Advances in Optics and Optoelectronics

Series Editor

Perry Ping Shum, Southern University of Science and Technology, Shenzhen, China

The Advances in Optics and Optoelectronics series focuses on the exciting new developments in the fast emerging fields of optics and optoelectronics. The volumes cover key theories, basic implementation methods, and practical applications in but not limited to the following subject areas:

AI Photonics
Laser Science and Technology
Quantum Optics and Information
Optoelectronic Devices and Applications
Fiber-Based Technologies and Applications
Near-infrared, Mid-infrared and Far-infrared Technologies and Applications
Biophotonics and Medical Optics
Optical Materials, Characterization Methods and Techniques
Spectroscopy Science and Applications
Microscopy and Adaptive Optics
Microwave Photonics and Wireless Convergence
Optical Communications and Networks

Within the scopes of the series are monographs, edited volumes and conference proceedings. We expect that scientists, engineers, and graduate students will find the books in this series useful in their research and development, teaching and studies.

Zhengjun Liu · Xuyang Zhou · Shutian Liu
Editors

Computational Optical Imaging

Principle and Technology

Springer

Editors
Zhengjun Liu
School of Physics
Harbin Institute of Technology
Harbin, China

Xuyang Zhou
School of Physics
Harbin Institute of Technology
Harbin, China

Shutian Liu
School of Physics
Harbin Institute of Technology
Harbin, China

ISSN 2731-6009 ISSN 2731-6017 (electronic)
Advances in Optics and Optoelectronics
ISBN 978-981-97-1454-4 ISBN 978-981-97-1455-1 (eBook)
https://doi.org/10.1007/978-981-97-1455-1

This Springer imprint is published by the registered company Springer Nature Singapore Pte Ltd.
The registered company address is: 152 Beach Road, #21-01/04 Gateway East, Singapore 189721, Singapore

Paper in this product is recyclable.

Preface

With the development of modern imaging technology, computational imaging technology has become a new form of imaging. It is an interdisciplinary product, involving optics, image and signal processing, mathematics, control, machinery, and other fields. In order to obtain high-quality imaging, some calculation methods and measurement techniques bring new opportunities to computational optical imaging in research and application. At the same time, computing technology can improve the intelligent and automatic elements of the imaging system, and greatly improve the performance of the optical imaging system. In view of this, many researchers have turned to this field to start their research work. With the development of computational imaging technology, the form of optical imaging system tends to be diversified.

At the basic principle level, mathematics is very important for the modeling of optical imaging systems. The algorithms can act as a bridge between the model and the actual measured image data. At the same time, the algorithms are the basis of some simulation research work. Computational optical imaging takes the measurement of light field data as the core research object. Because computational imaging technology is still in the process of development, there is no clear and strict definition at present. Perhaps, we can refer to systems involving computers and optical hardware as computational optical imaging systems. Although this definition is vague, it can include many imaging methods.

This book provides the readers with a comprehensive overview of principles, methodologies, and recent advances in computational optics imaging. This book is a valuable reference for scientists, engineers, scientific researchers, postgraduate students, and senior undergraduate students with majors in optics, image, signal, and video processing. To enlighten the reader, we would like to divide this book into 14 chapters:

Chapter 1—"Introduction".
Chapter 2—"Ghost Imaging".
Chapter 3—"Imaging Methods via Compressed Sensing".
Chapter 4—"Computerized Tomography".
Chapter 5—"Computer Generated Holography".
Chapter 6—"Non-line-of-Sight Imaging".

Chapter 7—“Coded Aperture Imaging”.
Chapter 8—“Phase Imaging”.
Chapter 9—“Light Field Camera”.
Chapter 10—“Adaptive Optical Imaging”.
Chapter 11—“Fluorescence Imaging”.
Chapter 12—“Synthetic Aperture Imaging”.
Chapter 13—“Mask-Modulated Lensless Imaging”.
Chapter 14—“Polarization Imaging”.

Harbin, China

Zhengjun Liu
Xuyang Zhou
Shutian Liu

Acknowledgements The editors would like to take this opportunity to express their sincere gratitude to the authors of the chapters for extending their wholehearted support in sharing some of their latest results. Without their significant contribution, this book volume could not have fulfilled its mission. All the authors are indebted to the editorial staff at Springer, for their hard working and selfless attitude in helping with our book plan.

Contents

Chapter 1
Introduction

Zhengjun Liu, Xuyang Zhou, and Shutian Liu

Computational optical imaging is a new imaging method recently. It integrates optics, imaging, computer, control and other multi-disciplinary knowledge. This technology brings new opportunities for the development of optical systems with various imaging scales. For example, with the support of computing technology, we can turn the optical systems or structures that cannot be imaged into an imaging tool. With the support of computing technology, the imaging system becomes more intelligent and the automatic information processing is enhanced.

1.1 Appearance

With the increasing maturity of the discrete recording mode, the connection between the imaging system and the computer becomes more natural. The projection data of the optical system can be easily transferred to the computer for processing. The charge coupled device (CCD) or optoelectronic chip plays an important role, as shown in Fig. 1.1. Within the optical system, discrete modulation devices also provide people with more choices of experiments and applications, such as digital micromirror device (DMD), spatial light modulator (SLM), and binary optical element (BOE). For the light source, the LED array can be regarded as a discrete light source, on the basis of which we can carry out space-time discrete illumination. Therefore, the optical imaging system is almost completely discretized. However, the photons transmitted as signals are still described by the continuous model. Of course, the pulse light source further cuts this continuous signal source.

Z. Liu (✉) · X. Zhou · S. Liu
School of Physics, Harbin Institute of Technology, Harbin 150001, China
e-mail: zjliu@hit.edu.cn

S. Liu
e-mail: stliu@hit.edu.cn

Z. Liu et al. (eds.), *Computational Optical Imaging*, Advances in Optics and Optoelectronics, https://doi.org/10.1007/978-981-97-1455-1_1

Fig. 1.1 Discretization of optical imaging system

On the basis of the discretization of the above hardwares, the theory and method of two-dimensional signal processing can be directly used in computational optical imaging systems. For example, discrete convolution, correlation, discrete Fourier transform and so on, can be directly applied. Even some pure mathematical transformations can be realized in such a discrete optical system, where signal light provides its fast signal processing characteristics.

1.2 Performance

With the rapid development of various-scale imaging technologies, resolution and field of view as important technical indicators have been paid great attention by researchers and important users. Due to the limitation of current hardware and micro-nano processing technology, the imaging resolution cannot be improved indefinitely. For the imaging field of view, if the imaging time is not limited, the scanning mode can infinitely expand the field of view. If we consider the synthesis of these two technical indicators, that is, spatial bandwidth product, it is a more effective amount of system performance evaluation to measure the ability of an imaging device to capture information flux.

For fast imaging or transient process imaging, the data acquisition time left to the imaging system is limited or very short. Therefore, the imaging speed is also an important technical index to consider the performance of the microscopic imaging system. They can be used to evaluate the ability of microscopic imaging system to obtain spatio-temporal information. It is an urgent research task to study the microscopic imaging system with fast and high-throughput imaging ability, which has important theoretical significance and economic benefits for practical applications such as observing the dynamic (transient) process of large-size targets and on-line detection.

In order to develop a fast and high-throughput microscopic imaging system, computational optical imaging technology is indispensable. It can connect super-resolution microscopic imaging theory and microscopic system hardware to achieve the combination of software and hardware to achieve high-quality microscopic imaging targets. With the unique advantages and potential of mathematics and algorithms, the application of computational optical imaging technology in computational optical imaging system is of great strategic significance for the rapid detection of high-end manufacturing industries such as chips and electronic panels.

In recent years, the space-frequency domain scanning scheme plays an important role in improving the field of view and resolution of computational microscopic imaging technology, and a lot of research results emerge accordingly. However, the scanning measurement method will affect the imaging speed, so a variety of parallel measurement methods have emerged, such as spatial multi-channel technology, multi-wavelength lighting and other methods to achieve a single exposure and obtain the projection information of the target through the optical microscopic imaging system at one time. Therefore, the parallel measurement method is a feasible technical route for the development of rapid computing microscopic imaging.

Fidelity is a key index of imaging system. It can be measured by the signal-to-noise ratio, and it mainly comes from the factors related to the aberration of the optical system. In the computational imaging system, we can rely on the digital compensation method to balance various unexpected interference factors, and finally output a result image as fidelity as possible. The sharpness of the image can be quantitatively estimated by the image quality evaluation function in the aspect of autofocus, which mainly depends on the summation of the absolute value of the image gradient. The sharpness evaluation function is a form of non-reference image evaluation function, which can be used to evaluate the simulation and experimental results without limitation.

Intelligence and automation is a natural attribute of computational imaging system. Computers can enable optical systems to achieve many flexible information processing functions. For example, the combination of image acquisition and image analysis, the automatic adjustment of the position and attitude of optical elements and so on. In addition, the imaging of extreme environment or complex media is an opportunity to display the ability of computational optical imaging technology, and it is possible to obtain low-resolution target image when it is difficult to image.

1.3 Progress

Here, the related research progress of computational optical microscopic imaging technology in high-flux imaging is listed and analyzed, and some related achievements in field of view, resolution and fast imaging are summarized.

(1) **Fourier ptychographic imaging technology and its development**

Spatial scanning and laminated scanning are of great help to the improvement of the field of view of the imaging system [1, 2]. When the surface information of the sample to be tested is constant or during the scanning period, the scanning method can obtain a large field of view, and the laminated scanning imaging can obtain the complex information of the light field on the target surface. Polarization sensitive Fourier ptychographic microscopy (FPM) has the ability of high resolution and wide field birefringence imaging [3]. In the design of portable optical microscopic system, the computational microscopic imaging system composed of smart phones can exceed the ordinary desktop microscope in terms of spatial bandwidth product, and has the ability of color complex imaging [4]. The computational microscopic system designed according to inclined lighting provides a new way to improve the field of view and resolution [5]. FPM technology which can remove artifacts [6] has the characteristics of non-iteration, high parallelism computing ability, strong fault tolerance and so on. Intelligent correction technology makes the performance of FPM system more stable [7, 8].

(2) **Parallel Measurement and Computing Technology**

Synthetic aperture interference [9] is an effective technical way to design high-resolution large field-of-view quantitative phase microscopic imaging. The technology of multi-beam array interference microscopy [10] overcomes the limitations of digital holographic microscopy, and its main advantage is that it can integrate more wavefronts and provide higher time stability. The combination of the inherent coherent gating characteristics of partially coherent light and the imaging mode of oblique illumination [11] can significantly increase the spatial frequency bandwidth and have a very considerable imaging field of view.

The lateral translation method [12] can generate high-resolution speckle lighting patterns in a large field of view, and the image reconstruction scheme with optimization algorithm can complete the necessary digital correction for defects such as unknown lighting patterns, system aberrations and pattern translation. Intensity diffraction tomography based on non-interference synthetic aperture [13] has the characteristics of high resolution, high contrast, strong tomography ability and so on. The test method of multi-view and multi-depth [14] is a parallel measurement method with great potential. The microscopic imaging system [15] designed by the combination of multi-core optical fiber and compressed sensing has a large spatial bandwidth product. Dual-amplitude imaging method based on photon sieve [16] uses Pancharatnam-Berry phase to realize polarization and wavelength control, and maintains the spatial bandwidth product unchanged.

(3) **High-flux Microscopic Imaging Techniques**

Based on the mixed bright-dark field intensity transfer equation, high-flux microscopy [17] can synthesize complex-valued images with high resolution and large field of view. With the combination of optical coherent microscopy and hybrid adaptive optics [18], a higher spatio-temporal bandwidth product can be obtained by introducing an ultra-fast scanning source. In the two-step phase shift interferometry, the series optical structure [19] can make full use of the spatial bandwidth product of the camera to improve the resolution. In the aspect of optical element integration, the design scheme based on optical invariants [20] can maximize the spatial bandwidth product of the microscopic imaging system. The plane-surface-plane microscopic imaging technology [21] has high spatio-temporal resolution and massive data processing ability, and the sample plane to be tested is magnified to a larger spherical image, so as to obtain the resolution and field of view that the ordinary imaging system cannot have and beyond the diffraction imaging limitation [22].

Pinhole holographic imaging technology [23] adopts moving array microlens and high resolution image array coding, which has considerable spatial bandwidth product and resolution. In order to achieve the best matching between the image sensor chip and the objective lens, the defocus image can make full use of the spatial bandwidth product of the objective lens [24]. The off-axis digital holographic microscopic imaging technique using single measurement [25] provides the optimization framework of the optical imaging system in terms of spatial bandwidth product. The off-axis holographic imaging technology [26] uses white light illumination to complete single exposure imaging at the expense of spatial resolution or field of view, and achieves the phase measurement target of high time domain bandwidth.

(4) **Microscopic imaging technology based on multi-wavelength illumination**

The super-resolution imaging method based on wavelength scanning [27] can replace the imaging hardware unit to realize physical displacement, improve the resolution of wide-field imaging system and significantly increase the spatial bandwidth product. The panchromatic light field microscopy based on single pixel imaging [28] can increase the spatial bandwidth product and improve the spatial resolution of the reconstructed light field. The intensity diffraction tomography based on single exposure [29] using ring programmable color light emitting diodes, it provides an oblique lighting space layout scheme that accurately matches the numerical aperture of the objective lens, with unmarked, high-throughput and fast imaging ability. The wide field of view quantitative phase imaging method based on wavelength scanning [30] provides a solution to overcome the aliasing of spatial information and forms a sufficient variety of diffraction information. The optically transparent subsurface composed of a patterned graphene sandwich structure and a radio frequency detector [31] can realize adaptive frequency manipulation to match the incident beam, providing a feasible light source design and control scheme for polychromatic lighting imaging.

(5) **Fast computational imaging methods**

In the aspect of image reconstruction, computational optical imaging technology provides technical support with strong flexibility and high plasticity for rapid image generation [32, 33]. The Kramers-Kronig relation [34] converts the intensity distribution into the phase distribution, realizes the non-iterative phase recovery calculation [34], expands the spatial bandwidth product [35], and reduces the computational complexity of image reconstruction [36]. The construction of generalized discrete Fourier-Hankel transform [37] can realize efficient basis vector decomposition and reduce the amount of calculation. It is suitable for the mathematical modeling of beam modulation and imaging optical elements such as Maxwell fisheye lens, spherical invisibility cloak, aspherical thin lens and cylindrical lens. The adaptive sampling calculation method [38] can reduce the computational complexity and make full use of the spatial bandwidth product information in the imaging system. The multi-wavelength phase recovery method [39] combines physical constraints and sparse prior information to ensure the microscopic imaging quality in the presence of noise in the measurement process. In the incoherent digital holographic microscopic system, the adaptive method based on random parallel gradient descent [40] has the advantages of parallel calculation of multiple aberrations and fast convergence.

1.4 Prospect

Computational optical imaging is not only a bridge of multi-disciplines, but also a tool for information acquisition. Computational imaging methods can break through the limitations of traditional direct optical imaging technology on precision adjustment of optical system, manufacturing process of optical elements, observation object, development cost and so on. It is a very active research hotspot in academic circles, and it represents the development trend of advanced optical imaging technology. It can be believed that this technology will play an important role in the future. In the aspect of image acquisition in extreme cases, we are expected to get clearer image results, faster imaging process and more information.

Acknowledgements This work was supported by the National Natural Science Foundation of China (Nos. 11874132, 61975044, 12074094, 12374271, and 62375065); Interdisciplinary Research Foundation of HIT (No. IR2021237).

References

1. Faulkner HML, Rodenburg JM. Movable aperture lensless transmission microscopy: a novel phase retrieval algorithm. Phys Rev Lett. 2004;93(2): 023903.
2. Zheng G, Horstmeyer R, Yang C. Wide-field, high-resolution Fourier ptychographic microscopy. Nat Photon. 2013;7(9):739–45.
3. Song S, Kim J, Hur S, Song J, Joo C. Large-area, high-resolution birefringence imaging with polarization-sensitive Fourier ptychographic microscopy. ACS Photon. 2021;8(1):158–65.
4. Lee KC, Lee K, Jung J, Lee SH, Kim D, Lee SA. A smartphone-based Fourier ptychographic microscope using the display screen for illumination. ACS Photon. 2021;8(5):1307–15.
5. Ma X, Yao M, Zhang Z, Peng J, Li S, Zhong J. Computational oblique illumination microscopy with isotropic high resolution. IEEE Trans Comput Imag. 2020;6:317–27.
6. Bianco V, Mandracchia B, Behal J, Barone D, Memmolo P, Ferraro P. Miscalibration-tolerant Fourier ptychography. IEEE J Sel Top Quant Electron. 2021;27(4):7500417.
7. Bian L, Zheng G, Guo K, Suo J, Yang C, Chen F, Dai Q. Motion-corrected Fourier ptychography. Biomed Opt Express. 2016;7(11):4543–53.
8. Zheng C, Zhang S, Yang D, Zhou G, Hu Y, Hao Q. Robust full-pose-parameter estimation for the LED array in Fourier ptychographic microscopy. Biomed Opt Express. 2022;13(8):4468–82.
9. Hu C, Kandel ME, Lee YJ, Popescu G. Synthetic aperture interference light (SAIL) microscopy for high-throughput label-free imaging. Appl Phys Lett. 2021;119(23): 233701.
10. Dong J, Yetisen AK, Zhao C, Dong X, Brändle F, Wang Q, Jakobi M, Saur D, Koch AW. Single-shot high-throughput phase imaging with multibeam array interferometric microscopy. ACS Photon. 2021;8(12):3536–47.
11. Duris M, Bouchal P, Rovenská K, Chmelík R. Coherence-encoded synthetic aperture for super-resolution quantitative phase imaging. APL Photon. 2022;7: 046105.

12. Yeh LH, Chowdhury S, Waller L. Computational structured illumination for high-content fluorescence and phase microscopy. Biomed Opt Express. 2019;10(4):1978–98.
13. Li J, Zhou N, Sun J, Zhou S, Bai Z, Lu L, Chen Q, Zuo C. Transport of intensity diffraction tomography with non-interferometric synthetic aperture for three-dimensional label-free microscopy. Light Sci Appl. 2022;11:154.
14. Ahar A, Chlipala M, Birnbaum T, Zaperty W, Symeonidou A, Kozacki T, Kujawinska M, Schelkens P. Suitability analysis of holographic vs light field and 2D displays for subjective quality assessment of Fourier holograms. Opt Express. 2020;28(24):37069–91.
15. Shin J, Bosworth BT, Foster MA. Compressive fluorescence imaging using a multi-core fiber and spatially dependent scattering. Opt Lett. 2017;42(1):109–12.
16. Frese D, Sain B, Zhou H, Wang Y, Huang L, Zentgraf T. A wavelength and polarization selective photon sieve for holographic applications. Nanophoton. 2021;10(18):4543–50.
17. Lu L, Li J, Shu Y, Sun J, Zhou J, Lam EY, Chen Q, Zuo C. Hybrid brightfield and darkfield transport of intensity approach for high-throughput quantitative phase microscopy. Adv Photon. 2022;4(5): 056002.
18. Liu S, Mulligan JA, Adie SG. Volumetric optical coherence microscopy with a high space-bandwidth-*time* product enabled by hybrid adaptive optics. Biomed Opt Express. 2018;9(7):3137–52.
19. Ibrahim DGA. Fast phase-shifting technique for 3-D surface micro-topography measurement. Measurement. 2019;135:106–11.
20. Bumstead JR, Park JJ, Rosen IA, Kraft AW, Wright PW, Reisman MD, Côté DC, Culver JP. Designing a large field-of-view twophoton microscope using optical invariant analysis. Neurophotonics. 2018;5(2): 025001.
21. Fan J, Suo J, Wu J, Xie H, Shen Y, Chen F, Wang G, Cao L, Jin G, He Q, Li T, Luan G, Kong L, Zheng Z, Dai Q. Video-rate imaging of biological dynamics at centimetre scale and micrometre resolution. Nat Photon. 2019;13(11):809–16.
22. Park J, Brady DJ, Zheng G, Tian L, Gao L. Review of bio-optical imaging systems with a high space-bandwidth product. Adv Photon. 2021;3(4): 044001.
23. Wang Z, Lv G, Feng Q, Wang A, Ming H. Enhanced resolution of holographic stereograms by moving or diffusing a virtual pinhole array. Opt Express. 2020;28(15):22755–66.
24. Wang H, GöRöcs Z, Luo W, Zhang Y, Rivenson Y, Bentolila LA, Ozcan A. Computational out-of-focus imaging increases the space-bandwidth product in lens-based coherent microscopy. Optica. 2016;3(12):1422–9.
25. Picazo-Bueno JA, Trusiak M, Mico V. Single-shot slightly off-axis digital holographic microscopy with add-on module based on beamsplitter cube. Opt Express. 2019;27(4):5655–69.
26. Shan M, Kandel ME, Majeed H, Nastasa V, Popescu G. White-light diffraction phase microscopy at doubled space-bandwidth product. Opt Express. 2016;24(25):29033–9.
27. Luo W, Zhang Y, Feizi A, Göröcs Z, Ozcan A. Pixel super-resolution using wavelength scanning. Light Sci Appl 2016;5: e16060.
28. Yao M, Cai Z, Qiu X, Li S, Peng J, Zhong J. Full-color light-field microscopy via single-pixel imaging. Opt Express. 2020;28(5):6521–36.
29. Wu X, Sun J, Zhang J, Lu L, Chen R, Chen Q, Zuo C. Wavelength-scanning lensfree on-chip microscopy for wide-field pixel-super-resolved quantitative phase imaging. Opt Lett. 2021;46(9):2023–6.
30. Zhou N, Li J, Sun J, Zhang R, Bai Z, Zhou S, Chen Q, Zuo C. Single-exposure 3D label-free microscopy based on color-multiplexed intensity diffraction tomography. Opt Lett. 2022;47(4):969–72.
31. Geng M, Yang X, Chen H, Bo X, Li M, Liu Z, Lu W. Optically transparent graphene-based cognitive metasurface for adaptive frequency manipulation. Photon Res. 2023;11(1):129–36.
32. Zuo C, Chen Q. Computational optical imaging: an overview. Inf Laser Eng. 2022;51(2):20220110.
33. Zhao W, Zhao S, Li L, Huang X, Xing S, Zhang Y, Qiu G, Han Z, Shang Y, Sun D, Shan C, Wu R, Gu L, Zhang S, Chen R, Xiao J, Mo Y, Wang J, Ji W, Chen X, Ding B, Liu Y, Mao H,

Song BL, Tan J, Liu J, Li H, Chen L. Sparse deconvolution improves the resolution of live-cell super-resolution fluorescence microscopy. Nat Biotechnol. 2022;40(4):606–17.

34. Baek Y, Park Y. Intensity-based holographic imaging via space-domain Kramers-Kronig relations. Nat Photon. 2021;15(5):354–60.
35. Baek Y, Lee K, Shin S, Park Y. Kramers-Kronig holographic imaging for high-space-bandwidth product. Optica. 2019;6(1):45–51.
36. Shen C, Liang M, Pan A, Yang C. Non-iterative complex wave-field reconstruction based on Kramers-Kronig relations. Photon Res. 2021;9(6):1003–12.
37. Venkatesh S, Schurig D. Computationally fast EM field propagation through axi-symmetric media using cylindrical harmonic decomposition. Opt Express. 2016;24(25):29246–68.
38. Zhang W, Zhang H, Jin G. Adaptive-sampling angular spectrum method with full utilization of space-bandwidth product. Opt Lett. 2020;45(16):4416–9.
39. Gao Y, Cao L. Projected refractive index framework for multi-wavelength phase retrieval. Opt Lett. 2022;47(22):5965–8.
40. Zhang W, Man T, Zhang M, Zhang L, Wan Y. Computational adaptive holographic fluorescence microscopy based on the stochastic parallel gradient descent algorithm. Biomed Opt Express. 2022;13(12):6431–42.

Chapter 2
Ghost Imaging

Wei Zhang, Shuo Ma, and Zhengjun Liu

2.1 Introduction

While traditional imaging methods utilize light reflected or scattered from an object for imaging, where light from the object plane is transmitted to the detector plane and the CCD receives the light field information for direct imaging, ghost imaging is a technique that uses optical correlation for imaging. Ghost imaging system is generally divided into a signal light path and reference light path, the signal light path using a bucket detector without spatial resolution, the reference light path using a CCD, the peculiarity is that any one of the detectors used to detect the signal cannot be imaged individually, the use of correlation between the signals of the two detectors in order to image.

Initially, it was found that the use of two-photon entangled state light source can be used for imaging, and the object to be measured is not in the light path of the imaging, people believe that this imaging method is like a ghostly quantum imaging, so called ghost imaging. The special feature of this imaging method is that the image is not obtained from the detector, but through the relevant calculations, this novel imaging method began to enter people's field of vision. At first, there has been no reasonable explanation about the mechanism of ghost imaging, and the related researches tend to think that ghost imaging is a property unique to quantum entangled light sources, and some statistical knowledge is used to prove that classical light sources cannot have some characteristics of quantum entangled light sources, and reasonable explanations are given. However, some people do not think so, and use the classical light source

W. Zhang · S. Ma · Z. Liu (✉)
School of Physics, Harbin Institute of Technology, Harbin 150001, China
e-mail: zjliu@hit.edu.cn

W. Zhang
e-mail: zw18894226727@163.com

S. Ma
e-mail: 2893753451@qq.com

Z. Liu et al. (eds.), *Computational Optical Imaging*, Advances in Optics and Optoelectronics, https://doi.org/10.1007/978-981-97-1455-1_2

in the experiment using a system similar to the quantum ghost imaging to realize the classical light source ghost imaging, also gave a reasonable explanation, that although the classical light source can only partially realize the characteristics of the quantum light source, but this is already enough for ghost imaging. Thereafter, there is still a discussion on the physical mechanism of ghost imaging, and it is considered that the quantum mechanics of nonlocal two-photon interference is what is necessary to quantitatively characterize thermo-optic ghost imaging. Subsequently, the introduction of spatial light modulators allowed a computational ghost imaging method requiring only a single-pixel detector to be proposed and experimentally verified as reasonable, denying non-local two-photon interference. Since then, the physical mechanism of ghost imaging is no longer discussed, and the more interesting question is how to improve the signal-to-noise ratio and acquisition time of ghost imaging. Subsequently, the single-pixel imaging method, which is similar to the ghost imaging technique and can be imaged by using only a single-pixel detector, has been increasingly discussed, although for the single-pixel imaging technique, more attention has been paid to the different illumination modulation methods.

The classical coherence theory of light as an electromagnetic wave is discussed, focusing on first-order and second-order coherence. First-order coherence reflects the correlation between the amplitudes of light fields in different time-space, e.g., for electromagnetic waves $E_1(\mathbf{r}_1, t_1)$ and $E_2(\mathbf{r}_2, t_2)$, define the classical first-order correlation function as $\langle E^*(\mathbf{r}_1, t_1)E(\mathbf{r}_2, t_2)\rangle$, where systematic $\langle\cdot\rangle$ represent average. Then define the classical first-order coherence function as follow

$$\gamma^{(1)}(x_1, x_2) = \frac{\langle E^*(x_1)E(x_2)\rangle}{\sqrt{\langle|E(x_1)|^2\rangle\langle|E(x_2)|^2\rangle}}. \tag{2.1}$$

The first-order coherence function $\gamma^{(1)}(x_1, x_2)$ can be used to characterize whether a light field is coherent or not, is a correlation function between the amplitudes of light fields, and cannot distinguish between light fields with different photon statistical properties. Second-order coherence, on the other hand, reflects the correlation between the intensity of the light field at two space-time points, i.e. $\langle I(x_1)I(x_2)\rangle$, called the second-order correlation function. The corresponding classical second-order coherence function is defined as

$$\gamma^{(2)}(x_1, x_2; x_2, x_1) = \frac{\langle I(x_1)I(x_2)\rangle}{\langle I(x_1)\rangle\langle I(x_2)\rangle} = \frac{\langle E^*(x_1)E^*(x_2)E(x_2)E(x_1)\rangle}{\langle|E(x_1)|^2\rangle\langle|E(x_2)|^2\rangle}. \tag{2.2}$$

The classical first-order and second-order coherence theories are given above, and the relevant quantum coherence theories are not analyzed here. Analyzing coherence is important, and ghost imaging using classical light sources basically involves analyzing the correlation between the signal light path and the detector signal of the reference light path to image or improve the quality of imaging.

2.2 Quantum Ghost Imaging

Entangled states are a peculiar phenomenon in quantum mechanics that cannot be explained by conventional theories. The classical example of a two-particle entangled state is the EPR absurd proposed by Einstein, Podelsky and Rosen in 1935 [1], stating that quantum mechanics allows two particles to be satisfied by their state: the position of each particle is indeterminate, but observing one particle at a certain position implies that the other particle must be found at a specific corresponding position. The ghost imaging method was first proposed by Pittman in 1995 and the ghost imaging system is shown in Fig. 2.1.

In nonlinear optics, spontaneous parametric down conversion (SPDC) is a commonly used process to generate a two-photon entangled state, where pump light irradiated on a nonlinear crystal produces a nonlinear polarization in the crystal, resulting in the emission of a pair of photons in an entangled state [2]

$$|\Psi\rangle = \sum_{s,i} \delta(\omega_s + \omega_i - \omega_p)\delta(k_s + k_i - k_p)|k_s\rangle \otimes |k_i\rangle. \tag{2.3}$$

which $\omega_j, k_j (j = s, i, p)$ refer to the angular frequencies, wave vectors, of the signal, idler and pump light, respectively. The phase matching condition needs to be satisfied for SPDC as follow

$$\omega_p = \omega_s + \omega_i, k_p = k_s + k_i, \tag{2.4}$$

that is conservation of energy and conservation of momentum. The signal light and pump light obtained from the conversion have a spatial position or momentum correlation, and the object and detector are illuminated with the signal light and the idle

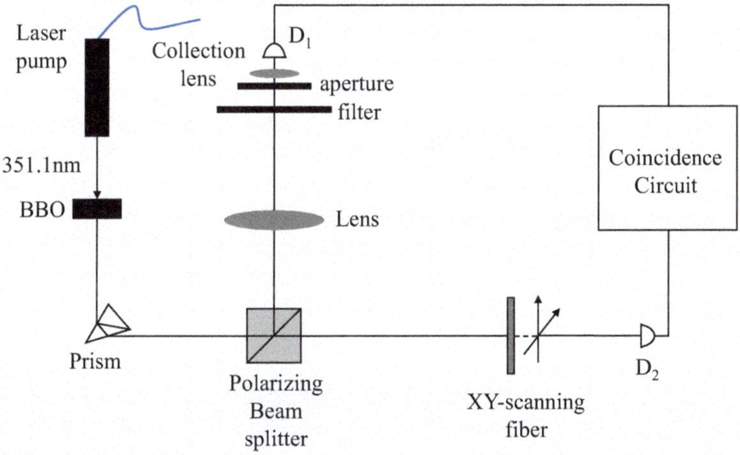

Fig. 2.1 Quantum imaging system

light, respectively, and the object information can be calculated based on the correlation between the signal light and the pump light. Pittman et al. do not give a mechanism for the imaging, but instead argue that the core of ghost imaging is the positional or momentum correlation satisfied by two-photons.

Due to the weak entangled photons produced in this experiment, a single-photon detector was used to probe the signal, and in the case of a large number of photon pairs, the imaging needs to utilize the correlation function of the fluctuations in the intensity of the signal and idler light. The unfolding of the system for quantum ghost imaging is shown in Fig. 2.2, when the distance from object to the imaging lens is S, from the fiber array detector to the imaging lens is S', we have (Fig. 2.3)

$$\frac{1}{S} + \frac{1}{S'} = \frac{1}{f}. \tag{2.5}$$

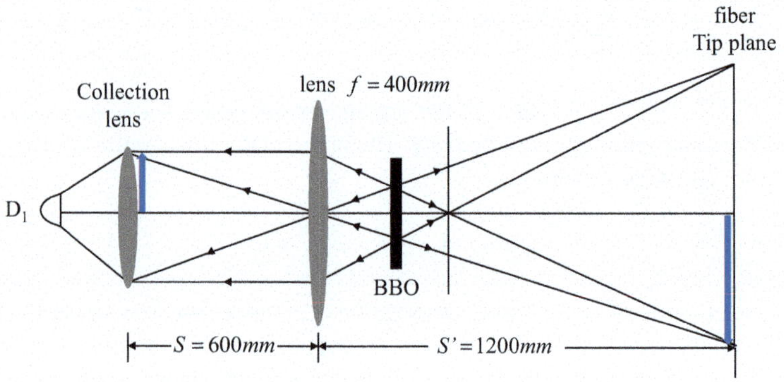

Fig. 2.2 "Unfold" figure of ghost imaging system

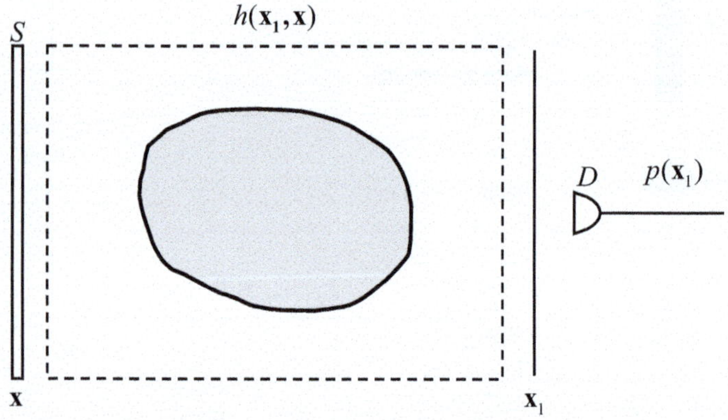

Fig. 2.3 The system of single-photon imaging

Quantum ghost imaging is based on quantum entanglement properties, and in principle, correlations based on classical light sources do not fully mimic quantum light sources based on entangled states, but can only be partially simulated, at which point the use of entangled photons in an imaging system can show results that cannot be mimicked by any other two-photon source [3]. At this point it is believed that classical light sources are not capable of ghost imaging. That is why it seems important to study the role of quantum entanglement in two-photon imaging, determining whether in principle ghost imaging can be realized with classical light sources. A specific analysis based on probability theory and quantum is shown below.

(1) Single-photon imaging

In order to analyze the quantum imaging system, the quantum single-photon imaging system and the basic equations are first analyzed.

Pure state single photon source

$$|\Psi\rangle = \int d\mathbf{x}\varphi(\mathbf{x})|1_{\mathbf{x}}\rangle, \tag{2.6}$$

where $|1_{\mathbf{x}}\rangle = 1/(2\pi)^2 \int e^{\mathrm{i}\mathbf{k}\cdot\mathbf{x}}|1_{\mathbf{k}}\rangle dk$ is the positional representation of the single photon state. $\varphi(\mathbf{x})$ is state probability amplitude, $h(\mathbf{x_1}, \mathbf{x})$ is the impulse response function through nonlinear medium. In the output plane, the probability density of the photon in position $\mathbf{x_1}$ is

$$p(\mathbf{x}_1) \propto |\int \mathrm{d}\mathbf{x}\varphi(\mathbf{x})h(\mathbf{x}_1, \mathbf{x})|^2. \tag{2.7}$$

If the single photon is in a mixed state described by the density operator

$$p(\mathbf{x}_1) \propto \int \mathrm{d}\mathbf{x}d\mathbf{x}'\gamma(\mathbf{x}, \mathbf{x}')h(\mathbf{x}_1, \mathbf{x})h^*(\mathbf{x}_1, \mathbf{x}'). \tag{2.8}$$

Then the probability density of the photon is

$$p(\mathbf{x}_1) \propto \int d\mathbf{x}d\mathbf{x}'\gamma(\mathbf{x}, \mathbf{x}')h(\mathbf{x}_1, \mathbf{x})h^*(\mathbf{x}_1, \mathbf{x}'). \tag{2.9}$$

(2) Two-photon imaging

Pure state two photon source

$$|\Psi\rangle = \iint \mathrm{d}\mathbf{x}\mathrm{d}\mathbf{x}'\varphi(\mathbf{x}, \mathbf{x}')|1_{\mathbf{x}}, 1_{\mathbf{x}'}\rangle. \tag{2.10}$$

where $|1_{\mathbf{x}}, 1_{\mathbf{x}'}\rangle$ is the two-photon state at transverse coordinates $\mathbf{x}$ and $\mathbf{x}'$, as shown in Fig. 2.4, the correlation probability density generated by a photon emitted by

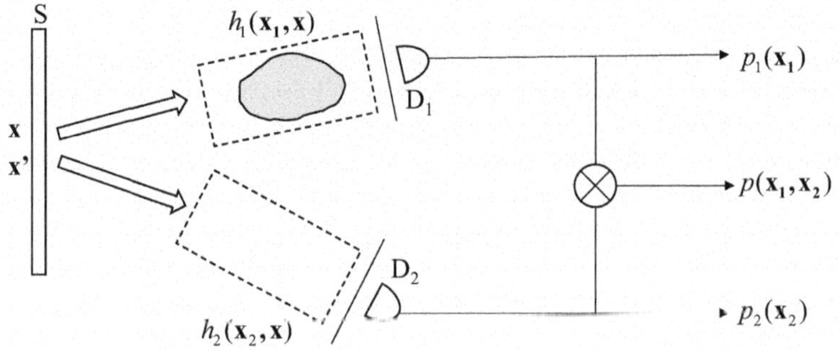

Fig. 2.4 The system of two-photo imaging

a quantum source after passing through a nonlinear optical system with impulse response functions of $h_1(\mathbf{x}_1, \mathbf{x})$ and $h_2(\mathbf{x}_1, \mathbf{x})$, respectively

$$p(\mathbf{x}_1, \mathbf{x}_2) \propto \left| \iint \mathrm{d}\mathbf{x}\mathrm{d}\mathbf{x}' \varphi(\mathbf{x}, \mathbf{x}') h_1(\mathbf{x}_1, \mathbf{x}) h_2(\mathbf{x}_2, \mathbf{x}') \right|^2 . \tag{2.11}$$

(3) Two distinct single-photon probability densities

Consider two distinguishable photon probability density functions associated with two-photon systems. The first is the probability density of observing one photon (regardless of whether the other photon is detected or not) at $\mathbf{x}_1$, This is called the single-photon probability density $p_1(\mathbf{x}_1)$. The second is the probability density of observing a photon at the output surface of System 1 in Fig. 2.4 and also observing a photon at any position on the output surface of System 2, called the fringe probability density $\overline{p}_1(\mathbf{x}_1)$. Similarly there are $p_2(\mathbf{x}_2)$ and $\overline{p}_2(\mathbf{x}_2)$. And Single-photon probability density on the output plane of the system j:

$$p_j(\mathbf{x}_j) \propto \int \mathrm{d}\mathbf{x}\mathrm{d}\mathbf{x}' \gamma_j(\mathbf{x}, \mathbf{x}') h_j(\mathbf{x}_j, \mathbf{x}) h_j^*(\mathbf{x}_j, \mathbf{x}'),\ j = 1, 2. \tag{2.12}$$

The fringe probability density of detecting a photon at detector D1 is:

$$\overline{p}_1(\mathbf{x}_1) = \int \mathrm{d}\mathbf{x}_2 p(\mathbf{x}_1, \mathbf{x}_2). \tag{2.13}$$

(4) Probability density of non-entangled sources

For a non-entangled two-photon source, the state probability amplitude can be factorized: $\varphi(\mathbf{x}, \mathbf{x}') = \varphi_1(\mathbf{x})\varphi_2(\mathbf{x}')$, then the joint probability density $p(\mathbf{x}_1, \mathbf{x}_2)$ can also be factorized, and then the measurement of the joint probability density will not have any significance because the photons arrive at different systems independently of each other. The photon arrivals to the imaging systems are independent, and each imaging system can be viewed as a coherent imaging system, at which point, which satisfies $\overline{p}_j(\mathbf{x}_j) = p_j(\mathbf{x}_j)$, is consistent with the probabilistic theory.

(5) Probability density of entangled sources

For entangled sources, the state probability amplitude $\varphi(x, x') = \varphi(\mathbf{x})\delta(\mathbf{x}-\mathbf{x}')$, then the joint probability density:

$$p(\mathbf{x}_1, \mathbf{x}_2) \propto \left| \int \mathrm{d}\mathbf{x}\varphi(\mathbf{x})h_1(\mathbf{x}_1, \mathbf{x})h_2(\mathbf{x}_2, \mathbf{x}) \right|^2. \tag{2.14}$$

And the single-photon probability density and the fringe probability density are satisfied, respectively:

$$p_j(\mathbf{x}_j) \propto \int \mathrm{d}\mathbf{x}|\varphi(\mathbf{x})|^2|h_j(\mathbf{x}_j, \mathbf{x})|^2,\ j = 1, 2, \tag{2.15}$$

$$\overline{p}_j(\mathbf{x}_j) \propto \int \mathrm{d}\mathbf{x}\mathrm{d}\mathbf{x}'\varphi(\mathbf{x})\varphi^*(\mathbf{x}')g_k(\mathbf{x}, \mathbf{x}')h_j(\mathbf{x}_j, \mathbf{x})h_j^*(\mathbf{x}_j, \mathbf{x}'),\ j, k = 1, 2, j \neq k. \tag{2.16}$$

It is obvious that $\overline{p}_j(\mathbf{x}_j) \neq p_j(\mathbf{x}_j)$, which Incompatible with classical probability theory. Based on this, it seems that entangled light sources possess some statistical properties that classical light sources cannot have, and that quantum entanglement is essential to realize ghost imaging.

2.3 Classic and Computational Ghost Imaging

Based on the analysis in the previous section, it seems that classical light sources do not have the conditions to realize ghost imaging, and the special requirements of quantum light sources will limit the development of ghost imaging technology and make it difficult to be applied. Subsequently, however, ghost imaging has been realized using a classical incoherent light source similar to the optical path of quantum ghost imaging [4] (as shown in Fig. 2.5). For the previously discussed problem $\overline{p}_j(\mathbf{x}_j) \neq p_j(\mathbf{x}_j)$, Ryan et al. pointed out that this is not a statistical property unique to quantum mechanics, and conjectured that it is due to the loss of photons on the way to the signal detector, then the corresponding $\overline{p}_j(\mathbf{x}_j) \neq p_j(\mathbf{x}_j)$. There is Established, if there is no gain or loss of the system photons, the impulse response function h_1 in system 1 of Fig. 2.4 Satisfies the normalization

$$g_k(\mathbf{x}, \mathbf{x}') = \int \mathrm{d}\mathbf{x}''h_1(\mathbf{x}'', \mathbf{x})h_1^*(\mathbf{x}'', \mathbf{x}') = \delta(\mathbf{x} - \mathbf{x}'), \tag{2.17}$$

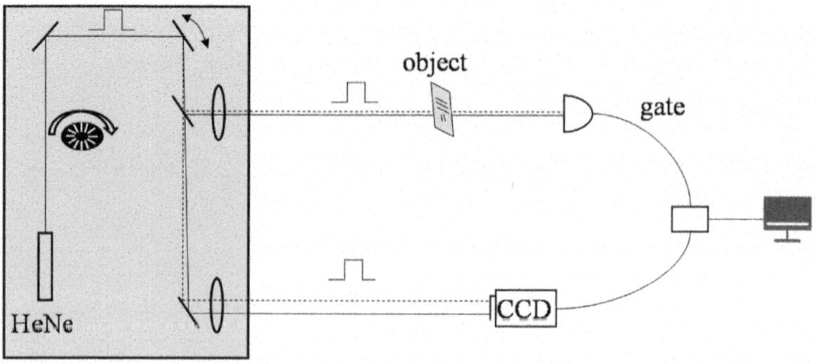

Fig. 2.5 Ghost imaging using a classical incoherent light source

Bringing the equation $g_k(\mathbf{x}, \mathbf{x}') = \delta(\mathbf{x}-\mathbf{x}')$ into the (K) equation, it can be derived that $\overline{p}_j(\mathbf{x}_j) = p_j(\mathbf{x}_j)$ to satisfy the requirements of classical probability theory.

In contrast to conventional quantum ghost imaging, Ryan et al. no longer use an entangled light source but a classical light source, which is generated using a chopper and a deflector that cuts the laser beam through the chopper, then deflects it to some degree through the deflector, and finally passes it through the beam splitter to obtain the classical light source.

Classical light sources can be modeled in the measurement and reference light paths by the probability of a random process with excitation fields P_n and distributions as $E_n^{(t)}$ and $E_n^{(r)}$. At this point, similar to the previous quantum light source, the probability of detection is respectively

$$p_1(x_1) = \sum_n P_n \left| \int \mathrm{d}x h_1(x_1, x) E_n^{(t)}(x) \right|^2, \tag{2.18}$$

$$p_2(x_2) = \sum_n P_n | \int \mathrm{d}x' h_2(x_2, x') E_n^{(r)}(x') |^2, \tag{2.19}$$

$$p(x_1, x_2) = \sum_n P_n \left| \int \mathrm{d}x h_1(x_1, x) E_n^{(t)}(x) \right|^2 \times \left| \int \mathrm{d}x' h_2(x_2, x') E_n^{(r)}(x') \right|^2. \tag{2.20}$$

If the probability distribution in the above equation is the same as the probability distribution in any entangled light source of the measurement and reference light paths, then there is

$$\varphi(x, x')\varphi^*(y, y') = \sum_n P_n E_n^{(t)}(x) E_n^{(r)}(x') E_n^{(t)*}(y) E_n^{(r)*}(y'). \tag{2.21}$$

For entangled quantum sources, the above equation is not satisfied because the right-hand side of the above equation cannot be decomposed as a function of (x, x') and (y, y'), so the classical sources cannot fully simulate pure-state entangled quantum sources. However, the classical light source only needs to generate the same joint probability distribution as the entangled light source, since the image formation is independent of the probability distribution of the single light path. For the reference light path, one can choose $E_s^{(r)}(x')$ to generate a diffractivity-limited spot at the reference plane position s, one has

$$|\int \mathrm{d}x' h_2(x_2, x') E_s^{(r)}(x')|^2 \approx \delta(x_2 - s). \tag{2.22}$$

then we choose $E_s^{(t)}(x) = \alpha \int \mathrm{d}x' h_2(s, x')\varphi(x, x')$, where $\alpha^{-2} \int \mathrm{d}s = 1$. At this point, the joint probability distribution of the classical light source is

$$p(x_1, x_2) = \sum_s P_s \alpha^2 \delta(x_2 - s) \left| \int \mathrm{d}x \mathrm{d}x' h_1(x_1, x) h_2(s, x') \varphi(x, x') \right|^2 \tag{2.23}$$

The classical joint distribution obtained is the same as that obtained for quantum sources.

Considering the similarity between the propagation behavior of the two-photon wave function and the mutual coherence function of thermal radiation, yet another classical ghost imaging system based on pseudo-thermal light has been proposed and experimentally and theoretically justified.

Theoretically, Gatti et al. verified the feasibility of realizing ghost imaging with a thermal light source by comparing the correlation measurement imaging scheme using a thermal light and an entangled light source [5]. As shown in Fig. 2.6, for the thermal light source correlation imaging system, the input–output relationship of the beam splitter is satisfied

$$b_1(\vec{x}) = ra(\vec{x}) + tv(\vec{x}), \quad b_2(\vec{x}) = ta(\vec{x}) + rv(\vec{x}). \tag{2.24}$$

The field in the plane of the detector is

$$c_i(\vec{x}_i) = \int \mathrm{d}\vec{x}_i' h_i(\vec{x}_i, \vec{x}_i') b_i(\vec{x}_i') + L_i(\vec{x}_i), \quad i = 1, 2. \tag{2.25}$$

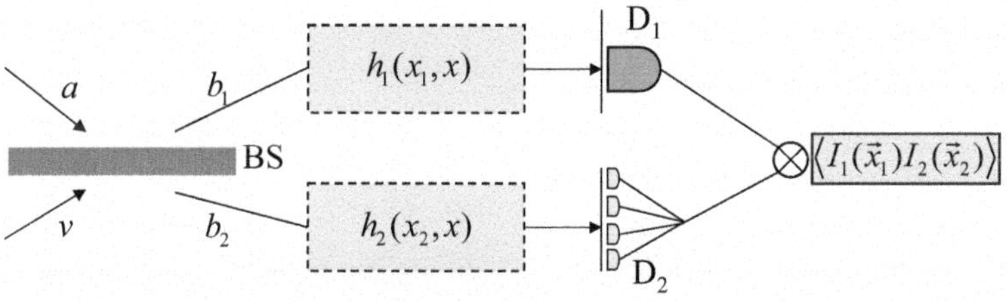

Fig. 2.6 Ghost imaging with incoherent thermal light

All information about the body is derived from a correlation function about intensity fluctuations

$$G(\vec{x}_1, \vec{x}_2) = \langle I_1(\vec{x}_1) I_2(\vec{x}_2)\rangle - \langle I_1(\vec{x}_1)\rangle\langle I_2(\vec{x}_2)\rangle, \tag{2.26}$$

where $\langle I_i(\vec{x}_i)\rangle = \langle c_i^{\dagger}(\vec{x}_i) c_i(\vec{x}_i)\rangle$. We can obtain

$$G_{\text{th}}(\vec{x}_1, \vec{x}_2) = |\text{tr}|^2 \left| \int \text{d}\vec{x}_1' \int \text{d}\vec{x}_2' h_1^*(\vec{x}_1, \vec{x}_1') h_2(\vec{x}_2, \vec{x}_2') \times \langle a^{\dagger}(\vec{x}_1') a(\vec{x}_2')\rangle \right|^2, \tag{2.27}$$

$$G_{\text{PDC}}(\vec{x}_1, \vec{x}_2) = \left|\left| \int \text{d}\vec{x}_1' \int \text{d}\vec{x}_2' h_1(\vec{x}_1, \vec{x}_1') h_2(\vec{x}_2, \vec{x}_2') \times \langle b_1(\vec{x}_1') b_2(\vec{x}_2')\rangle \right|\right|^2. \tag{2.28}$$

The comparison shows that the second-order thermal correlation $\langle a^{\dagger}(\vec{x}) a(\vec{x}\,')\rangle$ and the signal-idle photon correlation function $\langle b_1(\vec{x}) b_2(\vec{x}')\rangle$ have the same effect.

Analyzing a specific thermal ghost imaging system, as shown in Fig. 2.7, assuming that the object is described by a complex amplitude transmittance function $T(\vec{x})$, and that in the signal light path, the object is placed at the focal length of the lens, the impulse response function can be written as

$$h_1(\vec{x}_1, \vec{x}_1') = -\frac{\text{i}}{\lambda f} \exp\left(-\frac{2\pi i}{\lambda f} \vec{x}_1 \cdot \vec{x}_1'\right) T(\vec{x}_1'). \tag{2.29}$$

The test optical path impulse response function can be obtained when $z = f$

$$h_2(\vec{x}_2, \vec{x}_1') = -(\text{i}/\lambda f) \exp[-(2\pi i/\lambda f) \vec{x}_2 \cdot \vec{x}_1']. \tag{2.30}$$

Obtain the formula for the correlation function

$$G_{\text{th}}(\vec{x}_1, \vec{x}_2) \propto |\langle n(-\vec{x}_2 \frac{2\pi}{\lambda f})\rangle_{\text{th}} \widetilde{T}((\vec{x}_2 - \vec{x}_1) \frac{2\pi}{\lambda f})|^2. \tag{2.31}$$

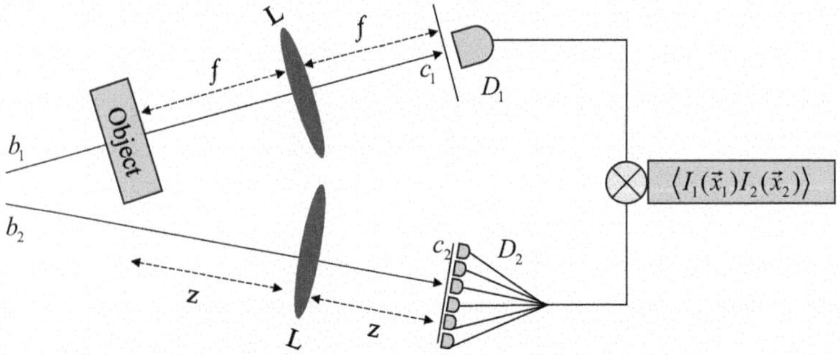

Fig. 2.7 One ghost imaging scheme with incoherent thermal light

which $\langle n(\vec{q})\rangle_{\mathrm{th}}$ is the expectation of the photon number in thermal mode. $\widetilde{T}(\vec{q}) = \int (\mathrm{d}\vec{x}/2\pi)e^{-\mathrm{i}\vec{q}\cdot\vec{x}}T(\vec{x})$ is the spectrum of objects.

The test optical path impulse response function can be obtained when $z = 2f$

$$h_2(\vec{x}_2, \vec{x}_2') = \delta(\vec{x}_2 + \vec{x}_2')\exp[-\mathrm{i}|\vec{x}_2|^2\pi/\lambda f]. \tag{2.32}$$

The correlation function obtained at this point satisfies

$$G_{\mathrm{th}}(\vec{x}_1, \vec{x}_2) \propto \left|\left\langle n(\vec{x}_1\frac{2\pi}{\lambda f})\right\rangle_{\mathrm{th}}\right|^2 |T(-\vec{x}_2)^2|. \tag{2.33}$$

Then the information about the object is derived from the light intensity correlation function.

In the experimental aspect, Alejandra et al. [6] used the optical path shown in Fig. 2.8 to realize the pseudo-thermal ghost imaging experiment for the first time, and verified that the pseudo-thermal ghost imaging can also satisfy the two-photon Gaussian thin lens equation similar to the quantum ghost imaging, but the more complex the information of the object, the visibility of the pseudo-thermal ghost imaging image realized by the pseudo-thermal light is lower, and the image is more blurred, and the practical application needs to satisfy the insensitivity of detection method to the background noise. At this time, some people believe that pseudo-thermal light ghost imaging is generated by non-local two-photon interference, and do not recognize the classical light source can achieve ghost imaging.

The computational ghost imaging method uses codable spatial light modulators and CCDs, which can digitize the data and realize ghost imaging with the help of computers. After the computational ghost imaging method was proposed, the spatial light modulator gradually replaced the random light source, the CCD replaced the photon detector, and people were no longer entangled in the ghost imaging mechanism whether it was a quantum light source or a classical light source, but focused on faster imaging and higher-quality imaging, and computational ghost imaging became a new computational optical imaging method.

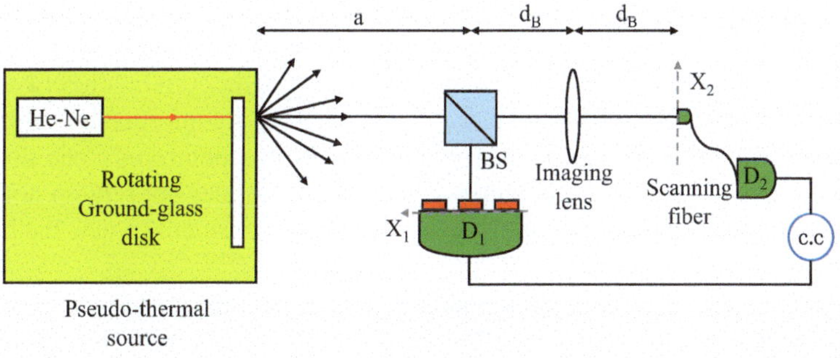

Fig. 2.8 Experimental diagram of the realization of ghost imaging by a thermal light source

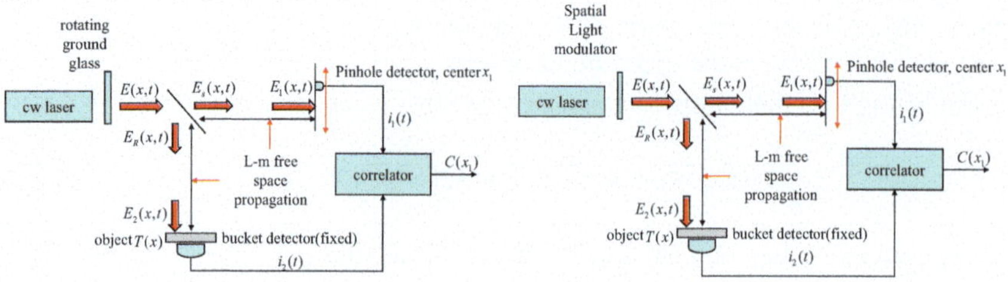

Fig. 2.9 Comparison of pseudo-thermal optical ghost imaging and ghost imaging realized by spatial light modulators

The concept of computational ghost imaging was first introduced by Jeffrey H. Shapiro. Considering that pseudo-thermal light sources can be well modeled as narrow-band classical Gaussian stochastic processes with coherent separable Gaussian-Schell model correlation structures, laser irradiation of rotating hair glass to generate thermal light sources was replaced by the use of spatial light modulators [7].

As shown in Fig. 2.9, the spatial light modulator loads a randomized phase $\exp[\mathrm{i}(\phi_{nm}(t))]$ at each pixel, Due to randomness, $\phi_{nm}(t)$ satisfies independent and identically distributed processes

$$\langle e^{i\phi_{nm}(t)}\rangle = 0 \text{ and } \langle e^{\mathrm{i}[\phi_{nm}(t_2)-\phi_{jk}(t_1)]}\rangle = \delta_{jn}\delta_{km}e^{-|t_2-t_1|/T_0}. \tag{2.34}$$

The central limit theorem gives $E_1(\rho, t)$ and $E_2(\rho, t)$ satisfy the joint Gaussian distribution. Thus, a spatial light modulator shown on the right side of Fig. 2.9 will generate ghost imaging with a spatial resolution of $\lambda_0 L/D$ in a spatial region of width $\lambda_0 L/d$. Noise generators can drive the spatial light modulator to generate random phases, but more widely used is strong sinusoidal modulation

$$\phi_{nm}(t) = \Phi\cos[(\Omega_0 + \Delta\Omega_{nm})t], \tag{2.35}$$

where in each pixel is random.

Since the modulation produced by the spatial light modulator is known, we can use diffraction theory to calculate the diffraction pattern of the free propagation of the light field. At this point, using the computational ghost imaging system shown in Fig. 2.10, the distribution of the object can be derived from the correlation between the information received by the bucket detector and the intensity information calculated in advance to realize the ghost imaging process. At this point, only one detector is used, so this method of computational ghost imaging does not rely on nonlocal two-photon interference in quantum mechanics, and is considered a new method of computational optical imaging.

Under the guidance of Jeffrey H. Shapiro's theory, Bromberg et al. completed the first experiment of computational ghost imaging [8], and the experimental optical path is shown in Fig. 2.11. The spatial light modulator is loaded with a random phase pattern $\varphi_r(x, y)$, and for each random phase, since the phase pattern is known, the

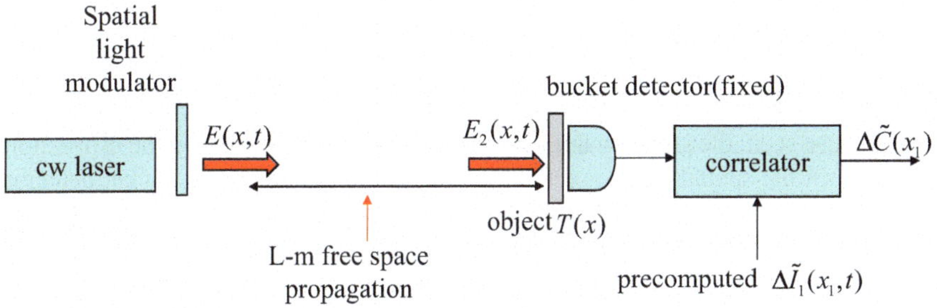

Fig. 2.10 Schematic of computational ghost imaging

field distribution after diffraction through free space is known. If the input field is $E_r(x, y, z = 0) = E_0 e^{i\varphi_r(x,y)}$, the field distribution after free propagation through a distance z can be computed using scalar diffraction theory

$$E_r(x, y, z) = \frac{\exp(i\lambda z)}{i\lambda z} \int d\xi d\eta E_r(x - \xi, y - \eta, 0) e^{i(\pi/\lambda z)(\xi^2+\eta^2)}. \tag{2.36}$$

The bucket detector is placed at a distance L from the object, and the light field arriving at the plane of the detector is $I_r = |E_r(x, y, z = L)|^2$, intensity measured by the bucket detector placed behind the object is

$$B_r = \int dx dy I_r(x, y, L) T(x, y). \tag{2.37}$$

Mutual correlation between the calculated light intensity I_r and the bucket detector intensity B_r is

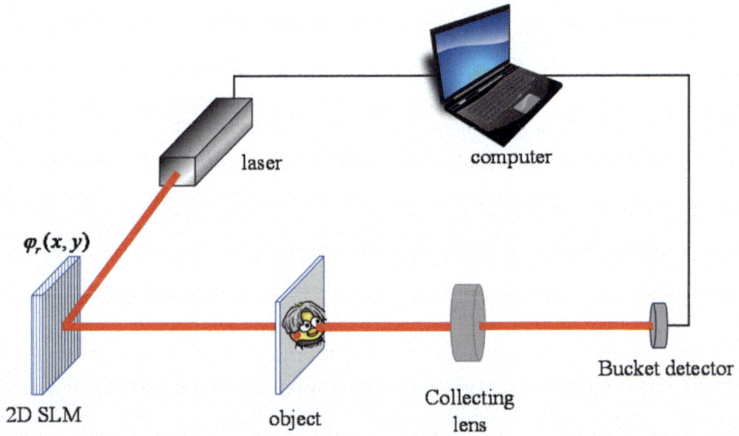

Fig. 2.11 Experimental setup of computational ghost imaging

$$G(x, y) \equiv \frac{1}{N}\sum_{r=1}^{N}(B_r - \langle B\rangle)I_r(x, y) = \langle BI(x, y)\rangle - \langle B\rangle\langle I(x, y)\rangle. \quad (2.38)$$

As can be seen, the ghost imaging results are obtained by summing the diffraction pattern I_r of the spatial light modulator computed weighted by the bucket detector intensity B_r. Theoretically, the greater the overlap between the randomly modulated diffraction pattern and the projected object, the greater the bucket detector intensity B_r will be, and the diffraction pattern I_r will be summed up with a greater weight. In summary, it is believed that the target image can be decomposed as a superposition of a series of random diffraction patterns, and the weights are the values detected by the bucket detectors.

Computational ghost imaging techniques require a long acquisition time to obtain a good signal-to-noise ratio, and this time can be greatly reduced by using compressed sensing techniques [9]. Compressed sensing is an advanced sampling and reconstruction technique where the main idea is to take into account the presence of redundancy in most natural signals or object structures and to reduce the redundancy in order to reduce the number of measurements needed for a realistic reconstruction. Computational ghost imaging techniques using compressive sensing can improve the signal-to-noise ratio of the reconstructed image while shrinking the acquisition time.

The compressive sensing algorithm searches for the sparsest image in the compressed basis, which contains the measured M < N random projections. The compressive sensing based reconstruction algorithm needs to solve a convex optimization problem to find the appropriate image $T_{CS}(x, y)$, minimizing the number of L1 modulus under the sparse basis

$$T_{CS} = T', \min\left\|\Psi\{T'(x, y)\}\right\|_{L_1}, \quad (2.39)$$

limited by $\int dxdy I_r(x, y)T'(x, y) = B_r, \forall_{r=1..M}$, which Ψ refer to transform operator for sparse bases. For the same number of random diffraction patterns, the computational ghost imaging method based on compressed perception can achieve better imaging quality compared with the computational ghost imaging side.

The method of compressed sensing can indeed greatly reduce the acquisition time required for imaging, but the improvement of signal-to-noise ratio of imaging is not too obvious, and previously, the ghost imaging technique is only applicable to some binary images, and it cannot achieve good results for complex images, so the practicality does not seem to be high. The proposed differential ghost imaging greatly improves the signal-to-noise ratio of ghost imaging and improves the quality of imaging.

Differential ghost imaging is an imaging method based on spatially correlated beams, which allows the projection function of an object to be calculated in absolute units, greatly improving the signal-to-noise ratio. If the bucket detector detects not the signal of a randomly scattered light field illuminating a transmissive object, but the signal of a fluctuating transmissive object, an analysis of the signal-to-noise ratio

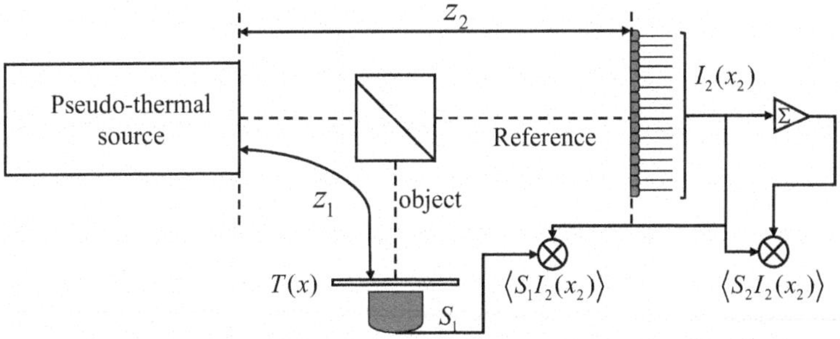

Fig. 2.12 Schematic of differential ghost imaging

of the imaging shows that this leads to a significant improvement in the signal-to-noise ratio of the imaging, and that the improvement in signal-to-noise ratio is very significant for highly transmissive objects [10].

The schematic of differential ghost imaging is shown in Fig. 2.12. Assuming that the detectors all detect a large number of photons, ignoring the effect of scattering noise, and considering that the reference and signal beams are perfectly correlated random variables, it is possible to make $I_1(\mathbf{x}_1) = \alpha I_2(\mathbf{x}_2)$, where α is a factor that takes into account the imbalance of the beam splitter and the detector between the two optical paths, in Fig. 2.12.

In the conventional ghost imaging method, the bucket detector detects the signal as follow

$$S_1 = \int_{A_{\text{beam}}} I_1(\mathbf{x}_1)T(\mathbf{x}_1)\mathrm{d}^2\mathbf{x}_1, \tag{2.40}$$

where $T(x, y)$ denotes the transmittance function of the object, a measurement obtained by conventional ghost imaging is

$$\langle O(\mathbf{x}_2)\rangle = \langle \delta S_1 \delta I_2(\mathbf{x}_2)\rangle, \tag{2.41}$$

$$\delta S_i = S_i - \langle S_i\rangle, \delta I_2(x_2) = I_2(x_2) - \langle I_2(x_2)\rangle. \tag{2.42}$$

Then after some treatment, there is

$$\langle \delta O^2(\mathbf{x}_2)\rangle = \langle \delta S_1^2\rangle\langle \delta I_2^2(\mathbf{x}_2)\rangle, \tag{2.43}$$

$$\langle O^2\rangle = A_{\text{coh}} A_{\text{beam}} \langle I_1\rangle^2 \langle I_2\rangle^2 \overline{T^2}. \tag{2.44}$$

The signal-to-ratio of conventional ghost imaging is obtained as below

$$(\mathrm{SNR})_{\mathrm{GI}} = \frac{[\Delta\langle O(\mathbf{x}_2)\rangle]^2_{\min}}{\langle \Delta O^2 \rangle} = \frac{m}{N_{\mathrm{speckle}}} \frac{\Delta T^2_{\min}}{\overline{T^2}}. \tag{2.45}$$

It can be seen that $SNR_{GI} \propto 1/\overline{T^2}$, for near-transparent objects ($\overline{T^2} \sim 1$), ghost imaging is more difficult to reconstruct the object and requires more random scattering patterns.

Differential ghost imaging methods can be used to solve this problem, imagine by using a new differential bucket detector:

$$S_{-} = \int_{A_{\mathrm{beam}}} I_1(\mathbf{x}_1)\delta T(\mathbf{x}_1)\mathrm{d}^2\mathbf{x}_1, \tag{2.46}$$

At this point the bucket detector detects not the signal of a random scattering pattern illuminating the object, but the signal of a random scattering pattern illuminating the object differentially. In order to find a way to measure this, define the bucket detection signal as follow:

$$S_2 = \int_{A_{\mathrm{beam}}} I_2(\mathbf{x}_2)\mathrm{d}^2\mathbf{x}_2, \tag{2.47}$$

and then we can obtained

$$S_{-} = S_1 - \frac{\langle S_1 \rangle}{\langle S_2 \rangle} S_2, \tag{2.48}$$

at this time, it is easy to measure S_{-}, and similarly, differential ghost imaging is measured as

$$\langle O_{-}(\mathbf{x}_2)\rangle = \langle \delta S_{-} \delta I_2(\mathbf{x}_2)\rangle = \langle S_1 I_2(\mathbf{x}_2)\rangle - \frac{\langle S_1 \rangle}{\langle S_2 \rangle}\langle S_1 I_2(\mathbf{x}_2)\rangle, \tag{2.49}$$

From the above two equations we can derive an expression for the differential ghost imaging signal-to-noise ratio obtained after mth independent measurements as follow

$$(\mathrm{SNR})_{\mathrm{DGI}} = \frac{[\Delta\langle O_{-}(\mathbf{x}_2)\rangle]^2_{\min}}{\langle \Delta O_{-}^2 \rangle} = \frac{m}{N_{\mathrm{speckle}}} \frac{\Delta T^2_{\min}}{\overline{\delta T^2}}, \tag{2.50}$$

Compared to the signal-to-noise ratio of conventional ghost imaging, there are

$$\frac{(\mathrm{SNR})_{\mathrm{DGI}}}{(\mathrm{SNR})_{\mathrm{GI}}} = 1 + \frac{1}{\sigma^2_{\mathrm{rel}}}, \tag{2.51}$$

where $\sigma^2_{\mathrm{rel}} = \overline{\delta T^2}/(\overline{T})^2$, for objects with a large transmittance, there are usually $\overline{\delta T^2} \ll (\overline{T})^2, \sigma^2_{\mathrm{rel}} \ll 1$, then the differential ghost imaging method has a significant improvement in signal-to-noise ratio compared to the conventional ghost imaging method.

In ghost imaging, some noise can affect the imaging quality, for example, instability of the laser light source or variations in the modulation efficiency of the spatial light modulator can affect the ghost imaging signal-to-noise ratio. In general, the larger the value measured by the bucket detector, which indicates the larger weight accumulated for that particular scattering field, the external noise affects the scattering weight and thus the overall reconstruction. Normalized ghost imaging is based on normalizing each individual measurement S and its average value according to a reference signal R to deal with the effect of noise sources on imaging.

Similar to differential ghost imaging, normalized ghost imaging uses two bucket detectors, the CCD is replaced by bucket detector in the reference optical path, and the measurements of the two bucket detectors are eventually correlated to derive the object transmittance function. The experimental schematic of normalized ghost imaging is shown in Fig. 2.13, and after the object, the bucket detector measurements can be expressed as

$$S = \int_{A_l} I(x_S, y_S) T(x_S, y_S) \mathrm{d}x_S \mathrm{d}y_S, \tag{2.52}$$

where A_l is the area of laser, $T(x_s, y_s)$ is the transmittance function of object. The bucket detector measurements of the reference optical path can be expressed as

$$R = \int I(x_R, y_R) \mathrm{d}x_R \mathrm{d}y_R \tag{2.53}$$

In conventional computational ghost imaging, the reconstruction result of an object $O(x, y)$ is obtained from S and computed $I(x, y)$ Object $O(x, y) = \langle O_i(x, y) \rangle$, the contribution to the reconstruction at each iteration is

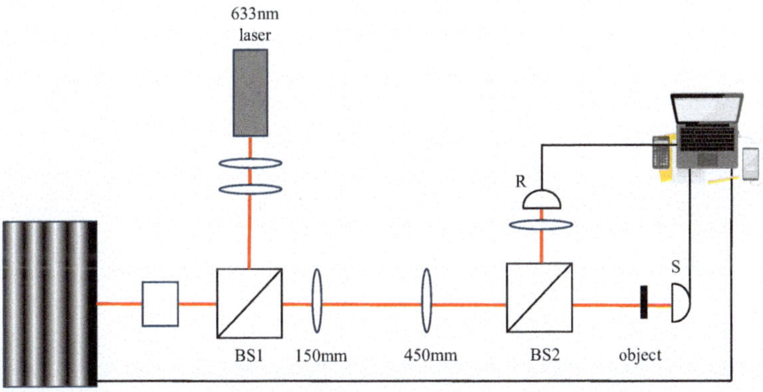

Fig. 2.13 Schematic of normalized ghost imaging

$$O_i(x, y) = (S - \langle S \rangle)(I(x, y) - \langle I(x, y) \rangle) \tag{2.54}$$

It can be seen that the weights used for reconstruction are $(S - \langle S \rangle)$. In differential ghost imaging, we have

$$O_i(x, y) = \left(S - \frac{\langle S \rangle}{\langle R \rangle} R \right)(I(x, y) - \langle I(x, y) \rangle), \tag{2.55}$$

The weight factor used for reconstruction becomes $(S - \langle S \rangle R / \langle R \rangle)$. At this point, the normalized ghost imaging method is considered, which normalizes each individual measurement S as well as the mean value according to the reference signal R to deal with time-varying noise more intuitively. In normalized ghost imaging,

$$O_i(x, y) = \left(\frac{S}{R} - \frac{\langle S \rangle}{\langle R \rangle} \right)(I(x, y) - \langle I(x, y) \rangle) \tag{2.56}$$

The weight factor used for reconstruction becomes $(S/R - \langle S \rangle / \langle R \rangle)$. A comparison shows that differential ghost imaging and normalized ghost imaging satisfied following equation

$$\langle O(x, y)_{NGI} \rangle = \frac{1}{\langle R \rangle} \langle O(x, y)_{DGI} \rangle \tag{2.57}$$

And normalized ghost imaging has been shown to have the same signal-to-noise ratio as differential ghost imaging.

The application of deep learning in optical imaging [11] exemplifies the advantages of the application of deep learning in solving various inverse problems in optical imaging. Deep learning can also be used in ghost imaging techniques. In ghost imaging, the signal-to-noise ratio of the reconstructed image is proportional to m/N_s, and the larger the number of random scattering patterns m used for illumination, the higher the quality of the reconstructed image is Often, traditional computational ghost imaging methods require multiple illumination, which is time consuming, and although compressed sensing improves the quality of the imaging, high-quality reconstruction is not possible even when m is too small. A ghost imaging method incorporating deep learning is proposed to achieve high quality imaging even with a small m value.

The flowchart of using deep learning ghost imaging is shown in Fig. 2.14, first, reconstruct the image with the traditional computational ghost imaging method using a small m value, which can be predicted that the reconstructed image O will be subjected to serious noise interference, and then use the reconstructed image O and the corresponding real dataset T as the training data and test data to train and test the deep neural network model, so that the deep neural network can realize the ghost imaging method with input reconstructed image O and output the real data T. In this way, the ghost imaging method with high quality reconstructed image can be realized

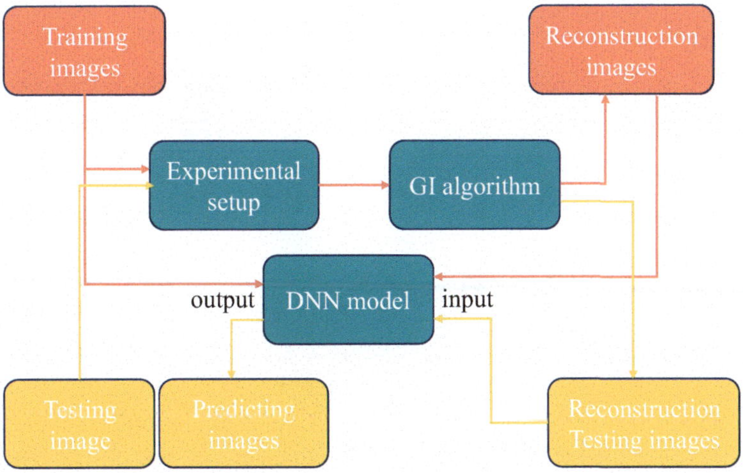

Fig. 2.14 Flowchart of ghost imaging using deep neural networks

with a small m value, and the robustness of the deep learning-based ghost imaging method to noise is stronger than the traditional ghost imaging method [12].

Mathematically, ghost imaging is analogous to an inverse problem in which image information is inferred from a time series of bucket detector measurements and reference light intensity. Positive and negative ghost imaging methods can simply obtain an image of an object from the raw reference detector data without involving the operations associated with traditional ghost imaging, so they are also known as correspondence imaging [13]. Positive–negative ghost imaging can obtain a normal image as well as an opposite negative image, and the quality of the image obtained after the same amount of time is much better than that of conventional ghost imaging.

The schematic diagram of positive and negative ghost imaging is shown in Fig. 2.15. In positive and negative ghost imaging, the bucket detector measurements are no longer used as a weighting factor to sum up all scattering patterns, but as a logical trigger to select the appropriate random scattering pattern. Traditional ghost imaging proposed that in ghost imaging, summing up all the random scattering patterns does not get any information of the image to be measured due to randomness, but if the random scattering pattern that meets certain conditions of the random scattering pattern, for example, the random scattering in the signal light path through the object detected by the bucket detector is greater than the average of all the detected values, then the information of the image can be obtained.

Analyze the principle of positive and negative ghost imaging. Considering that the object and the reference light path detector are equidistant from the light source, the light field distributions are also equal, and the light transmitted through the object constitutes a bucket detector signal that fluctuates at its mean value during each recording, and the corresponding reference light path detection values for that part of the object fluctuate in concert. If the object $\{\vec{x}_0\}$ is divided into two regions: X_1 the part that transmits light and X_2 the part that does not transmit light. When the

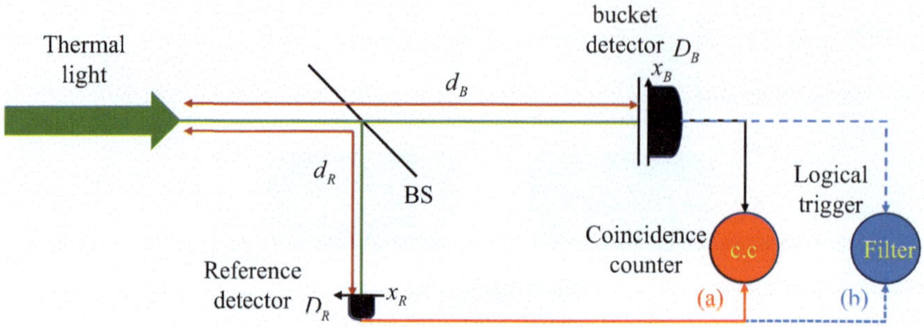

Fig. 2.15 Schematic of correspondence imaging

light field in the region fluctuates greatly, the bucket detector output will be larger, and the corresponding reference detector value in the region will also be larger. At this time, choose the part of the backet detection signal is large random scattering, and use it to sum up to derive the object information.

The bucket detector strength can be written as

$$I_B(t_i) = \sum_{\{\vec{x}\}} I_R(\vec{x}, t)|T(\vec{x})|^2, \tag{2.58}$$

Considering the randomness, the average of all reference signals can be written as

$$\overline{I}_R(\vec{x}_R) = \frac{1}{N} \sum_{\{t_i|i=1,\dots,N\}} I_R(\vec{x}_R, t_i) = \overline{I}_R \tag{2.59}$$

Then we can get $\overline{I}_B = \overline{I}_R \sum_{\{\vec{x}\}} |T(\vec{x})|^2$, in a narrow range away from $\overline{I}_B$, centered at I_B^+, make the approximation shown below

$$I_B(t_i) \simeq \sum_{\{\vec{x}|\vec{x}\in X_1\}} I_R(\vec{x}, t_i)|T(\vec{x})|^2 \tag{2.60}$$

If N_B^+ is the number of $I_B(t_i)$ in the range I_B^+, then the function can be written

$$\begin{aligned} G_+(\vec{x}) &\equiv \frac{1}{N_{B+}} \sum_{\{t_i|I_B(t_i)\approx I_B^+\}} I_B(t_i) I_R(\vec{x}, t_i) \\ G_+(\vec{x}) &\approx \frac{I_B^+}{N_{B+}} \sum_{\{t_i|I_B(t_i)\approx I_B^+\}} I_R(\vec{x}, t_i) \equiv \frac{I_B^+}{N_B^+} I^+(\vec{x}) \end{aligned} \tag{2.61}$$

Correlation functions can also be written as

$$G_{+}(\vec{x}) \approx \overline{I}_B \overline{I}_R + \Delta_R^2 |T(\vec{x})|^2, \Delta_R^2 = \frac{1}{N_B^+} \sum_{\{t_i | I_B(t_i) \approx I_B^+\}} \left[I_R(\vec{x}, t_i) - \overline{I}_R \right]^2 \quad (2.62)$$

Comparing the two equations gives

$$I^{+}(\vec{x}) \propto C_b + |T(\vec{x})|^2, \quad (2.63)$$

where $C_b = \overline{I}_B \overline{I}_R / \Delta_R^2$ is a constant due to background noise. Similarly, satisfying $I_B(t_i) \approx I_B^-$ can derive negative image information. To eliminate the noise, the negative image can be subtracted using the positive image. In addition to $\overline{I}_B$, the bucket detection values can be divided into more parts using I_N and I_P, and different weights can be set to obtain a better quality of the reconstructed image. This corresponding ghost imaging method because it does not utilize the relevant operations, strictly speaking, cannot be regarded as a ghost imaging technology, is inspired by the method of ghost imaging, the use of statistical ideas and theories and imaging of the new technology, after the simulation and experimentation of the accidental ground can be achieved good imaging results.

2.4 Single-Pixel Imaging

Single-pixel imaging is similar in principle to ghost imaging. In 2008 Duarte et al. pioneered the technique of single-pixel imaging [14]. The experimental optical pathway, as shown in Fig. 2.16, incorporates compressed sensing techniques, where only a single-pixel detector can be used to obtain the image information of the object with the use of a modulated illumination method. Subsequently, various modulated illumination methods have been proposed for high quality single pixel imaging.

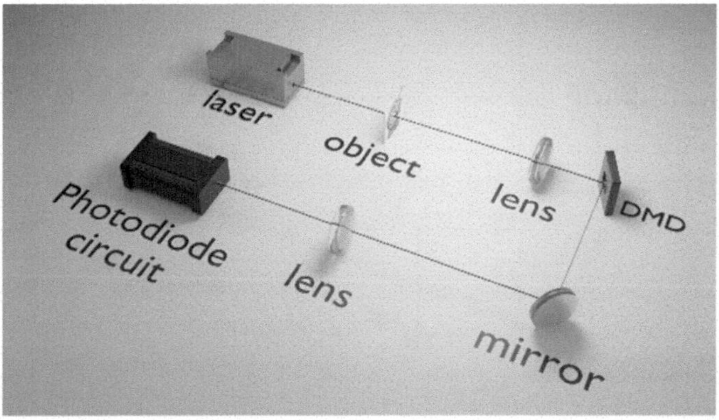

Fig. 2.16 Schematic of single-pixel camera[14]

Considering the natural property that the spectrum of an object is sparse, the image can be decomposed into a linear superposition of sinusoidal functions of different frequencies, then most of the spectrum corresponds to a coefficient of 0, and the sinusoidal pattern is also easy to generate, then consider the use of tuned sinusoidal signals to irradiate the object, the sinusoidal transformation of the single-pixel imaging is the use of the analyzed gray scale sinusoidal harmonic pattern, rather than the random scattering of the ghost imaging or binary pattern, to obtain the object's spectral information rather than the object's null-domain information, the spectrum is obtained by the use of the inverse Fourier transform can be reconstructed object [15].

The experimental schematic diagram of the sinusoidal transform single-pixel imaging technique is shown in Fig. 2.17. A sinusoidal pattern is used to illuminate the object, a single-pixel detector is used to collect transmitted or reflected light, and the Fourier coefficients of the object can be obtained by using a four-step phase-shift method, which is analyzed as shown below:

$$P_{\phi}\left(x, y; f_x, f_y\right) = a + b \cdot \cos\left(2\pi f_x x + 2\pi f_y y + \phi\right). \tag{2.64}$$

The total reflected intensity produced by a structural light source when it illuminates an object is

$$E_{\phi}\left(f_x, f_y\right) = \iint_{\Omega} R(x, y) P_{\phi}\left(x, y; f_x, f_y\right) \mathrm{d}x\mathrm{d}y. \tag{2.65}$$

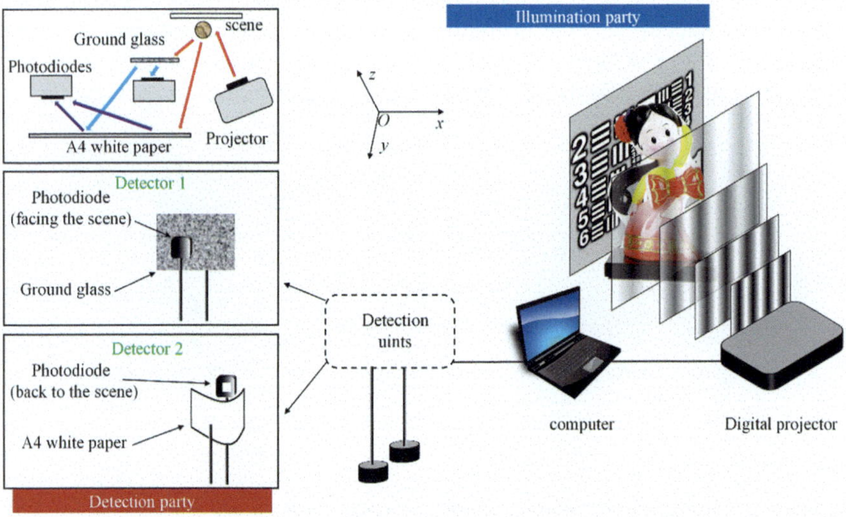

Fig. 2.17 Experimental schematic diagram of sinusoidal single-pixel imaging [15]

If a single-pixel detector is used, the single-pixel detector collects only a portion of the reflected light, and the single-pixel detector responds in total

$$D_\phi(f_x, f_y) = D_n + kE_\phi(f_x, f_y). \tag{2.66}$$

The complex Fourier coefficients $C(f_x, f_y)$ can be derived using the four-step phase-shifting method, and the DC term of the background illumination can be removed at the same time

$$\begin{aligned}\mathcal{F}\{R(x, y)\} &= \left[D_0(f_x, f_y) - D_\pi(f_x, f_y)\right] + \mathrm{i}\left[D_{\pi/2}(f_x, f_y) - D_{3\pi/2}(f_x, f_y)\right] \\ &= \left[E_0(f_x, f_y) - E_\pi(f_x, f_y)\right] + \mathrm{i}\left[E_{\pi/2}(f_x, f_y) - E_{3\pi/2}(f_x, f_y)\right] \\ &\propto \iint\limits_\Omega R(x, y)\exp\left[\mathrm{i}2\pi(f_x x + f_y y)\right]\mathrm{d}x\mathrm{d}y.\end{aligned} \tag{2.67}$$

Then the final object $R(x, y)$ can be derived from the Fourier inverse transform

$$R(x, y) = \mathcal{F}^{-1}\left\{\left[D_0(f_x, f_y) - D_\pi(f_x, f_y)\right] + \mathrm{i}\left[D_{\pi/2}(f_x, f_y) - D_{3\pi/2}(f_x, f_y)\right]\right\}. \tag{2.68}$$

The Fourier transform single-pixel imaging method utilizes the sparsity and conjugate prior of the image spectrum to achieve fast single-pixel imaging with very high efficiency and low computational cost. Compared with the sinusoidal transform single-pixel imaging method, the Fourier transform single-pixel method performs significant sampling in the frequency domain and does not need to sample all the Fourier coefficients [16]. The centrosymmetric conjugate property of the natural image spectrum is also introduced into the graphical strategy, and only two steps of the phase-shift method are required for each spatial frequency.

The illumination strategy of the Fourier-transform single-pixel method is different from that of sinusoidal-transform single-pixel imaging, and in order to illustrate the illumination strategy in detail, the practical significance of the single-pixel measurements in the frequency domain should be analyzed first. According to the Fourier transform theory, a two-dimensional image I can be represented as $I = \sum_{\forall i} c_i B_i$ and the illumination pattern can be represented as $P = \sum_{\forall j} \hat{c}_j B_j$.

The single-pixel measurement d can be represented as

$$\begin{aligned}s &= \left|\sum\nolimits_{\forall m,n} I(m, n)P(m, n)\right| \\ &= |\sum\nolimits_{\forall m,n} [\sum\nolimits_{\forall i} c_i B_i(m, n)][\sum\nolimits_{\forall j} \hat{c}_j B_j(m, n)]| \\ &= \left|\sum\nolimits_{\forall i,j} c_i \hat{c}_j [\sum\nolimits_{\forall m,n} B_i(m, n) B_j(m, n)]\right|,\end{aligned} \tag{2.69}$$

where (m, n) represent 2D spatial coordinates, utilizing the orthogonality of the Fourier basis

$$f(x) = \begin{cases} \sum_m \sum_n B_i(m,n)B_j(m,n) = 0,\ i \neq j, \\ \sum_m \sum_n B_i(m,n)B_j(m,n) = 1,\ i = j. \end{cases} \tag{2.70}$$

We can obtain

$$s = \left| \sum_j c_j \hat{c}_j \right|. \tag{2.71}$$

Let $c_1 = a_0 + ib_0, c_2 = a_0 - ib_0, c_3 = d_0$ denote the three nonzero Fourier coefficients of the target image I, and $\hat{c}_1 = a_1 + ib_1, \hat{c}_2 = a_1 - ib_1, \hat{c}_3 = d_1$ denote the coefficients of the sinusoidal pattern used for illumination, which leads to

$$\begin{aligned} s &= |c_1\hat{c}_1 + c_2\hat{c}_2 + c_3\hat{c}_3| \\ s &= |(a_0 + ib_0)(a_1 + ib_1) + (a_0 - ib_0)(a_1 - ib_1) + d_0 d_1| \\ s &= 2(a_0 a_1 - b_0 b_1) + d_0 d_1 \end{aligned} \tag{2.72}$$

Based on the above derivation, obtaining the Fourier coefficients transforms to calculating a_0 and b_0, and the calculation of the coefficients can be realized by projecting three patterns. The first one is an equal-intensity uniform pattern, which is equal to the average pixel value of the projected pattern P, and is used to calculate d_0, the other two patterns Fourier coefficients satisfy the sinusoidal pattern of $\{a_1 = 1/2, b_1 = 0, d_1 = 1\}$ and $\{a_1 = 0, b_1 = 1/2, d_1 = 1\}$, then a_0 and b_0 can be calculated simply. According to the above analysis, we can calculate all the Fourier coefficients we want to get, and finally use the Fourier inverse transform to get the spatial image. In addition, there is also the Hadamard transform single-pixel imaging similar to the Fourier transform single-pixel imaging, based on the Hadamard transform [17], the Hadamard spectrum of the object image is obtained, and finally the inverse Hadamard transform is utilized to derive the object image.

2.5 Basic Structures of Different Ghost Imaging Systems

2.5.1 Quantum Ghost Imaging System

Quantum ghost imaging is a type of imaging system that uses entangled photon pairs to create an image of an object. Entangled photons are pairs of photons that are generated in such a way that their properties, such as polarization or phase, are intrinsically linked, even when they are separated by large distances. A typical example of an entangled state is the two-photon state generated from spontaneous parametric down conversion (SPDC), the nonlinear optical process in which a laser pump incident on a crystal leads to the emission of a pair of photons in an entangled state.

As the initial version of ghost imaging, quantum ghost imaging uses the second-order correlation property of entangled photon pairs produced by SPDC to carry out "ghost" imaging. In the experiment, the entangled photon pairs generated by SPDC are divided into two paths by polarization beam splitter: the object optical path and the reference optical path. The two separated photons propagate along their respective paths. In the light path with an object, photons interact with the object and are then detected by a bucket detector with no spatial resolution. At the same time, the photons that do not interact with the object in the reference optical path are detected by the detector which can distinguish the spatial position. Then, the measurement signals from the two optical paths are input into the coincidence counting circuit. After a large number of randomly distributed entangled photon pairs in space, the object image is reconstructed iteratively according to the measured results.

Pittman et al. [2] completed the quantum ghost imaging experiment for the first time. As shown in the Fig. 2.18, entangled photon pairs are produced in a nonlinear beta barium borate (BBO) crystal pumped by a 351.1 nm laser beam. The pair of entangled photons emitted from the crystal is separated by a polarization beam splitter so that the two entangled photons propagate along different paths. The transmission object is placed in one of the paths, and a lens is generally placed between the polarization beam splitter and the object to constrain the spatial position of the photons.

Photons passing through the object are collected by a lens placed in front of the detector and then detected by a single-photon sensitive single-pixel detector. At the same time, the photons propagating in another optical path and orthogonal polarization with the signal path do not interact with the object and are detected by the scanning detector after passing through a lens. The scanning detector consists of a multimode fiber and another single photon detector. One end of the multimode fiber is connected to the single photon detector, and the other end scans the image plane

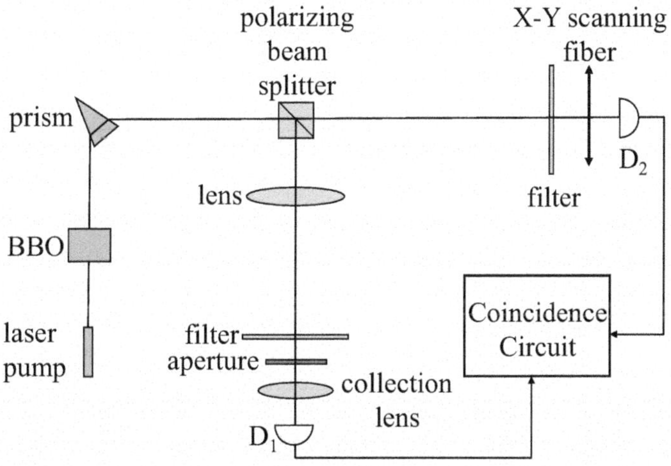

Fig. 2.18 The experimental setup of ghost imaging

of the lens to collect photons so that the position of photons can be determined in real time. The output signals of two single photon detectors are sent to a coincidence counting circuit of a narrow bandwidth acceptance window. The coincidence count is carried out by referring to the optical fiber plane coordinate function in the beam, and then the object image is reconstructed with the coincidence result.

It is worth mentioning that the scanning process at high resolution (low fiber diameter) will lead to lower detection efficiency. However, in order to improve the sampling efficiency, the scan detector can be replaced by an area array camera with high quantum efficiency.

In addition, quantum ghost imaging can not only image amplitude objects, but also pure phase objects. Abouraddy et al. [18] realize the imaging of phase-only objects using entangled photons for the first time. The only difference between their experimental device and that of Pittman et al. is that a small hole is placed in front of the two detectors located on the object optical path and the reference optical path. Due to the existence of the small holes, the system is partially coherent in space, so that the intensity correlation of the two beams is related to their phase difference. Before their work, any coincidence imaging technology using barrel detectors could not realize the imaging of phase objects, and their work opened the prelude to the imaging of phase objects by ghost imaging.

The arrangement used in their experiment is shown schematically in Fig. 2.19. The source of light is spontaneous parametric down-conversion generated in a nonlinear crystal (NLC) pumped by a laser beam. The down-converted photons are emitted in pairs in a collinear configuration; they are subsequently separated into two beams (each of which is spatially incoherent). One of the beams impinges on the object, upon reflection from the object, the beam is detected by a single detector D_1 fitted with a fixed small pinhole P_1. The other beam does not interact with the object, and is simply directed to a small pinhole P_2 and a detector D_2. The coincidence rate of the photons detected by the two detectors is measured as detector D_2 (together with pinhole P_2) is scanned while P_1 is held fixed. If the intervening optical system is appropriately designed, a coherent image of the phase object may be obtained.

2.5.2 Ghost Imaging System with Classical Light Sources

Since the original ghost imaging experiment was based on entangled photons, ghost imaging was considered to be essentially a quantum phenomenon at that time. But soon, Abouraddy et al. [3] questioned this, thinking that ghost imaging can also be achieved through classical light sources. Later, Bennink et al. [4] officially made the first ghost imaging experiment using classical angle-dependent laser pulses. From then on, some scientists began to think that the only requirement for ghost imaging is that there is a spatial correlation between the two light fields, whether it is a quantum field or a classical field. The ghost imaging system using laser pulse as light source is shown in Fig. 2.20. This imaging system is completely similar to the Quantum ghost imaging system, except that the classical source is used instead of the entangled

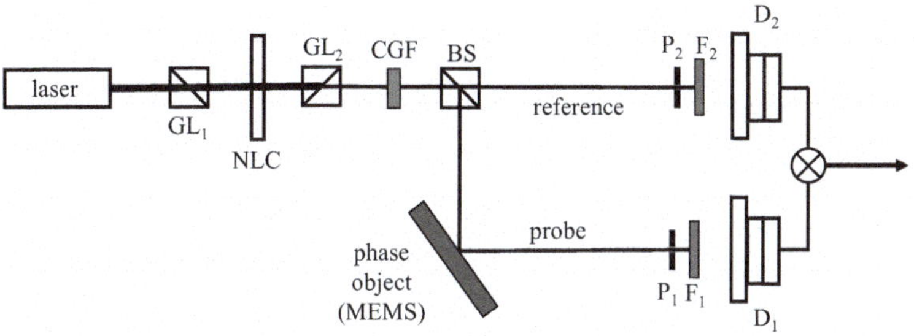

Fig. 2.19 Experimental arrangement for the quantum imaging of a reflective pure phase object (MEMS). NLC represents the nonlinear crystal, CGF stands for a colored glass filter, GL_1 and GL_2, are two orthogonally oriented Glan-Laser polarizing beam splitters, BS is a nonpolarizing beam splitter, P_1 and P_2, are vertical slits, F_1 and F_2, are interference filters, D_1 and D_2, are single photon detectors, and ⊗ represents an electronic coincidence circuit

source. The classic light source is made by cutting the laser beam, deflecting it by a variable amount, and then passing it through the beam splitter.

However, there are still some scientists who insist that ghost imaging is a quantum phenomenon, so that the discussion about whether ghost imaging needs quantum entanglement has been going on for a long time. Valencia et al. [6] built a ghost imaging device using classical incoherent thermal light as light source for the first time. In their experiment, the thermal light source is divided into two channels by using a beam splitter to simulate the entanglement characteristics of entangled photon pairs, and then they compare the role of thermal light and entangled photons in ghost imaging. The experimental results and analysis show that the classical thermal light source can image ghosts in both the near field and the far field. At the same time, their work further overturns the idea that ghost imaging is a quantum phenomenon.

The ghost imaging experiment setup with thermal light source is shown in Fig. 2.21. The laser emitted by the He–Ne laser irradiates on the rotating ground glass, destroys the phase correlation of the laser, and simulates the light field statistics

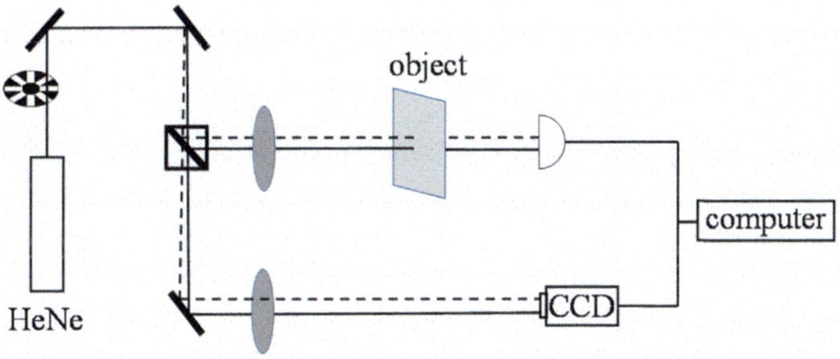

Fig. 2.20 The experimental setup used to perform ghost imaging with a classically correlated source

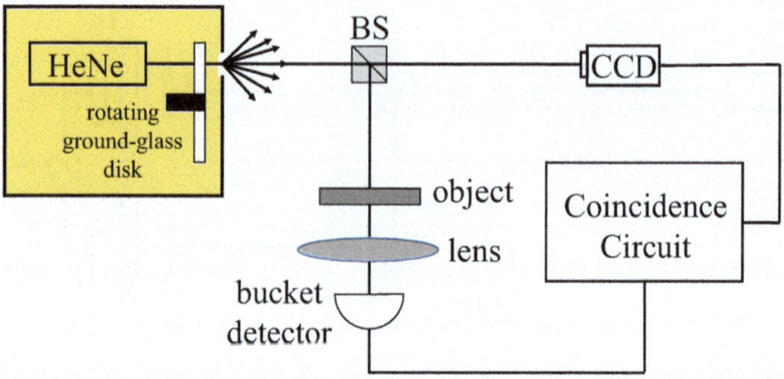

Fig. 2.21 The ghost imaging experiment setup with thermal light source

of the true thermal light to get the pseudo-thermal light source. Then the pseudo-thermal light is divided into two paths by a beam splitter, one of which is detected by an area array camera after transmission, and the spatial distribution of the light field is recorded, which is used as a reference optical path. The rest of the path is used as a signal optical path, which is transmitted at the same distance to the surface of the object to be imaged, and the light passing through the object is detected by the bucket detector under the collection of the lens. In addition, there is a synchronous trigger signal between the camera and the barrel detector to ensure that both have the same exposure time. As the spatial distribution of the light field changes with the rotation of the ground glass, several groups of light field distribution and bucket detection signals recorded by the camera can be obtained in the whole data acquisition process. The object information can be reconstructed by correlating and calculating the detection values of the reference optical path and the signal optical path.

After the quantum property of ghost imaging has been gradually overturned, scientists' enthusiasm for ghost imaging has not diminished, which has greatly promoted the in-depth development of ghost imaging. A few years later, ghost imaging based on pseudo-thermal light source was popularized and became a typical representative. On the one hand, it completely proves that ghost imaging can be realized by classical light source, not a pure quantum phenomenon; on the other hand, pseudo-thermal light source is relatively easy to obtain, which is helpful to promote ghost imaging to practical application.

Then the researchers compare the classical light ghost imaging and quantum ghost imaging through theory and experiment, and think that the main difference between the two kinds of ghost imaging results is that there is meaningless background noise in the classical ghost imaging, but not in the two-photon ghost imaging. The background noise in the classical light-ghost imaging results affects the imaging resolution and signal-to-noise ratio. Although the signal-to-noise ratio of classical light ghost imaging is lower than that of quantum ghost imaging, it is easier to achieve thermal ghost imaging because the classical thermal light source is easy to obtain.

2.5.3 Computational Ghost Imaging System

Based on the classic light ghost imaging experiment, Shapiro of the Massachusetts Institute of Technology used a spatial light modulator (SLM) instead of rotating ground glass to customize a series of pseudo-thermal light fields irradiated on the object plane. Thus, there is no need to use the beam splitter and the area array camera with spatial resolution in the experiment, and the real single-pixel imaging is realized. As a new version of ghost imaging, he defines this single-channel custom light field distribution of ghost imaging as computational ghost imaging (CGI) [7].

Bromberg et al. proposed ghost imaging using only a single detector to study CGI experimentally [8]. Using the spatial light modulator instead of the ground glass, only the signal light path is used, and the light intensity is also calculated to restore the object image, but the light field intensity of the reference light path is obtained by calculation rather than experimental detection. Figure 2.22 shows the computational ghost imaging setup. A pseudothermal light beam is generated by applying controllable phase masks with a spatial light modulator. The light emitted by the laser is illuminated on the object to be measured after phase or intensity modulation, and the total light intensity is received by the single pixel detector (SPD). Finally, by calculating the total light intensity sequence collected by SPD and the known reference light intensity for ghost imaging correlation calculation, the reconstructed object information can be recovered.

Computational ghost imaging simplifies the traditional ghost imaging experiment, reduces the experimental cost and threshold, and makes the new imaging technology of ghost imaging more practical. The traditional ghost imaging requires the correlation of two signals, which can easily introduce noise. In contrast, computational ghost imaging only needs to process a single signal, resulting in less signal noise. The signal used in computational ghost imaging is a random light field generated by a spatial light modulator, which produces higher resolution and finer details. In contrast, the signal used in traditional ghost imaging is a coherent beam generated by a beam splitter, which may introduce coherent artifacts and other limitations in

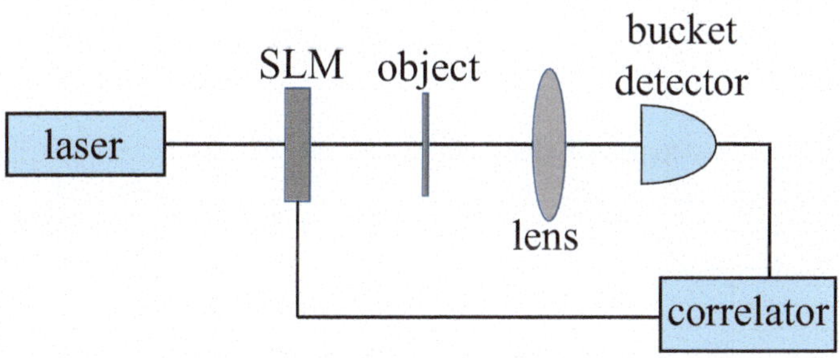

Fig. 2.22 The computational ghost imaging setup

resolution. Generally speaking, compared with traditional ghost imaging, computational ghost imaging has lower hardware cost, higher time resolution, lower signal noise and higher resolution, so it has greater advantages in some applications.

2.5.4 Temporal Ghost Imaging System

In the field of optics, there are dual characteristics of time and space, and many technologies are only proposed to be applied in the spatial domain at the beginning, but after a period of precipitation, they can often be extended to the time domain and achieve unexpected results. Temporal ghost imaging is proposed after people have a full understanding of spatial ghost imaging. Spatial ghost imaging reconstructs the spatially resolved image by using a single pixel detector without spatial resolution combined with a series of known spatial structure coding. Temporal ghost imaging uses slow detectors without time resolution combined with a series of known time structure coding to reconstruct fast signals with time characteristics. In addition, the development of temporal ghost imaging is basically the same as that of spatial ghost imaging, from the original dual-path classical temporal ghost imaging to single-optical path computational temporal ghost imaging.

When light propagates, there is a duality in space and time, that is, the diffraction of light in space is equivalent to the dispersion of light in time domain. Temporal ghost imaging evolved from spatial ghost imaging according to this theory. Ryczkowski et al. [19] completed temporal ghost imaging for the first time in experiments, and their experimental device is shown in Fig. 2.23.

First of all, the light wave fluctuating with time is emitted by the incoherent light source in time domain. Then, let the light wave pass through an optical fiber beam splitter to get two beams with the same fluctuation characteristics. On one path, the time object is placed, and after the time object interacts with the light wave, the integral intensity over a period of time is recorded by the slow detector, while on the other path, the trend of the light wave intensity changing with time is accurately recorded by the fast detector.

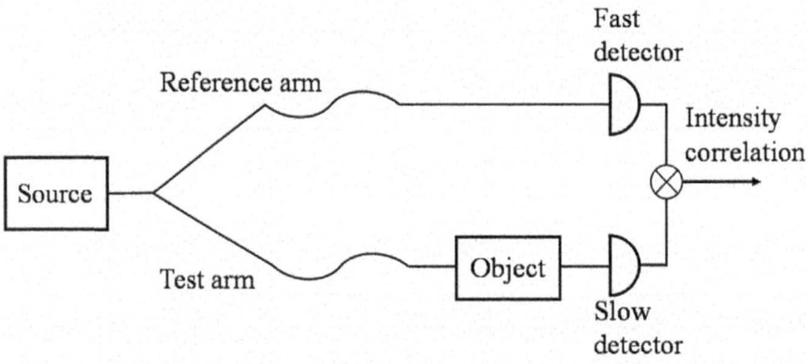

Fig. 2.23 Temporal ghost imaging experiment setup with classical non-stationary pulsed light

Similar to the spatial ghost imaging, when the light intensity fluctuation of the light wave is similar to the time object signal, the light intensity recorded by the slow detector will be larger, which will give more weight to the time-resolved signal recorded by the fast detector. On the contrary, the slow detector will get a smaller light intensity value, which will provide a smaller weight value for the time-resolved signal recorded by the fast detector.

After several times of weighted summation, that is, time correlation operation, the imaging result will gradually approach the time object. In the experiments of Ryczkowski et al., they used slow detectors to realize the imaging of time objects with 50 ps time resolution. Inspired by the computational ghost imaging in the spatial domain, Devaux et al. [20] used liquid crystal display to generate random signals for time domain coding of spatial multiplexing in the same year, and the signals interacting with time objects were collected by the area array camera, thus realizing the computational temporal ghost imaging with a single shot.

2.6 Applications of Ghost Imaging

Ghost imaging has garnered widespread attention due to its many unique characteristics that are not typically found in conventional imaging methods. The separation of the object path and reference path in ghost imaging systems provides greater flexibility. Furthermore, the inability of the detector in the object path to have spatial distribution allows for high-resolution imaging to be achieved. Ghost imaging is also able to function under challenging optical conditions, such as atmospheric turbulence and environments containing scattering media. Given these advantageous imaging characteristics, ghost imaging technology has found applications in various fields.

2.6.1 Optical Encryption Based on Ghost Imaging

The advantages of optical encryption include high-speed operation and the ability to hide data in multiple dimensions (such as phase, wavelength, spatial frequency, or polarization), making it difficult for unauthorized parties to access protected information. Additionally, optical procedures are a natural medium for dealing with images or holograms and benefit from continuous advancements in electro-optic devices.

Clemente applied ghost imaging technology to optical image encryption [21]. Figure 2.24 illustrates the encryption process. Alice encrypts the image using computational ghost imaging system, where a series of random phase $\varphi_i(x, y)$ are introduced into the spatial light modulator to simulate a thermal light source. A spatially coherent monochromatic laser beam passes through the SLM and the diffracted light field $I_i(x, y)$ illuminates the object. The light passing through the object is recorded by a single pixel detector to obtain a series of intensity values as ciphertext, $\{B_i\}$. This operation is repeated N times, each time with a different phase distribution, resulting

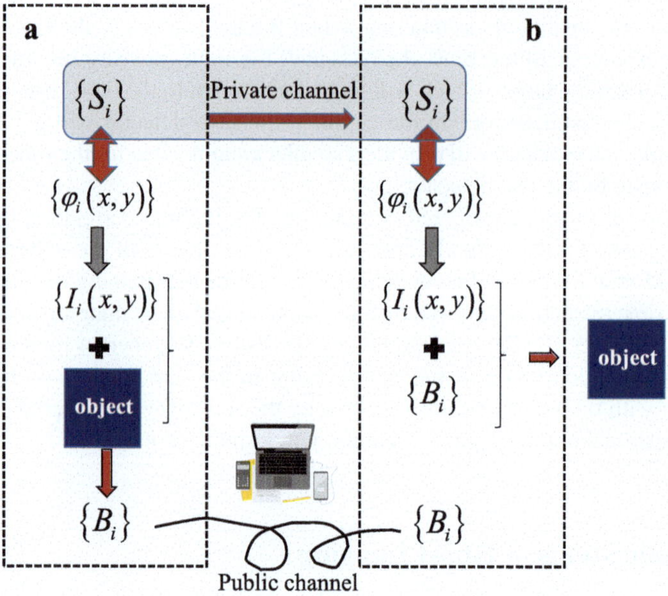

Fig. 2.24 Encryption and decryption process of encryption method based on CGI

in a set of N key components. Therefore, the object information is encoded in a vector of N components, with each component containing a corresponding intensity value detected by a single pixel detector. The encoded object information is then transmitted to Bob through a public channel. Meanwhile, the set of random phases introduced by the SLM, $\{S_i\}$, is transmitted to Bob via a secure channel as the encryption key, who shall decrypts the information through the key.

According to the CGI algorithm, the object information can be recovered by the following linear operations

$$O(x, y) = \frac{1}{N} \sum_{i=1}^{N} [B_i - \langle B \rangle] I_i(x, y), \tag{2.73}$$

where $\langle B \rangle$ is the average value of the measured intensity values $\{B_i\}$. $I_i(x, y)$ represents the Fresnel diffraction corresponding to the phase distribution i,

$$I_i(x, y) = |E_{in}(x, y) \exp[j\varphi_i(x, y)] \otimes h_z(x, y)|^2 \tag{2.74}$$

Here, $E_{in}(x, y)$ is the complex field of the coherent light beam, $h_z(x, y)$ is the Fresnel propagation kernel at a distance z, and $\otimes$ denotes the 2D convolution operation. Because the phase distribution is known, Bob can easily calculate the intensity distribution and decrypt the object information.

Compared with the traditional optical encryption scheme, the ciphertext based on computational ghost imaging is the light intensity sequence value, which greatly simplifies the storage space for storing ciphertext. The information recorded by this method is in the form of intensity, which simplifies the system. This research promotes the application of ghost imaging in optical information security. But the result of this paper is the numerical simulation of a binary image.

Tanha et al. used this method to process grayscale images and color images, and carried out experiments [22]. Besides the application of color object CGI in color imaging and processing, it can also significantly improve the security level of encrypted ghost images. The main idea behind using color objects is to distribute the information used to encrypt monochromatic images on at least three information channels, namely, red-green-blue (RGB). In three-channel encryption using CGI, each color refers to only a portion of the object information. In fact, in order to faithfully retrieve ghosting images through decryption, all three channels are needed. The experimental settings using virtual objects and three-channel reflected fluorescence CGI are shown in Fig. 2.25.

In traditional CGI, a coherent laser beam produces a speckle pattern after passing through a spatial light modulator, and the speckle pattern passes through the transparent area of the object mask. In order to reconstruct the ghosting image of the object mask, a single pixel detector placed behind the object measures the total reception

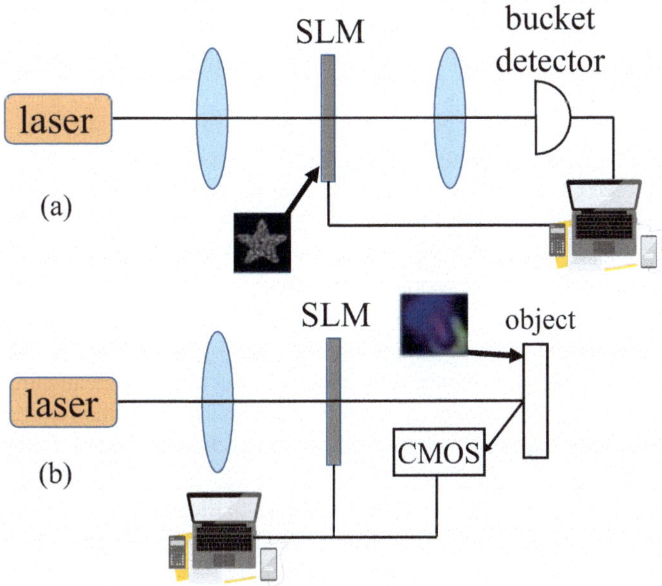

Fig. 2.25 Experimental setup with a virtual object and three channel reflective fluorescent CGI: **a** object is a gray-scale picture of a star that has been multiplied to a random speckle pattern and then loaded into the SLM, and **b** reflective CGI with an object made from florescent material in three different colors and placed approximately perpendicular to the laser beam

intensity of each random speckle pattern. The intensity detected by a single-pixel detector directly depends on the speckle pattern. Through experiments, the random speckle pattern is multiplied by the transmission function of the image to simulate the object mask on the SLM plane. In fact, the transmitted light through the SLM contains information about the speckle pattern and transmission function of the object (grayscale image) collected by the converging lens on a single pixel detector. Removing the transmission mask will compress the CGI setup and provide greater flexibility in the encryption of grayscale images without rebuilding its transmission mask.

In order to encrypt the ghost image of color objects, SLM is used to generate speckle patterns on fluorescent materials. The fluorescent material used comes from the yellowish green spectrum rather than pure green. Although the contrast of the ghost image retrieved from each channel is reduced, this sharing is invalid when using all channels to retrieve the final ghost image. The experimental results of eavesdropping on single-channel and three-channel CGI show that unauthenticated users are more likely to make errors in decrypting ghost images, and the percentage of key eavesdropping is smaller. For three-channel CGI, the security is obviously improved. Compared with traditional CGI encryption using transmission mask, these two methods are more robust to attacks.

In the two methods, the key percentage that unauthorized users can decrypt ghosts is different. Attackers will need to obtain more valid key components to decrypt ghost images reconstructed from grayscale CGI. In the case of color object CGI, it is even more robust to attacks. To sum up, by combining the bucket data of three channels with color objects, we can not only improve the security of encrypted ghosts, but also expand the application of CGI.

Zhao et al. propose a high performance optical encryption (OE) scheme based on computational ghost imaging (CGI) with QR code and compressive sensing (CS) technique, named QR-CGI-OE scheme [23]. The combination of QR code and polarization digital holography expands the key space and further improves the security level of the optical encryption system. At the same time, the optical information encryption scheme based on incoherent superposition has a higher security level and better robustness to the noise attack of QR code. On the other hand, compressed sensing, as a modern sampling compression technology, combined with computational ghost imaging, greatly shortens the measurement time and greatly improves the quality of the reconstructed image. Therefore, it seems a good method to combine QR code and CS technology into the optical encryption scheme based on computational ghost imaging.

The diagram of this encryption scheme is shown in Fig. 2.26. The N random phase distributions generated by Alice are a key and are shared with its authorized user Bob. First, Alice encodes the information with a QR code, and then encrypts the image encoded by the QR code with the help of the computational ghost imaging optical system. Here, the measurement result of the bucket detector of the ghost imaging system is encrypted and transmitted to the Bob. Using the key, Bob decrypts the encrypted information, uses GI and CS technology to obtain QR-coded image, and then recovers the information by QR decoding.

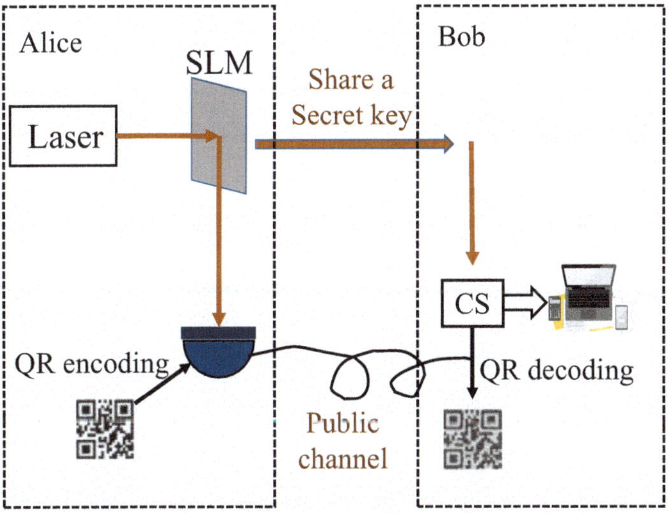

Fig. 2.26 The optical encryption scheme based on computational ghost imaging with CS technique and QR code (QR-CGI-OE). SLM is spatial light modulator. The authorized users, Alice and Bob, are sharing a secret key, and Alice sends the measurement results to Bob in a public channel

This optical encryption scheme has many advantages. Firstly, the scheme uses the computational ghost imaging technology in the field of quantum optics. The encrypted version of the target image is not a complex matrix, but only an intensity vector, which greatly reduces the number of bits sent by Alice to Bob. Secondly, through the use of QR code, the security of the scheme is greatly improved. The results show that even if the eavesdropper eavesdropped on 60% of the key components of the original image, they could not get any information about the original image. In addition, the proposed scheme provides a practical solution for optical encryption because QR codes are fast readable and can be used by almost anyone. Moreover, the CS restoration algorithm is used to reduce the number of measurements and improve the quality of the reconstructed image.

Wu et al. propose an optical multi-image encryption scheme based on computational ghost imaging and position multiplexing [24]. This optical multi-image encryption scheme is shown in Fig. 2.27.

In the traditional pseudothermal ghost imaging, pseudothermal light is generated by a laser beam through a rotating diffuser. The resulting speckle beam is divided into a signal beam and a reference beam. The signal beam passes through the object, which is located at a distance z, is detected by a bucket detector with no spatial resolution. After the propagating distance z', the reference beam is captured by a CCD. Here, the CCD camera is synchronized with the barrel detector to ensure that the captured speckle image is related to the total intensity. If the distance z and z' are equal, the image of the object is obtained by cross correlation between the speckle field and the total intensity after many times of capture and detection. When the difference between z and z' is large enough, the resolution of the ghost image is

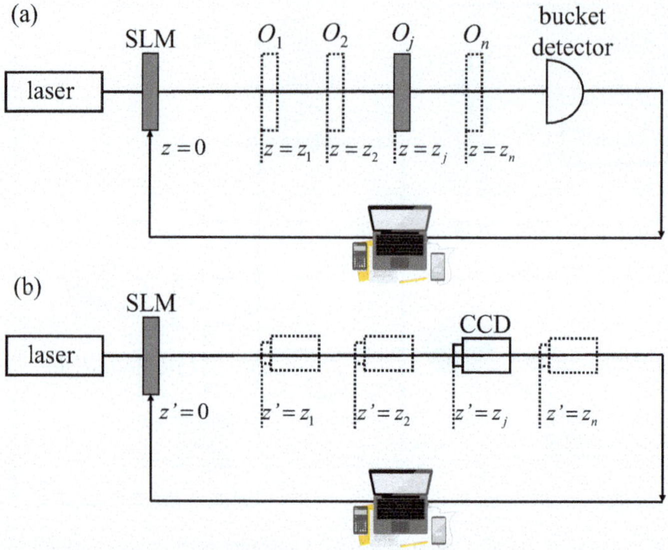

Fig. 2.27 Optical setup for proposed multi-image encryption scheme. **a** Encryption procedure. **b** Decryption procedure. SLM represents spatial light modulator

equal to the size of the object, and the image result will be blurred. This multi-image encryption method is based on this characteristic of ghost imaging.

In the process of encryption, the virtual image is calculated by using different diffraction distances, and each ordinary image is encrypted into an intensity vector. All the intensity vectors are superimposed together to generate the final ciphertext. Different from the general multi-image cryptosystem, the ciphertext in this scheme is only an intensity vector, not a complex amplitude. The final ciphertext is an intensity vector, and the result that needs to be detected and recorded in the encryption process is also an intensity value. Therefore, the detection process of this scheme is more convenient than other multi-image encryption systems. This is the characteristic of the proposed multi-image encryption scheme.

Ghost imaging optical encryption technology has good anti-interference performance against external optical interference, noise, etc., which can ensure the stability and reliability of encrypted data. Ghost imaging optical encryption technology can enhance the encryption strength by increasing the optical path length, adjusting the phase, and other means, which has good scalability. The encryption and decryption speed of ghost imaging optical encryption technology is very fast, suitable for situations that require high-speed encryption and decryption.

2.6.2 Optical Object Authentication with Ghost Imaging

Optical object authentication refers to the process of verifying the authenticity and integrity of an optical object. It involves confirming that the optical object has not been manipulated, altered, or tampered with in any unauthorized way. With the rise of digital image editing tools, it has become easier to modify images, leading to concerns about the trustworthiness and reliability of visual information. Optical object authentication techniques aim to address these concerns by providing methods to verify if an image is genuine and has not been subjected to any unauthorized modifications. Optical object authentication is important in fields such as forensics, journalism, and digital archives, where the integrity and trustworthiness of visual information are critical. By ensuring the authenticity of optical objects, it helps to maintain trust, prevent fraud, and ensure the accuracy of information conveyed through visual media.

In ghost imaging, one beam illuminates the test object and produces the scatterer collected by the bucket detector, while the other beam does not interact with the object (as a reference arm). By correlating the intensity distribution obtained by the bucket detector with the intensity distribution at the reference arm, the information of the object can be reconstructed.

Although ghost imaging has been achieved through the use of various configurations (such as quantum, classical and computational), the main purpose is still to correlate the photocurrent obtained by the separate detectors illuminated by two highly correlated beams. It is well known that in many cases, a large number of realizations (such as 5000 or even 20000) are usually required to extract objects. Therefore, the question is whether a significant number of implementations can be used to identify objects in ghost imaging which can lead to reduced acquisition time and efficient use of photons under low-light imaging conditions.

Chen et al. [25] demonstrate how the reconstructed object can be authenticated using the significantly small number of realizations (i.e., less than 5% of Nyquist limit) in computational ghost imaging. The main objective for this study is to authenticate the object reconstructed by using the significantly small number of realizations rather than to extract the high-fidelity object.

Figure 2.28 shows a schematic diagram of the experimental setup. A series of random phase contours are embedded in the spatial light modulator (SLM) in turn. The light source is divided into object beam and reference beam, which are collected by bucket detector (without spatial resolution) and CCD respectively. The intensity patterns at the reference arm can be precomputed (indicated by the dashed box in Fig. 2.28), which is one main advantage of computational ghost imaging. In the object path, the beam illuminates the object, and the scatters are collected by a bucket detector just placed behind the object. Series of intensity distributions are collected by the bucket detector.

For the object reconstruction, the intensity patterns in the reference arm are correlated with the intensity distributions in the object path. Chen et al. apply the nonlinear correlation algorithm to authenticate the reconstructed object when a significantly

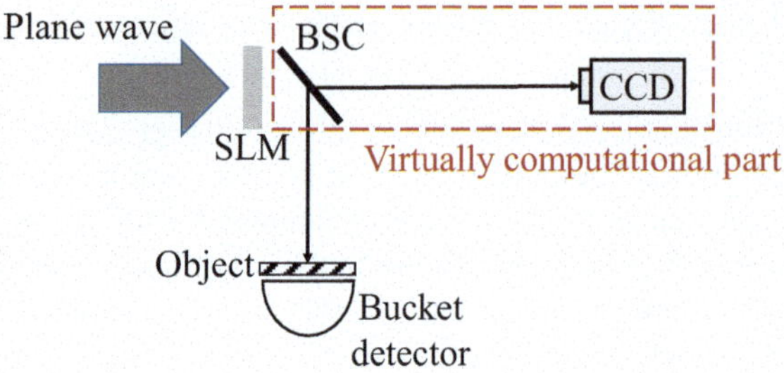

Fig. 2.28 Schematic experimental setup: SLM spatial light modulator; BSC, nonpolarizing beam splitter cube; CCD, 2D charge-coupled device

small number of realizations (i.e., less than 5% of Nyquist limit) are used. A flow chart is shown in Fig. 2.29 to illustrate object authentication process aforementioned. Because only a significantly small number of realizations are used, the reconstructed object is just like speckle pattern, and does not clearly render information about the original object. The authentication between the reconstructed object and the original object is based on the nonlinear correlation algorithm. If a remarkable and sharp peak is generated over the noisy background, then the reconstructed object is effectively authenticated even without clear visualization of the object. If there is no peak, the image is false.

Chen et al. [26] report grayscale object authentication based on ghost imaging using binary signals. It is found that binary signals can be applied at reference and object beam arms, which provides a promising alternative to explore the potential of ghost imaging for some applications. Reducing the data size with only two quantization levels is beneficial to the storage and transmission. In the ghost imaging, a large number of measurements, such as 10000 or more, are usually requested. When

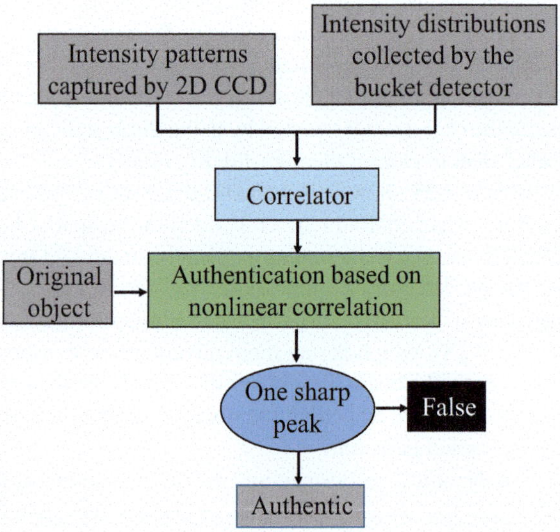

Fig. 2.29 Flow chart for the optical object (information) authentication in computational ghost imaging using a significantly small number of realizations

only two quantization levels are applied to compress the recorded information, data sizes can be dramatically reduced for facilitating the storage and transmission.

The reference beam arm assumes the presence of only two quantization levels, specifically 0 and 1, which can be achieved through data compression. The recorded signals can be adjusted to binary form by flexibly designing the threshold. Similarly, in the object beam arm, binary signals consisting of only 0 and 1 are also utilized through data compression. Utilizing the principles of classical optical coherence theory, the object can be reconstructed by utilizing intensity correlation. For this study, a limited number of measurements are utilized, and ghost imaging incorporates the use of binary signals. It is demonstrated that during the reconstruction process, the object's visibility is not clear, and an algorithm based on nonlinear correlation can be employed for object authentication.

Chen et al. [27] have further improved the optical object authentication based on ghost imaging using a "compressive sensing algorithm" instead of a "classical correlation algorithm in computational ghost imaging." As a result, the realizations for object authentication were further reduced from 5% of the Nyquist limit to 3% of the Nyquist limit. The object authentication algorithm is still the nonlinear correlation algorithm, but the key difference is the object reconstructed algorithm. The object reconstructed algorithm in their work is the compressive sensing algorithm rather than the classical correlation algorithm.

Figure 2.30 shows a diagram of compressive ghost imaging. In this diagram, a 2D digital micromirror device (DMD) was employed. The computer-controlled DMD was illuminated by a laser with a Gaussian intensity distribution to generate speckles.

A series of light intensity patterns in the reference arm can be calculated, and a bucket detector is used to measure the total light intensity transmitted through the object. A compressive sensing (CS) algorithm is employed to reconstruct the object. The CS algorithm is used to search for the most sparse image in the compressible basis that fulfilled the $M < N$ random projections, where M is the total number of measurements, and N is the total number of resolution cells in the image. A number of realizations, such as 15 or 30% of the Nyquist limit, may be still required to obtain a clear ghost image, although the compressive sensing technique is applied. However, the reconstructed object can be authenticated by the nonlinear correlation algorithm when a significantly small number of realizations (i.e., less than 3% of the Nyquist limit) were used.

Object authentication based on ghost imaging can reconstruct the image of the object through a small amount of measurement data, so as to improve the efficiency of authentication. Ghost imaging technology has a certain anti-jamming ability for optical noise, scattering and interference, and can be reliably authenticated in complex environments. Besides, ghost imaging does not require direct contact with the certified object, so it can avoid damage or pollution to the object. In the future, object authentication based on ghost imaging will not only be applied in the field of ghost imaging but also in image analysis and pattern recognition.

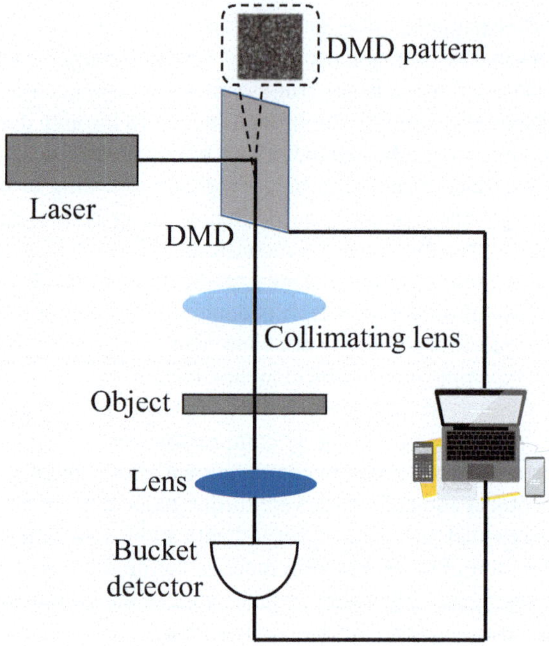

Fig. 2.30 Diagram of compressive ghost imaging

2.6.3 Ghost Imaging Lidar

In recent decades, lidar has emerged as a prominent tool for remote sensing, playing a crucial role in detection. Traditional lidar systems can be categorized into two types based on their measurement modes: scanning imaging lidar and non-scanning imaging lidar, such as flash radiography. Scanning imaging lidar utilizes a pulsed laser to scan the target region point-by-point, generating a real-space image. However, this approach faces challenges in imaging high-speed moving targets. On the other hand, non-scanning imaging lidar employs a high-resolution imaging system coupled with a pulsed flash laser, enabling the capture of the entire field of the target in a single exposure. Nevertheless, the detection sensitivity of non-scanning imaging lidar is compromised due to the division of reflected light intensity into small pixels of a charge-coupled device (CCD) camera. Moreover, the detection distance of non-scanning imaging lidar is limited by the signal-to-noise ratio (SNR) of received photons distributed across the aperture of the imaging system and the entire imaging plane. Additionally, the imaging resolution of scanning imaging lidar is constrained by the Rayleigh criterion of the emitting aperture, while for non-scanning imaging lidar, it is determined by the numerical aperture of the receiving system.

To enhance the information acquisition in remote sensing, it is beneficial to employ high-resolution imaging techniques. Zhao et al. [28] expand the utilization of pseudo-thermal light ghost imaging to remote imaging, introducing a practical ghost imaging lidar via sparsity constraints (GISC lidar) system. In the realm of remote sensing applications, the ghost imaging methodology employs speckle-to-speckle correlation in conjunction with classical imaging systems. Due to the impracticality of matching the distance between the light source and the reference detection plane with the actual remote sensing distance (the distance between the light source and the target), the imaging systems can be tailored to manipulate the light field, making the miniaturization of the reference optical system in ghost imaging remote sensing devices more feasible. Moreover, the incorporation of sparsity constraints with the ghost imaging approach using speckle-to-speckle correlation can be more directly achieved.

The configuration of the proposed GISC lidar is illustrated in Fig. 2.31. The light source consists of a 532 nm solid-state pulsed laser with a pulse width of 10 ns, a rotating ground glass disk, and a set of field lenses. These elements combine to generate a speckle field at stop 2, also known as the field stop. A beam splitter (BS) divides the light emitted from the source into two paths: the object path and the reference path. In the object path, the light passes through the objective lens f_0 and then reaches the target. The photons reflected by the target are collected by a light concentrator, specifically a Cassegrain telescope with an aperture of 420 mm and a focal length of 5 m. After passing through an interference filter with a half-width of 1 nm, the photons are detected by a photomultiplier tube (PMT). In the reference path, the light travels through the reference lens f1 and then reaches a CCD camera. Stop 2 is strategically positioned on the conjugate plane of both the target and the CCD camera, allowing for control of the field of view (FOV) on the target plane. The field lens is utilized to maximize the utilization of light energy and create a visual aperture stop, referred to as stop 3 in Fig. 2.31. To ensure consistent transverse size of the light beam at the objective and reference lenses, stop 1 is employed. This stop also guarantees that the entrance pupil is identical for both lens f_0 and f_1. Additionally, the PMT converts the reflected light signals from the target into electrical signals, while the interference filter suppresses background light.

By exploiting sparsity constraints of the image, nonlinear sparsity reconstruction techniques usually yield, as predicted by theoretical analysis and confirmed by experiments, better results when the target is sparse in a representation basis. The reconstruction of GISC lidar is formulated in the sparsity reconstruction framework. By reshaping each of the speckle intensity distribution recorded by the CCD camera into a row vector, the measurement matrix is obtained after observations. Meanwhile, the intensities recorded by the PMT in the object path are arranged as a column vector. The target's image can be reconstructed by solving a convex optimization program.

In comparison to traditional imaging lidar systems, GISC lidar combines the benefits of both scanning and non-scanning imaging lidar. Similar to scanning imaging lidar, GISC lidar achieves high detection efficiency for long remote sensing distances as all photons collected by the concentrator are directed towards the same PMT. Conversely, like non-scanning imaging lidar, GISC lidar emits laser pulses that cover the entire detection field, enabling target reconstruction without the need for target scanning. This feature allows for high-speed imaging of targets, even with a lower sampling number than what is required by conventional imaging systems based on

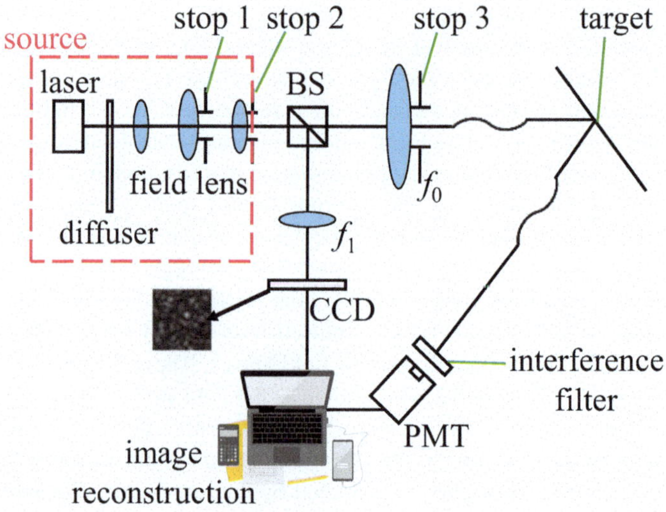

Fig. 2.31 Experimental setup for GISC lidar

Nyquist criteria. Additionally, GISC lidar mitigates the impact of atmospheric turbulence on imaging quality to some extent, an advantage over traditional imaging lidar. In summary, GISC lidar offers advantages such as long remote sensing distances, high imaging speeds, and super-resolution imaging capabilities.

The field of three-dimensional (3D) remote imaging has garnered significant attention due to its ability to capture detailed target characteristics. Despite the advancements made in 3D remote imaging techniques such as scanning imaging lidar and pulsed floodlight-illumination imaging lidar, these methods still have limitations in terms of detection range and application flexibility. However, a promising approach known as Ghost Imaging via Sparsity Constraint (GISC) has emerged. This technique allows for the reconstruction of a high-resolution two-dimensional image using a significantly reduced number of measurements. Building upon the GISC framework, Gong et al. [29]. have developed a 3D GISC lidar system that incorporates time-resolved measurements to capture depth information of targets. Through experimental validation, they have demonstrated the system's ability to reconstruct a stable 3D scene at a range of approximately 1.0 km, even when using measurements below the Nyquist limit.

For ghost imaging, the light fields within the range of axial correlation depth are always spatially correlated, thus all the targets positioned in this range can be clearly imaged even though the detection plane in the reference path is fixed, which provides a necessary condition for 3D GISC. In order to obtain the depth information of the target, similar to lidar, a narrow pulsed laser is used as the source of 3D GISC with pseudo-thermal light and a time-resolved technique is used to measure the target's reflection signals. Therefore, when the reflection signals from the target are recorded by a time-resolved single-pixel bucket detector, the target's 3D image can be obtained

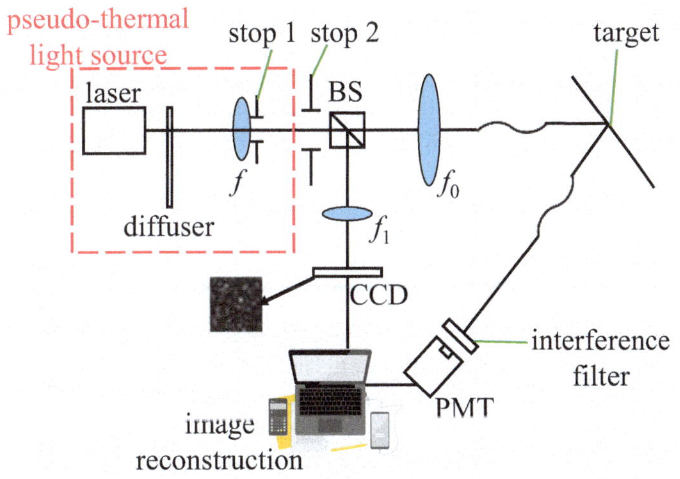

Fig. 2.32 Experimental setup of 3D GISC lidar system with pseudo-thermal light

by a structured image reconstruction method of 3D GISC. To demonstrate 3D GISC lidar, the optical setup is illustrated in Fig. 2.32.

By employing a rotating diffuser to modulate the laser, a sequence of independent and random speckle patterns can be generated. In the measurement framework of 3D GISC lidar, the recorded random speckle patterns captured by the CCD sensor are utilized to construct the sensing matrix. Correspondingly, the one-dimensional time-resolved intensity distributions, detected by the PMT, are employed as the measurement data. By leveraging the inherent characteristics of the target as constraints, the 3D GISC method enables the restoration of both the target's 3D image and its tomographic image at each time delay (or depth). In this scenario, the target's images at different depths can be sparsely represented using a suitable basis transform, ensuring that there is no spatial overlap among the images. Furthermore, the target's gray distribution is always real and nonnegative. The structured sparse reconstruction algorithm is employed to recover the target's 3D images, which can be accomplished within a few minutes on a standard desktop computer.

3D GISC lidar offers both high detection sensitivity and efficient information extraction, distinguishing it from traditional 3D imaging lidar. Similar to scanning imaging lidar, 3D GISC lidar achieves high detection sensitivity for long remote sensing distances by directing all collected photons towards the same PMT. This allows for a larger detection range compared to pulsed floodlight-illumination imaging lidar. Conversely, like pulsed floodlight-illumination imaging lidar, the laser pulse emitted by 3D GISC lidar covers the entire detection field, enabling target reconstruction without the need for scanning. Consequently, 3D GISC lidar overcomes the challenge of imaging moving targets at high speeds, which is a limitation of scanning imaging lidar. Moreover, the image reconstruction of a surface 3D scene with $n \times n \times n$ pixels can be achieved by 3D GISC lidar using far fewer than n^2 pulses.

In contrast, both scanning imaging lidar and pulsed floodlight-illumination imaging lidar require samples at or beyond the Nyquist limit for surface 3D scene reconstruction. Additionally, 3D GISC lidar can be applied to non-invasive optical imaging in medical diagnosis, and a similar 3D microscopy technique can be employed to observe the surface morphology of materials and biomedical specimens.

2.7 4X-Ray Ghost Imaging

The application of X-ray crystallography has played a crucial role in advancing scientific and technological developments over the last century. By providing insights into the internal structures of various materials, including biomolecules and nanomaterials, X-ray crystallography has become an invaluable tool in the field of materials analysis. However, its limitations lie in its applicability only to crystalline structures, rendering it ineffective in resolving non-crystalline materials. As a result, the structure information of significant molecular materials, such as membrane proteins, remains inaccessible due to challenges in growing them into macroscopic crystals.

In a groundbreaking study, Yu et al. [30] have successfully demonstrated the application of hard X-ray Fourier-transform ghost imaging (FGI) using an innovative experimental technique. This experiment was based on the same fundamental principle as visible light FGI. By dividing a light beam from a spatially incoherent source into two beams—a testing beam and a reference beam—after passing through a beam splitter, they were able to achieve FGI in the hard X-ray range. However, the challenges posed by hard X-rays make the realization of FGI particularly difficult. Unlike visible light, there is no ideal beam splitter capable of producing twin beams for hard X-rays. To overcome this obstacle, the researchers employed an alternative approach. Instead of splitting the thermal light into two beams, they utilized a controllable pseudothermal light source and a fixed panel detector. By shuttling the sample in and out of the beam during a single pseudo-coherent duration (a period of stable light source), they were able to obtain the desired results. When the sample was inserted into the beam, the signals detected by a single pixel of the panel detector served as those detected with a point detector in the testing beam. Conversely, when the sample was moved out of the beam, the signals detected by the panel detector served as those detected in the reference beam. Consequently, by completing both measurements within a single pseudo-coherent duration, the intensity fluctuations could be acquired using the same panel detector.

Figure 2.33 shows the experimental setup. A pseudothermal X-ray source, which can generate a controllable chaotic speckle pattern fluctuation to emulate the behavior of a spatially incoherent source, was used to illuminate the sample. The pseudothermal source is produced by a monochromatic X-ray beam passing through a slit array and a movable porous gold film deposited on a Si_3N_4 substrate. The monochromatic X-ray beam was produced by passing the x rays emitted from the synchrotron source through a double crystal monochromator with a certain energy resolution.

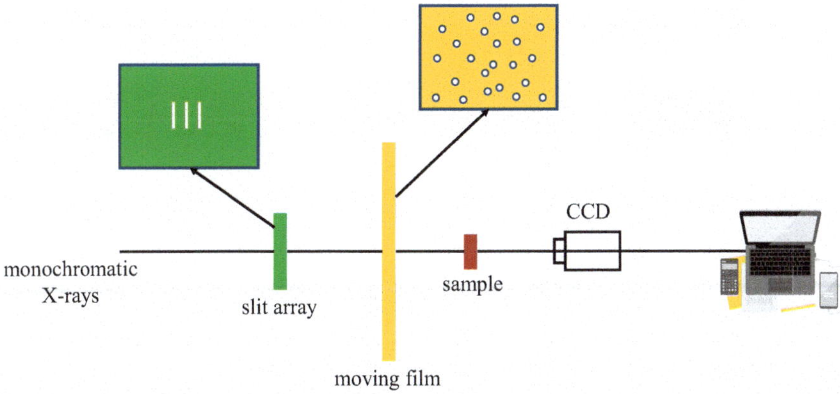

Fig. 2.33 Experimental setup for X-ray FGI using a pseudothermal X-ray source. A monochromatic X-ray beam passes through a slit array and a moving porous gold film to generate a controllable pseudothermal X-ray speckle pattern fluctuation, which is used in the experiment as the pseudothermal X-ray source. The sample is moved in and out of the beam to get the intensity fluctuation, as required by the scheme of FGI without a beam splitter

By employing a two-stage process for phase-retrieval image reconstruction based on Fourier-transform ghost imaging (FGI), the amplitude component of the sample's transmittance is first extracted from the Fourier-transform patterns. Subsequently, the retrieved amplitude component is combined with the Fourier-transform pattern to retrieve the phase component of the sample's transmittance. This innovative approach expands the application of X-ray crystallography to non-crystalline samples. As a lens-less imaging technique, the spatial resolution of X-ray FGI is fundamentally limited only by the wavelength. An important advantage of X-ray FGI is its independence from highly coherent X-ray sources for achieving X-ray diffraction imaging. Consequently, it offers a practical means to obtain high-resolution images of non-crystalline samples using commonly available laboratory X-ray sources. Moreover, this advancement opens up the possibility of applying FGI to not only bosons but also fermions, such as neutrons and electrons, potentially revolutionizing current neutron and electron scattering methods widely utilized in materials science and biomedicine research.

In the same year, Pelliccia et al. [31] achieved a proof-of-principle experiment for ghost imaging in the hard-x-ray energy range. They utilized a synchrotron X-ray beam and employed a thin crystal in Laue diffraction geometry to split the beam. By utilizing an ultrafast imaging camera, they were able to capture images of X-rays emitted by individual electron bunches. At this temporal scale, the synchrotron emission process introduced shot noise that manifested as speckles, resulting in speckle correlation between the two split beams. To recover the ghost image of an object, the researchers split the beam into two spatially separated locations on the camera screen and placed the object in one of the beams. By establishing an appropriate intensity correlation between the two speckle beams, the ghost image of the object was successfully reconstructed. This achievement holds great significance

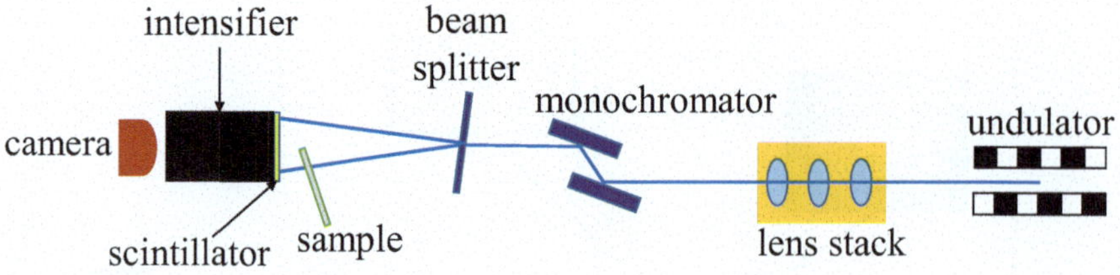

Fig. 2.34 Scheme of the experimental setup (not to scale). X-ray beam propagation is from right to left. X-ray pulses from the undulator are focused by a refractive lens stack and then monochromatized by a double-bounce Si monochromator. The beam splitter, working in Laue diffraction, is located at the focal position of the lenses. Both transmitted and diffracted beams are imaged on the ultrafast camera coupled with a scintillator and an image intensifier

as ghost imaging with hard X-rays exhibits a remarkable characteristic. Due to the intensity correlation between separate beams, ghost imaging is remarkably robust against turbulence in either beam. This property has led to applications in atmospheric imaging, making ghost imaging with hard X-rays an area of great interest.

The experimental setup is depicted in Fig. 2.34. The beam from the undulator was focused by a Be refractive lens stack to a focal spot at the beam splitter position. The monochromator, a pair of Si crystals located at 140 m from the source, monochromatized the beam at an energy of 20 keV. The beam splitter was aligned to select one of the Laue reflections in the forward direction. At the same time, the crystal enabled a portion of the undiffracted beam to be transmitted in the forward direction. The camera was placed 20 cm downstream of the beam splitter to enable detection, within its field of view, of both diffracted and transmitted beams.

To generate two identical copies of the beam, a beam splitter was employed, and both copies were simultaneously captured by a high-speed camera. The camera's frame rate was sufficiently high to effectively isolate the X-rays emitted by a single bunch or the average of two consecutive bunches. Placing an X-ray opaque Cu wire in one of the beams, its image was integrated spatially to form a point (bucket) detector. By utilizing the intensity correlation between the bucket signal and the image of the empty beam, the ghost image of the wire was successfully reconstructed under two different configurations. The integrated transmitted intensity from a sample positioned in the first beam was correlated with the spatially resolved intensity measured in the second, empty beam to extract the shadow of the sample. This experimental demonstration of direct X-ray ghost imaging holds significant promise for potential applications in medical imaging and ultrafast X-ray studies using free electron lasers. In both scenarios, ghost imaging could potentially reduce the radiation dose on the sample by utilizing a suitably weak bucket beam while maintaining the speckle correlation with the reference beam.

2.8 General Conclusions

In this chapter, the principle of ghost imaging is introduced. ghost imaging is a technique that uses optical correlation for imaging, its system is generally divided into a signal light path and reference light path, the signal light path using a bucket detector without spatial resolution, the reference light path using a CCD, the peculiarity is that any one of the detectors used to detect the signal cannot be imaged individually, the use of correlation between the signals of the two detectors in order to image.

The evolution of ghost imaging is examined, encompassing its historical progression, current status, and future prospects. This encompasses the transition from early quantum ghost imaging to the utilization of classical light sources, as well as the advancement from traditional two-arm ghost imaging to the emergence of one-arm computational ghost imaging. Furthermore, the progression from computational ghost imaging to the refinement of single-pixel imaging is also considered.

Ultimately, the fundamental configurations of diverse ghost imaging systems and the utilization of ghost imaging across various domains are delineated. This encompasses the deployment of ghost imaging in optical encryption, optical image authentication, lidar technology, and the advancement of X-ray ghost imaging.

Acknowledgements This work was supported by the National Natural Science Foundation of China (Nos. 11874132, 61975044, 12074094, 12374271, and 62375065); Interdisciplinary Research Foundation of HIT (No. IR2021237).

References

1. Einstein PB, Rosen N. Can quantum-mechanical description of physical reality be considered complete? Phys Rev A. 1935;47(10):777.
2. Pittman TB, Shih YH, Strekalov DV, Sergienko AV. Optical imaging by means of two-photon quantum entanglement. Phys Rev A. 1995;52(5):R3429–32.
3. Abouraddy AF, Saleh BEA, Sergienko AV, Teich MC. Role of entanglement in two-photon imaging. Phys Rev Lett. 2001;87(12):123602–12.
4. Bennink RS, Bentley SJ, Boyd RW. "Two photon" coincidence imaging with a classical source. Phys Rev Lett. 2002;89(11): 113601.
5. Gatti A, Brambilla E, Bache M, Lugiato LA. Correlated imaging, quantum and classical. Phys Rev A. 2003;70(1):235–8.
6. Valencia SG, D'Angelo M, Shih Y. Two-photon imaging with thermal light. Phys Rev Lett. 2005;94(6): 063601.
7. Shapiro JH. Computational ghost imaging. Phys Rev A. 2008;78(6): 061802.
8. Bromberg Y, Katz O, Silberberg Y. Ghost imaging with a single detector. Phys Rev A. 2009;79(5): 053840.
9. Katz BY, Silberberg Y. Compressive ghost imaging. Appl Phys Lett. 2009;95: 131110.
10. Ferri F, Magatti D, Lugiato LA, Gatti A. Differential ghost imaging. Phys Rev Lett. 2010;104(25): 253603.
11. Horisaki R, Takagi R, Tanida J. Learning-based imaging through scattering media. Opt Express. 2016;24(13):13738–43.

12. Lyu M, Wang W, Wang H, Wang HC, Li GW, Chen N, Situ GH. Deep-learning-based ghost imaging. Sci Rep. 2017;7(1):17865.
13. Luo KH, Huang BQ, Zheng WM, Wu LA. Nonlocal imaging by conditional averaging of random reference measurements. Chin Phys Lett. 2012;29(7): 074216.
14. Duarte MF, Davenport MA, Takhar D, Laska JN, Sun T, Kelly KF, Baraniuk RG. Single-pixel imaging via compressive sampling. IEEE Signal Proc Mag. 2008;25(2):83–91.
15. Zhang Z, Ma X, Zhong J. Single-pixel imaging by means of Fourier spectrum acquisition. Nat Commun. 2015;6(1):6225.
16. Bian L, Suo J, Hu X, Chen F, Dai QH. Efficient single pixel imaging in Fourier space. J Opt. 2016;18(8): 085704.
17. Wang L, Zhao S. Fast reconstructed and high quality ghost imaging with fast Walsh-Hadamard transform. Photonics Res. 2016;4(6):240–4.
18. Abouraddy AF, Stone PR, Sergienko AV, Saleh BEA, Teich MC. Entangled-photon imaging of a pure phase object. Phys Rev Lett. 2004;93(21): 213903.
19. Ryczkowski P, Barbier M, Friberg AT, Dudley JM, Genty G. Ghost imaging in the time domain. Nat Photon. 2015;10(3):167.
20. Devaux F, Moreau PA, Denis S, Lantz E. Computational temporal ghost imaging. Optica. 2016;3(7):698–701.
21. Clemente P, Durán V, Torres-Company V, Tajahuerce E, Lancis J. Optical encryption based on computational ghost imaging. Opt Lett. 2010;35(14):2391–3.
22. Tanha M, Kheradmand R, Ahmadi-Kandjani S. Gray-scale and color optical encryption based on computational ghost imaging. Appl Phys Lett. 2012;101(10): 101108.
23. Zhao SM, Wang L, Liang WQ, Cheng WW, Gong LY. High performance optical encryption based on computational ghost imaging with QR code and compressive sensing technique. Opt Commun. 2015;353:90–5.
24. Wu JJ, Xie ZW, Liu ZJ, Liu W, Zhang Y, Liu ST. Multiple-image encryption based on computational ghost imaging. Opt Commun. 2016;359:38–43.
25. Chen W, Chen XD. Object authentication in computational ghost imaging with the realizations less than 5% of Nyquist limit. Opt Lett. 2013;38(4):546–8.
26. Chen W, Chen XD. Grayscale object authentication based on ghost imaging using binary signals. Europhys Lett. 2015;110(4):44002.
27. Chen ZP, Shi JH, Zeng GH. Object authentication based on compressive ghost imaging. Appl Opt. 2016;55(30):8644–50.
28. Zhao CQ, Gong WL, Chen ML, Li ER, Wang H, Xu WD, Han SS. Ghost imaging lidar via sparsity constraints. Appl Phys Lett. 2012;101(14): 141123.
29. Gong WL, Zhao CQ, Yu H, Chen ML, Xu WD, Han SS. Three-dimensional ghost imaging lidar via sparsity constraint. Sci Rep. 2016;6:26133.
30. Yu H, Lu RH, Han SS, Xie HL, Du GH, Xiao TQ, Zhu DM. Fourier-transform ghost imaging with hard X-rays. Phys Rev Lett. 2016;117(11): 113901.
31. Pelliccia D, Rack A, Scheel M, Cantelli V, Paganin DM. Experimental X-ray ghost imaging. Phys Rev Lett. 2016;117(11): 113902.

Chapter 3
Imaging Methods via Compressed Sensing

Guancheng Huang, Xuyang Zhou, Zhengjun Liu, and Shutian Liu

3.1 Introduction

Prior to the introduction of the theory of Compressive Sensing (CS), traditional signal acquisition relied upon the Nyquist theorem, which dictated that signals should be sampled at a frequency twice that of their bandwidth. The resulting signal was then compressed, stored, and transmitted before being reconstructed. This theorem ensured that the sampled signal retained all the information of the original signal to facilitate accurate reconstruction [1]. However, CS theory challenges this conventional framework, as the signal's sampling frequency is no longer entirely determined by its bandwidth, but instead by its structure and content. For signals that are sparse, a lower sampling frequency than that required by Nyquist's theorem can be used, and the original signal can be reconstructed with high probability using reconstruction algorithms [2, 3].

CS theory was first proposed in 2006 by Donoho et al. [4] and has since undergone extensive development and refinement. The core of the theory is founded on three key elements: (1) sparse representation of signals, which assumes that the original signal is sparse or can be represented through a sparse basis expansion; (2) selection of an optimal measurement matrix, which is crucial in measuring the signal to obtain compressed results that retain a minimal amount of original information. The quality of the measurement matrix is assessed based on its adherence to the characteristics of incoherence or Restricted Isometry Property (RIP); (3) the signal reconstruction algorithm, which is essential in the retrieval of the sparse original signal from the measured results.

G. Huang · X. Zhou · Z. Liu (✉) · S. Liu
School of Physics, Harbin Institute of Technology, Harbin 150001, China
e-mail: zjliu@hit.edu.cn

S. Liu
e-mail: stliu@hit.edu.cn

Z. Liu et al. (eds.), *Computational Optical Imaging*, Advances in Optics and Optoelectronics, https://doi.org/10.1007/978-981-97-1455-1_3

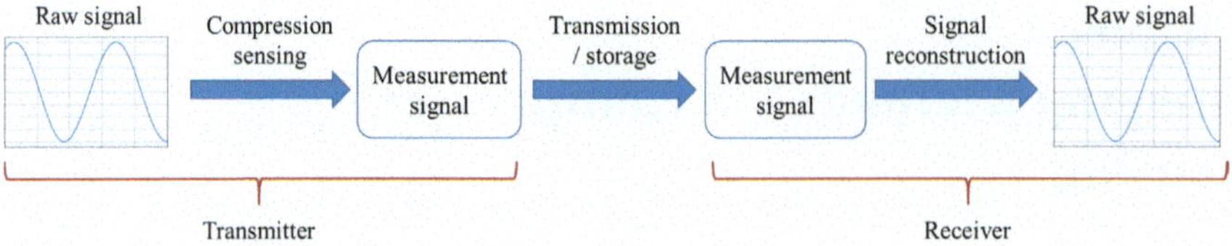

Fig. 3.1 The signal processing process of compression sensing

CS theory surpasses the limitations of traditional signal processing methods, making significant strides in signal processing and data transmission. It provides new avenues and concepts for information acquisition and processing and has gained widespread attention and application in diverse fields, such as medical imaging [5–7], remote sensing [8, 9], high-speed imaging [10], and wireless communication [11]. Based on the sparsity of the signal, CS can sample the signal at a rate lower than that of the Nyquist sampling theorem and restore the original signal with high probability, making CS a valuable tool in computational imaging. The signal processing process of compression sensing is illustrated in Fig. 3.1.

3.2 Basic Principle

The traditional signal processing approach involves high-speed signal sampling at the sender, followed by quantization of the discrete signal, coding, decoding, and ultimately signal recovery at the receiver. However, this signal processing flow is based on the Nyquist sampling theorem, which mandates sampling at twice the highest frequency and results in exorbitant equipment costs and wastage of resources. In contrast, CS theory improves on the traditional sampling theorem by exploiting the sparsity of signals. This enables data compression and sampling through the multiplication of the measurement matrix, culminating in signal reconstruction at the receiver.

The compression sensing algorithm involves three primary steps. First, signal compression is achieved through the sparse representation of signals. Second, the measurement matrix is designed to acquire observation values. Finally, signal reconstruction is performed to restore the original signal. Figure 3.2 illustrates the specific process of CS.

CS theory posits that a signal can be sampled at a frequency lower than the Nyquist limit, provided that the signal is sparse and the observation matrix satisfies the RIP [12–14]. The high probability recovery of the original signal can then be achieved via a reconstruction algorithm. Specifically, a vector $\boldsymbol{\alpha}$ is defined as sparse if it satisfies Eq. (3.1), where $p \in [0, 2)$ and $R > 0$.

$$\boldsymbol{\alpha}_p = \left(\sum_i |\alpha_i|^p \right)^{1/p} \leq R. \tag{3.1}$$

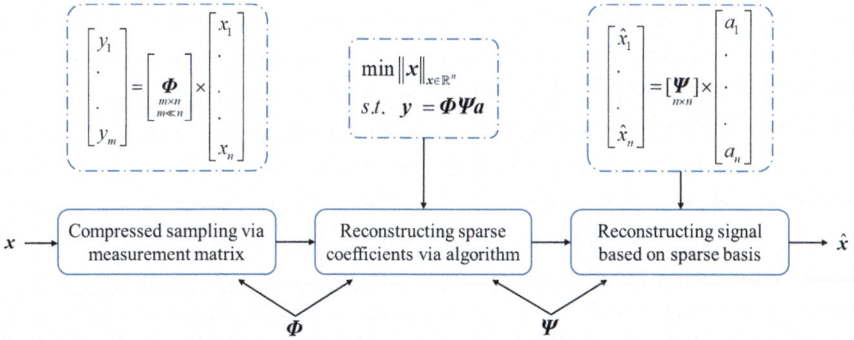

Fig. 3.2 The solving process of CS

As the majority of signals themselves are not sparse, CS is commonly applied to the $\boldsymbol{\alpha}$ representation of a signal under a certain transformation $\boldsymbol{\Psi}$. Specifically, CS can be expressed through

$$y = \boldsymbol{\Phi x} = \boldsymbol{\Phi \Psi}^{-1}\boldsymbol{\alpha} = \boldsymbol{\Theta\alpha}. \tag{3.2}$$

wherein the transformation coefficient a of signal $\mathrm{x} \in \mathbb{R}^n$ is K-sparse (with $K \ll n$) under the transformation $\alpha = \boldsymbol{\Psi}x$, and $\mathrm{y} \in \mathbb{R}^m$ denotes the measured value of signal $\boldsymbol{x}$ through the measurement matrix $\boldsymbol{\Phi} \in \mathbb{R}^{m\times n}$. In addition, $\boldsymbol{\Theta} = \boldsymbol{\Phi\Psi}^{-1}$ is called sensing matrix. The accurate restoration of the original signal $\boldsymbol{x}$ can be achieved by utilizing the observation value $\boldsymbol{y}$, given that the measurement matrix satisfies the RIP condition, which equates to the statement that the measurement matrix $\boldsymbol{\Phi}$ is uncorrelated with the transformation matrix $\boldsymbol{\Psi}$. It is further required that the number of measurements $m > O(K\log(n))$.

The process of signal reconstruction involves the restoration of low-dimensional measurements $\boldsymbol{y}$ to their original high-dimensional signal $\boldsymbol{x}$. However, this problem is known to have multiple solutions, rendering it an NP-Hard problem. To address this issue, the following constrained optimization problem can be employed as a viable solution

$$\boldsymbol{x} = \arg\min_{x} \|\boldsymbol{a}\|_0 \ \text{s.t.}\ \boldsymbol{y} = \boldsymbol{\Theta\alpha}, \tag{3.3}$$

where the symbol $\|\cdot\|$ represents the l_0 norm, which presents significant challenges for direct solutions. However, under specific conditions, the l_0 norm and l_1 norm exhibit equivalence, with the latter being relatively more tractable. To effectively alleviate the computational complexity, the aforementioned formula may be reformulated as follows

$$\boldsymbol{x} = \arg\min_{x} \|\boldsymbol{a}\|_1 \ \text{s.t.}\ \boldsymbol{y} = \boldsymbol{\Theta\alpha}. \tag{3.4}$$

The design of CS reconstruction algorithm is based on Eq. 3.4 with prior or various constraints. Different prior or constraints have different effects on the reconstruction performance of the algorithm.

Next, we will introduce the sparse representation, measurement matrix, and reconstruction algorithm in CS theory in detail.

3.3 Signal Sparse Representation

The sparse representation of signals is a fundamental concept in modern image processing that enables images to be stored and transmitted using small coefficients. This approach plays a vital role in various applications of image processing, such as image compression and de-noising. Furthermore, sparse representation is a key theoretical foundation of CS, which has a significant impact on signal sampling and reconstruction. In traditional image processing, the conventional sparse representation method can only process fixed data distributions and cannot adaptively handle different types of signals. Therefore, the construction of a sparse dictionary tailored to the data characteristics is required to facilitate the adaptive processing of various signals.

3.3.1 Signal Sparsity

The sparsity of a signal refers to the property where only a few elements in the signal have non-zero values, while the majority of the elements are either zero or close to zero. In such cases, the original signal can be compressed and measured directly. The sparsity of a signal is measured by the number of non-zero elements in the sparse signal, and the higher the sparsity, the more difficult it is to compress. Most signals in nature are non-sparse, but these non-sparse signals can become sparse in a certain transform domain, where they can be represented as a finite set of sparse coefficients under an appropriate sparse basis. In this case, the signal can be compressed and stored more efficiently, without losing much of its original information.

In accordance with the theory of harmonic analysis, a one-dimensional discrete-time signal f of length N can be expressed as a linear combination of a set of orthonormal basis functions

$$f = \boldsymbol{\Psi} a = \sum_{i=1}^{N} \psi_i a_i, \tag{3.5}$$

where $\boldsymbol{\Psi} = [\psi_1, \psi_2, \ldots, \psi_N] \in \mathbb{R}^{N \times N}$ is the orthonormal basis, ψ_i is the column vector, the weighted coefficient sequence of $\boldsymbol{f}$ is denoted by $\boldsymbol{a} \in \mathbb{R}^{N \times 1}$ and serves as the equivalent representation of $\boldsymbol{f}$ on the standard basis. Figure 3.3 illustrates that

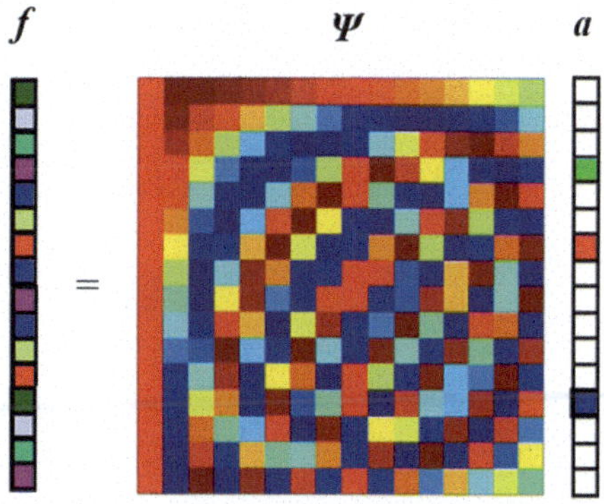

Fig. 3.3 The process of sparsely representing the original signal (with 3-sparse)

when only K elements of $\boldsymbol{a}$ are non-zero, the representation of signal $\boldsymbol{f}$ under base $\boldsymbol{\Psi}$ is sparse. Therefore, $\boldsymbol{a}$ is referred to as the K-sparse representation of $\boldsymbol{f}$.

3.3.2 Sparse Representation

Sparse representation is a widely used technique for signal processing, which aims to represent a signal by a linear combination of a set of orthogonal basic signals. Given a signal $\mathrm{y} \in \mathbb{R}^m$ and a dictionary $\mathrm{A} \in \mathbb{R}^{m \times q}$, where each column of $\boldsymbol{A}$ is called a basis vector, typically with $m < q$, i.e., the dictionary is over-completed, the signal $\boldsymbol{y}$ can be represented as a linear combination of the basis vectors in $\boldsymbol{A}$. The coefficients of $\boldsymbol{y}$ on the dictionary $\boldsymbol{A}$ are denoted as $\boldsymbol{x}$. Since the dictionary $\boldsymbol{A}$ is over-completed, the representation is not unique.

The objective of sparse representation is to find a representation with the maximum number of zero coefficients. This means that the signal is represented using only a small number of the basis vectors, and the other coefficients are close to zero. Such a representation is highly desirable since it simplifies the analysis and processing of signals. However, finding the sparsest representation is a challenging problem and requires sophisticated algorithms.

Figure 3.4 illustrates the basic principle of sparse representation. The figure shows a signal $\boldsymbol{y}$ and its representation as a linear combination of a set of basis vectors. The sparsest representation of $\boldsymbol{y}$ is obtained by selecting only a few basis vectors, which have non-zero coefficients.

In the coefficient matrix, non-zero values represent the essential attributes and inherent structure of signal data, with actual physical meaning. Large non-zero coefficients indicate that the energy of the signal is concentrated on a small number of basis vectors, where these vectors with non-zero coefficients represent the most fundamental characteristics of the signal. Finding the vectors corresponding to these large coefficient values is equivalent to identifying the essential structure of the signal. It is necessary to construct a sparse representation that is compatible with the

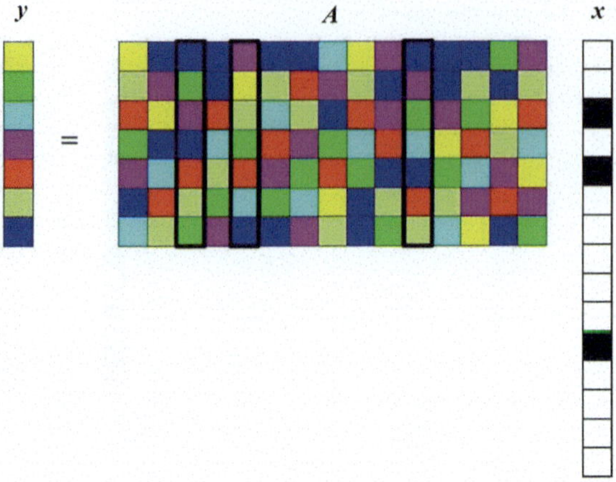

Fig. 3.4 Sparse representation model

specific signal characteristics. Several commonly used sparse representation methods in CS theory include: sparse representation based on orthogonal transformation basis, sparse representation based on signal multi-scale geometric analysis, and sparse representation based on redundant dictionaries.

3.3.3 Sparsification Based on Orthogonal Transformation Basis

(1) Fourier transform

Fourier transform is a mathematical tool and one of the most commonly used in signal processing techniques for the field of imaging. In Fourier transform, a signal is decomposed into a sum of sine and cosine functions, and the amplitude and phase of each function are determined. These functions are called Fourier basis functions, and their linear combination forms the original signal. Fourier transform transforms a signal from the time domain to the frequency domain, which is used to analyze the distribution of signal energy in various frequency components, in order to perform specific analysis of the phase and amplitude of the signal.

When the time function $f(t)$ satisfies

$$\int_{R} |f(t)|^2 \mathrm{dt} < \infty, \tag{3.6}$$

the set of all $f(t)$ forms a space $L^2(\mathbb{R})$, where the functions in this space have square-integrable energy. This space is also known as the space of finite energy. A finite energy function $f(t)$ can be represented as a sum of sinusoidal waves, which is inverse Fourier transform as

$$f(t) = \frac{1}{2\pi} \int_{-\infty}^{+\infty} F(\omega) e^{i\omega t} d\omega, \tag{3.7}$$

where each sinusoidal wave's amplitude $F(\omega)$ is equal to the correlation between the wave and the function $f(t)$. This is, Fourier transform is as follows

$$F(\omega) = \int_{-\infty}^{+\infty} f(t) e^{-i\omega t} dt \tag{3.8}$$

The Fourier transform is a fundamental mathematical tool used to represent a function in terms of its frequency components. It achieves this by decomposing a function into a sum of sinusoidal waves with varying frequencies, where the amplitudes and phases of the constituent waves are determined by the function's original values. The resulting representation of the function in the frequency domain provides valuable insights into its underlying structure and properties.

Specifically, the Fourier transform enables the extraction of the frequency domain characteristics of signals, which can be particularly useful in applications such as signal processing, communication systems, and image analysis. By analyzing a signal's frequency content, one can gain a better understanding of its behavior and properties, and develop techniques for manipulating and processing it.

The Fourier transform is traditionally defined for continuous-time functions. However, in the context of digital signal processing, it is common to work with signals that are only available in a discrete form, obtained by sampling a continuous-time signal at regular intervals. This discreteness introduces a fundamental difference in the mathematical properties of the signal, requiring the use of discrete Fourier transform techniques to extract frequency-domain information. The forward and inverse discrete Fourier transforms are shown below

$$X(k) = \sum_{n=0}^{N-1} x(n) \exp\left(-i\frac{2\pi}{N}nk\right), 0 \le k \le N-1, \tag{3.9}$$

$$x(n) = \frac{1}{N} \sum_{k=0}^{N-1} X(k) \exp\left(i\frac{2\pi}{N}nk\right), 0 \le n \le N-1. \tag{3.10}$$

The discrete Fourier transform represents the sampled signal as a vector of length N, where N is the number of samples taken. By decomposing the signal into its constituent sinusoidal components, the discrete Fourier transform generates a set of complex coefficients, which can be thought of as representing the amplitude and phase of the signal at specific frequency bins. These coefficients, known as discrete Fourier transform coefficients in the frequency domain.

The Fourier transform provides a powerful framework for sparse representation, which is a widely used technique in many areas of signal processing. In this approach, a signal is represented as a linear combination of a small number of basis functions, where the basis functions are typically chosen to be the sine and cosine functions at different frequencies. By selecting an appropriate subset of basis functions, one can represent the signal with high accuracy using only a small number of coefficients, leading to more efficient processing and analysis.

It is worth noting that the regularity of a function plays an important role in determining its sparse representation using the Fourier transform. Specifically, as a function becomes more regular, the amplitude of its Fourier coefficients decays faster as the frequency increases, making it easier to approximate the function using a small number of low-frequency coefficients. Thus, the Fourier transform provides a principled and efficient way to represent and analyze a wide range of functions and signals.

(2) Wavelet transform

The wavelet transform, like the Fourier transform, is a powerful tool for time-frequency analysis of signals. Unlike the Fourier transform, which represents a signal in terms of its frequency content alone, the wavelet transform allows for the decomposition of a signal into sub-signals of different frequency ranges while simultaneously providing information about the temporal occurrence of these sub-signals. This is achieved through the use of a set of basis functions, known as wavelets, which are defined by a specific frequency and time range. The decomposition of a signal using wavelets provides a representation of the signal that has good locality, allowing for the efficient representation of local transient structures with relatively few coefficients. This is in contrast to the Fourier transform, which requires a large number of coefficients to accurately represent transient features in a signal. As a result, the wavelet transform has become an important tool in many areas of science and engineering, including signal processing, image compression, and data analysis.

Given a basic function $\psi(t)$, let

$$\psi_{a,b}(t) = \frac{1}{\sqrt{a}}\psi\left(\frac{t-b}{a}\right) \tag{3.11}$$

where a and b are positive constants.

The given equation reveals that $\psi_{a,b}(t)$ is obtained as a result of applying translation and scaling operations to the fundamental function $\psi(t)$. The scaling parameter a and the time-shift parameter b enable the generation of a family of basic functions $\psi_{a,b}(t)$ as they are continuously varied. Therefore, $\psi(t)$ is referred to as the mother wavelet, which exhibits the decay property and is a compactly supported function that remains locally non-zero. Notably, when the frequency parameter ω equals zero, the Fourier transform of $\psi(t)$ also attains a zero value.

The family of functions $\psi_{a,b}(t)$ generated by applying translation and scaling operations to the mother wavelet are referred to as wavelet bases. It is worth highlighting that the basis functions of wavelet transform are waveform functions that exhibit a finite width and vary in both frequency and position. Hence, wavelet transform offers a versatile approach for signal analysis, as it facilitates the decomposition of complex signals into a set of wavelet coefficients that capture both temporal and spectral information.

The direct-inverse wavelet transform of a square integrable signal $x(t) \in L^2(\mathbb{R})$ can be expressed as follows,

$$WT_x(a,b) = \frac{1}{\sqrt{a}} \int_{-\infty}^{+\infty} x(t)\psi_{a,b}^{*}\left(\frac{t-b}{a}\right)\mathrm{d}t, \tag{3.12}$$

$$x(t) = \frac{1}{C_\psi} \int_{-\infty}^{\infty}\int_{-\infty}^{\infty} WT_x(a,b)\psi\left(\frac{t-b}{a}\right)\frac{\mathrm{d}a\mathrm{d}b}{a^2}, \tag{3.13}$$

where * represents complex conjugation. C_ψ is the normalization constant of wavelet function.

Based on the equation, it can be inferred that the expression $WT_x(a,b)$ is a function of two independent variables, namely a and b. The wavelet transform can be interpreted as the inner product between a given signal $x(t)$ and a family of wavelet basis functions. The scaling factor a is utilized to adjust the scale of the mother wavelet, whereby larger values of a result in a greater time support range of $\psi(t/a)$. On the other hand, the time shift parameter b is employed to determine the temporal location of the signal analysis as $\psi(t-b)$. By jointly manipulating a and b, the window parameters for signal analysis can be established.

In the definition of the traditional wavelet transform, the variables a, b, and t are considered continuous, which presents a challenge when implementing the transform on a computer. In order to compute the wavelet transform of a signal, it is necessary to discretize a, b, and t, which involves setting a and b to discrete values and sampling t at discrete intervals. Discretization of a is usually done by setting $a = 2^j$, discretely sampling a at powers of 2, while also discretizing b. By discretizing the scale-shift plane (time-frequency plane) formed by a and b, the discrete binary wavelet can be obtained as

$$\psi_{j,k}(t) = 2^{-j/2}\psi\left(2^{-j}t - k\right), \tag{3.14}$$

where $j, k \in \mathbb{Z}$. Then the discrete wavelet transform of signal x is shown as

$$WT_x(j,k) = \left\langle x(t), \psi_{j,k}(t)\right\rangle. \tag{3.15}$$

The wavelet transform has the ability to extract both local and global information from a signal by decomposing it into different scales of frequency components. The high-frequency components capture the details and local variations of the signal, while the low-frequency components represent the overall trends and long-term behavior. This characteristic makes the wavelet transform a powerful tool for signal sparse representation, where the goal is to represent a signal with a small number of non-zero coefficients.

By decomposing a signal into wavelet coefficients at multiple scales, the wavelet transform provides a multi-resolution analysis of the signal, allowing us to extract both fine-grained and coarse-grained features of the signal. The high-frequency wavelet coefficients tend to be sparse, meaning that they contain only a small amount of energy and contribute relatively little to the overall structure of the signal. On the other hand, the low-frequency coefficients tend to be more energy-rich and dominant in representing the global behavior of the signal.

Based on these properties, in wavelet transform, the high-frequency coefficients can be discarded or truncated without significant loss of information. By selectively retaining the most significant coefficients, the wavelet transform can produce a sparse representation of the signal, which not only reduces storage and computational complexity but also facilitates further analysis and processing.

In addition to the well-known Fourier and wavelet transforms, there exist a plethora of orthogonal transform bases-based sparse representation methods that exhibit notable sparsity performance in various signal processing applications, including but not limited to the cosine transform and the Walsh-Hadamard transform. Due to space constraints, a comprehensive elaboration on each of these techniques is beyond the scope of this discussion.

3.3.4 Sparsification Based on Multi-scale Geometric Analysis

Observing natural objects at different scales reveals different features, and selecting an appropriate scale is crucial for recognizing and understanding the target [15]. However, because images generally contain objects of different sizes, it is difficult to predefine an optimal scale for processing and analyzing images. As a result, the need to analyze image features at multiple scales has gradually emerged.

The image representation method used in multiscale geometric analysis simulates and approximates the human visual system to a certain extent. Typically, when humans observe a visual scene, they first focus on the overall structure of the scene, and then shift their attention to local details of interest. Furthermore, the human visual system is particularly sensitive to directionality. Multiscale geometric analysis is directional, and through multiscale transformations, it can capture global information and local detail information at different scales. In terms of image signals, there is a functional similarity between the human visual system and multiscale geometric analysis. Therefore, multiscale geometric analysis can be successful in the field of image analysis and representation.

For any given signal $f(t) \in V_0$, it is possible to decompose it into an approximate component V_1 and a detailed component W_1. Subsequently, the approximate component V_1 can be subjected to further decomposition, and the procedure can be iterated on the resulting approximate components, enabling the decomposition of both approximate and detail components at any desired scale. Figure 3.5 illustrates the fundamental theoretical principle and framework of multiscale analysis, as

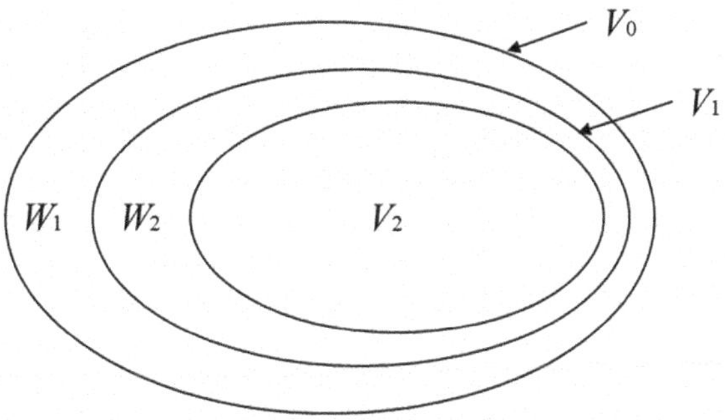

Fig. 3.5 Frame of the orthogonal sub-space, V_j is scale space and W_j is detail space

$$\mathbf{V}_0 = \mathbf{V}_1 \oplus \mathbf{W}_1 = \mathbf{V}_2 \oplus \mathbf{W}_2 \oplus \mathbf{W}_1 = \mathbf{V}_3 \oplus \mathbf{W}_3 \oplus \mathbf{W}_2 \oplus \mathbf{W}_1 = \cdots. \tag{3.16}$$

Currently, the mainstream methods for multiscale geometric analysis include ridgelet [16], bandelet [17], curvelet [18], contourlet [19], and shearlet [20] transforms. Depending on whether image features need to be combined, these methods can be classified into two categories: adaptive and non-adaptive. Adaptive methods, such as the bandelet and wedgelet transform, require edge detection methods to utilize the edge detection information of the decomposed image for optimal representation. Non-adaptive methods, such as the ridgelet, curvelet, and contourlet transform, decompose the image based on a given set of supporting bases and do not depend on the geometric features of the image. The latest method for scale geometric analysis, the shearlet transform, uses shear filters to achieve multidirectional decomposition of the image.

(1) Curvelet transform

For objective functions with smooth singularities in their curves, the curvelet transform offers a stable, efficient, and nearly optimal representation approach. Its central concept is to decompose an image into multiple subbands of varying scales and orientations, followed by a local, reversible transformation on each subband to achieve a representation with improved compressibility and separability. This approach is highly desirable in image processing applications, as it can efficiently capture the complex features of images and allow for accurate analysis and manipulation.

Let $\varphi_j(x)$ be the mother curvelet, and let $\hat{\varphi}_j(\omega) = U_j(\omega)$ be its Fourier transform. Then, all curvelets at scale 2^{-j} can be obtained through translations and rotations of $\varphi_j(x)$. To this end, we introduce the equidistant rotation angle sequence $\theta_l = 2\pi \times 2^{-\lfloor j/2 \rfloor} \times l$, $(l = 0, 1, \ldots, 0 \le \theta_l \le 2\pi)$ and the displacement parameter sequence $k = (k_1, k_2) \in \mathbb{Z}^2$. Thus, the curvelet located at scale 2^j, direction angle θ, and position $x_k^{(j,l)} = R_{\theta_l}^{-1}(k_1 \times 2^{-j}, k_2 \times 2^{-j/2})$ can be expressed as

$$\varphi_{j,l,k}(x) = \varphi_j\left[R_{\theta_l}\left(x - x_k^{(j,l)}\right)\right]. \tag{3.17}$$

Then the curvelet transform is defined as

$$c(j, l, k) = \langle f, \varphi_{j,l,k} \rangle = \int_{R^2} f(x)\varphi_{j,l,k}(x)\mathrm{d}x. \tag{3.18}$$

According to the Plancherel theorem, the transformation of a curvelet in the frequency domain can be defined as follows

$$c(j, l, k) = \frac{1}{2\pi^2}\int \hat{f}(\omega)\varphi_{j,l,k}(\omega)d\omega = \frac{1}{2\pi^2}\int \hat{f}(\omega)U_j\left(R_{\theta_l}\omega\right)\exp\left[i\left\langle x_k^{(j,l)}, \omega\right\rangle\right]\mathrm{d}\omega. \tag{3.19}$$

Similar to other transforms used in multiscale analysis, curvelet exhibits different components that correspond to different scales, including coarse and fine scales. The concept of a low-pass window, denoted by W_0, is introduced, and the curvelet at coarse scales is defined as

$$\begin{cases} \varphi_{j_0,k}(x) = \varphi_{j_0}\left(x - 2^{-j_0}k\right), \\ \varphi_{j_0}(\omega) = 2^{-j_0}W_0\left(2^{-j_0}|\omega|\right). \end{cases} \tag{3.20}$$

Based on the structural characteristics of the support space of curvelets in the frequency domain and the aforementioned definitions, it becomes apparent that curvelets at coarse scales lack directionality. Specifically, the complete curvelet transform comprises of isotropic wavelets $(\varphi_{j_0,k})_k$ at coarse scales and directional elements $\varphi_{j,l,k}$, $(j \geq j_0, l, k)$, at fine scales. This structural composition is illustrated in Fig. 3.6.

The discrete curvelet transform exhibits linearity, which renders it amenable to processing input images $f[t_1, t_2]$, $(0 < t_1, t_2 < n)$ represented in the Cartesian coordinate system. Its output is expressed in terms of a set of coefficients $c^D(j, k, l)$ depending on a basis of functions defined on a multi-scale and multi-directional grid,

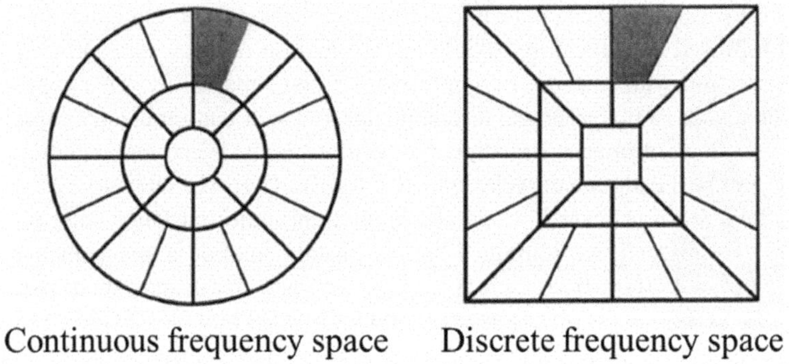

Fig. 3.6 Sketch map of curvelet transform in frequency domain. The wedges with color are regions of support respectively in continuous situation and discrete situation

$$c^D(j,k,l) = \sum_{0 \le t_1, t_2 < n} f[t_1, t_2] \varphi^D_{j,l,k}[t_1, t_2]. \tag{3.21}$$

Here, $\varphi^D_{j,l,k}[t_1, t_2]$ represents the discrete curvelet transform framework, j, k, and l are the scale, direction, and position parameters, while t_1 and t_2 denote the discrete coordinates. In the discrete domain, concentric slanted regions at different angles need to be used as the support domain instead of the continuous sector region in the continuous domain, as shown in Fig. 3.6.

(2) Contourlet transform

The contourlet transform possesses multi-scale and multi-directional properties, as well as excellent time-frequency localization characteristics. It only requires a small number of non-zero coefficients to effectively capture the contour information of images. By utilizing an iterative filter bank, the contourlet transform can perform a multidirectional decomposition on high-frequency components, thus obtaining detailed information on the texture and contours of the image in arbitrary directions. As a result, it has good direction sensitivity and excellent nonlinear approximation performance.

The overall idea of the Contourlet transform is to first use the Laplacian Pyramid (LP) to perform a multiscale decomposition of the image to "capture" singular points, and then use a Directional Filter Bank (DFB) to perform a multidirectional decomposition on the high-frequency sub-band image obtained from the decomposition, connecting similar singular points to form contour segments.

Figure 3.7 shows the curve representations of wavelet bases and contourlet bases. At different resolutions, the support intervals of the wavelet transform are represented by squares of different sizes, and as the resolution decreases, the support interval sizes also decrease until they become points. In contrast, the support intervals of the contourlet transform are represented by rectangles, and the final curve representation is similar to approximating the original image using a line segment basis structure. When the resolution is the same, the number of support domains in the contourlet transform is significantly less than that in the wavelet transform. Therefore, the contourlet transform can sparsely represent curves using fewer basis functions.

The contourlet transform has a dual-iterative filter structure. The LP decomposition process is shown in Fig. 3.8a. First, the original image is subjected to low-pass filtering and down-sampling to obtain a low-frequency component c. Then, the low-frequency component is subjected to one-time up-sampling and low-pass filtering to obtain an approximation that is the same size as the original image. Then this approximation is subtracted from the original image to obtain the high-frequency component d. Through the LP transformation at multiple scales, a low-frequency approximation component and multiple band-pass components in different directions can be obtained.

DFB synthesizes singular points distributed in the same direction into one coefficient $c^{l_j}_{j,k}[n]$ by using a conjugate mirror filter bank with a fan-shaped structure, that is, performs l_j-level binary tree decomposition on the image direction, and divides the bandpass image spectrum of LP decomposition into 2^{l_j} wedge-shaped frequency

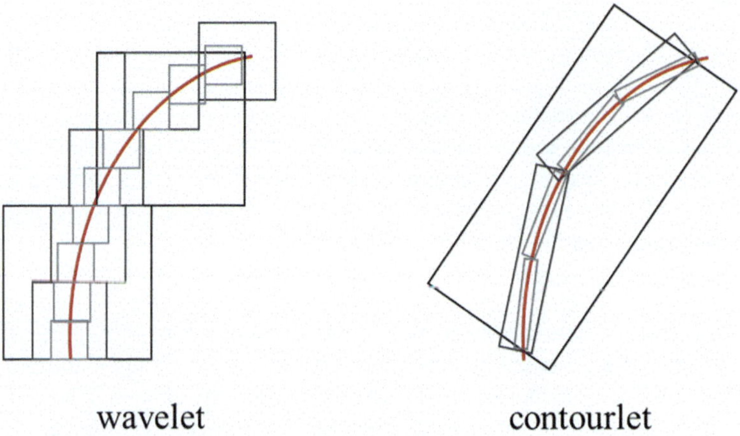

Fig. 3.7 Illustration of curve with wavelet and contourlet

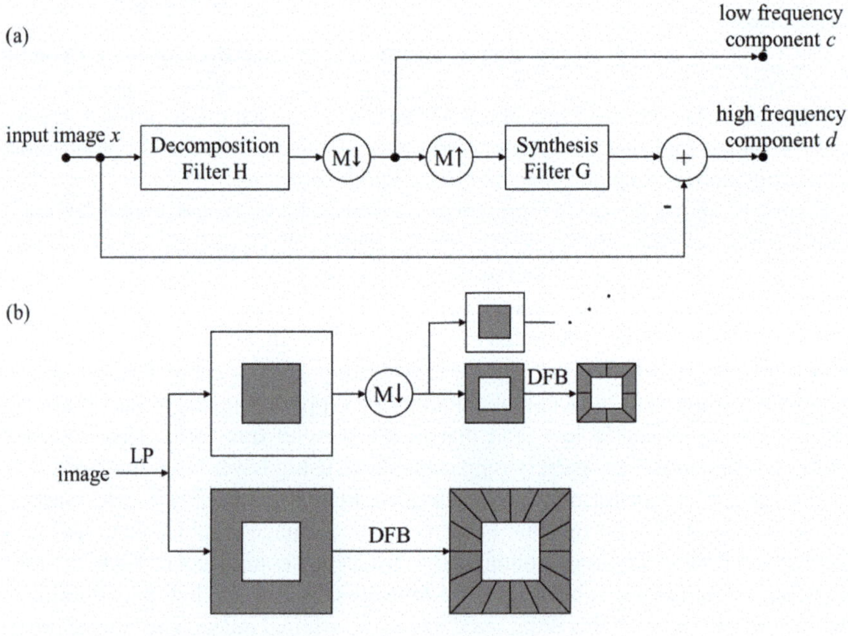

Fig. 3.8 Decomposition process diagram of contourlet transform

sub-bands $c_{j,k}^{l_j}[n]$, so as to complete the direction decomposition on each scale, as shown in Fig. 3.8b, the decomposition process is expressed as

$$x = \sum_{n} a_J[n]\phi_{J,n} + \sum_{j}\sum_{n}\sum_{k} c_{j,k}^{l_j}[n]\lambda_{j,k,n}^{l_j}. \tag{3.22}$$

Here, $a_J[n] = \langle x, \phi_{J,n} \rangle$, $c_{j,k}^{l_j}[n] = \langle x, \lambda_{j,k,n}^{l_j} \rangle$, $x \in L^2(\mathbb{Z}^2)$ and $n \in \mathbb{R}^2$. The scale parameter j represents the series of scale decomposition, the position parameter n represents the two-dimensional coordinate position, the direction parameter k represents the direction of the band-pass sub-band, the number of levels in the band-pass sub-band decomposition is denoted by l_j, and J is a positive integer being >2.

In order to eliminate the spectral aliasing phenomenon, enhance directional selectivity, and achieve translation invariance in the contourlet transform, the non-subsampled contourlet transform (NSCT) was proposed. Similar to the contourlet transform, the NSCT separates multiscale analysis and directional analysis. However, the NSCT differs in that it employs a double filter bank structure composed of a non-subsampled pyramid (NSP) and non-subsampled directional filter bank (NSDFB) filters, instead of the LP and DFB filters in the contourlet transform.

By exploiting the shift-invariance property of the Z-transform, the NSCT removes the down-sampling and up-sampling operations in the LP and DFB decompositions and instead up-samples the corresponding filters for analysis and synthesis filtering of the signal. The elimination of down-sampling and up-sampling allows the NSCT to achieve translation invariance. Its framework is shown in Fig. 3.9a, which divides the two-dimensional frequency domain into wedge-shaped sub-bands as shown in Fig. 3.9b.

(3) Shearlet transform

Shearlet transform is a nearly optimal, multidimensional sparse representation method constructed by an affine system with synthetic dilation. It represents the latest advances in multiscale geometric analysis and can achieve sparse representation of multidimensional signals and anisotropic information at multiple scales and multiple directions.

In addition to its superior error approximation order, the shearlet transform has the advantage of being subdivided in a layer-by-layer manner in the frequency domain,

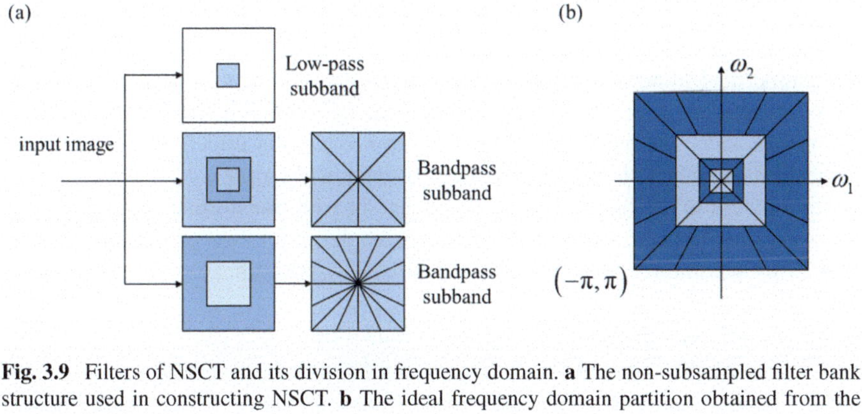

Fig. 3.9 Filters of NSCT and its division in frequency domain. **a** The non-subsampled filter bank structure used in constructing NSCT. **b** The ideal frequency domain partition obtained from the structure in (**a**)

which enhances its ability to represent images accurately. Furthermore, it has a relatively simple mathematical structure, requiring only the generation of basis functions by scaling, translating, and rotating a single function. The size, direction, and number of support sets can be customized to meet specific requirements, enabling true multi-scale and multi-directional decomposition of signals and providing a more sensitive tool for the geometric representation of high-dimensional data.

Using the classical theory of affine systems with synthesis and expansion properties, we construct shearlet by combining geometric transformations and multi-scale analysis. In two-dimensional space, the definition of the shearlet transform is expressed as follows

$$SH_{\psi}f(j,o,l) = \langle f, \psi_{j,o,l} \rangle. \tag{3.23}$$

Here, $SH_{\psi}f(j,o,l)$ represents the coefficient of the shearlet transform, j, o and l represent the scale parameter, direction parameter and translation parameter of the shearlet transform respectively, f represents a function, and $\psi_{j,o,l}$ represents the shearlet. The definition of shearlet $\psi_{j,o,l}$ is expressed as

$$\begin{aligned} \psi_{j,o,l} &= \left|\det \boldsymbol{M}_{j,o}\right|^{-\frac{1}{2}} \psi\left[\boldsymbol{M}_{j,o}^{-1}(x-l)\right], \\ \boldsymbol{M}_{j,o} &= \boldsymbol{B}_o \boldsymbol{A}_j,\ \boldsymbol{B}_o = \begin{bmatrix} 1 & -o \\ 0 & 1 \end{bmatrix},\ \boldsymbol{A}_j = \begin{bmatrix} j & 0 \\ 0 & \sqrt{j} \end{bmatrix}, \end{aligned} \tag{3.24}$$

where det represents the calculation of the determinant, $\boldsymbol{A}_j$ represents the anisotropic dilation matrix, and $\boldsymbol{B}_o$ represents the non-expansive shear matrix. $\boldsymbol{A}_j$ and $\boldsymbol{B}_o$ are invertible 2×2 matrices, and $|\det B| = 1$.

The shearlet functions are a type of synthetic wavelet that can implement scale and geometric transformations, and has the characteristics of multi-resolution analysis, enabling the construction of tight-supported frames at different scales, orientations, and positions. The frequency domain decomposition of shearlet is expressed in Fig. 3.10.

The support base of each element of the shearlet is a pair of trapezoids that are symmetrical along the origin of the coordinate. The size of each trapezoid is approximately $2^{2j} \times 2^{j}$, and its directional slope is 2^{-j}. Compared to wavelets and contourlets, shearlets have a more optimal approximation effect on the details in images. It is beneficial to design compressed sensing algorithm.

The process of decomposing images using shearlet transformation is shown in Fig. 3.11. The first step is to use the non-subsampled Laplacian pyramid transform (NSLP) to decompose image f into two parts: low-pass image f_l^1 and high-pass image f_h^1.

The second step is to calculate the Fourier transform value $\hat{f}_h^1$ of f_h^1 on the pseudo-polar lattice to obtain the matrix Pf_h^1. The third step is to process the matrix Pf_h^1 using a bandpass filter to obtain the Fourier transform values $\{\hat{f}_{h,k}^1\}_{k=1}^{N_1}$ of the sub-band coefficients in each direction in the first level scale, where N_1 represents the

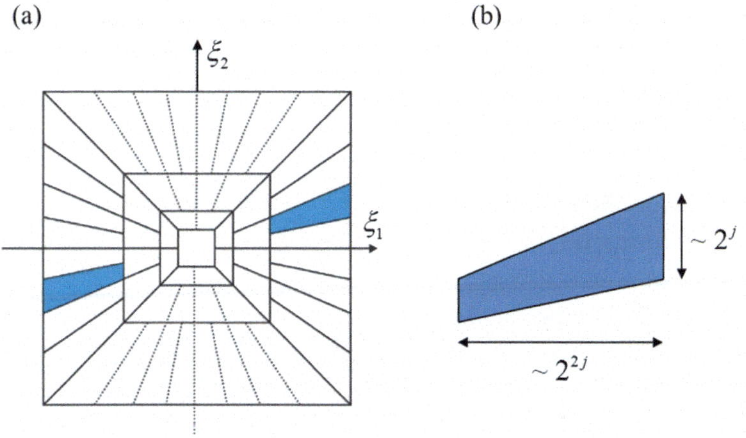

Fig. 3.10 Frequency domain decomposition diagram of shearlet. **a** Frequency domain decomposition diagram. **b** Frequency domain support base

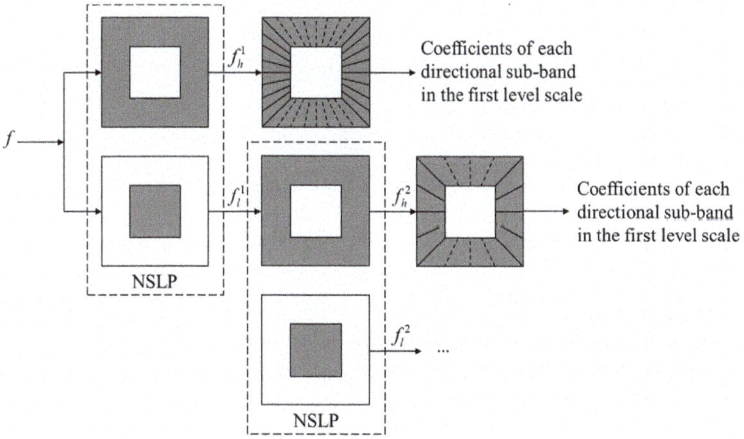

Fig. 3.11 Schematic diagram of image decomposition using shear wave transformation

number of all directions in the first level scale. The fourth step is to use the inverse Fourier transform to process $\{\hat{f}_{h,k}^1\}_{k=1}^{N_1}$ and obtain the coefficients $\{f_{h,k}^1\}_{k=1}^{N_1}$ of each directional sub-band in the first level scale of the shearlet. By replacing f with f_l^1 in the first step and repeating the work from step 1 to step 4, the coefficients of each directional sub-band in the second level scale of the shearlet can be obtained. By using a similar method, the coefficients of various directional sub-bands at other levels of scale can be obtained.

By understanding several commonly used multi-scale geometric analysis methods, theoretical guidance is provided for multi-scale sparse representation of images, and reliable bases are provided for the regularization of CS reconstruction.

In addition to the three multiscale analysis transformations mentioned above, there are still many studies on multiscale analysis methods in the field of signal multiscale analysis, and we will not elaborate on them here.

3.3.5 *Sparse Representation Based on Redundant Dictionaries*

The mathematical description of signal sparse decomposition based on a redundant dictionary is as follows: Let $D = |\boldsymbol{d}_i,\ i = 1,\ 2,\ \ldots,\ L|$ be a set, where $\boldsymbol{d}_i$ represents unit vectors that can span the entire Hilbert space. The set D is called a dictionary, and each column in the dictionary is an atom. The signal has a length of N, and because $L \gg N$, D is referred to as a redundant dictionary. For a signal $f \in \mathbb{R}^N$, K atoms are selected from the dictionary D to approximate f.

$$f_k = \sum_{i \in I_k, |I_k| = K} \langle f, \boldsymbol{d}_i \rangle \boldsymbol{d}_i = \sum_{i \in I_k, |I_k| = K} \alpha_i \boldsymbol{d}_i. \tag{3.25}$$

In this equation, I_k represents the index of the sequence $\boldsymbol{d}_i$.

The approximation error is

$$\varepsilon = \inf_{f_k} \| f - f_k \|. \tag{3.26}$$

Since $K \le N$, f_k is a sparse approximation of f. The coefficients α_i are not unique, and under the error condition, the goal is to select a set of atoms with the sparsest decomposition coefficients from various possible combinations to approximate the signal f. Sparsity can be measured using the l_0 norm.

Therefore, finding the sparsest representation of a signal is equivalent to solving the l_0 norm problem as

$$\min \|\boldsymbol{\alpha}\|_0 \text{ s.t. } f = \sum_{i \in I_k, |I_k| = K} \alpha_i \boldsymbol{d}_i \sqrt{a^2 + b^2}. \tag{3.27}$$

In this equation, $\|\boldsymbol{\alpha}\|_0$ is the number of non-zero elements in the sequence α_i.

All the atoms from the dictionary are used as column vectors to form an $N \times L$ matrix (where $N < L$), denoted as $\boldsymbol{\Phi}$, as

$$\min \|\boldsymbol{\alpha}\|_0 \text{ s.t. } f = \boldsymbol{\Phi}\boldsymbol{\alpha}. \tag{3.28}$$

The above equation represents the sparse representation model in matrix form. However, solving for x for a redundant dictionary D is an NP-hard problem.

Dictionary design is a core issue in sparse representation theory. Whether a signal can be represented sparsely essentially depends on the performance of the dictionary itself. The degree of matching between the structure of the signal and the atoms in the dictionary determines the ease of sparse representation. When atoms match the structure of the signal better, the signal is more easily represented sparsely.

Common methods for constructing redundant dictionaries include:

(1) Using existing analytical methods such as wavelet orthogonal transforms, multiscale geometric analysis, and tight frame systems.
(2) Combining the above methods to obtain even sparser representations. This approach has been widely used in image processing.
(3) For specific applications, dictionaries can be obtained through training, such as K-SVD and online learning [21, 22]. The main approach is to learn a sparse representation dictionary for training samples, and this dictionary can be used to represent other input signals with similar features sparsely.

3.4 Construction of Measurement Matrix

The measurement matrix plays a crucial role in CS, and its performance impacts the compression and reconstruction of the original signal. The quality of the measurement matrix is determined by the RIP. In CS theory, for precise reconstruction of a K-sparse signal x, the measurement matrix $\boldsymbol{\Phi}$ must satisfy the following defined RIP [23].

The RIP parameter δ_K of the measurement matrix $\boldsymbol{\Phi}$ is defined as the minimum value that satisfies the following equation

$$(1-\delta)\|\boldsymbol{x}\|_2^2\|\boldsymbol{\Phi}_x\|_2^2(1+\delta)\|\boldsymbol{x}\|_2^2, \tag{3.29}$$

where x is a K-sparse signal. If $\delta_K < 1$, it is said that the measurement matrix $\boldsymbol{\Phi}$ satisfies the K-order RIP.

Although the RIP condition has a perfect form and rich implications, it cannot be practically used to guide the construction of a specific measurement matrix or to effectively determine whether a given measurement matrix meets the required properties. Therefore, the concept of coherence theory [24] was introduced.

An equivalent condition for RIP is that the measurement matrix and the sparse representation basis are uncorrelated, meaning that the measurement matrix $\boldsymbol{\Phi}$ and the transformation matrix $\boldsymbol{\Psi}$ are uncorrelated. The correlation is defined as:

$$\mu(\boldsymbol{\Phi}, \boldsymbol{\Psi}) = \sqrt{n} \max_{1 \le k, j \le n} \left|\langle \varphi_k, \boldsymbol{\psi}_j \rangle\right|, \tag{3.30}$$

where n is the number of columns in the matrices, φ_k, $\boldsymbol{\psi}_j$ respectively represent column vectors of the measurement matrix and the transformation matrix. A smaller μ indicates a lower correlation.

Since the inception of CS theory, various methods have been developed for constructing measurement matrices. The general form of the measurement matrix is as follows

$$\Phi = \begin{bmatrix} \varphi_{11} & \varphi_{12} & \cdots & \varphi_{1n} \\ \varphi_{21} & \varphi_{22} & \cdots & \varphi_{2n} \\ \ddots & \ddots & \ddots & \ddots \\ \varphi_{m1} & \varphi_{m2} & \cdots & \varphi_{mn} \end{bmatrix}. \tag{3.31}$$

These matrices can generally be categorized into two main types: random measurement matrices and deterministic measurement matrices. Let us delve into further details below.

3.4.1 Random Measurement Matrix

(1) Gaussian random matrix

The elements of a Gaussian random measurement matrix are independent random variables with an expected value of 0 and a variance of $1/m$, and are expressed as follows

$$\Phi_{i,j}^{\text{Gaussian}} \sim N\left(0, \frac{1}{\sqrt{M}}\right). \tag{3.32}$$

The coherence parameter between the Gaussian random measurement matrix and the orthogonal sparse matrix is small, and it can satisfy the RIP property well [25]. It only requires a small number of observations m, $m \geq cK\log(n/K)$ (where c is a constant), to accurately reconstruct sparse signals. Therefore, the Gaussian matrix is widely used.

(2) Bernoulli random matrix

Each element in a Bernoulli random measurement matrix follows a Bernoulli distribution, which is an independent probability distribution and a special case of the binomial distribution. This matrix has only binary values and is represented as follows

$$\Phi_{i,j}^{\text{Bernoulli}} = \begin{cases} +\frac{1}{\sqrt{M}}, & P = \frac{1}{2}. \\ -\frac{1}{\sqrt{M}}, & P = \frac{1}{2}. \end{cases} \tag{3.33}$$

The Bernoulli random measurement matrix exhibits good randomness. With a number of measurements m, $m \geq cK\log(n/K)$ (where c is a constant), it can satisfy the RIP property well, and there is a high probability of accurately reconstructing the original signal [26].

(3) Chaotic random matrix

The construction of a Chaotic random measurement matrix can be derived from its logical mapping function [27]

$$\varphi_{k+1} = \mu\varphi_k(1 - \varphi_k), \tag{3.34}$$

where $\mu \in (0, 4)$ and $\varphi_k \in (0, 1)$. In the special case of $\mu = 4$, the solution of system is

$$\varphi_k = \frac{1}{2}\left[1 - \cos\left(2\pi\theta \cdot 2^k\right)\right], \tag{3.35}$$

where $\theta \in (0, \pi)$ satisfies

$$\varphi_0 = \frac{1}{2}\cos(2\pi\theta). \tag{3.36}$$

It is well-known that chaotic systems can produce highly complex sequences. The chaotic random measurement matrix can be given by the following equation

$$\Phi_{\text{Chaotic}} = \begin{bmatrix} \varphi_0 & \varphi_m & \cdots & \varphi_{m(n-1)} \\ \varphi_1 & \varphi_{m+1} & \cdots & \varphi_{m(n-1)+1} \\ \vdots & \vdots & \ddots & \vdots \\ \varphi_{m-1} & \varphi_{2m-1} & \cdots & \varphi_{mn-1} \end{bmatrix}. \tag{3.37}$$

In addition to these, the random measurement matrix also includes partial Fourier matrices [28], partial orthogonal matrices [29], sparse random matrices [30], and so on. These random matrices exhibit good independence and to some extent satisfy RIP. However, due to their randomness, the hardware implementation of these matrices can be complex. Therefore, scholars have researched deterministic matrix construction algorithms to achieve the compression observation of the original signal through deterministic matrices [31].

3.4.2 Deterministic Measurement Matrix

(1) Edge expansion graph adjacency matrix

The edge expansion graph adjacency matrix utilizes the adjacency relationship of an unbalanced bipartite graph and is applied in the compression observation process [25]. As shown in Fig. 3.12, the unbalanced bipartite graph directly simulates CS observation process. The graph is divided into two parts: the left subgraph is defined as A_1, and the right subgraph is defined as A_2, containing two types of nodes. Following the numerical description of coding theory, nodes in A_1 are defined as variable nodes, while nodes in A_2 are defined as parity-check nodes. The number of nodes in A_1 is n, corresponding to the original signal x, and the number of nodes in A_2 is m, corresponding to the signal y after compression observation. Assuming $m \leq n$, there are edges connecting nodes between A_1 and A_2, and the connection relationship corresponds to the adjacency matrix $\boldsymbol{C}$

$$\boldsymbol{C}_{ij} = \begin{cases} 1, & \text{right i node links to left } j \text{ node.} \\ 0, & \text{other.} \end{cases} \tag{3.38}$$

where $i = 1, 2, ..., m$; $j = 1, 2, ..., n$.

In the case of $1 \le K \le n/2$, if the (K, ε)-edge expansion graph has a degree d in the left subgraph and a size m in the right subgraph:

$$\begin{cases} d = O\{[\log(n/K)]/\varepsilon\}, \\ m = O\{[K\log(n/K)]/\varepsilon^2\}. \end{cases} \tag{3.39}$$

Then its adjacency matrix $\boldsymbol{C}$ satisfies the RIP property for any K-sparse signal.

(2) Polynomial deterministic matrix

Based on the principles of CS theory and finite field theory, deterministic polynomial matrices are constructed using polynomial mapping matrices [32]. Let us assume a prime number, denoted as p, where the elements in the finite field $\boldsymbol{F}$ are integers modulo p. Therefore, the order of the finite field $\boldsymbol{F}$ is p, meaning that there are p elements in the $\boldsymbol{F}$ field, which can be represented as $\{0, 1, \ldots, p-1\}$. We create a $p \times p$ matrix, denoted as $\boldsymbol{E}$, which contains p^2 elements. The assignment of elements in $\boldsymbol{E}$ is as follows.

Fig. 3.12 A (K, ε)-edge expansion graph

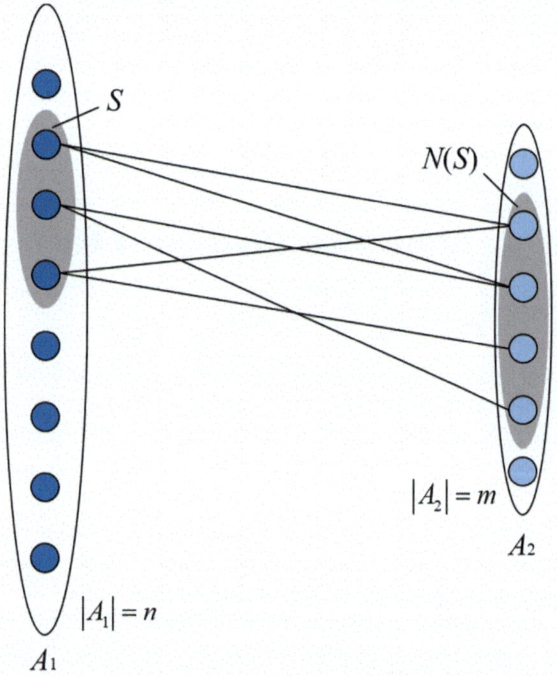

Let us define an integer $r(0 < r < p)$. Then, $\mathbf{P}_r$ represents the set of polynomials with degrees not exceeding r,

$$\boldsymbol{P}_r = \{Q(t) = a_0 + a_1 t + \cdots + a_n t^r | a_n \in F\}. \tag{3.40}$$

The number of polynomials in $\boldsymbol{P}_r$ is p^{r+1}. For any $Q \in P_r$, considering Q as a mapping function $\boldsymbol{F} \to \boldsymbol{F}$, then $\boldsymbol{E}$ serves as the mapping matrix $t \to Q(t)$, satisfying

$$Q(t) = \boldsymbol{E}t^*. \tag{3.41}$$

Here, t^* represents the set of all possible values of t.

The position index of matrix $\boldsymbol{E}$ is defined as follows

$$\begin{bmatrix} (0,0) & (0,1) & \cdots & (0,p-1) \\ (1,0) & (1,1) & \cdots & (1,p-1) \\ \vdots & \vdots & & \vdots \\ (p-1,0) & (p-1,1) & \cdots & (p-1,p-1) \end{bmatrix}. \tag{3.42}$$

In matrix $\boldsymbol{E}$, the element at the position indexed by $(t, Q(t))$ is set to 1, while all other positions are set to 0. $\boldsymbol{E}$ is then transformed into a $p^2 \times 1$ vector $\mathbf{v}_Q$, where every p elements in $\boldsymbol{v}_Q$ contain a single 1, resulting in a total of p 1 s. Since the polynomial coefficients have different values, they generate different vectors. Therefore, when all coefficients are considered, we can construct p^{r+1} vectors, treating each vector as a column to create a $p^2 \times p^{r+1}$ matrix.

The deterministic matrices constructed in the aforementioned manner satisfy the RIP under the condition where the sparsity level $K < p/r + 1$. This ensures the exact reconstruction of the original sparse signal.

(3) Code matrix

Inspired by the concept of finite fields, researchers have expanded the application of existing coding techniques by using Optical Orthogonal Codes (OOC) to construct deterministic measurement matrices for CS [33]. Optical orthogonal codes find wide application in the design of code-division multiple access channels in optical communication. By leveraging cyclic shifts of OOC and nested Hadamard matrices, a ternary real matrix with excellent CS performance can be constructed.

An (n, ω, λ) OOC is a set of binary sequences with a total of n elements, denoted as $\{\boldsymbol{v}^{(i)} | 0 \le i \le n-1\}$. Each sequence contains ω non-zero elements, and between any two sequences, the relationship is as follows

$$\forall \tau \in [0, n-1], \theta_{v(i),v(j)}(\tau)\lambda. \tag{3.43}$$

$$\theta_{s^{(i)},s^{(j)}}(\tau) = \sum_{t=0}^{n-1} \boldsymbol{v}_t^{(i)} \boldsymbol{v}_{t \oplus \tau}^{(j)}, \tag{3.44}$$

where ⊕ represents modulo-n addition.

The construction of OOC utilizes a cyclic difference set group $\boldsymbol{G}$, where $G(n, \omega)$ consists of ω non-zero modulo-n integers with distinct pairwise differences. A characteristic sequence $\boldsymbol{a} = \{a_t | t \in (0, 1, \ldots, n-1)\}$ is defined, such that if $t \in G$, then $a_t = 1$, otherwise, $a_t = 0$. This sequence belongs to $(n, \omega, 1)$ OOC.

Building on the foundation of OOC, the construction of the measurement matrix is as follows:

(a) Generate S 0–1 sequences $\boldsymbol{v}^{(i)}(0 \le i \le n-1)$, belonging to the (n, ω, λ) OOC, and cyclically shift each sequence $\boldsymbol{v}^{(i)}$ to generate nS sequences. Each sequence is used as a column vector in matrix $\boldsymbol{B}$.
(b) Construct a $v \times v\,(v = \omega + \delta,\ 0 \le \delta \le \omega)$ Hadamard matrix $\boldsymbol{H}$. For each column in $\boldsymbol{B}$, replace the 1 elements with different rows from $\boldsymbol{H}$ and expand the 0 elements into rows of length v, where all elements are 0. This results in an $n \times vnS$ matrix B_e.
(c) Normalize the matrix to generate the measurement matrix $\boldsymbol{D} = (1/\sqrt{\omega})\boldsymbol{B}_e$.

The measurement matrix $\boldsymbol{D}$ constructed using the above method satisfies a tight frame structure. Compared to other tight frame structure measurement matrices, it requires less storage space and fewer parameters. Its column coherence coefficient is $O(1/\sqrt{\omega})$. For a K-sparse signal x, if it satisfies $K = O[m/\log(n)]$, then the original signal can be accurately reconstructed with a probability of $1 - n^{-1}$.

In addition to this, various coding techniques such as discrete Chirp coding [34], algebraic coding [35], Reed-Muller coding [36], Berlekamp-Justesen coding [37], as well as various low-density parity-check codes [38, 39], have been proven to serve as CS measurement matrices, enabling the precise reconstruction of the original signal. These techniques are not elaborated on here.

3.4.3 *Structural Design of Measurement Matrix*

As the algorithms for constructing CS measurement matrices continue to evolve, scholars have proposed methods to design measurement matrix structures by optimizing them, combining the advantages of different algorithms, and creating more reasonable measurement matrices. Common structural design methods for measurement matrices mainly include Toeplitz circulant matrices, block matrices, and nested matrices.

(1) Toeplitz circulant matrices

Toeplitz matrices are constructed using vector circulations, which have evident structural characteristics [40, 41]. Therefore, they are easy to implement in hardware, require fewer generated random variables compared to independently and identically distributed random matrices, have low computational complexity, and relatively simple reconstruction algorithms. As a result, the circulant matrix construction method has attracted widespread attention.

In the construction process, vectors $g = [g_1, g_2, \ldots, g_n]$ with elements of ± 1 following a Bernoulli distribution are first generated. This vector is used as the first row vector of the Toeplitz matrix. Subsequently, the other row vectors of matrix are constructed through circulation, and finally, a column-wise normalization is performed, defining the normalized elements as u. The resulting matrix is as follows

$$\boldsymbol{U} = \begin{bmatrix} u_n & u_{n-1} & \cdots & u_2 & u_1 \\ u_1 & u_n & \cdots & u_3 & u_2 \\ \vdots & \vdots & & \vdots & \vdots \\ u_{m-1} & u_{m-2} & \cdots & & u_m \end{bmatrix}. \tag{3.45}$$

When the elements in the matrix independently follow a specific distribution $P(u)$, and $m \geq K^3 \ln(n/K)$, the RIP is satisfied. Toeplitz matrices can be combined with specific sequences to generate sequence matrices with Toeplitz structures, such as discrete Chirp Toeplitz matrices, and so on.

(2) Block matrices

A large number of measurement matrices that satisfy the RIP exhibit a dense matrix structure, which demonstrates excellent performance in reconstructing the original signal. However, these dense matrices pose challenges in terms of storage and computational complexity during their construction. To overcome these limitations, scholars have proposed structural design methods based on block matrices [42].

One such design is the block-diagonal matrix structure, where a sub-Gaussian random matrix construction method is employed to achieve compressive observations of sparse signals. The model can be described as follows

$$\begin{bmatrix} y_1 \\ y_2 \\ \vdots \\ y_m \end{bmatrix} = \underbrace{\begin{bmatrix} \boldsymbol{\Phi}_1 & & & \\ & \boldsymbol{\Phi}_2 & & \\ & & \ddots & \\ & & & \boldsymbol{\Phi}_j \end{bmatrix}}_{\boldsymbol{\Phi}} \begin{bmatrix} x_1 \\ x_2 \\ \vdots \\ x_n \end{bmatrix}, \tag{3.46}$$

where the measurement matrix $\boldsymbol{\Phi}$ is an $m \times n$ matrix. If the elements in each matrix block $\boldsymbol{\Phi}_j$ are different, it is referred to as a distinct block diagonal (DBD) matrix. If the elements are the same, it is referred to as a repeat block diagonal (RBD) matrix. Both DBD and RBD matrices, while ensuring an adequate number of measurements, can satisfy the RIP and achieve measurement performance almost equivalent to high-density sub-Gaussian random matrices.

Additionally, block matrix structures, when combined with concepts like cyclic or sparse matrices, lead to the creation of structures like block circulant matrices and sparse block circulant matrices, which can also meet the RIP to a certain extent.

(3) Nested matrices

Column coherence is a crucial metric for assessing the performance of measurement matrices. For some binary matrices, constructing matrices with low coherence is particularly challenging due to the non-negativity of their elements. This limitation imposes significant constraints on matrix dimensions. The construction method of nested matrices allows for an extension of matrix dimensions while maintaining low column coherence, thereby enhancing the flexibility of constructing deterministic matrices [43].

We assume a binary matrix $\boldsymbol{V}_{m\times n_1}$, where each column in matrix $\boldsymbol{V}$ has the same non-zero elements (p in number), and the column coherence coefficient is $\mu(\boldsymbol{V})$. Additionally, let matrix $\boldsymbol{W}$ be a $p \times n_2$ matrix in which all elements have the same absolute value. With these conditions, it is possible to construct a nested matrix $\mathbf{Z}_{m\times(n_1 n_2)}$, where the column coherence coefficient satisfies $\mu(\mathbf{Z}) \leq \max(\mu(\boldsymbol{V}), \mu(\boldsymbol{W}))$.

The matrix construction process is illustrated in Fig. 3.13, where the i-th non-zero element in matrix $\boldsymbol{V}$ is replaced by the i-th row of matrix $\boldsymbol{W}$, and zero elements are extended to the same dimension until all non-zero elements in the matrix have been replaced.

By nesting matrices with low column coherence coefficients, it is possible to extend the matrix dimensions while keeping the column coherence coefficient unchanged. Examples include nested Hadamard matrices, discrete Fourier transform matrices, and so on.

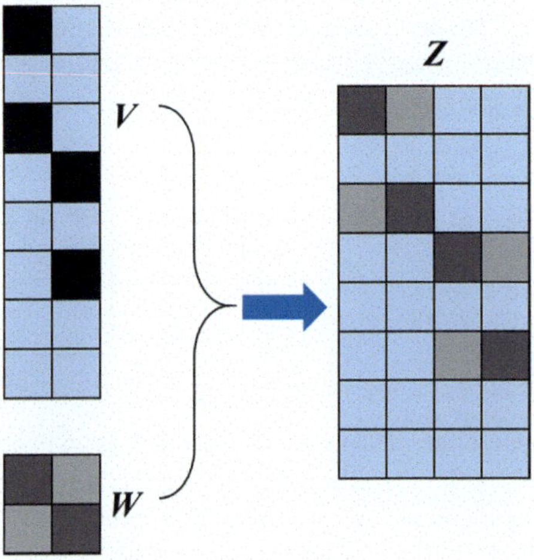

Fig. 3.13 Nested matrices

3.5 Reconstruction Algorithm

CS employs a global non-adaptive measurement approach, where the number of observations is significantly lower than the signal length, thus greatly reducing data acquisition, at the cost of computational complexity in signal reconstruction algorithms. The reconstruction algorithm is at the core of CS theory, as it determines the process of recovering the original signal from the observed values. Being an asymmetric algorithm, the complexity of CS reconstruction is much higher than that of sampling and compressing the signal. Therefore, the quality of the reconstruction algorithm directly impacts the success of CS.

Differing from the linear sensing problem based on Nyquist theory, CS faces the challenge of solving an underdetermined system of equations due to the significantly smaller number of observations m compared to the signal length n. When the signal x is sparse or compressible, solving the underdetermined system of equations can be translated into a problem of minimizing the l_0 norm.

$$\boldsymbol{x} = \arg\min_{x} \| \boldsymbol{\Psi x} \|_0 \ \text{s.t.}\, \boldsymbol{y} = \boldsymbol{\Phi x}. \tag{3.47}$$

However, statistical theory and combinatorial optimization theory tell us that combinatorial optimization is an NP-hard problem, which cannot be efficiently solved numerically when N is large and has poor resistance to noise. When the measurement matrix $\boldsymbol{\Phi}$ satisfies the RIP constraint, the combinatorial optimization problem (l_0 constraint optimization problem) can be transformed into a numerically tractable convex optimization problem with l_1 constraints.

$$\boldsymbol{x} = \arg\min_{x} \| \boldsymbol{\Psi x} \|_1 \ \text{s.t.}\, \boldsymbol{y} = \boldsymbol{\Phi x}. \tag{3.48}$$

Reconstruction algorithms can be roughly divided into the following categories.

The first category is greedy iterative algorithms, proposed for combinatorial optimization problems. These algorithms primarily focus on finding a more effective or non-zero way to link the signal to the atomic dictionary as measurement atoms (coefficients). The basic principle is to iteratively search for the support set of a sparse vector and reconstruct the signal using a constrained support least squares estimation. This category of algorithms includes: matching pursuit (MP), orthogonal matching pursuit [44] (OMP), stagewise orthogonal matching pursuit [45] (StOMP), Regularized Orthogonal Matching Pursuit [46] (ROMP), compressive sampling matching pursuit [47] (CoSaMP), and iterative hard thresholding [48] (IHT), among others. The complexity of these algorithms is mostly determined by the number of iterations required to find the correct support set. These algorithms are computationally fast but require more measurement data and have lower precision.

The second category is convex optimization algorithms or optimization approximation methods. These methods find an approximation of the signal by transforming non-convex problems into convex ones. The most commonly used method in this category is the basis pursuit (BP) algorithm. This algorithm proposes replacing the l_0 norm with the l_1 norm to solve the optimization problem, making it amenable to linear programming techniques. Another algorithm is the focuss algorithm, which uses l_p norms ($p \leq 1$) instead of the l_0 norm to solve the optimization problem. Additionally, other researchers have proposed solving the problem by minimizing the smooth transformation of the l_0 norm, known as the smoothed l_0 method [49]. These algorithms have slower computation speed (computational complexity is n^3), but require fewer measurements ($O[K \log(n/K)]$) and offer higher precision. Another common convex relaxation algorithm in this category is the gradient projection for sparse reconstruction (GPSR) algorithm [50]. The GPSR algorithm solves the bounded-constraint optimization problem using gradient descent, and it requires projections onto the feasible region to ensure feasibility during the iteration process.

The third category of algorithms is based on information prior for reconstruction. Instead of treating sparsity as a constraint in the objective function, these methods further refine the reconstruction algorithm based on different characteristics of the signal.

In consideration of the temporal correlation of signals, Bayesian-based reconstruction algorithms have been proposed. When the signal exhibits a strong temporal correlation, these algorithms can provide superior reconstruction accuracy compared to other reconstruction methods. This category includes expectation-maximization (EM) algorithm [51], Bayesian CS (BCS) algorithm [52], sparse Bayesian learning (SBL) algorithm [53] based on the single measurement vector (SMV) model, and multiple measurement vectors (MMV) model-based multiple sparse Bayesian learning (MSBL) algorithm [54].

Based on the local smoothness properties of natural images, the total variation (TV) reconstruction algorithm has been proposed, using total variation as the regularization term. The TV model assumes that images are locally smooth and can effectively remove noise in images and recover information from smooth regions of the original image under the TV regularization. Subsequently, algorithms like two-step iterative shrinkage thresholding (TwIST) algorithms [55] and TV minimization by augmented Lagrangian and alternating direction algorithms (TVAL3) [56] based on TV have also been developed. Building upon the TV prior model, scholars have proposed the reweighted total variation (ReTV) model [57], which assigns different weights to different gradient values, with smaller weights for larger gradient values, thereby preserving edges to some extent.

Based on non-local self-similarity (NSS) in images, combined with sparse models, various image recovery models have been derived. In the context of CS reconstruction, using the non-local characteristics of images for reconstruction, considering non-local similar image blocks instead of individual pixels, and characterizing image details through weighted means, the non-local total variation (NLTV) model has been proposed [58]. Recognizing that the TV model can lead to over-smoothing, models like total variation recovery by nonlocal regularization (TVNLR) have been developed, which combine local total variation sparsity with non-local self-similarity to reduce the staircase effect caused by image smoothing [59].

Now, let us delve into an analysis of commonly used reconstruction algorithms.

3.5.1 Greedy Algorithm

The basic principle of the MP algorithm is to first establish a fundamental library of functions D to analyze the signal. It does not require that all the basic functions $g_i(t)$ (also known as atoms) in the library be mutually orthogonal, but it does require that their l_2 norm $||g_i(t)||_2 = 1$. Therefore, this set of functions is not mutually independent; it contains redundancy. The task of matching pursuit is to decompose the signal $f(t)$ into a linear combination of a group of members $g_i(t)$ $(i = 1, 2, ..., T)$ from the library during each matching, with a goal of minimizing T. Essentially, it is a method that approximates the signal by progressively searching for its sparse representation through approximation with a series of individual atoms.

Although the MP algorithm is simple, it finds suboptimal solutions, leading to slow convergence and relatively poor sparsity in the approximation results. The OMP algorithm overcomes these drawbacks by projecting the signal vector onto a subspace spanned by selected atoms. An improvement in the OMP algorithm is that during the decomposition of the signal into its orthogonal and residual components, when selecting an atom from the atomic library in each iteration, it undergoes Schmidt orthogonalization with all the previously selected and used atoms from earlier steps. This ensures that the algorithm obtains orthogonal residuals with respect to the atoms chosen in each iteration, preventing the repeated selection of already chosen atoms and making it easier to avoid suboptimal solutions in favor of optimal ones. This orthogonalization process makes the OMP algorithm converge faster under the same accuracy requirements, enabling the reconstruction of the original signal with higher efficiency.

The specific workflow of the OMP algorithm for estimating the original signal is as follows:

Input:
Measurement matrix $\boldsymbol{\Phi}$
Observation vector $\boldsymbol{y}$
Sparsity level K

Output:
Approximate solution of the original signal $\hat{\boldsymbol{x}}$ (K-sparse)
Initialization:
Residual vector $\boldsymbol{r}_0 = \boldsymbol{y}$
Index set $\boldsymbol{\Lambda}_0 = \emptyset$ (empty set)
Iteration variable $t = 1$

Step 1:
Calculate the index λ that corresponds to the maximum inner product between the residual $\boldsymbol{r}$ and the column vector of sensing matrix $\boldsymbol{\varphi}_j$.

$$\lambda_t = \underset{j=1,2,\cdots,n}{\arg\max} \left|\langle \mathbf{r}_{t-1}, \varphi_j \rangle\right|. \tag{3.49}$$

Step 2:

Update the index set $\mathbf{\Lambda}_t = \mathbf{\Lambda}_{t-1} \cup \{\lambda_t\}$, and simultaneously update the set of reconstructed atoms $\Phi_t = [\Phi_{t-1}, \varphi_{\lambda_t}]$.

Step 3:

Using the least-squares algorithm, obtain the approximate solution of the original signal after t iterations as

$$\hat{\mathbf{x}}_t = \arg\min_{\hat{\mathbf{x}}} \|\mathbf{y} - \Phi_t \hat{\mathbf{x}}\|_2. \tag{3.50}$$

Step 4:

Update the residual $\mathbf{r}_t = \mathbf{y} - \Phi_t \hat{\mathbf{x}}_t$ and increment the iteration variable t by 1.

Step 5:

Check if $t > K$. If true, stop the iteration; if false, go back to Step 1 and continue with the next iteration.

End.

In the first iteration, the OMP algorithm can select the correct columns from the measurement matrix. However, due to the small random correlations that may exist between the column vectors of the matrix, it is challenging to analyze the subsequent iterations of the OMP algorithm. Furthermore, ROMP ensures that too many erroneous elements are not selected by choosing the maximum element of the alternative signal and using regularization. It is faster and provides a more balanced and stable reconstruction result compared to the OMP algorithm.

The specific workflow of the ROMP algorithm for estimating the original signal is as follows:

Input:

Measurement matrix $\boldsymbol{\Phi}$

Observation vector $\boldsymbol{y}$

Sparsity level K

Output:

Approximate solution $\hat{x}$ for the original signal $\hat{\boldsymbol{x}}$ (K-sparse)

Initialization:

Residual vector $\boldsymbol{r}_0 = \boldsymbol{y}$.

Index sets $\boldsymbol{\Lambda}_0 = \emptyset,\ \boldsymbol{J} = \emptyset$.

Iteration count $t = 1$.

Step 1:

Calculate the correlation coefficients

$$\lambda_t = \arg\max_{j=1,2,\cdots,n} \left|\left\langle \boldsymbol{r}_{t-1}, \boldsymbol{\varphi}_j \right\rangle\right|, \tag{3.51}$$

select the top K indices to form the set $\boldsymbol{J}$.

Step 2:

Perform regularization of the correlation coefficients corresponding to the selected indices based on $|\lambda(i)|2|\lambda(j)|$ ($\boldsymbol{J}_0 \subset \boldsymbol{J}$, $i, j \in \boldsymbol{J}_0$), and update the results into $\boldsymbol{J}_0$.

Step 3:

Update the support set $\boldsymbol{\Phi}_\Lambda$, where $\boldsymbol{\Lambda}_t = \boldsymbol{\Lambda}_{t-1} \cup \boldsymbol{J}_0$, update the approximate solution

$$\hat{\boldsymbol{x}}_t = \arg\min_{\hat{\boldsymbol{x}}} \|\boldsymbol{y} - \Phi_\Lambda \hat{\boldsymbol{x}}\|_2. \tag{3.52}$$

Step 4:

update the residual $\boldsymbol{r}_t = \boldsymbol{y} - \Phi_\Lambda \hat{\boldsymbol{x}}_t$.

Step 5:

If $\boldsymbol{\Lambda} \geq 2K$, stop the iteration; otherwise, increment t by 1 and return to Steps 1–5.

End.

Additionally, an improvement over the OMP algorithm is the multi-candidate orthogonal matching pursuit (MOMP) algorithm [60]. In each iteration, while OMP selects only one candidate column to add to the atomic set, MOMP selects multiple candidate columns to add to the optimal atomic set. This reduces the number of iterations and lowers the computational complexity of reconstructing the signal.

The StOMP algorithm simplifies OMP by using a thresholding method to find alternative signals, sacrificing approximation accuracy to further improve computational speed. It is better suited for solving large-scale problems.

The tree-based orthogonal matching pursuit (TOMP) algorithm constructs a sparse tree and tracks important coefficients within the tree [61]. TOMP takes into account the relationship between sparse signal positions in various singular subbands when considering multiscale signal decomposition. As a result, it builds a faster and higher-precision algorithm compared to OMP.

Different from OMP algorithms that precisely select a single atom in each iteration, the CoSaMP algorithm, in each iteration, chooses multiple atoms from the atom library that are more correlated and simultaneously removes some atoms to improve algorithm efficiency. The CoSaMP algorithm ensures that the candidate atom set will not exceed 3 K atoms at most, and in each iteration, the support set consists of 2 K atoms while at most K atoms are removed. Unlike the ROMP algorithm where atoms selected in previous iterations are retained, in CoSaMP, the atoms selected in one iteration may be discarded in the next iteration.

Matching pursuit algorithms have a simple workflow. By selecting matching atoms, they can quickly reconstruct the original signal using least-squares estimation. However, they are essentially heuristic greedy algorithms, which may converge to local optima and have a higher computational complexity. Their accuracy may not be as high as that of convex optimization-based reconstruction algorithms.

3.5.2 Convex Optimization Algorithms

Greedy algorithms all have a notable drawback, which is the need for matrix inversion operations in the projection matrix when projecting the residual vector onto the current subspace. This results in a high algorithmic complexity and substantial computational costs in practical usage. Based on the convex relaxation model, signal reconstruction shifts from solving the l_0-norm to solving the l_1-norm, which can be addressed using convex optimization algorithms. Classical convex optimization algorithms include the BP algorithm, iterative shrinkage thresholding (IST) algorithm [48, 62], alternative direction multiplier method (ADMM) [63], and so on.

The BP algorithm replaces the l_0-norm with the l_1-norm and solves it using linear programming, making it the most commonly used convex optimization algorithm. The minimization of the l_1-norm linear programming problem can be formulated as

$$\min \boldsymbol{c}^{\mathrm{T}}\boldsymbol{x} \text{ s.t.} \boldsymbol{A}\boldsymbol{x} = \boldsymbol{b},\ \boldsymbol{x} \geq 0. \tag{3.53}$$

Here, $\boldsymbol{c}^{\mathrm{T}}\boldsymbol{x}$ is the objective function, and $\boldsymbol{c}$ is a constant vector. Introducing the following auxiliary variables

$$\begin{gathered} \boldsymbol{A} \Leftrightarrow [\boldsymbol{\Phi}, -\boldsymbol{\Phi}],\ \boldsymbol{b} \Leftrightarrow \boldsymbol{y}, \\ \boldsymbol{c} \Leftrightarrow \begin{bmatrix} 1 \\ 1 \end{bmatrix}, \boldsymbol{x} \Leftrightarrow \begin{bmatrix} \boldsymbol{u} \\ \boldsymbol{v} \end{bmatrix}, \end{gathered} \tag{3.54}$$

where $\boldsymbol{u}$ and $\boldsymbol{v}$ are non-negative vectors separating the positive and negative components of $\boldsymbol{x}$, the constraint conditions can be expressed as

$$[\boldsymbol{\Phi}, -\boldsymbol{\Phi}]\begin{bmatrix} \boldsymbol{u} \\ \boldsymbol{v} \end{bmatrix} = \boldsymbol{y}. \tag{3.55}$$

Based on the norm definition, the objective function can be written as

$$\|\boldsymbol{x}\|_1 = \sum_i |\boldsymbol{x}_i| = \sum_i |\boldsymbol{u}_i - \boldsymbol{v}_i|. \tag{3.56}$$

Based on the optimization theory of linear programming, highly accurate solutions can be obtained, albeit with slower computational speed. For complex convex relaxation problems, specific solutions can be obtained using optimization tools like CVX.

The IST algorithm combines the concepts of approximation error and sparsity based on the gradient descent approach. It uses the following iterative formula,

$$\boldsymbol{x}^{n+1} = H_K\left[\boldsymbol{x}^n + \boldsymbol{A}^*(\boldsymbol{y} - \boldsymbol{A}\boldsymbol{x}^n)\right], \tag{3.57}$$

$$H_K(\boldsymbol{x}_i) = \begin{cases} 0, & |\boldsymbol{x}_i| < \lambda_{n,K}. \\ \boldsymbol{x}_i, & |\boldsymbol{x}_i| \geq \lambda_{n,K}. \end{cases} \tag{3.58}$$

Here, $H_K(\boldsymbol{x}_i)$ represents the hard thresholding function, and $\lambda_{n,K}$ is known as the hard threshold, which is a predetermined constant. This method selects the top K absolute maximum elements at each step, ensuring sparsity in the solution. Additionally, this approach takes into account measurement conditions, ensuring reconstruction accuracy.

The soft thresholding algorithm is a variation of the hard thresholding algorithm. In this variation, the hard thresholding function in the former is replaced with a soft thresholding function, resulting in the following iterative formula

$$\boldsymbol{x}^{n+1} = S_K\left(\boldsymbol{x}^n + \boldsymbol{A}^*\left(\boldsymbol{y} - \boldsymbol{A}\boldsymbol{x}^n\right), \lambda\right), \tag{3.59}$$

where the soft thresholding function is defined as

$$S_K(\boldsymbol{x}_i) = \begin{cases} \boldsymbol{x}_i + \lambda, & \boldsymbol{x}_i < -\lambda. \\ 0, & |\boldsymbol{x}_i| \leq \lambda. \\ \boldsymbol{x}_i - \lambda, & \boldsymbol{x}_i > \lambda. \end{cases} \tag{3.60}$$

The ADMM is an algorithm that combines dual ascent methods and augmented Lagrangian methods. It is typically used to solve models like the following

$$\min f_1(\boldsymbol{x}) + f_2(\boldsymbol{x}) \text{ s.t.} \boldsymbol{A}\boldsymbol{x} + \boldsymbol{B}\boldsymbol{x} = \boldsymbol{y}. \tag{3.61}$$

The augmented Lagrangian function for the objective problem is

$$L_\rho(\boldsymbol{x}, u, \lambda) = f_1(\boldsymbol{x}) + f_2(\boldsymbol{x}) + \lambda(\boldsymbol{A}\boldsymbol{x} + \boldsymbol{B}\boldsymbol{u} - \boldsymbol{y}) + \frac{\rho}{2}\|\boldsymbol{A}\boldsymbol{x} + \boldsymbol{B}\boldsymbol{u} - \boldsymbol{y}\|_2^2. \tag{3.62}$$

The iterative formula used by this algorithm is

$$\begin{cases} \boldsymbol{x}^{k+1} = \arg\min\limits_{x} L_\rho\left(\boldsymbol{x}, \boldsymbol{u}^k, \lambda^k\right), \\ z^{k+1} = \arg\min\limits_{z} L_\rho\left(\boldsymbol{x}^k, \boldsymbol{u}, \lambda^k\right), \\ \lambda^{k+1} = \boldsymbol{y}^k + \rho\left(\boldsymbol{A}\boldsymbol{x}^{k+1} + \boldsymbol{B}\boldsymbol{u}^{k+1} - \boldsymbol{y}\right). \end{cases} \tag{3.63}$$

To incorporate prior knowledge of the original signal, researchers have proposed a weighted model, where each component of the objective function is multiplied by a weight. This weighted model considers the following weighted l_1 optimization format

$$\min \sum_{i=1}^{n} \omega_i |\boldsymbol{x}_i| \text{ s.t.} \boldsymbol{A}\boldsymbol{x} = \boldsymbol{y}, \tag{3.64}$$

where ω_i is a positive number representing the weight for the corresponding component of the original signal.

The weighted l_1 optimization model can be viewed as a convex relaxation model of the weighted l_0 optimization model

$$\min|\boldsymbol{W}\boldsymbol{x}_i| \text{ s.t.} \boldsymbol{A}\boldsymbol{x} = \boldsymbol{y}, \tag{3.65}$$

where $\boldsymbol{W} = \text{diag}(\omega_1, \omega_2, \ldots, \omega_n)$ is a diagonal matrix.

When $\omega_i \neq 0$, $(i = 1, 2, \ldots, n)$, the solutions of the weighted l_1 optimization model and the l_0 norm optimization model are consistent. There are various methods for selecting the weights, with one classical approach being to set the weights as the reciprocals of the actual signal values.

$$\omega_i = \begin{cases} 1/|\boldsymbol{x}_i|, & \boldsymbol{x}_i \neq 0. \\ \infty, & \boldsymbol{x}_i = 0. \end{cases} \tag{3.66}$$

Using the alternating iterative algorithm, weights are first set using prior information, and then the weighted l_1 optimization model is solved with fixed weights. The iterative algorithm can be summarized as follows

$$\begin{cases} \omega_i^{(0)} = 1, \; i = 1, 2, \cdots, n, \\ \boldsymbol{x}^{(l)} = \arg\min \left\| \boldsymbol{W}^{(l)}\boldsymbol{x} \right\|_1, \text{ s.t.} \boldsymbol{A}\boldsymbol{x} = \boldsymbol{y}, \\ \omega_i^{(l+1)} = 1 \Big/ \left(\left| \boldsymbol{x}_i^{(l)} \right| + \varepsilon \right). \end{cases} \tag{3.67}$$

This method is simple and reliable, taking into account the distribution of the support set of the original signal. Numerous experiments have shown that this weighted approach can effectively improve various performance aspects of existing algorithms. This weighted solving approach can be further refined into two categories: weighted reconstruction methods and iterative weighted methods. The difference lies in the former applying weights once, with the weights remaining unchanged during the iterative process, while the latter effectively solves a series of problems similar to the former. In each iteration, it uses the amplitude characteristics of the previously obtained solution to reset the weights, thus obtaining further results.

Weighted l_1 optimization models have several advantages. With the introduction of prior information, more usable information becomes available, making it favorable for signal recovery. Moreover, compared to traditional l_1 optimization models, it can provide accurate results with fewer samples. In other words, by using the weighted l_1 optimization model, it is possible to achieve precise recovery with a less complete (i.e., smaller) linear measurement set. Additionally, compared to the l_0 optimization model, it remains a convex optimization model, making it easy to solve.

3.5.3 Information-Prior-Based Algorithms

With the introduction of prior information, higher reconstruction quality can be achieved through sparser sampling. In the field of computational imaging, prior models based on local smoothness and non-local self-similarity properties are widely used for optimization in imaging. By introducing prior information in the framework of CS imaging, and enhancing sparse reconstruction constraints, it is possible to achieve global reconstruction with fewer measurement data.

As one of the classical prior information algorithms, the TV reconstruction algorithm utilizes total variation as the regularization term in problem solving and has become an important branch in CS reconstruction [64]. The TV constraint model for images is as follows

$$\min \mathrm{TV}(\boldsymbol{f}), \text{ s.t.} \boldsymbol{y} = \boldsymbol{\Phi f}, \tag{3.68}$$

where the objective function $\mathrm{TV}(\boldsymbol{f})$ represents the sum of the discrete gradients of the image, that is:

$$\mathrm{TV}(\boldsymbol{f}) = \sum_{a,b} \sqrt{\left(f_{a+1,b} - f_{a,b}\right)^2 + \left(f_{a,b+1} - f_{a,b}\right)^2} = \sum_i \|D_i \boldsymbol{f}\|_2, \tag{3.69}$$

where $f_{a,b}$ represents the pixel values of the image.

As a representative TV reconstruction algorithm, the TVAL3 algorithm solves the sparse reconstruction problem regularized by total variation using the augmented Lagrangian method and alternating direction transformation. It has the advantages of speed and high reconstruction quality. Additionally, the algorithm supports various measurement matrices and constraints, making it highly flexible. The computational framework of the TVAL3 algorithm is as follows.

The augmented Lagrangian function for the total variation regularization constraint problem is as follows

$$\widetilde{\mathcal{L}}_A(f) = \sum_i \|D_i \boldsymbol{f}\|_2 - \lambda^T(\boldsymbol{\Phi x} - \boldsymbol{y}) + \frac{\mu}{2}\|\boldsymbol{\Phi x} - \boldsymbol{y}\|_2^2, \tag{3.70}$$

where λ and μ are Lagrange multipliers. By introducing the relaxation variable $\boldsymbol{w}$, the TV regularization model becomes

$$\min_{\boldsymbol{w}_i, \boldsymbol{f}} \sum_i \|\boldsymbol{w}_i\|_2 \text{ s.t.} \boldsymbol{\Phi x} = \boldsymbol{y} \text{ and } D_i \boldsymbol{f} = \boldsymbol{w}_i. \tag{3.71}$$

And its corresponding augmented Lagrangian problem is

$$\min_{w_i, f} \sum_i \left(\|w_i\|_2 - v_i^T (D_i f - w_i) + \frac{\beta}{2} \|D_i f - w_i\|_2^2 \right)$$
$$-\lambda^T (\Phi x - y) + \frac{\mu}{2} \|\Phi x - y\|_2^2, \tag{3.72}$$

where $\mathbf{v}_i^T$ represents the Lagrange multiplier vector.

The alternating minimization scheme is applied to solve this problem. For a fixed f, the minimum value of w_i for all i can be obtained using the following formula

$$w_i = \max\left\{ \left\| D_i f - \frac{\mathbf{v}_i}{\beta} \right\|_2 - \frac{1}{\beta}, 0 \right\} \frac{D_i f - \mathbf{v}_i/\beta}{\|D_i f - \mathbf{v}_i/\beta\|_2}. \tag{3.73}$$

On the other hand, when w_i is held constant, we perform an approximate minimization of the quadratic function concerning f. This is achieved by taking a single steepest descent step using a step size determined through a back-tracking non-monotone line search approach, which initiates from a Barzilai-Borwein (BB) step length. Following each steepest descent iteration, we adjust w_i and repeat this procedure until we have achieved an approximate solution to the augmented Lagrangian problem within a specified tolerance.

Consider $\hat{f}$ and $\hat{w}_i$ as an approximate solution. The multipliers are subsequently updated for all I, as

$$\mathbf{v}_i \leftarrow \mathbf{v}_i - \beta\left(D_i \hat{f} - \hat{w}_i\right),$$
$$\lambda \leftarrow \lambda - \mu(\Phi \hat{f} - y). \tag{3.74}$$

Then, we update the multipliers until the results converge.

In addition to the TVAL3 algorithm representing the total variation prior reconstruction algorithm, there are other reconstruction algorithms based on prior information such as non-local similarity prior reconstruction algorithms and signal temporal correlation prior reconstruction algorithms. These algorithms have various representation methods and optimization models in different CS application domains, significantly enhancing the efficiency of CS reconstruction. Due to space limitations, further elaboration on this topic will not be provided.

3.6 CS in Computational Imaging

CS theory, as an excellent signal processing framework, allows for signal sampling at rates below the Nyquist sampling theorem and employs reconstruction algorithms to highly probabilistically recover the original signal. Now, we will review some applications of CS in the field of optical imaging, demonstrating its practical application value and advantages.

3.6.1 Single-Pixel Camera

The practical implementation of CS theory began with the development of a single-pixel camera [65, 66]. This single-pixel camera is an optical system comprising two converging lenses, a digital micromirror device (DMD), an analog-to-digital (A/D) converter, and a single photodiode. It operates by sequentially projecting structured light patterns onto the object being imaged and measuring the reflected light.

The operational procedure of a single-pixel camera unfolds as follows: An incident light source is utilized to illuminate the targeted object. Subsequently, the initial lens serves to focus the resultant reflected light onto a DMD. The DMD, comprising a grid array of microscopic mirrors in a resolution of 1024×768, affords the capability for individual mirror adjustment, enabling discrete 'on' and 'off' states. Light reflected from the 'on' micro-mirrors is then directed towards a second lens, which further focuses it onto a solitary photodiode. The photodiode performs an integration of these incoming light measurements, thereby producing a linear representation of the object under observation. This measured voltage signal is subsequently subjected to amplification prior to digitization via an A/D converter.

To capture diverse information, the orientation of the DMD's micromirrors is programmed in a pseudo-randomized fashion for each measurement iteration. Iterating this process yields a succession of pseudo-random projections, each of which is duly digitized and transmitted to a receiver. The ultimate objective of this multi-step process is the subsequent reconstruction of the original image through the application of CS reconstruction techniques.

Single-pixel cameras have found extensive utility across a spectrum of imaging domains, encompassing applications in the realms of infrared imaging [67] and terahertz (THz) imaging [68], microscopy imaging [69], and various other imaging paradigms. These cameras offer an advantageous juxtaposition of simplicity and efficiency over their complex array detector counterparts.

In the context of THz imaging, a comprehensive system configuration emerges, comprising a THz transmitter/receiver pairing, converging lenses, and a planar screen endowed with a stochastic pattern [70, 71]. This screen effectively samples the THz beam, ushering in spatial modulation as a replacement for traditional raster scanning methodologies, thereby culminating in the reduction of scanning durations. To facilitate the creation of randomized illumination patterns in THz imaging, a repertoire of methodologies has been employed, ranging from the utilization of rotating disks to the employment of random binary patterns. Metamaterials have emerged as instrumental assets for the precise manipulation and control of THz radiation, imparting versatility and precision to THz imaging systems [72].

Moreover, single-pixel camera technology has not only been seamlessly incorporated into compressive microscopy, thus offering a cost-effective alternative to conventional methodologies [73], but has also been instrumental in the design of streamlined infrared cameras, circumventing the intricacies inherent to traditional hardware configurations [74].

Innovations in this domain include the advent of an active illumination single-pixel camera, which eschews the DMD in favor of a video projector, thus diversifying the technological landscape [75]. Furthermore, ongoing endeavors are firmly directed towards the development of novel imaging systems predicated on single photodiodes, coalescing with the principles of CS, which hold promising prospects across a multiplicity of diverse applications [76].

Single-pixel imaging, grounded in the principles of CS, is confronted by a set of pertinent challenges. Foremost among these is the intricate balance between acquisition time and the number of random projections, which directly influences the ultimate quality of the reconstructed images. This dilemma underscores the need for an astute consideration of the trade-off inherent in the process, as a protracted acquisition time might enhance image quality but inversely impact the real-time applicability of this technology. Additionally, the prerequisite for stationary objects during measurements imposes inherent limitations on the imaging of dynamic scenes, a constraint intrinsic to the CS-based single-pixel paradigm. The endeavor to extend the capabilities of single-pixel cameras to facilitate the real-time capture of video sequences confronts practical constraints, signifying the exigency for innovative methodologies in this regard.

However, the single-pixel camera, founded on CS principles, has undeniably exhibited remarkable versatility and efficiency across a spectrum of imaging applications. Its allure lies in the inherent simplicity of its design and the potential for substantial cost reduction vis-à-vis conventional imaging systems. Notwithstanding these merits, it is crucial to acknowledge that the aforementioned challenges persist as active domains of research and development. The relentless pursuit of novel techniques and methodologies, aimed at surmounting these impediments, continues to be at the forefront of contemporary research in the field of single-pixel imaging. In sum, single-pixel cameras constitute a promising avenue for cost-effective and versatile imaging solutions, yet the persistent quest to reconcile the trade-off between acquisition time and image quality remains an enduring focal point within the research community.

3.6.2 Coded Aperture

In traditional cameras, lenses are employed to focus on objects, ensuring that each sensor corresponds to a pixel in the resulting image. However, when a scene contains information at varying depths, typical cameras yield blurry images. This blurriness arises because light from objects not in the focal plane converges outside the plane of focus. A solution to this problem involves using a coded mask placed at the lens plane, comprising transparent and opaque areas. Recently, coded apertures have been adapted to the CS framework [77].

Notably, Scholars designed a coded aperture camera specifically for the precise reconstruction of high-resolution images from single low-resolution measurements using CS techniques [78] Their approach utilized Toeplitz structure matrices, which satisfy RIP with high probability, to construct coded aperture masks suitable for both conventional coded aperture and Fourier imaging setups. These coded aperture masks enable simultaneous capture of all CS measurement values, significantly reducing data acquisition time. Subsequently, CS techniques are applied to the randomly sampled linear measurements of the scene obtained through the coded aperture, facilitating the reconstruction of a super-resolved image. In a related work, researchers utilized pseudo-random coded masks to reconstruct super-resolved videos from sequences of low-resolution coded frames using CS techniques [79]. They harnessed temporal information to expedite the overall reconstruction process, employing two approaches: the first leveraged the previous frame's reconstructed image as the initial approximation for the next frame, while the second used CS techniques to concurrently reconstruct multiple frames, exploiting the similarities between consecutive frames to reduce reconstruction time considerably.

One of the most promising applications of coded apertures lies in the design of camera architectures for compressive spectral imaging. Spectral imagers capture scenes with two spatial coordinates and one wavelength coordinate, forming a 3D structure often referred to as a data cube. Conventional approaches to data acquisition in spectral imaging involve slow processes, utilizing liquid crystal tunable filters or light sources to capture several narrowband spectra at each spatial point. In this context, CS has demonstrated substantial potential. Scholars designed a single-shot spectral imager based on the CS framework [80]. This new class of spectral imagers records 2D-coded spatial and spectral projections of the 3D data cube. Instead of capturing every voxel within the desired 3D data cube, this architecture acquires a limited number of coded projections and employs them to estimate spatial and spectral information within the data cube using CS reconstruction methods. The key advantage of these CS-based snapshot spectral imagers over conventional ones is their adaptability when spectral resolution takes precedence over spatial resolution. By taking multiple measurements using multiple coded apertures and focal plane measurement sets, this approach is further expanded. Additionally, researchers demonstrated compressive temporal imaging using a shifted coded aperture, where each temporal frame was modulated through mechanical translation of the coded aperture [81]. This approach integrates all spectral planes; however, frames can be isolated by distinguishing their spectral planes based on their local code structure, akin to code division multiple access.

CS-based imaging can also be achieved by employing a random phase mask placed on the lens. Random projections are obtained by convolving the image with this random phase mask [82].

3.7 General Conclusions

This chapter introduces the theoretical background and mathematical models of CS, and reviews the three main components of the CS framework: sparse representation of signals, construction of observation matrices, and relevant theoretical methods for reconstruction algorithms under sparse constraints. Finally, it elaborates on some specific applications of CS in the field of optical imaging. This chapter provides an overview of the theoretical framework of CS systems, guiding readers to gain a comprehensive understanding of how CS operates in computational imaging.

Acknowledgements This work was supported by the National Natural Science Foundation of China (Nos. 11874132, 61975044, 12074094, 12374271, and 62375065); Interdisciplinary Research Foundation of HIT (No. IR2021237).

References

1. Candès EJ, Wakin MB. An introduction to compressive sampling. IEEE Signal Proc Mag. 2008;25:21–30.
2. Tsaig Y, Donoho DL. Extensions of compressed sensing. Signal Process. 2006;86:549–71.
3. Khosravy M, Cabral TW, Luiz MM, Gupta N, Crespo RG. Random acquisition in compressive sensing: a comprehensive overview. Int J Ambient Comput. 2021;12:140–65.
4. Donoho DL. Compressed sensing. IEEE Trans Inform Theory 2006;52:1289–1306.
5. Mardani M, Gong EH, Cheng JY, Vasanawala SS, Zaharchuk G, Xing L, Pauly JM. Deep generative adversarial neural networks for compressive sensing MRI. IEEE Trans Med Imaging. 2019;38:167–79.
6. Haris K, Hedstrom E, Kording F, Bidhult S, Steding-Ehrenborg K, Ruprecht C, Heiberg E, Arheden H, Aletras AH. Free-breathing fetal cardiac MRI with doppler ultrasound gating, compressed sensing, and motion compensation. J Magn Reson Imaging. 2020;51:260–72.
7. Matcuk GR, Gross JS, Fields BKK, Cen S. Compressed sensing MR imaging (CS-MRI) of the knee: assessment of quality, inter-reader agreement, and acquisition time. Magn Reson Med Sci. 2020;19:254–8.
8. Tomei S, Bacci A, Giusti E, Martorella M, Berizzi F. Compressive sensing-based inverse synthetic radar imaging imaging from incomplete data. IET Radar Sonar Nav. 2016;10:386–97.
9. Xue J, Zhao YQ, Liao WZ, Chan JCW. Nonlocal tensor sparse representation and low-rank regularization for hyperspectral image compressive sensing reconstruction. Remote Sens. 2019;11:11020193.
10. Liu Y, Yuan X, Suo JL, Brady DJ, Dai QH. Rank minimization for snapshot compressive imaging. IEEE Trans Pattern Anal. 2019;41:2990–3006.
11. Choi JW, Shim B, Ding Y, Rao B, Kim DI. Compressed sensing for wireless communications: useful tips and tricks. IEEE Commun Surv Tut. 2017;19:1527–50.
12. Candes EJ, Tao T. Decoding by linear programming. IEEE Trans Inform Theory. 2005;51:4203–15.
13. Baraniuk R, Davenport M, DeVore R, Wakin M. A simple proof of the restricted isometry property for random matrices. Constr Approx. 2008;28:253–63.
14. Candès EJ. The restricted isometry property and its implications for compressed sensing. Cr Math. 2008;346:589–92.
15. Lindeberg T. Scale-space: a framework for handling image structures at multiple scales. Cern Report. 1996;96:27–38.
16. Do MN, Vetterli M. The finite ridgelet transform for image representation. IEEE Trans Image Process. 2003;12:16–28.
17. Le Pennec E, Mallat S. Bandelet image approximation and compression. Multiscale Model Sim. 2005;4:992–1039.
18. Candès E, Demanet L, Donoho D, Ying LX. Fast discrete curvelet transforms. Multiscale Model Sim. 2006;5:861–99.

19. Do MN, Vetterli M. The contourlet transform: an efficient directional multiresolution image representation. IEEE Trans Image Process. 2005;14:2091–106.
20. Easley G, Labate D, Lim WQ. Sparse directional image representations using the discrete shearlet transform. Appl Comput Harmon A. 2008;25:25–46.
21. Aharon M, Elad M, Bruckstein A. K-SVD: an algorithm for designing overcomplete dictionaries for sparse representation. IEEE Trans Signal Proces. 2006;54:4311–22.
22. Mairal J, Bach F, Ponce J, Sapiro G. Online learning for matrix factorization and sparse coding. J Mach Learn Res. 2010;11:19–60.
23. Candès EJ, Romberg J, Tao T. Robust uncertainty principles: exact signal reconstruction from highly incomplete frequency information. IEEE Trans Inform Theory. 2006;52:489–509.
24. Tropp JA. Greed is good: algorithmic results for sparse approximation. IEEE Trans Inform Theory. 2004;50:2231–42.
25. Raginsky M, Jafarpour S, Harmany ZT, Marcia RF, Willett RM, Calderbank R. Performance bounds for expander-based compressed sensing in Poisson noise. IEEE Trans Signal Process. 2011;59:4139–53.
26. Yu NY. Indistinguishability and energy sensitivity of Gaussian and Bernoulli compressed encryption. IEEE Trans Inf Foren Sec. 2018;13:1722–35.
27. Can HP, Xiao S, Liu F. Chaotic binary sensing matrices. Int J Bifurcat Chaos. 2019;29:19501219.
28. Xu GW, Xu ZQ. Compressed sensing matrices from Fourier matrices. IEEE Trans Inform Theory. 2015;61:469–78.
29. Rousseau S, Helbert D. Compressive color pattern detection using partial orthogonal circulant sensing matrix. IEEE Trans Image Process. 2020;29:670–8.
30. Kühn R. Spectra of sparse random matrices. J Phys A Math Theor. 2008;41: 295002.
31. Zeng L, Zhang XW, Chen L, Cao TY, Yang JB. Deterministic construction of toeplitzed structurally chaotic matrix for compressed sensing. Circ Syst Signal Pr. 2015;34:797–813.
32. DeVore RA. Deterministic constructions of compressed sensing matrices. J Complexity. 2007;23:918–25.
33. Yu NY, Zhao N. Deterministic construction of real-valued ternary sensing matrices using optical orthogonal codes. IEEE Signal Process Lett. 2013;20:1106–9.
34. Applebaum L, Howard SD, Searle S, Calderbank R. Chirp sensing codes: deterministic compressed sensing measurements for fast recovery. Appl Comput Harmon A. 2009;26:283–90.
35. Li SX, Gao F, Ge GN, Zhang SY. Deterministic construction of compressed sensing matrices via algebraic curves. IEEE Trans Inform Theory. 2012;58:5035–41.
36. Calderbank R, Howard S, Jafarpour S. Construction of a large class of deterministic sensing matrices that satisfy a statistical isometry property. IEEE J Sel Top Signal Process. 2010;4:358–74.
37. Amini A, Marvasti F. Deterministic construction of binary, bipolar, and ternary compressed sensing matrices. IEEE Tans Inform Theory. 2011;57:2360–70.
38. ten Brink S, Kramer G, Ashikhmin A. Design of low-density parity-check codes for modulation and detection. IEEE Trans Commun. 2004;52:670–8.
39. Dimakis AG, Smarandache R, Vontobel PO. LDPC codes for compressed sensing. IEEE Trans Inform Theory. 2012;58:3093–114.
40. Gray RM. Toeplitz and circulant matrices: a review. Found Trends Commun Inform Theory. 2006;2:155–239.
41. Haupt J, Bajwa WU, Raz G, Nowak R. Toeplitz compressed sensing matrices with applications to sparse channel estimation. IEEE Trans Inform Theory. 2010;56:5862–75.
42. Sun JM, Wang S, Dong Y. Sparse block circulant matrices for compressed sensing. IET Commun. 2013;7:1412–8.
43. Amini A, Montazerhodjat V, Marvasti F. Matrices with small coherence using p-Ary block codes. IEEE Trans Signal Proces. 2012;60:172–81.
44. Sahoo SK, Makur A. Signal recovery from random measurements via extended orthogonal matching pursuit. IEEE Trans Signal Process. 2015;63:2572–81.

45. Wang J, Kwon S, Shim B. Generalized orthogonal matching pursuit. IEEE Trans Signal Process. 2012;60:6202–16.
46. Needell D, Vershynin R. Uniform uncertainty principle and signal recovery via regularized orthogonal matching pursuit. Found Comput Math. 2009;9:317–34.
47. Needell D, Tropp JA. CoSaMP: iterative signal recovery from incomplete and inaccurate samples. Appl Comput Harmon A. 2009;26:301–21.
48. Blumensath T, Davies ME. Iterative hard thresholding for compressed sensing. Appl Comput Harmon A. 2009;27:265–74.
49. Mohimani H, Babaie-Zadeh M, Jutten C. A fast approach for overcomplete sparse decomposition based on smoothed l0 norm. IEEE Trans Signal Process. 2009;57:289–301.
50. Figueiredo MAT, Nowak RD, Wright SJ. Gradient projection for sparse reconstruction: application to compressed sensing and other inverse problems. IEEE J Sel Top Signal Process. 2007;1:586–97.
51. Do CB, Batzoglou S. What is the expectation maximization algorithm? Nat Biotechnol. 2008;26:897–9.
52. Ji SH, Xue Y, Carin L. Bayesian compressive sensing. IEEE Trans Signal Process. 2008;56:2346–56.
53. Wipf DP, Rao BD. Sparse Bayesian learning for basis selection. IEEE Trans Signal Process. 2004;52:2153–64.
54. Wipf DP, Rao BD. An empirical Bayesian strategy for solving the, simultaneous sparse approximation problem. IEEE Trans Signal Process. 2007;55:3704–16.
55. Bioucas-Dias JM, Figueiredo MAT. A new twist: two-step iterative shrinkage/thresholding algorithms for image restoration. IEEE Trans Image Process. 2007;16:2992–3004.
56. Li CB, Yin WT, Jiang H, Zhang Y. An efficient augmented Lagrangian method with applications to total variation minimization. Comput Optim Appl. 2013;56:507–30.
57. Candès EJ, Wakin MB, Boyd SP. Enhancing sparsity by reweighted l1 minimization. J Fourier Anal Appl. 2008;14:877–905.
58. Liu Q, Gao XB, He LH, Lu W. Single image dehazing with depth-aware non-local total variation regularization. IEEE Trans Image Process. 2018;27:5178–91.
59. Zhang HY, Liu L, He W, Zhang LP. Hyperspectral image denoising with total variation regularization and nonlocal low-rank tensor decomposition. IEEE Trans Geosci Remote. 2020;58:3071–84.
60. Huang SS, Zhu JB. Recovery of sparse signals using OMP and its variants: convergence analysis based on rip. Inverse Probl. 2011;27: 035003.
61. Jost P, Vandergheynst P, Frossard P. Tree-based pursuit: algorithm and properties. IEEE Trans Signal Process. 2006;54:4685–97.
62. Beck A, Teboulle M. A fast iterative shrinkage-thresholding algorithm for linear inverse problems. SIAM J Imaging Sci. 2009;2:183–202.
63. Yuan XM. An improved proximal alternating direction method for monotone variational inequalities with separable structure. Comput Optim Appl. 2011;49:17–29.
64. Li B, Salucci M, Tang W, Rocca P, Massa A. An accelerated total-variation compressive sensing approach to field strength reconstruction. IEEE Trans Antenn Propag. 2020;68:8244–8.
65. Duarte MF, Davenport MA, Takhar D, Laska JN, Sun T, Kelly KF, Baraniuk RG. Single-pixel imaging via compressive sampling. IEEE Signal Proc Mag. 2008;25:83–91.
66. Edgar MP, Gibson GM, Padgett MJ. Principles and prospects for single-pixel imaging. Nat Photon. 2019;13:13–20.
67. Edgar MP, Gibson GM, Bowman RW, Sun B, Radwell N, Mitchell KJ, Welsh SS, Padgett MJ. Simultaneous real-time visible and infrared video with single-pixel detectors. Sci Rep. 2015;5:10669.
68. Lu Y, Wang XK, Sun WF, Feng SF, Ye JS, Han P, Zhang Y. Reflective single-pixel terahertz imaging based on compressed sensing. IEEE Trans Thz Sci Techn. 2020;10:495–501.
69. Ma YY, Lee Y, Best-Popescu C, Gao L. High-speed compressed-sensing fluorescence lifetime imaging microscopy of live cells. Proc Natl Acad Sci USA. 2021;118: e2004176118.

70. Chan WL, Moravec ML, Baraniuk RG, Mittleman DM. Terahertz imaging with compressed sensing and phase retrieval. Opt Lett. 2008;33:974–6.
71. Chan WL, Charan K, Takhar D, Kelly KF, Baraniuk RG, Mittleman DM. A single-pixel terahertz imaging system based on compressed sensing. Appl Phys Lett. 2008;93(12): 121105.
72. Shrekenhamer D, Watts CM, Padilla WJ. Terahertz single pixel imaging with an optically controlled dynamic spatial light modulator. Opt Express. 2013;21:12507–18.
73. Woringer M, Darzacq X, Zimmer C, Mir M. Faster and less phototoxic 3D fluorescence microscopy using a versatile compressed sensing scheme. Opt Express. 2017;25:13668–83.
74. Chen HZ, Xi N, Song B, Chen LL, Zhao JG, Lai KWC, Yang RG. Infrared camera using a single nano-photodetector. IEEE Sens J. 2013;13(3):949–58.
75. Magalhaes F, Araújo FM, Correia MV, Abolbashari M, Farahi F. Active illumination single-pixel camera based on compressive sensing. Appl Opt. 2011;50:405–14.
76. Soldevila F, Salvador-Balaguer E, Clemente P, Tajahuerce E, Lancis J. High-resolution adaptive imaging with a single photodiode. Sci Rep. 2015;5:14300.
77. Fu C, Don ML, Arce GR. Compressive spectral imaging via polar coded aperture. IEEE Trans Comput Imag. 2017;3:408–20.
78. Chen J, Wang YT, Wu HX. A coded aperture compressive imaging array and its visual detection and tracking algorithms for surveillance systems. Sensors. 2012;12:14397–415.
79. Yang XL, Wu HZ, Yin YK, Meng XF, Peng X. Multiple-image encryption base on compressed coded aperture imaging. Opt Lasers Eng. 2020;127: 105976.
80. Wang Z, Yi S, Chen A, Zhou M, Luk TS, James A, Nogan J, Ross W, Joe G, Shahsafi A, Wang KX, Kats MA, Yu ZF. Single-shot on-chip spectral sensors based on photonic crystal slabs. Nat Commun. 2019;10:1020.
81. Llull P, Liao XJ, Yuan X, Yang JB, Kittle D, Carin L, Sapiro G, Brady DJ. Coded aperture compressive temporal imaging. Opt Express. 2013;21:10526–45.
82. Stern A, Javidi B. Random projections imaging with extended space-bandwidth product. J Disp Technol 2007;3: 315–320.

Chapter 4
Computerized Tomography

Xuyang Zhou and Zhengjun Liu

4.1 Introduction

Computed tomography (CT) imaging is a non-invasive diagnostic technique widely used in medical imaging. By combining X-ray and computer technology, CT imaging provides high-resolution, cross-sectional images of the internal structures of the human body, which provide physicians with important clinical information to aid in the diagnosis and treatment of disease. In this chapter, we will systematically introduce CT imaging technology from the origin of X-rays, including the principles of CT imaging, classification, development and application of CT imaging equipment. CT imaging algorithms are categorized and typical reconstruction algorithms are introduced. Finally, the new technology and future development direction of CT imaging are envisioned.

Nowadays, computed tomography (CT) technology has a wide range of applications in the field of medical imaging and continues to evolve and innovate. As a non-invasive, non-invasive examination method, CT imaging, through the fusion of X-ray and computer technology is capable of obtaining detailed structural images of the internal tissues of the human body at high resolution, providing physicians with important clinical diagnostic information.

This chapter will provide a comprehensive introduction to the history, principles, and application areas of CT imaging, providing insight into its advantages and limitations. We will examine the structure and composition of CT scanning devices, as well as their operating principles and parameter settings. In addition, the basics of CT image acquisition and reconstruction will be emphasized, including different reconstruction algorithms and image enhancement techniques. As technology continues

X. Zhou · Z. Liu (✉)
School of Physics, Harbin Institute of Technology, Harbin 150001, China
e-mail: zjliu@hit.edu.cn

X. Zhou
e-mail: xyzhou0508@163.com

Z. Liu et al. (eds.), *Computational Optical Imaging*, Advances in Optics and Optoelectronics, https://doi.org/10.1007/978-981-97-1455-1_4

to advance, we will also delve into advanced CT image reconstruction methods, such as spiral CT, multilevel CT, sparse reconstruction, and statistical reconstruction algorithms, as well as the application of deep learning in CT image reconstruction.

In addition to reconstruction techniques, we will investigate methods for CT image alignment and deformation, as well as techniques for image segmentation and feature extraction. These key steps lay the foundation for subsequent CT image analysis and applications covering image classification, quantization, measurement, simulation, and virtual reality.

Finally, we will look ahead to new technologies and future directions in CT imaging, exploring hardware-based technological innovations, the future of algorithms and methods, and the potential application of artificial intelligence in CT imaging. With the continuous development of science and technology, CT imaging will surely continue to provide more accurate and reliable information for medical diagnosis and protect the health of patients.

4.1.1 From X-ray to CT Imaging

The birth of X-rays can be traced back to an accidental experiment in 1895. At that time, the German physicist Wilhelm Conrad Röntgen was conducting research on a high vacuum glass tube known as a Crookes tube, which was used to study cathode rays [1]. During his experiments, Röntgen noticed that even when the tube was sealed and most of the light was blocked, it emitted a mysterious type of radiation. This radiation passed through a layer of black paper and caused a nearby screen to glow, creating a fascinating and unknown phenomenon.

Intrigued by this discovery, Röntgen began further investigations into the nature of this radiation. He called this new form of radiation "X-rays," with the "X" representing the unknown, as the nature of these rays was not yet fully understood.

In subsequent experiments, Röntgen found that X-rays could penetrate soft tissues in the human body but were blocked by denser structures such as bones. This discovery sparked his curiosity to explore the potential medical applications of X-rays.

In early 1896, Röntgen published his research findings, and it caused a sensation in the scientific community. His discovery quickly spread, and doctors and scientists began using X-rays to examine the internal structures of the human body, including bones and foreign objects. The birth of X-rays marked a groundbreaking breakthrough in medical imaging, providing a completely new means and perspective for medical diagnosis. Röntgen's discovery paved the way for the development of medical imaging and radiology. In recognition of his outstanding contributions to the discovery and application of X-rays, Röntgen was awarded the Nobel Prize in Physics in 1901 [2]. His work laid a solid foundation for the subsequent advancements in medical imaging and radiology.

4.1.2 *Principle of CT Imaging*

The basic principle of CT imaging is to use X-ray beams to scan a certain thickness of the human body to examine the level of the site, by the detector to receive X-rays through the level of the visible light, converted to electrical signals by the photo-electric converter, and then by the analog/digital converter (analog/digital converter) into digital signals, entered into a computer for processing. Figure 4.1 shows the CT imaging systems.

The image formation is processed as if the selected layers were divided into a number of rectangles of the same volume, called voxels. The information from the scan is calculated to obtain the X-ray attenuation or absorption coefficients for each voxel, which are then arranged in a matrix, the digital matrix. The digital matrix can be stored on a disk or CD-ROM. Each number in the digital matrix is converted by a digital/analog converter into small squares of varying shades of gray from black to white, i.e., pixels, which are arranged in a matrix to form the CT image.

The principle of CT imaging is based on the use of X-rays and advanced computer processing techniques to create detailed cross-sectional images of the human body. The CT imaging process involves the following key steps:

X-ray Source and Detectors: The CT scanner consists of an X-ray tube and an array of detectors positioned opposite each other. The X-ray tube emits a narrow beam of X-rays that pass through the body during the scanning.

Rotational Scanning: The patient lies on a movable table, which is then slowly advanced into the gantry of the CT scanner. The X-ray tube and detector array rotate around the patient in a complete 360° arc, capturing multiple X-ray projections from different angles.

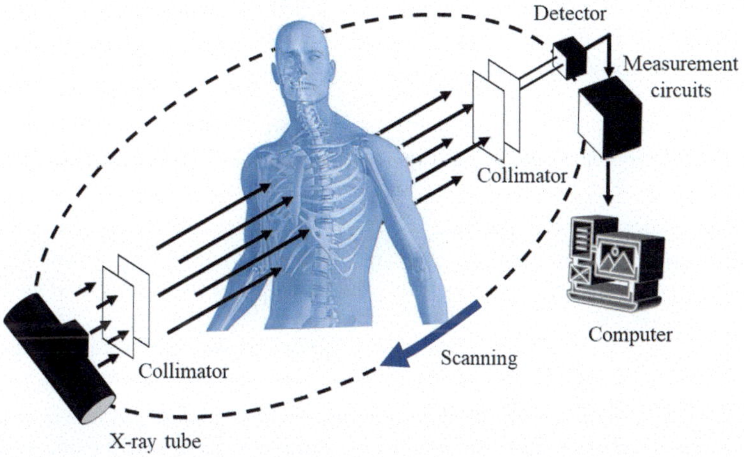

Fig. 4.1 CT imaging systems

Data Acquisition: As the X-ray tube rotates, the detectors measure the amount of X-rays that pass through the body at each angle. This data is collected and digitized, resulting in a series of raw X-ray attenuation measurements.

Image Reconstruction: The raw data is sent to a powerful computer, which uses specialized algorithms to process the data and reconstruct cross-sectional images, or "slices," of the body. These images represent the X-ray attenuation properties of the internal structures at different depths within the body.

Hounsfield Units: The reconstructed CT images are represented in Hounsfield Units (HU), a scale that quantifies the X-ray attenuation of each pixel in the image. Air is set as –1000 HU, water as 0 HU, and bone as positive values. This scale allows for differentiation of various tissues and structures based on their density.

Image Display: The CT images are displayed on a computer monitor, allowing radiologists and physicians to analyze the cross-sectional anatomy and identify any abnormalities or pathology.

The principle of CT imaging relies on the fact that different tissues in the body have different X-ray attenuation properties due to variations in density. By collecting X-ray attenuation measurements from multiple angles and using advanced mathematical algorithms, CT imaging can create detailed and highly informative images of internal body structures, providing valuable information for diagnostic purposes, treatment planning, and monitoring of various medical conditions.

4.1.3 Development of CT

In 1974, Robert Ledley, an engineer at George Town Medical Center, designed the first whole-body CT scanner, and in 1975, Prof. Ralph Alfidi of Case Western Reserve University also took the lead in researching CT imaging of the abdomen, and with their efforts, the scope of CT was expanded to include the chest, abdomen, spine, and extremities, creating the era of CT tomography for whole-body scanning. Extremities, creating a whole-body scanning CT tomography era, this era is divided into the first generation of CT machines and the second generation of CT machines in two phases [3, 4].

The first generation CT machine in Fig. 4.2 adopts the rotate/translate mode to scan and collect information. Due to the use of a pen-shaped X-ray beam and only 4.2 detectors, the data collected was small, the time required was long, and the image quality was poor.

The scanning method of the second generation CT machine in Fig. 4.3 has not changed with the previous generation, only the X-ray beam was changed to fan, the detector was increased to 30, the scanning range was enlarged, the acquisition data was increased, and the image quality was improved, but it still can't avoid the artifacts caused by the patient's physiological movement.

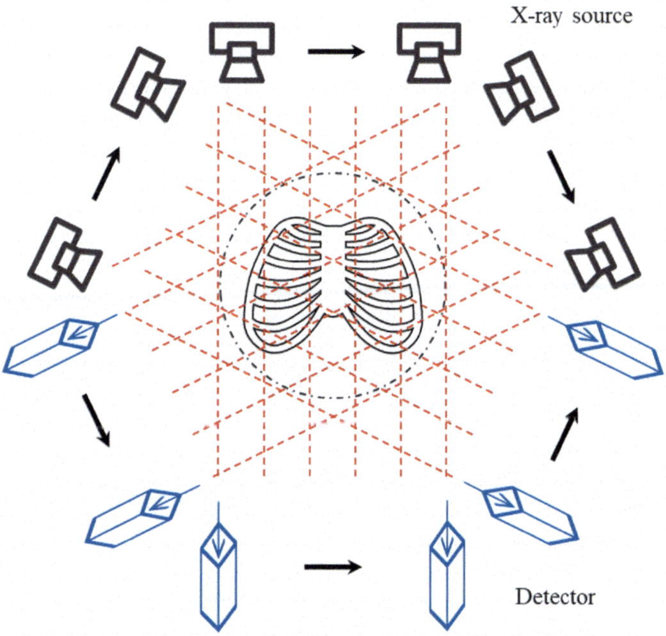

Fig. 4.2 First-generation CT with parallel beam scanning

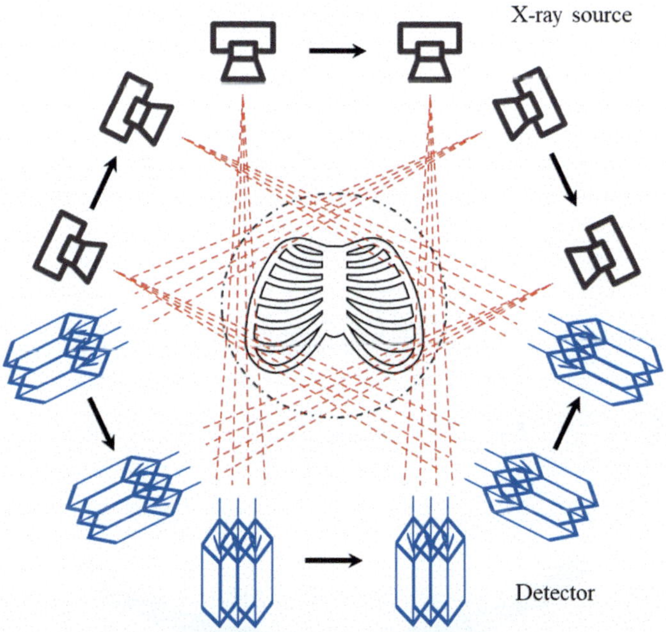

Fig. 4.3 Second-generation CT with fan sector beam scanning and multi-detector

In 1989, with the application of slip ring technology, a new CT scanning technology different from tomography mode was born-helical scan mode in Fig. 4.4. 1990, the first clinical research results of helical CT were published, which initiated the first volumetric scanning revolution, and brought about a wide range of clinical applications of dynamically enhanced CT. The development of CT technology into the third generation is a qualitative leap and an important milestone in the history of CT development [5].

In order to increase the scanning speed, the third and fourth generation CT machines were developed on the basis of the second generation CT machines, of which the third generation CT machines have since gradually become mainstream. The third-generation CT machine has a surge of 300~800 controllers and only rotates with the relative X-ray tube (rotate/rotate mode), which collects more data, and the scanning time is within 5 s. The artifacts are greatly reduced, and the quality of the image is significantly improved.

The problem of cardiac scanning has led to the gradual evolution of CT with the development of four and five generations of CT machines. The fourth generation

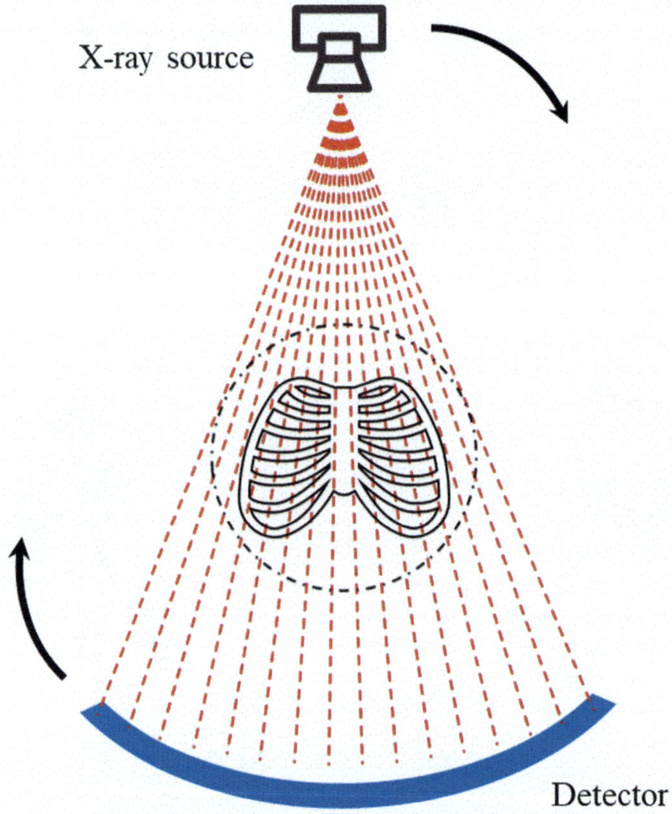

Fig. 4.4 Third-generation CT with fan beam scanning and rotating detector

of CT machines increased to 1,000–2,400 controllers and arranged in a ring and stationary, only the X-ray tube rotates around the patient, i.e., rotate/stationary mode, which has a fast scanning speed and high image quality.

The fifth generation of CT machines has reduced the scanning time to 50 ms, addressing the cardiac scan, an electron beam generated by an electron gun directed at a ring-shaped tungsten target, with a ring arrangement of detectors to collect the information. 64-slice CT can obtain images of 64 layers of a patient's body in only 0.33 s, with a spatial resolution of less than 0.4 mm, which improves the quality of the images, especially for the pulsing heart. heart. However, the fourth and fifth generation CT machines were not commercialized as Figs. 4.5 and 4.6.

Based on the third generation of CT machines, CT technology has evolved in this era mainly in two directions: faster and wider, as Fig. 4.5. In the aspect of fast imaging (improving time resolution), the concept of triple-source CT was proposed and patented in 1979 (as shown in Fig. 4.7), the birth of dual-source CT in 2006, and the birth of Philips iCT in 2009, which, based on the leading air cushion bearing technology, firstly increased the physical rotation speed of CT to 0.27 s. And in the aspect of wide (improving the coverage range), from the 1992 The first double-row spiral CT (Elscint Twin, the first realization of the sub-mm imaging, CT from now on to truly realize to meet the isotropic three-dimensional imaging) began, the width

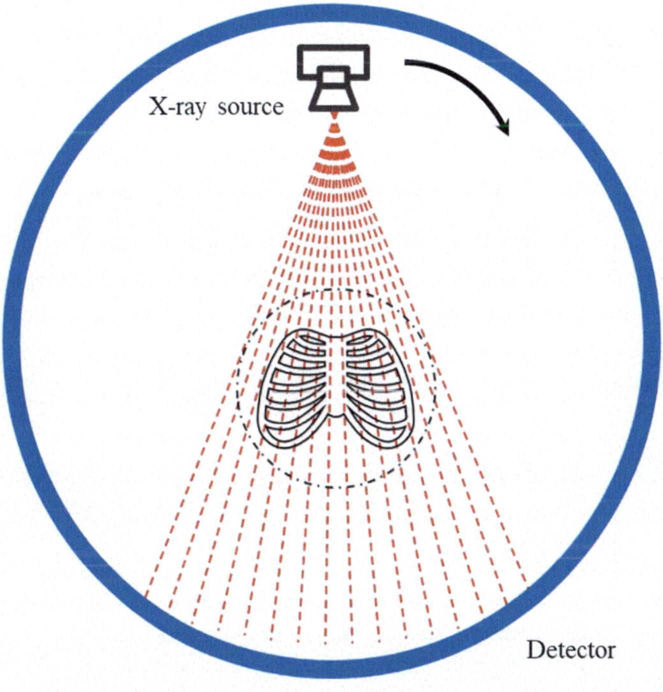

Fig. 4.5 Fourth-generation CT with fan beam scanning and loop detector

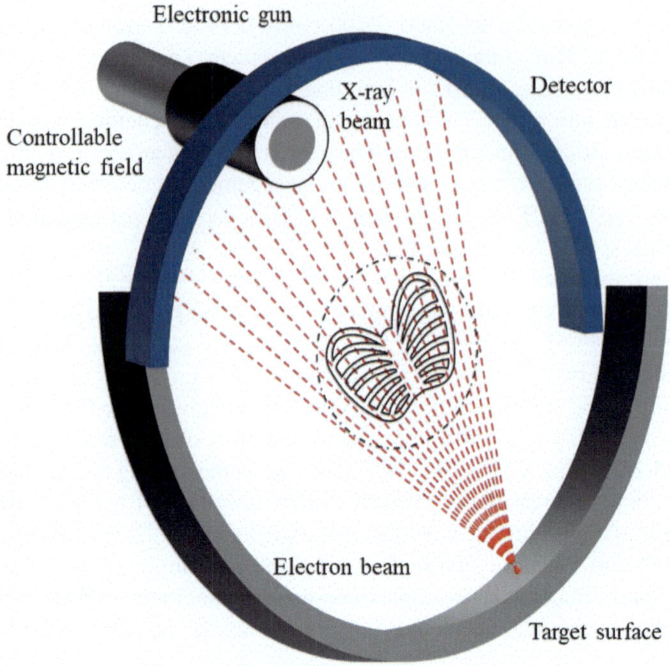

Fig. 4.6 Fifth-generation CT with electron beam scanning

of the detector after more than 20 years of rapid development, and ultimately the development of 16 cm coverage of the wide-body detector CT [6].

In 2006, the birth of Tube-Based Dual Energy CT solved the problem of cardiac scanning, and also reopened another direction of CT development—CT energy imaging. The use of two sets of tube-detector system in the two energy points respectively after imaging in the image data domain for material decomposition, thus realizing the identification of different material components, from the CT realized from the simple physical characteristics of the analysis to the complex chemical characteristics of the analysis. This became the beginning of the CT energy spectrum era. Later, with the birth of energy spectrum CT further realized the material decomposition in the projection data domain [6].

In 2016, the world's first dual-layer detector-based spectral CT was born, which receives high- and low-energy X-ray photons through spatially-equivalent upper and lower detectors, respectively, to achieve energy resolution and color spectral imaging at the detector end [7, 8].

In 2021, the technology is further developed to 100 kVp spectral imaging and 80 cm large aperture and new spherical wide-body spectral detector. The new detector further improves the ability to calculate and analyze X-ray energy at the detector end, and accurately differentiate between high and low energy levels. Another important milestone is that another detector-based spectral CT technology, semiconductor

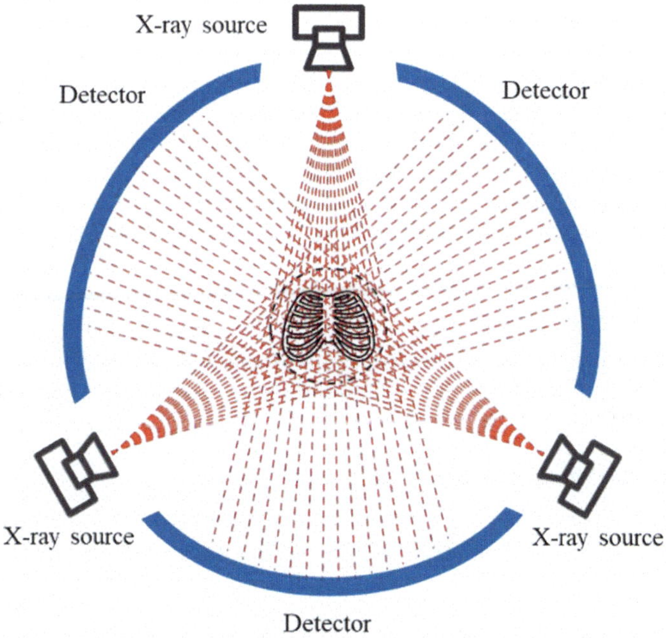

Fig. 4.7 Triple-source CT (Based on the third-generation CT, it consists of three independent bulb-detector systems in the same rack.)

photon counting CT, has begun to move towards the clinic. The imaging advantages of low noise and high spatial resolution were realized [9], allowing for multi-contrast agent examination and molecular imaging exploration.

In early 2022, researchers at the Technical University of Munich applied dark-field X-ray imaging methods to clinical CT for the first time [10]. Spectrum's wide aperture CT and advanced image reconstruction algorithms, combined with Kromek's high performance Cadmium Zinc Telluride (CZT) digital detector, increased the energy range of high-energy isotopes up to 400 keV. In the same year, Canada's KA Imaging introduced its first mobile dual-energy DR (digital radiography). It supports dual-energy subtraction, provides bone and tissue differentiation through a single exposure, and simultaneously acquires three images (DR, bone and soft tissue dual-energy X-ray images). Philips introduces the first Angio CT with a spectral detector—the Spectral Angio CT suite—bringing its breakthrough spectral CT imaging technology to an integrated composite Angio CT solution.

4.1.4 Application Areas of CT Imaging

The following are some of the major application areas of CT imaging:

Medical diagnosis

Radiology: CT scans are used to detect and diagnose a variety of diseases and conditions such as tumors, fractures, strokes, cardiovascular disease, organ damage and infections.

Surgical Planning: CT imaging is used for surgical planning to help surgeons pinpoint lesions, organs and blood vessels for more precise procedures.

Physical Examination: CT scans are also used for general physical examinations for early detection of health problems.

Radiation therapy planning

CT imaging can be used in cancer radiation treatment planning to help doctors determine the location and size of lesions to more accurately target radiation treatments.

Dental and craniofacial surgery

Diagnostic Dentistry: CT scans are used in dentistry to diagnose dental and maxillofacial problems such as root infections, periodontal disease, and implant placement planning.

Craniofacial Surgery: CT imaging is used in the planning of craniofacial surgical procedures such as skull reconstruction and facial plastic surgery.

Medical research

Scientific Research: CT scans are used in medical research to study different disease models, anatomy and pathology.

Drug development: CT imaging can help evaluate the efficacy of new drugs, as well as conduct studies on the distribution and metabolism of drugs in animal models.

Industrial and materials science

Non-destructive testing: CT imaging can be used to examine the internal structure of components, welded joints and materials in industrial and materials science without destructive testing.

Quality Control: CT scans can be used for quality control to ensure that manufactured products meet specifications and standards.

Geology and archaeology

Geology: CT imaging is used in geological research to study the structure of underground rocks and minerals.

Archaeology: CT scanning is used in the study of archaeological specimens, such as ancient artifacts and fossils, to non-destructively obtain internal information.

Aerospace engineering

Materials Testing: CT imaging is used to examine the structure and integrity of materials used in aerospace engineering.

Component Inspection: CT scans can be used to inspect components of airplanes and spacecraft to ensure they are free of internal defects.

Overall, CT imaging has a wide range of applications in medicine, scientific research and engineering, providing detailed information about internal structures to help diagnose diseases, plan treatments, study materials and carry out non-destructive testing.

4.1.5 Advantages and Limitations of CT Imaging

(1) Advantages

High-resolution images: CT scans provide high-resolution images that show details of the body's internal structures, helping doctors to more accurately diagnose and treat diseases.

Three-dimensional reconstruction: CT imaging produces three-dimensional images that allow doctors to view organs and tissues at all angles, providing more comprehensive information.

Non-invasive: CT scanning is a non-invasive test that does not require surgery or endoscopy. Patients usually do not need to undergo painful tests or procedures.

Rapid imaging: CT scans are usually very rapid, with the entire scan taking only a few minutes, which is important for diagnosis in emergency situations.

Multi-position imaging: CT machines can capture images of the patient in different positions, which helps doctors evaluate organs and tissues more fully.

Applicable to different tissues: CT imaging is excellent for different types of tissues, such as bones, soft tissues and blood vessels, and therefore provides useful information in a wide range of medical situations.

Repeatability: CT scans can be performed as many times as needed without harming the patient, which helps monitor disease progression and treatment outcomes.

For radiation treatment planning: CT imaging can be used to accurately plan cancer radiation treatments, helping doctors better locate tumors and pinpoint treatment areas.

For research and education: CT imaging is also useful for medical research and education, providing detailed information on anatomy and pathology.

Wide range of applications: In addition to medicine, CT imaging has applications in engineering, geology, archaeology, and other fields, where it is used to examine materials, components, and underground structures.

In summary, the high resolution, non-invasive, multi-site imaging and applicability to different tissues make CT imaging a valuable imaging technique that helps provide accurate diagnosis and treatment planning.

(2) Limitations

Although computed tomography (CT) imaging has many advantages, it also has some disadvantages and limitations, including the following:

Radiation Exposure: CT scans use X-rays to generate images and therefore expose patients to radiation. While many improvements have been made to modern CT equipment to minimize the radiation dose, radiation exposure is still a consideration for certain populations such as pregnant women and children, as well as for patients who require frequent CT exams.

Need for Contrast: If a patient is having an enhanced CT exam a contrast agent (usually a contrast agent containing iodine) is needed to enhance the visualization of the image, but the contrast agent may trigger an allergic reaction, especially in those patients who are allergic to iodine.

Not for specific tissues: CT imaging may not be as advantageous as other imaging techniques, such as magnetic resonance imaging (MRI) or positron emission tomography (PET-CT), for certain tissues, such as brain and lung tissue.

Bone Tissue Artifacts: CT scans may produce bone tissue artifacts in the images, making diagnosis of other tissues difficult. These artifacts may interfere with accurate diagnosis in some cases.

Limited resolution for soft tissues: Although CT is excellent for imaging bone, resolution may be relatively low for certain soft tissues, such as muscle and fat.

Not suitable for continuous monitoring: CT scans are usually used for one-time imaging and are not suitable for continuous monitoring of disease progression or treatment effects.

Inability to detect certain medical conditions: Some medical conditions may not be amenable to diagnosis by CT scanning, such as neurodegenerative diseases or certain lung diseases.

Limited ability to image blood vessels: Although CT angiography can be used to examine blood vessels, it may not provide vascular imaging comparable to digital subtraction angiography (DSA) or magnetic resonance angiography (MRA).

Expensive: CT equipment is expensive to purchase and maintain and therefore may lead to higher healthcare costs.

In conclusion, CT imaging is a useful imaging technique, but it is not a universal solution for all situations. Doctors usually choose the most appropriate imaging technique based on the patient's specific situation and needs. In addition, CT scans should be used with caution in radiation-sensitive patients and appropriate radiation protection measures should be taken when necessary.

4.2 Fundamentals of CT Image Acquisition and Reconstruction

The acquisition and reconstruction of CT images is a highly complex field of technology that plays an important role not only in medical diagnosis but also has a wide range of applications in many other fields. In this chapter, we will take an in-depth look at the CT image acquisition process and the key steps in image reconstruction.

4.2.1 CT Image Acquisition Process

In the context of medical imaging applications, CT imaging uses X-rays to obtain detailed images of the inside of the human body. The steps involved in acquiring a CT image are as follows:

X-ray source: CT scans use an X-ray source, which produces a high-energy beam of X-rays.

Patient positioning: The patient is usually asked to lie on the scanning bed to ensure that the desired area is in the focus of the scanner.

X-Ray detector: Located on the opposite side of the scanner, the detector is used to measure the x-ray beam that passes through the patient. The detector usually consists of a number of small units to collect data from incoming X-rays from different directions.

Rotation scanning: The CT scanner usually rotates around the patient while the X-ray source and detector rotate to acquire X-ray data at different angles.

Data acquisition: At each rotational position, the x-ray beam passes through the patient and is measured by the detector. This data, which includes the degree of X-ray absorption, is used to construct the image.

Reconstruction filtering: The acquired data is calculated and mathematically algorithmized to remove artifacts and noise, which is then used for image reconstruction.

4.2.2 Overview of CT Image Reconstruction Algorithms

In the process of CT scan projection and data reconstruction, it mainly involves the mutual transformation of projection data and image data between object space, Radon space and Fourier space as Fig. 4.8.

Image reconstruction is an important branch of image processing, its significance is to obtain the internal structure of the detected object without causing any physical damage to the object. The quality of the reconstruction algorithm is directly related

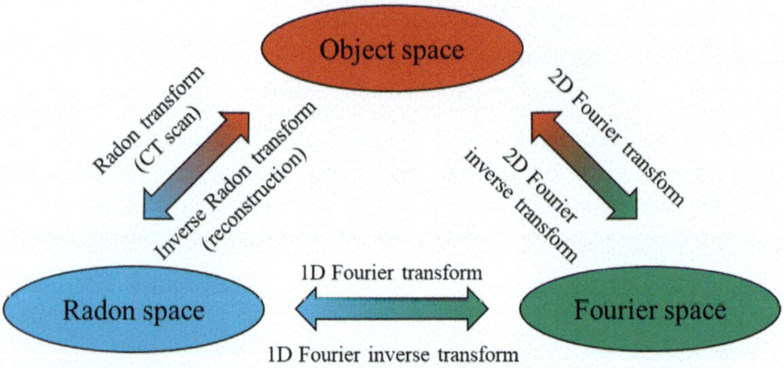

Fig. 4.8 The relationship between the three domains (2D Imaging)

to the quality of the reconstructed image, reconstruction speed and anti-interference ability, which is the key and difficult point in the whole CT processing process [11].

The Radon transform and the central slice theorem are the theoretical basis of analytic reconstruction. The distribution of the physical characteristic parameters of a section can be calculated by the linear integral value of the physical characteristic parameters in each direction of the section.

Projection is an important concept of CT imaging, and projection is also called "ray sum" or line integral:

$$p(s,\theta) = \int_{-\infty}^{\infty}\int_{-\infty}^{\infty} f(x,y)\delta(x\cos\theta + y\sin\theta - s)\mathrm{d}x\mathrm{d}y, \tag{4.1}$$

where, $f(x, y)$ is the density function defined on the plane x–y; δ is a Dirac function.

A sinogram is one of the methods used to describe a projection A sinogram is a map of the projected distribution of a layer of CT acquired prior to reconstruction, and the space corresponding to the sinogram is called the sinusoidal space The vertical axis of the sinusoidal space represents the angle of projection, and the horizontal axis represents the rays (detector units) of each projection. Chordal plots can be used to detect whether the CT system is working properly or not when one of the detector units fails, and the chordal plot contains a vertical straight line.

The Radon transform reveals the relationship between functions and projections. Assuming the function is $f(x, y)$ and taking projections of it at different angles then the relational equation for $p(t, \theta)$ and $f(x, y)$ is

$$p(t,\theta) = \int_{-\infty}^{\infty}\int_{-\infty}^{\infty} f(x,y)\delta(x\cos\theta + y\sin\theta - t)\mathrm{d}x\mathrm{d}y. \tag{4.2}$$

The inverse transformation is expressed as

$$f(x, y) = \frac{1}{2\pi^2} \int_0^{\pi} \int_{-\infty}^{\infty} \frac{1}{\sqrt{x^2 + y^2} \cos[\tan^{-1}(y/x) - \theta) - t} \frac{\partial p(t, \theta)}{\partial t} \mathrm{d}t \mathrm{d}\theta. \tag{4.3}$$

That is, the distribution of relative linear decay coefficients within an infinitely thin slice is uniquely determined by the set of all its line integrals.

In 2D imaging, CT reconstruction is the process of transforming object space and Radon space into each other. The central slice theorem for 2D images reveals that the Fourier space is a bridge between the Radon space and the object space, the theorem states: The parallel bundle projection of the 2D function $f(x, y)$ at angle θ is $p(t, \theta)$. The one-dimensional Fourier transform in the t-direction of the detector is $\hat{p}(t, \theta)$. The slice of the 2D Fourier space $\hat{f}(\xi, \theta)$ of $f(x, y)$ parallel to the detector direction over the origin is denoted as

$$\hat{p}(\xi, \theta) = \hat{f}(\xi \cos\theta, \xi \sin\theta). \tag{4.4}$$

4.3 Parallel Beam Image Reconstruction

4.3.1 Direct Fourier Reconstruction Algorithm

By the central slice theorem $S_\theta(\omega) = F(\omega \cos\theta, \omega \sin\theta)$, it follows that if the detector is rotated at least 180° around the object, the projection is acquired, then $\{S_\theta(\omega)\}$ can be calculated. That is, the frequency domain values on each angle in the 2D Fourier transform space of the image can be known, and the complete Fourier transform function $F(u, v)$ can be obtained. As shown in Fig. 4.9, once $F(u, v)$ is a known function, the original image function $f(x, y)$ can be reconstructed by 2D Fourier inverse transform recovery [12].

4.3.2 Filtered Back Projection

Filtered back projection (FBP) is one of the widely used image reconstruction algorithms today. The FBP algorithm is implemented by transforming the parameters and redefining the limits of integration as

$$f(x, y) = \int_{-\infty}^{\infty} \int_{-\infty}^{\infty} F(u, v) e^{i2\pi(ux + vx)} \mathrm{d}u \mathrm{d}v. \tag{4.5}$$

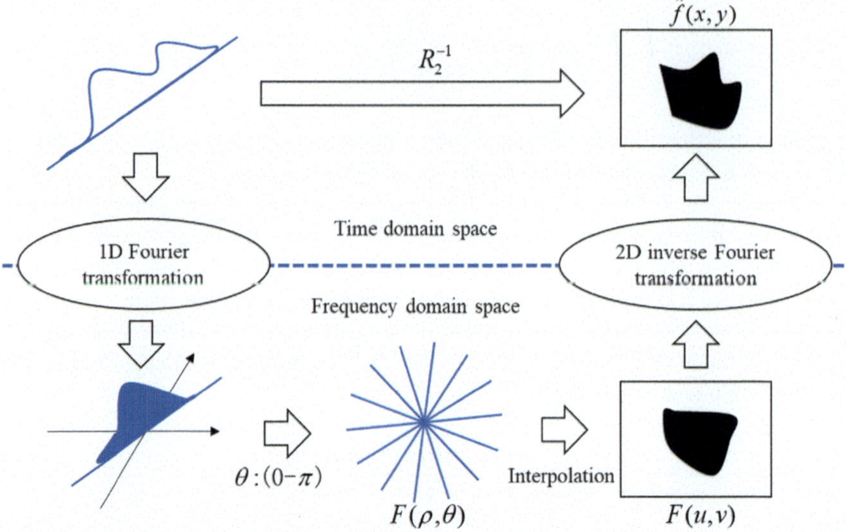

Fig. 4.9 Fourier reconstruction algorithm diagram

The transformation of (u, v) from a right-angle coordinate system to a polar coordinate system is expressed in the form of (ω, θ). The relationship is established as follows

$$\begin{cases} u = \omega \cos\theta, \\ v = \omega \sin\theta. \end{cases} \tag{4.6}$$

By the centre slice theorem

$$f(x, y) = \int_0^{\pi} \int_{-\infty}^{\infty} S_\theta(\omega)\exp[i2\pi\omega s(x\cos\theta + y\sin\theta)]|\omega|\mathrm{d}\omega\mathrm{d}\theta \tag{4.7}$$

where, $|\omega|$ achieves the effect of high-pass filtering of the projection data.

The filtered back-projection reconstruction process for 2D parallel beam projection first filters the projection data for each viewpoint θ, and then back-projects the cumulative to compute the function $f(x, y)$. $q_\theta(s)$ denotes the filtered projection data for the projection angle θ, and its filtering operator is denoted by $|\omega|$ in the frequency domain space.

If different mathematical deformations of $q_\theta(s)$ are performed, they will lead to different physical interpretations, corresponding to different filtered backprojection reconstruction algorithms, which are generally classified into two categories: convolutional backprojection algorithms and Radon inverse transform backprojection algorithms.

(1) Convolutional backprojection algorithms: Assuming $H(\omega) = |\omega|$, Its corresponding 1D Fourier inverse transform function is $h(s)$. The filtered data is $q_\theta(s)$. According to Fourier transform theory, the frequency domain product is equivalent to the time domain convolution.
(2) Radon inverse transformation backprojection algorithm: Decomposition of the transfer function of a slope filter into two parts: the derivative operation and the Hilbert transform

$$H(\omega) = \left|\omega\right| = \omega \mathrm{sgn}\omega, \tag{4.8}$$

where, $\mathrm{sgn}(\omega)$ is a symbolic function.

The inverse projection does the hardest work in image reconstruction, it takes the projection data measured from different angles and turns it into an image. This image is similar to the one we want to get, just a little blurred. The addition of the filter solves the problem of blurring in the reconstructed image and enables the imaging system to achieve the desired imaging effect.

4.4 Sector Beam Image Reconstruction

Parallel beam CT imaging scanning modalities shows in Fig. 4.10 are characterized by slow acquisition speeds, low ray utilization, and high mechanical accuracy requirements. While fan-beam CT imaging shows in Fig. 4.11 is more common in daily applications, we will introduce the image reconstruction algorithm for fan-beam in the following.

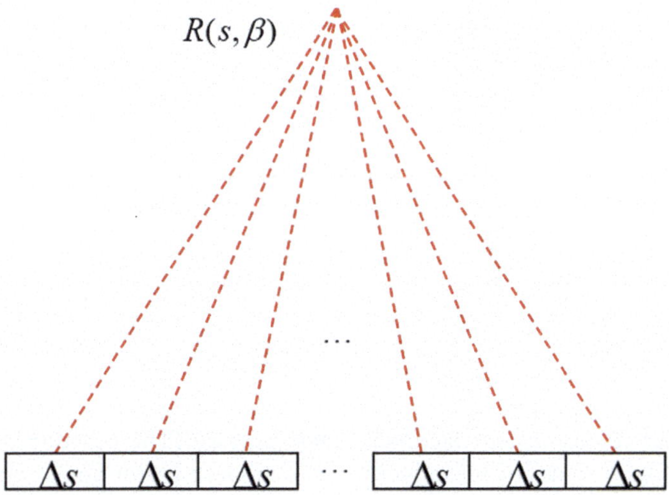

Fig. 4.10 Fan beam and linear detectors

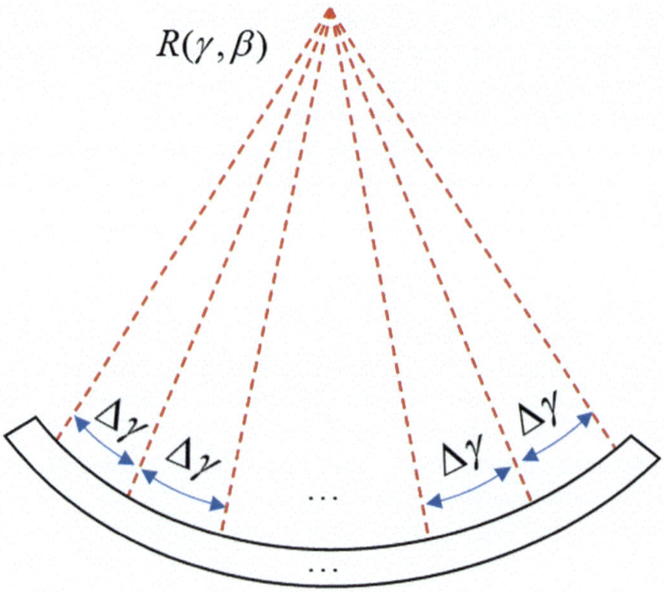

Fig. 4.11 Fan beam and arc detectors

On the basis of the parallel beam reconstruction algorithm and the equation relationship $\theta = \gamma + \beta$, $t = D \sin \gamma$, the variables of the parallel bundle are replaced with those of the fan-shaped bundle. The fan-bundle image reconstruction algorithm can be expressed as

$$f(r, \varphi) = \frac{1}{2} \int_{0}^{2\pi} \int_{-\gamma m}^{\gamma m} R(\gamma, \beta) h(D' \sin(\gamma' - \gamma)) D \cos \gamma \mathrm{d}\gamma \mathrm{d}\beta. \tag{4.9}$$

4.5 Cone-Beam Image Reconstruction Algorithm

Cone beam X-ray source has the advantages of high ray utilisation and high acquisition efficiency, which is of great significance for reducing the radiation dose and improving the reconstruction efficiency; at the same time, the 3D reconstruction algorithm can be used to obtain three-dimensional body data, which solves the problem of low inter-layer resolution of the traditional two-dimensional reconstruction. Cone beam image reconstruction algorithms mainly include FDK reconstruction algorithm, Grangeat reconstruction algorithm, Katsevich reconstruction algorithm and BPF reconstruction algorithm.

4.5.1 FDK Reconstruction Algorithm

Feldkamp et al. propose an approximation algorithm based on the reconstruction of circular orbital cone beams [13], This is now our FDK algorithm. Currently, FDK is the most practical algorithm for cone beam reconstruction of circular orbits. When the cone angle is less than 5 degrees, good reconstruction results can be obtained, but as the cone angle increases, FDK will bring more and more serious artefacts.

The FDK algorithm is actually a three-dimensional extension of the two-dimensional fan-beam filtered backprojection algorithm, which first weights the projection data that deviates from the centre plane, and then reconstructs the object using a filtered backprojection similar to the fan-beam algorithm. Its specific process can be divided into the following two steps: first weighted filtering of the projection data

$$p(\beta, a, b) = \left(\frac{R}{\sqrt{R \cdot R + a \cdot a + b \cdot b}} \cdot p(\beta, a, b) \right) * g^{p}(a), \tag{4.10}$$

$$\begin{aligned} \frac{R}{\sqrt{R \cdot R + a \cdot a + b \cdot b}} &= \frac{R}{\sqrt{R \cdot R + a \cdot a}} \cdot \frac{\sqrt{R \cdot R + a \cdot a}}{\sqrt{R \cdot R + a \cdot a + b \cdot b}} \\ &= \cos \lambda \cdot \cos k, \end{aligned} \tag{4.11}$$

where, $g^{p}(a)$ is a one-dimensional filter, R is the radius of the track and β indicates the angle of rotation. The weighted filtered projection data is then reconstructed by back-projection

$$f_{FDK}(x, y, z) = \int_{0}^{2\pi} \frac{R^2}{U(x, y, \beta)} p(\beta, a(x, y, z, \beta), b(x, y, z, \beta)) \mathrm{d}\beta. \tag{4.12}$$

4.5.2 Grangeat Reconstruction Algorithm

In the field of accurate reconstruction, the Grangeat algorithm [14] is a landmark. The Grangeat algorithm mainly reconstructs images using the 3D Radon inversion formula. In 3D imaging, the Radon transform is an area division of a 3D object. The Radon inversion formula needs to find the second-order derivative of the area fraction obtained by the Radon transform along the direction normal to the integration plane, and then carry out the 3D Radon inverse projection. Grangeat found the equivalence between the differential of the area fraction and the integral derivative of the cone-beam projection data on the line of the detector, and through which the first-order derivative of the area fraction can be calculated. However, the Grangeat algorithm is not an FBP-type algorithm, and embedding the image into the Radon space during the reconstruction process requires interpolation operations, which introduces a large error and cannot reconstruct long objects.

The implementation of the Grangeat algorithm is divided into the following four main steps

(1) Calculating the intermediate function F, which converts the cone beam projection into the derivative of the Radon transform.

$$r_{\lambda}^{(1)}\left(\vec{\beta}\right) = F\left(\vec{\beta}, \vec{\Phi}(\lambda) \cdot \vec{\beta}\right) = -2\pi \int_{S^2} g(\lambda, \vec{\alpha}) h_D(\vec{\alpha} \cdot \vec{\beta}) \mathrm{d}\vec{\alpha}. \tag{4.13}$$

(2) Handling data redundancy

$$r_{\lambda}^{(2)} = r_{\lambda}^{(1)} \frac{\left|\vec{\Phi}'(\lambda) \cdot \vec{\beta}\right|}{M_{\lambda}(\vec{\beta})}. \tag{4.14}$$

(3) Calculating the filtered cone beam projection

$$g_{\lambda}^{(1)}(\vec{\alpha}) = \int_{S^2/2} h_D(\vec{\alpha} \cdot \vec{\beta}) \gamma_{\lambda}^{(2)}(\vec{\beta}) \mathrm{d}\vec{\beta}. \tag{4.15}$$

(4) Obtaining the reconstructed object by 3D weighted inverse projection

$$f(\vec{x}) = \frac{-1}{2\pi} \int_{\Lambda} \frac{1}{\left\|\vec{x} - \vec{\Phi}(\lambda)\right\|^2} g_{\lambda}^{(1)}\left(\frac{\vec{x} - \vec{\Phi}(\lambda)}{\left\|\vec{x} - \vec{\Phi}(\lambda)\right\|}\right) \mathrm{d}\lambda. \tag{4.16}$$

4.5.3 *Katsevich Reconstruction Algorithm*

For any point in the region to be reconstructed, there is and only one line segment (PI line) passing through the point and having two points of intersection with the spiral trajectory [15]. Tam demonstrated that the minimum projection data required for an accurate reconstruction of an object inside a helix is the data inside the Tam window [16]. The Tam window is the area between the two projected lines on the detector between the upper half spiral orbital arc and the lower half spiral orbital arc where the light source point illuminates its location as shown in Fig. 4.12.

Katsevich proposed a one-dimensional filter function using translation invariance. The redundant data are normalised and weighted by the choice of filter direction. Only a specific region of data is needed to achieve accurate reconstruction of a single point, and the algorithm can effectively solve the longitudinal truncation reconstruction problem. However, the range of detector data used in Katsevich's algorithm is much

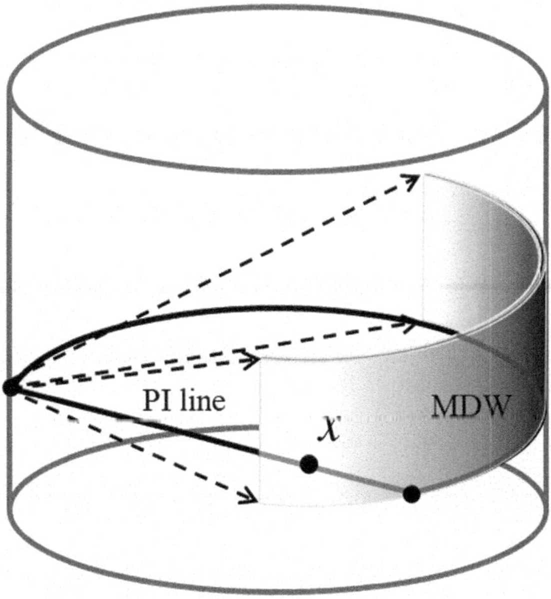

Fig. 4.12 The illustration of PIline and Tam-Danielsson. MDE: Minimum detection window

larger than that of the Tam window, and all data on the filtering path need to participate in the reconstruction, which cannot solve the transverse truncation problem [17].

The formulation of the cone-beam spiral Katsevich reconstruction algorithm can be decomposed as follows

$$f(\vec{x}) = -\frac{1}{4}\int_{I_{PI(\vec{x})}} \frac{1}{\|\vec{x} - y(s)\|^2}\Phi(s, \beta(s, \vec{x}))\mathrm{d}s, \tag{4.17}$$

where,

$$\Phi(s, \beta(s, \vec{x})) = \int_0^{2\pi} \frac{\partial}{\partial q}\Delta f(y(q), \Theta(s, \vec{x}, \gamma))|_{q=s}\frac{\mathrm{d}\gamma}{\sin\gamma}, \tag{4.18}$$

which is regarded as $e(s, \beta) \cdot \beta = 0$, $|e(s, \beta)| = 1$, then $\beta = (\cos\theta, \sin\theta)$ and $e(s, \beta) = (-\sin\theta, \cos\theta)$, $\beta, e(s, \beta) \in \Pi(s_2)$,

$$\Phi(s, \beta) = \int_0^{2\pi} \frac{\partial}{\partial q}Df(y(q), (\cos(\theta + \gamma), \sin(\theta + \gamma)))\bigg|_{q=5}\frac{\mathrm{d}\gamma}{\sin\gamma}. \tag{4.19}$$

4.5.4 *BPF Reconstruction Algorithm*

Zou et al. [18] proposed a reconstruction algorithm in the form of backprojection filtration (BPF) based on Katsevich's algorithm based on PI lines in 2004. The BPF algorithm is mainly based on the relationship between the differential inverse projection $g(r)$ of the projected data and the Hilbert transform $H(\vec{r})$ of the object to be reconstructed,

$$g(\vec{r}) = -2\pi H(\vec{r}). \tag{4.20}$$

The algorithm reconstructs the information corresponding to each PI line within the spiral trajectory through angle-by-angle differentiation, weighted inverse projection, Hilbert transform along PI lines, and resampling in Cartesian coordinate system to complete the reconstruction of 3D objects. Katsevich algorithm can only solve the problem of longitudinal truncation, while BPF algorithm utilises the theoretically minimal projection data for reconstruction, and it can solve the problem of transverse truncation reconstruction effectively. It is of great significance.

4.6 Iterative Reconstruction Algorithm

Iterative class algorithms are another important class of CT image reconstruction algorithms. Although transform domain methods such as filtered back-projection (FBP) are widely used in production due to their advantages of fast computation speed, high imaging accuracy, etc., they are not well suited to meet the requirements under some special application conditions. For example, when the projection data is insufficient or the projection angle is unevenly distributed, the application of the transform domain method often fails to obtain a better quality reconstructed image, while the iterative class of algorithms may be a better choice. According to the difference of data space utilised in iterative computation, iterative reconstruction techniques are divided into 3 categories.

(1) Iteration only in the image data space

After the original data are reconstructed according to the traditional FBP method, the obtained FBP image is then filtered according to the noise model to reduce noise and artefacts. This method is faster and the computation time is only slightly longer than that of the FBP method, but it inevitably has the limitations of the FBP method due to the processing based on the image space.

(2) Partial iterations are performed in both projected data spaces

Firstly, the projection data is reconstructed by the FBP method, and the obtained image data is compared and corrected with a statistically based ideal noise model that takes into account both photon and electronic noise, and then the corrected image is updated with the original projection data by orthographic projection, and then the calculation is carried out for the next iteration. This is done for several iterations to obtain the final image, which is also faster to reconstruct, but also has limitations due to the fact that only the noise model is considered.

(3) Complete iteration in the projected data space

Multiple models of the entire X-ray optical acquisition process of the X-ray beam from the focal point to the detector, where factors such as the focal point, the X-ray beam, the voxel, and the geometry of the detector are equally factored into the model, are complex and correspondingly computationally intensive, and the reconstruction time required is relatively long.

The most classical method for iterative reconstruction is the algebrare construction technique (ART) proposed by Gorden et al. The algorithm is based on the projection method first proposed by Kak et al. [12]. The basic idea is to discretise the problem. The reconstructed image area is divided into $n \times n$ equal squares, which are considered as pixels. $f(x, y)$ is the discrete value and is constant inside each pixel. This is followed by a combination of imaging physical processes and corresponding mathematical modelling. A system of algebraic equations to reconstruct the relationship between the image and the projected image is created. The algorithm is resistant to noise interference and can be combined with a priori knowledge to solve the problem [19]. The projection diagram is shown in Fig. 4.13.

The formula for the ART algorithm is

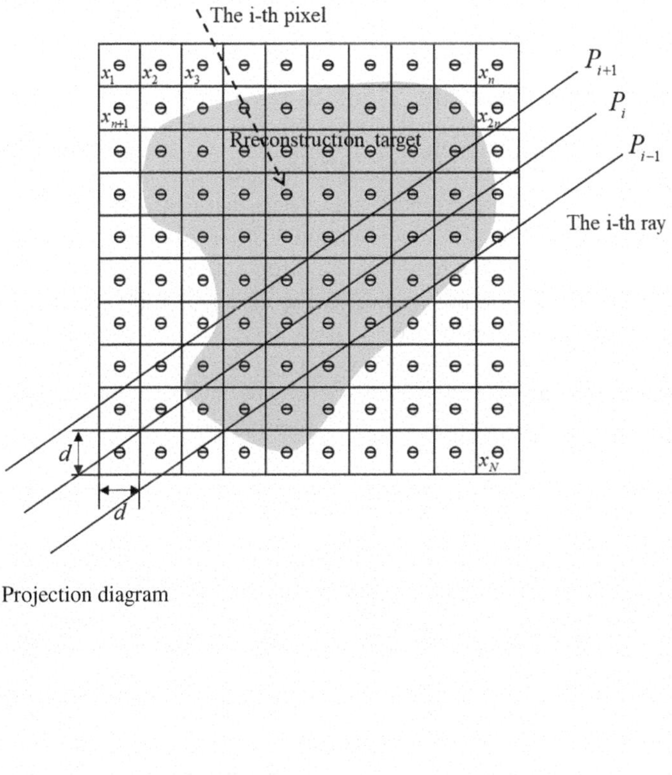

Fig. 4.13 Projection diagram

$$x_j^{(k+1)} = x_j^k + \lambda^{(k)} \frac{p_i - \sum_{m=1}^{M} a_{im} x_m}{\sum_{m=1}^{M} a_{im}^2} a_{ij}, \tag{4.21}$$

where, k denotes the number of iterations. The range of values of i is $1 \leq i \leq N$, λ is the relaxation factor and $0 < \lambda < 2$.

Simultaneous algebraic reconstruction technique (SART) is proposed as a major improvement of the algebraic reconstruction algorithm (ART). Since then, SART has become a powerful tool for iterative reconstruction. The SART algorithm, on the other hand, is much closer to the actual image generation process. Instead of viewing each ray as a separate unit, it views a projection angle as a correlated system. The formula for the SART algorithm is

$$x_j^{(k+1)} = x_j^{(k)} + \lambda^{(k)} a_{ij} \left(p_i - \sum_{m=1}^{M} a_{im} x_m \right) \left(\sum_{m=1}^{M} a_{im} \right)^{-1} \left(\sum_{i \in I_s} a_{ij} \right)^{-1}, \tag{4.22}$$

where, k is the number of iterations, $1 \leq i \leq N$, $1 \leq j \leq M$, λ is the relaxation factor and $0 < \lambda < 2$, I_θ is the set of projection indices under a projection angle.

The above describes the most commonly used reconstruction algorithms in CT imaging, and a summary of their basic steps and algorithmic features is shown in the Table 4.1.

4.7 Other Reconstruction Algorithms

The reconstruction algorithms described in this section have been developed to address specific CT imaging problems and needs that can improve image quality, reduce radiation dose, increase computational efficiency, or meet specific clinical application needs. Researchers and engineers in the field of medical imaging are constantly working to improve and develop these algorithms to meet different medical diagnostic and treatment needs. These algorithms include incomplete Angle reconstruction algorithm, local reconstruction algorithm and extended field of view reconstruction algorithm.

4.7.1 Incomplete Angle Reconstruction Algorithm

Incomplete angle reconstruction algorithm is a method used to reconstruct high quality images from finite Angle projection data. Typically, in traditional CT imaging, it is necessary to obtain full 360-degree Angle projection data, but in some cases, such

Table 4.1 Comparison of commonly used reconstruction algorithms

Algorithm names	Elementary steps	Specificities
Iterative method	>Set the initial matrix >Form a projection with the initial matrix >Compare the projection of the object to be reconstructed with the simulated projection >The iteration stops until the error meets the requirements	>Calculations are accurate >Time-consuming >Improved iterative algorithms are a hot topic in image reconstruction
Fourier transform reconstruction algorithm	>The 1D Fourier transform is first applied to each measured projection data >The 1D transform function obtained at different projection angles can form a complete 2D Fourier transform function in the frequency domain >By inverting this 2D transformation function, the density function in the required spatial domain is obtained	>It is computationally intensive due to the interpolation >There may be noticeable distortion in the high frequency section
Back projection (BP)	>The measured projections are distributed equally to each point according to its original path >The projected values in each direction are back-projected and superimposed at the image to infer the original image	>The absorption coefficient is highest at the centre. The further away from the centre the lower the value, resulting in a loss of sharpness at the edges of the image >Counter-projection causes the edges of the image to be unsharp
Backprojection filtration (BPF)	>After obtaining the projection function (a one-dimensional function) at a certain projection angle, it is filtered >The corrected projection function is then back-projected to produce the desired density function	Only the one-dimensional Fourier transform needs to be computed. The filtered inverse projection method significantly shortens the image reconstruction time

as rapid scanning or low-dose imaging, it may not be possible to obtain full projection data. Incomplete Angle reconstruction algorithm can generate better image quality by using missing projection data, adopting iterative algorithm or other mathematical techniques to fill in the missing information.

The compressed sensing theory proposed by Candes et al. [20] in 2006 is a theory to collect, encode and decode signal data under the condition that the known signal has sparsity or compressibility. In terms of the hardware implementation of compressed sensing theory, MIT has successfully developed MRIRF pulse devices and coded aperture cameras, and Rice University has developed single-pixel cameras and A/I converters. In the field of X-ray CT, compressed sensing algorithm based on iterative method can not only reduce radiation dose, but also ensure image quality [21].

From the perspective of optimization, finding the sparsiest signal representation is essentially an l0 minimization problem, and this problem is an NP-hard problem. For this reason, researchers have proposed many pioneering algorithms, which can basically be classified as greedy algorithm [22], l1 minimization and Total Variation. TV) minimization algorithm [23]. Recent studies show that compared with greedy algorithm and l1 minimization algorithm, TV minimization algorithm can more accurately preserve the edge details of image reconstruction, which is very important in image processing. Another advantage of the TV minimization algorithm is that in addition to processing sparse signals and images, it can also successfully process gradient-sparse signals and images. In recent years, many scholars have studied sparse representation model of CT image reconstruction and made some important research results. The common practice is to establish a suitable objective function first, and then find a fast algorithm to solve the objective function. Zhu et al. [24] added wavelet transform to the objective function to make the reconstructed image edges clearer. Deng Luzhen et al. [25] introduced the Contourlet transform of multi-scale geometric analysis and proposed an algorithm based on the combination of Countourlet transform and separating Bregman. Nien and Fessler [26] divided the system matrix into ordered subsets, then adopted the augmenting Lagrangian method to solve the subsets, and proposed the order-linear augmenting Lagrangian (OS-LALM) algorithm. Later, an improved SLack-OS-LALM algorithm was proposed [27]. Hao et al. [28] added prior information into the objective function and proposed a Bayesian reconstruction algorithm based on low-volume CT images.

Some of these algorithms can recover the whole information of the image well, but there is a small range of artifacts. Some reconstructed images have no obvious artifacts, but the reconstruction efficiency is reduced. The shortcomings of the algorithm still need to be improved.

4.7.2 Local Reconstruction Algorithm

In many engineering applications, due to hardware limitations or to reduce radiation dose, only certain region of interest (ROI) can or need to be imitated, which presents a local reconstruction problem. Local reconstruction algorithm is a method to reconstruct different regions or volumes of CT images independently. This is useful in situations where large data sets or complex scans need to be processed, as it can improve computational efficiency. Local reconstruction algorithms allow different parts of an image to be processed separately and then combined into a complete image.

Compared with traditional global reconstruction, local reconstruction has the following advantages:

(1) For large-size measured objects, the X-ray beam only needs to cover the ROI during CT scanning, which can greatly reduce the radiation dose and the need for detector size;
(2) It can greatly improve the speed of CT data scanning and acquisition, reduce the amount of projected data, and then improve the efficiency of image reconstruction, making local reconstruction have obvious advantages in dynamic imaging;
(3) It can use a finite size detector, combined with the geometric magnification property of X-ray beam projection, to achieve high-resolution three-dimensional imaging. These advantages make local reconstruction technology become one of the important development directions of the new generation of practical CT.

The solution to the problem of local CT imaging includes establishing the approximation function related to the object tomography image. Based on this, Smith et al. put forward a Lambda Tomography local reconstruction algorithm [29, 30], which uses local projection data to reconstruct a function with the same singularity as the ROI density function. By reconstructing part of the Hilbert transform of the density function to replace the original function [31], Katsevich proposes a local reconstruction algorithm of Pseudolocal Tomography.

However, because these functions can not replace the real object tomography image, it is difficult to meet the needs of practical engineering applications, which greatly reduces the significance of local CT imaging for practical engineering applications. In the ensuing long period of time, the research on imaging technology for object local ROI has been stalled and can not be well solved.

FBP algorithm can solve the problem of long object reconstruction well. In the case of vertical projection data truncation, the algorithm can still accurately reconstruct the scanned object image, and has great advantages in reconstruction speed compared with the iterative class reconstruction algorithm. This is a new idea in the history of image local reconstruction algorithm. Spiral cone-beam CT in BPF form based on PI lines [31–33] makes the reconstruction algorithm more accurate [32].

Defrise et al. made further achievements on the basis of BPF algorithm and relaxed the restriction of PI line. Using the relationship between the inverse projection function of the projection data derivative and the finite Hilbert transform of the attenuation coefficient function of the reconstructed object, based on the solution of the inverse transformation of the finite Hilbert transform and the unique certainty of the analytic function, it is proved that when only one end point of the PI segment is outside the support of the reconstructed object, The image on the PI line segment can still be accurately reconstructed from the Projection data passing through the PI line segment [33], that is, the iterative algorithm based on the Projection Onto Convex Sets (POCS) method.

For the local reconstruction problem of arbitrary truncation of projected data, Dennerlein proposed an Approximate Truncation Resistant Algorithm for Computed Tomography (ATRACT) based on Laplace operator [34]. It allows arbitrary truncation of projection data and has a reconstruction accuracy comparable to that of the method of completing projection data [35].

4.7.3 Extended Field of View Reconstruction Algorithm

The extended field reconstruction algorithm is used to expand the field of view of CT imaging to a larger area, so as to cover a wider range of anatomical structures. This is useful for specific clinical applications, such as cardiac imaging, where images of the heart and chest need to be captured simultaneously. These algorithms can expand the field of view by merging multiple sub-scans or through other techniques to obtain more comprehensive image information.

In cone-beam CT detection, due to the limited detector size and X-ray beam cone Angle, it is often encountered that the size of the object exceeds the FOV of cone-beam CT imaging. In this case, the projection of the object obtained will produce a projection truncation in the axial or horizontal direction, and the usual CT reconstruction algorithm requires that the object be completely covered by the beam. In the case of projection truncation, there will be large artifacts in the reconstruction results, so the problem of how to use small size detector to detect large size objects is proposed, that is, the problem of expanding the cone-beam CT imaging field.

Cone-beam CT imaging field extension can be simply divided into axial field extension, transverse field extension and bidirectional field extension. Axial field of view extension means that the field of view expands in the vertical direction in the detector plane, transverse field of view extension means that the field of view expands in the horizontal direction in the detector plane, and two-way field of view extension means that the axial and transverse field of view expand simultaneously.

A typical algorithm for axial field expansion is the helical FDK algorithm. Helical FDK algorithm is an approximate reconstruction algorithm of cone-beam helical trajectory based on circular trajectory FDK algorithm. Although it is an approximate reconstruction algorithm, compared with accurate reconstruction algorithm, approximate reconstruction algorithm usually has the advantages of low algorithm complexity, high computational efficiency, easy parallel acceleration, etc., and has good reconstruction effect in the case of small cone Angle. Therefore, the use of approximate reconstruction algorithm in cone-beam CT is usually the mainstream. The FDK algorithm has high efficiency, good robustness to noise and good reconstruction effect in the case of small cone Angle, so the spiral FDK and its improved algorithm still have a high use rate in cone-beam spiral CT.

There are many scanning methods that can realize transverse field expansion in cone-beam CT, mainly including bias scanning and rotation-translation (RT) scanning. The biasing scan needs to scan the object once, and the transverse field of view expansion rate cannot be more than 2 times. The RT scanning method in Fig. 4.14 is more flexible, requiring at least two scans, but at the same time can achieve more than twice the horizontal field of view through multiple scans (as shown in Fig. 4.13). In practice, the modified eccentric FDK algorithm is used to reconstruct the offset scanning. Due to the data loss caused by the scanning mode itself, the reconstructed results will have truncation artifacts in the horizontal direction.

Imaging field of view and imaging resolution are a pair of contradictory relations. When high-resolution imaging is performed on the detected object, the imaging field

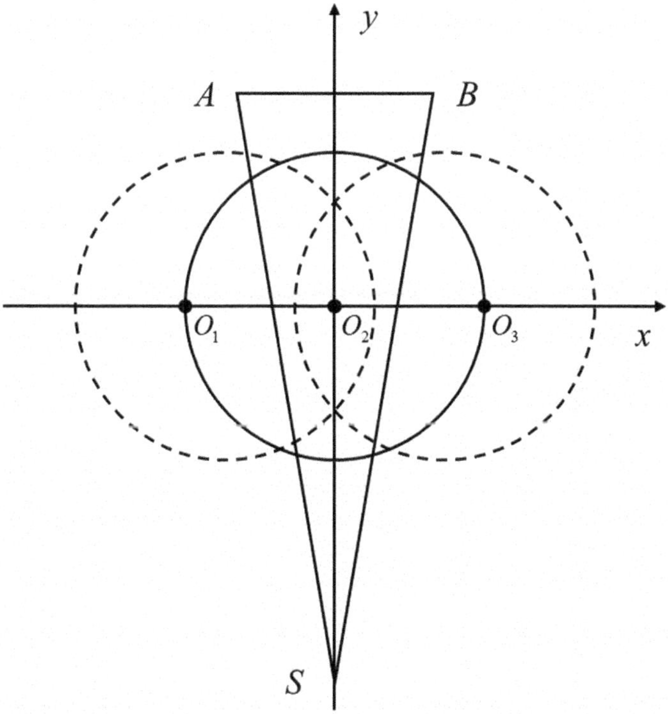

Fig. 4.14 RT scanning schematic

of view will become small, and the object will easily exceed the imaging field of CT in both vertical and horizontal directions. To image the whole object, it is necessary to expand the imaging field of cone-beam CT in both vertical and horizontal directions. At present, there are few researches on the bidirectional field expansion reconstruction algorithm of cone-beam CT. Cone-beam helical CT semi-coverage scanning method [36] and cone-beam CT double helix scanning method [37] can realize the simultaneous expansion of cone-beam CT vertical field of view and horizontal field of view. However, in the horizontal field of view expansion, both scanning methods can only achieve no more than 2 times the field of view expansion rate.

4.8 New Technology and Development of CT Imaging

4.8.1 Innovation of CT Imaging Technology Based on Hardware

Detector technology: Improved X-ray detector technology for CT systems to improve spatial resolution, sensitivity, and signal-to-noise ratio. The new generation of detectors may use more advanced materials, such as silicon, germanium or semiconductors, to increase detection efficiency.

Line scan CT: Increase the number of lines of the CT machine to achieve faster scanning speed and higher anatomical coverage, resulting in reduced artifacts and improved image quality.

Photon Counting detectors: Photon counting detectors can more accurately measure the number of photons of an incident X-ray, resulting in improved image quality and radiation dose management. This technique is expected to improve low-dose imaging, especially for children and long-term monitoring.

High-energy X-ray tubes: A new generation of high-energy X-ray tubes can produce a stronger X-ray beam to improve tissue contrast and reduce noise levels.

Rapid rotation system: Improve the mechanical design of CT systems to allow them to rotate more quickly, thereby reducing scan time, especially for patients in emergency situations and in motion.

Low-dose imaging: Develop more efficient radiation filters, control algorithms, and reconstruction techniques to enable low-dose imaging and reduce patient exposure to radiation.

High-resolution imaging: Improved X-ray beam transmission and receiving systems to improve the spatial resolution of CT images to better show tiny anatomical structures and lesions.

Mobile CT equipment: Design and develop mobile CT equipment to meet the needs of field medicine, emergency rescue and mobile medical diagnostics.

4.8.2 Future Trends of Algorithms and Methods

Deep learning and artificial intelligence: Deep learning techniques will continue to play a key role in CT imaging for tasks such as image reconstruction, noise reduction, lesion detection, and segmentation. Neural networks and deep learning algorithms will further improve image quality, increase automation, and improve the accuracy of clinical diagnosis.

Low-dose imaging: Algorithms will continue to be optimized to reduce radiation doses while still producing high-quality images. This is important for long-term monitoring and imaging of children.

Multispectral CT: The development of multispectral imaging will require higher-level algorithms to efficiently process data at multiple energy levels, providing richer tissue contrast and more accurate material identification.

Enhance image quality: Further improve the reconstruction algorithm to reduce artifacts, noise and other imaging artifacts and improve image quality. This helps to show tiny lesions and tissue structures more clearly.

Fast imaging: Algorithms will continue to improve to support faster scans for patients in emergency situations and in motion. This includes faster image acquisition and faster reconstruction algorithms.

Automated analysis: Further development of automated analysis algorithms to help physicians more easily identify and quantify abnormalities, such as tumors, vascular diseases, and more. This will improve the efficiency and accuracy of clinical decision making.

Personalized Medicine: Algorithms are customized to the individual characteristics of patients to enable personalized medicine, including optimizing scan parameters and image processing to better meet the needs of patients.

Multimodal imaging: Integrating data from different imaging modes, such as CT, MRI, and PET, to provide more comprehensive information and facilitate more precise diagnosis and treatment planning.

Cloud computing and distributed processing: Image data is uploaded to the cloud platform to provide faster and more efficient image analysis and diagnosis support through distributed processing and sharing.

4.8.3 Prospect of Artificial Intelligence in CT Imaging

Artificial intelligence holds great promise in CT imaging and will have a profound impact on medical imaging and patient care. Combined with the actual needs and shortcomings, the application prospects of artificial intelligence in CT imaging are summarized as follows:

Automated image analysis: AI algorithms can automatically analyze CT images to help doctors detect abnormalities, mark lesions, measure the dimensions of organs and structures, and track the progression of disease. This will improve the efficiency of clinical decision making, reduce the burden on physicians, and reduce diagnostic errors.

Disease screening and early diagnosis: AI can be of potentially great value in the early diagnosis of tumors, lung disease, cardiovascular disease, and other diseases. By analyzing large-scale CT image data, AI can identify tiny abnormalities and risk factors that can help advance intervention and treatment.

Personalized Medicine: AI can customize CT scan parameters and image processing based on a patient's individual characteristics and medical history to provide personalized medical recommendations and treatment options.

Low-dose imaging: AI algorithms can achieve high-quality imaging at lower radiation doses through noise reduction and image enhancement. This is very useful for long-term monitoring and imaging of children.

Image synthesis: AI can synthesize high-resolution CT images to produce sharper images from limited data or low-resolution scans, helping to improve image quality.

Intelligent assisted decision making: AI systems can provide doctors with information about diagnosis and treatment recommendations, helping them make more informed decisions based on large-scale clinical databases and case data.

Long-term monitoring and patient management: AI can automatically analyze and track a patient's disease progression, providing continuous monitoring and prediction to support patient management and treatment planning.

Fault detection and maintenance: AI can be used to monitor and maintain CT equipment to help predict equipment failures, optimize performance, and reduce downtime.

In conclusion, the promise of artificial intelligence in CT imaging is very broad, helping to improve the accuracy, efficiency and quality of clinical diagnosis, promoting the development of personalized medicine, improving patient care, and driving innovation in the field of medical imaging. With the further development of technology and the expansion of clinical applications, AI will continue to play an increasingly important role in medical imaging.

Acknowledgements This work was supported by the National Natural Science Foundation of China (Nos. 11874132, 61975044, 12074094, 12374271, and 62375065), Interdisciplinary Research Foundation of HIT (No. IR2021237).

References

1. Röntgen WC. On a new kind of rays. Nature. 1896;53:274–6.
2. Nobel A. The Nobel prizes in physics and chemistry for 1901 Nobel lectures. Phys. 1967; 1901–21
3. Richmond C. Sir Godfrey Hounsfield engineer who invented computed tomography and won the Nobel prize for medicine. BMJ Brit Med J. 2004;329:687.

4. Alfidi RJ, MacIntyre WJ, Meaney TF, Chernak ES, Janicki P, Tarar R, Levin H. Experimental studies to determine application of CAT scanning to the human body. Am J Roentgenol Radium Ther Nucl Med. 1975;124(2):199–207.
5. Rigauts H, Marchal G, Baert AL, Hupke R. Initial experience with volume CT scanning. J Comput Assist Tomogr. 1990;14(4):675–82.
6. Siegel MJ, Kaza RK, Bolus DN, Boll DT, Rofsky NM, De Cecco CN, Foley WD, Morgan DE, Schoepf UJ, Sahani DV, Shuman WP, Vrtiska TJ, Yeh BM, Berland LL. White paper of the society of computed body tomography and magnetic resonance on dual-energy CT, part 1: technology and terminology. J Comput Assist Tomogr. 2016;40(6):841–5.
7. Forghani R, De Man B, Gupta R. Dual-energy computed tomography: physical principles, approaches to scanning, usage, and implementation: Part 1. Neuroimaging Clin N Am. 2017;27(3):371–84.
8. Forghani R, De Man B, Gupta R. Dual-energy computed tomography: physical principles, approaches to scanning, usage, and implementation: part 2. Neuroimaging Clin N Am. 2017;27(3):385–400.
9. Higashigaito K, Euler A, Eberhard M, Flohr TG, Schmidt B, Alkadhi H. Contrast enhanced abdominal CT with clinical photon-counting detector CT: assessment of image quality and comparison with energy-integrating detector CT. Acad Radiol. 2022;29(5):689–97.
10. Viermetz M, Gustschin N, Schmid C, Haeusele J, Teuffenbach M, Meyer P, Bergner F, Lasser T, Proksa R, Koehler T, Pfeiffer F. Dark-field computed tomography reaches the human scale. Proc Natl Acad Sci USA. 2022;119(8): e2118799119.
11. Seeram E. Computed tomography: a technical review. Radiol Technol. 2018; 89(3): 279CT–302CT.
12. Kak AC, Slaney M, Wang G. Principles of computerized tomographic imaging. Med Phys. 2002;29(1):107–107.
13. Feldkamp LA, Davis LC, Kress JW. Practical cone-beam algorithm. J Opt Soc Am A. 1984;1(6):612–9.
14. Grangeat P. Mathematical framework of cone beam 3D reconstruction via the first derivative of the radon transform. In: Herman GT, Louis AK, Natterer F, editors. Mathematical methods in tomography. Lecture notes in mathematics, vol. 1497. Springer, Berlin, Heidelberg
15. Danielsson PE, Edholm P, Eriksson J, Magnusson M. Towards exact reconstruction for helical cone-beam scanning of long objects. A new detector arrangement and a new completeness condition. In: Proc. 1997 meeting on fully 3D image reconstruction in radiology and nuclear medicine. Pittsburgh; 1997. p. 141–4
16. Tam KC. Three-dimensional computerized tomography scanning method and system for imaging large objects with smaller area detectors: U.S. Patent 5,390,112[P]. 1995–2–14, 1995
17. Katsevich A. Analysis of an exact inversion algorithm for spiral cone-beam CT. Phys Med Biol. 2002;47(15):2583–97.
18. Zou Y, Pan X. Exact image reconstruction on PI-lines from minimum data in helical cone-beam CT. Phys Med Biol. 2004;49(6):941–59.
19. Mueller K, Yagel R, Wheller JJ. Fast implementations of algebraic methods for three-dimensional reconstruction from cone-beam data. IEEE Trans Med Imaging. 1999;18(6):538–48.
20. Candes EJ, Romberg J, Tao T. Robust uncertainty principles: exact signal reconstruction from highly incomplete frequency information IEEE Trans Inform Theory. 2006; 52(2): 489–9
21. Hashemi S, Beheshti S, Gill PR, Paul NS, Cobbold RS. Accelerated compressed sensing based CT image reconstruction. Comput Math Method Med. 2015;2015: 161797.
22. Tropp JA, Wright SJ. Computational methods for sparse solution of linear inverse problems. Proc IEEE. 2010;98(6):948–58.
23. Wang YL, Yang JF, Yin WT, Zhang Y. A new alternating minimization algorithm for total variation image reconstruction. SIAM J Imag Sci. 2008;1(3):248–72.
24. Zhu Z, Wahid K, Babyn P, Cooper D, Pratt I. Carter Y (2013) Improved compressed sensing-based algorithm for sparse-view CT image reconstruction. Comput Math Methods Med. 2013;18:84–104.

25. Deng LZ, Feng P, Chen MY, He P, Vo QS, Wei B. A CT reconstruction algorithm based on non-aliasing Contourlet transform and compressive sensing. Comput Math Methods Med. 2014; 23(5):751–9
26. Nien H, Fessle JA. Fast X-ray CT image reconstruction using a linearized augmented Lagrangian method with ordered subsets. IEEE Trans Med Imaging. (2014); 34(2):388–99
27. Nien H, Fessler JA. Relaxed linearized algorithms for faster X-Ray CT image reconstruction. IEEE Trans Med Imaging. 2016;35(4):1090–8.
28. Zhang H, Han H, Liang Z, Hu Y, Liu Y, Moore W, Ma J, Lu H. Extracting information from previous full-dose CT scan for knowledge-based Bayesian reconstruction of current low-dose CT images. IEEE Trans Med Imaging. 2016;35(3):860–70.
29. Smith KT, Keinert F. Mathematical foundations of computed tomography. Appl Opt. 1985;23:3950–7.
30. Faridani A, Ritman EL, Smith KT. Local tomography SIAM Appl Math. 1992;52:459–84.
31. Katsevich A, Ramm AG. Pseudolocal tomography SIAM Appl Math. 1996;56:167–91.
32. Zou Y, Pan XC, Sidky E. Theory and algorithms for image reconstruction on chords and within regions of interest. J Opt Soc Am A. 2005;22(11):2372–84.
33. Defrise M, Noo F, Clackdoyle R, Kudo H. Truncated Hilbert transform and image reconstruction from limited tomographic data. Inv Prob. 2006;22:1037–53.
34. Dennerlein F. Cone-beam ROI reconstruction using the Laplace operator. In: 11th international meeting on fully three-dimensional image reconstruction in radiology and nuclear medicine. 2011. pp.80–3
35. Hsieh J, Chao E, Thibault EJ, Grekowicz B, Horst A, Mcolash S, Myers TJ. A novel reconstruction algorithm to extend the CT scan field-of-view. Med Phys. 2004;31(9):2385–91.
36. Zou XB, Zeng L. Half-cover scanning and reconstruction for helical cone beam CT. Opt Prec Eng. 2010; pp. 434–42
37. Zou XB, Zeng L, Li ZJ. Dual helical cone-beam CT for inspecting large object. J X Sci Technol. 2009;17(3):233–52.

Chapter 5
Computer Generated Holography

Xiu Wen and Zhengjun Liu

Abstract Computational holography combines traditional optical imaging with computational techniques, which can improve the capabilities of a camera, and reduce the cost and size of camera elements. In this chapter, the theoretical basis and application of the computer generated holography is introduced. First, principles and reconstruction algorithms of digital holographic imaging are introduced. Then it compares several holographic computing models. And the lensless holographic imaging method is introduced, which focuses on phase retrieval algorithms and pixel super-resolution imaging. Then principles of 3D tomographic holographic imaging, structured light 3D imaging are introduced.

5.1 Introduction

Digital holography is a technique that allows the capture and reconstruction of three-dimensional (3D) images using digital sensors and computational algorithms. It was first introduced in the late 1960s and has since evolved significantly. Digital holography [1, 2] replaces the holographic plate or film with a digital sensor, such as a charge-coupled device (CCD) or a complementary metal–oxide–semiconductor (CMOS) sensor. The interference pattern created by the object and reference beams is captured by the sensor, which converts it into a digital hologram. To reconstruct the 3D image from the digital hologram, computational algorithms are used. These algorithms perform complex calculations to extract the phase and amplitude information from the hologram and generate a 3D representation of the object. This reconstructed image can then be viewed on a computer screen or other display devices.

X. Wen
School of Physical Science and Technology, Tiangong University, Tianjin 300387, China
e-mail: wenxiu@tiangong.edu.cn

Z. Liu (✉)
School of Physics, Harbin Institute of Technology, Harbin 150001, China
e-mail: zjliu@hit.edu.cn

Z. Liu et al. (eds.), *Computational Optical Imaging*, Advances in Optics and Optoelectronics, https://doi.org/10.1007/978-981-97-1455-1_5

Digital holography offers several advantages over traditional holography. It allows for real-time capture and reconstruction of 3D images, making it suitable for applications such as microscopy, metrology, and biomedical imaging. It also enables the manipulation and analysis of holographic data using digital processing techniques, opening up new possibilities for image enhancement and post-processing.

Computational optical imaging is a fascinating field that combines the principles of optics and computational algorithms to enhance and improve imaging techniques. It involves the use of mathematical models and algorithms to process and analyze optical data, allowing for the extraction of valuable information from images.

In computational optical imaging, various techniques are employed to overcome limitations in traditional imaging systems. For example, by using advanced algorithms, it is possible to enhance image resolution beyond the physical limits of the imaging system. This can be achieved through techniques such as super-resolution imaging, which reconstructs high-resolution images from multiple low-resolution images. Another important aspect of computational optical imaging is the ability to manipulate light and control its propagation. This can be done using techniques like wavefront shaping, which allows for the focusing of light through complex media or the creation of specific light patterns for imaging purposes. Furthermore, computational optical imaging also involves the use of machine learning and artificial intelligence algorithms to improve image analysis and interpretation. These algorithms can be trained to recognize patterns, detect anomalies, or even classify objects within images, enabling automated and more accurate image analysis.

Overall, computational optical imaging is a rapidly evolving field with numerous applications in various domains, including biomedical imaging, remote sensing, and industrial inspection. It holds great potential for advancing our understanding of the world around us and improving the quality and capabilities of imaging systems.

5.2 Digital Holographic Imaging

Digital holographic imaging has various applications, including microscopy, metrology, and non-destructive testing. It allows for high-resolution imaging of transparent or semi-transparent objects, providing depth information and enabling quantitative analysis.

5.2.1 Principles of Digital Holographic Imaging

In the traditional holography process, a laser beam is split into two parts: the object beam and the reference beam. The object beam illuminates the object, and the light scattered from the object combines with the reference beam to create an interference pattern on a holographic plate or film. When this plate or film is illuminated with a laser, a 3D image of the object is reconstructed.

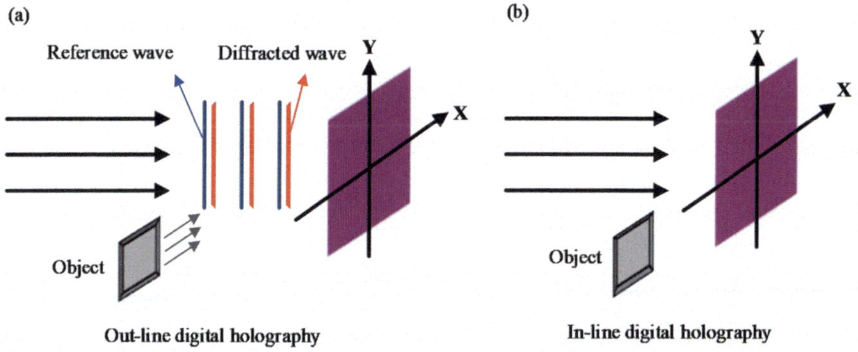

Fig. 5.1 Digital holographic imaging model: **a** out-line digital holography, **b** in-line digital holography

The principles of digital holographic imaging involve capturing and reconstructing holograms using digital techniques. Digital holographic imaging model is shown in Fig. 5.1. In digital holography, a hologram is created by recording the interference pattern between a reference beam and an object beam. This interference pattern contains information about the object's shape, size, and other optical properties. To capture a digital hologram, a coherent light source, such as a laser, is split into two beams. One beam, called the reference beam, is directed onto a reference mirror, while the other beam, called the object beam, illuminates the object of interest. The object beam interacts with the object and carries its information.

The reference beam $R(x, y)$ and the object beam $O(x, y)$ are then recombined, and their interference pattern is recoded by a digital sensor. This recorded interference pattern is the digital hologram as.

$$H(x, y) = |O(x, y)|^2 + |R(x, y)|^2 + O(x, y)R^*(x, y) + R(x, y)O^*(x, y) \quad (5.1)$$

To reconstruct the hologram and obtain a 3D image of the object, the recorded hologram is numerically processed using algorithms. These algorithms perform a Fourier transform on the hologram, separating the object information from the reference beam.

5.2.2 Digital Holographic Reconstruction Algorithm

The digital holographic reconstruction algorithm is the second key issue in digital holography research. Fresnel transform method, convolution method and angular spectrum method are the three most commonly used reconstruction algorithms. They are all based on the scalar diffraction theory and Fourier transform method of optics. They use computers to numerically simulate the diffraction process of optics, so as to obtain the digital reconstructed light wave field. There are several algorithms that can be used for this purpose, depending on the specific requirements and constraints of the application.

(1) Fresnel transform-based method

Fresnel transform is the used algorithm. It involves performing a numerical simulation of the diffraction process using the Fresnel approximation. This algorithm takes into account the interference patterns recorded in the hologram and reconstructs the wavefront of the object or scene.

The Fresnel transformation reconstruction method is based on the Fresnel diffraction theory and needs to meet the Fresnel approximation condition.

$$z_0 \geq \left\{ \frac{\pi}{4\lambda} \left[(x_0 - x)^2 + (y_0 - y)^2 \right]^2 \right\}_{\max}^{1/3} \tag{5.2}$$

where z_0 is the distance between the observation plane x–y and the diffraction aperture plane $x_0 - y_0$.

According to the Fresnel diffraction formula and the definition of Fourier transform, if the distribution of the reconstructed light wave on the camera plane is $C(x, y)$, then the complex amplitude of the diffracted light wave on the image plane is shown

$$\begin{aligned} U(x_i, y_i) &= \frac{\exp(ikz_i)}{i\lambda z_i} \int_{-\infty}^{\infty} \int_{-\infty}^{\infty} C(x, y) H(x, y) \exp\left\{ \frac{ik}{2z_i} \left[(x_i - x)^2 + (y_i - y)^2 \right] \right\} \\ \mathrm{d}x\mathrm{d}y &= \frac{\exp(ikz_i)}{i\lambda z_i} \exp\left[\frac{ik}{2z_i} \left(x_i^2 + y_i^2 \right) \right] \\ &\mathcal{F}\left\{ \int_{-\infty}^{\infty} \int_{-\infty}^{\infty} C(x, y) H(x, y) \exp\left[\frac{ik}{2z_i} \left(x^2 + y^2 \right) \right] \right\}. \end{aligned} \tag{5.3}$$

In the Fresnel transform reconstruction algorithm, only one Fourier transform is required to obtain the distribution of the image field.

(2) Convolution method

Convolution reconstruction method is based on Rayleigh-Sommerfeld diffraction integral formula and Linear system theory. According to convolution method, the complex amplitude of the above diffraction field is expressed as

$$U = \mathcal{F}^{-1}[\mathcal{F}(H \cdot C) \cdot \mathcal{F}(g)] \tag{5.4}$$

From Eq. (5.4), it can be seen that convolution method requires three Fourier transforms, while the Fresnel approximation reconstruction method only requires one Fourier transform, resulting in a faster reconstruction speed. In addition, if the hologram plane is considered as a spatial domain, the reconstructed image is transformed into the frequency domain. So the pixel size of the reconstructed image plane is shown

$$\Delta x_i = \frac{\lambda z_i}{N_x \Delta x}, \ \Delta y_i = \frac{\lambda z_i}{N_x \Delta y} \tag{5.5}$$

where λ is the lambda of the illumination and z_i is the distance between the object plane and the image plane. Δx and Δy are the resolution size in the object plane. This interval is consistent with the resolution of the reconstructed image. The transformation represented by Eq. (5.5) ultimately returns to the spatial domain, so the pixel size of the reconstructed image surface using convolution method is shown:

$$\Delta x_i = \Delta x, \ \Delta y_i = \Delta y \tag{5.6}$$

According to the holographic principle, the resolution of digital holographic imaging is determined by the numerical aperture of the hologram during recording. Therefore, recording a high numerical aperture hologram can achieve high-resolution imaging. However, the sampling interval of the reconstructed image plane is equal to the pixel size of the camera. If the convolution method is used for reproduction, the fine structure on the object plane smaller than the pixel size of the camera in actual reproduction cannot be reproduced.

(3) Angular spectrum method

The angular spectrum theory is a plane wave theory of diffraction. It strictly follows the Helmholtz equation of scalar diffraction. Therefore, it is an accurate description of diffraction phenomena in the frequency domain.

According to the theory of angular spectrum, the image plane $x_i \sim y_i$ with a distance of z_i from the hologram plane can be obtained the spectrum of the reconstructed light field on x–y is obtained by performing an inverse Fourier transform to obtain the complex amplitude distribution of the reconstructed light wave field as

$$U = \mathcal{F}^{-1}\{\mathcal{F}[C \cdot H] \cdot G_B\} \tag{5.7}$$

The reconstruction of holograms using the angular spectrum method requires a forward Fourier transform and an inverse Fourier transform, so the pixel size of the reconstructed image plane is also equal to the pixel size of the CCD.

It is important to note that the choice of algorithm depends on factors such as the quality of the hologram, computational resources available, and the desired level of accuracy in the reconstruction.

5.3 Computational Holographic Imaging

5.3.1 *Computational Holography Theory*

Computational holographic imaging is a cutting-edge technique in the field of optics and imaging that involves the use of digital processing to reconstruct 3D images of objects from the interference patterns created by light waves [3, 4]. It builds upon the principles of holography, which is a technique that allows the recording and reconstruction of the complete wavefront of light scattered by an object.

Traditional holography relies on the interference between a reference beam and an object beam to create an interference pattern on a holographic plate or film. When this pattern is illuminated with coherent light, it produces a 3D image of the object. However, this conventional method requires precise alignment of optical elements, stable setups, and sensitive recording materials, making it cumbersome and challenging for real-time applications.

Computational holographic imaging [5, 6], on the other hand, leverages the computational power of computers and advanced algorithms to overcome these limitations. A digital sensor captures the interference pattern generated when the object wavefront interacts with a reference wavefront. The recorded hologram is digitized and undergoes complex numerical processing to extract information about the object's phase and amplitude. Various computational techniques like Fourier transforms, diffraction calculations, and iterative algorithms are employed here. By utilizing the extracted information, the computational system reconstructs a high-resolution 3D image of the object. This reconstructed image can be visualized and explored from different angles, enabling detailed examination of the object's structure.

One of the significant advantages of computational holographic imaging is its ability to capture dynamic and time-varying scenes. Since the recording is done digitally, the holograms can be stored and processed further at different time points, making it possible to observe the changes in the object over time. This technology finds applications in various fields, including biomedical imaging, industrial inspection, non-destructive testing, and even in virtual reality and augmented reality displays.

Computational holographic imaging is an active and evolving research area, and ongoing advancements in sensor technology, computational algorithms, and data processing are continually improving its capabilities and expanding its potential applications.

5.3.2 *Sampling and Spatial Bandwidth Product*

The sampling of 3D objects refers to the sampling of continuous object information on the surface into discrete object information. In order not to lose information, according to the Nyquist–Shannon sampling theorem, the number of sampling points of objects is not less than four times of the object space bandwidth product as.

$$M_o N_o = 4\mathrm{w}_o h_o \Delta f_{mo} \Delta f_{mo} \tag{5.8}$$

where M_o and N_o are the sampling points of the object in the x and y directions, and w_o are h_o the physical dimensions of the object in the x and y directions. Δf_{mo} and Δf_{no} are the frequency band width of the object in the x and y directions, respectively. The bandwidth of an object is related to its precision. For sampling an object, it is necessary to sample it according to the resolution of the human eye, and the final displayed 3D image can meet the viewing needs of the human eye. When the sampling interval is too small and the object sampling is too dense, the increase in data volume will increase the calculation time. When the object sampling is too sparse, the final display cannot meet the requirements of the human eye, resulting in loopholes and a sense of granularity.

Similar to object sampling, the sampling points of a hologram are four times the spatial bandwidth product of the hologram, expressed as

$$M_h N_h = 4\mathrm{w}_h h_h \Delta f_{mh} \Delta f_{nh} = 16 w_h h_h f_{mh\,\max} f_{nh\,\max} \tag{5.9}$$

Among them, M_h and N_h are the sampling points of the hologram in the x and y directions, and w_h are h_h the physical dimensions of the hologram in the x and y directions. Δf_{mh} and Δf_{nh} are the bandwidth of the hologram in the x and y directions. $f_{mh\,\max}$ and $f_{nh\,\max}$ are the maximum spatial frequency of the hologram in the x and y directions.

5.3.3 Holographic Computing Model

(1) Holographic algorithm based on point cloud

As shown in Fig. 5.2, a point cloud is a collection of data points in a three-dimensional space, where each point represents a specific position and often includes additional information, such as color, intensity, or normal vector. Point clouds method are commonly generated using 3D scanning techniques, such as structured light scanning, and find applications in various fields like computer graphics, computer vision, and augmented reality.

In holographic computing, an object is regarded as a collection of numerous luminous points, where it is a luminous point on a three-dimensional object. The total complex amplitude distribution on the holographic surface is the superposition of the complex amplitudes of light waves emitted from all points, which can be expressed as:

Fig. 5.2 Holographic computing model based on point cloud

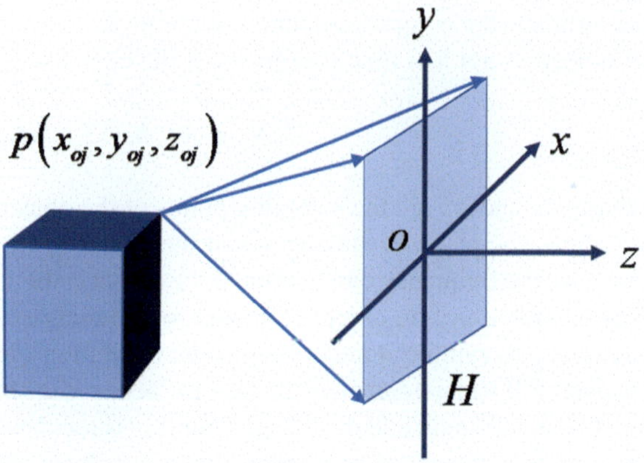

$$U(x_h, y_h) = \sum_{j=1}^{J} \frac{A_j}{z_{oj} + 2z_{oj}^{-1}\left[(x_h - x_{oj})^2 + (y_h - y_{oj})^2\right]} \times \exp(ikz_{oj}) \exp\left[ik\frac{(x_h - x_{oj})^2 + (y_h - y_{oj})^2}{2z_{oj}}\right]. \tag{5.10}$$

This algorithm requires each object point to be calculated on the holographic surface, which requires a huge amount of computation.

(2) Holographic algorithm based on layered

Figure 5.3 shows the principle of layered holographic algorithm. A 3D object is divided into multiple planes along the z direction. z_1 is the distance from the first plane to the holographic surface. And z_J is the distance from the Jth plane to the holographic surface. The complex amplitude distribution of each plane passes through Fresnel diffraction. The complex amplitude distribution of the light field on the holographic surface is obtained by diffraction. In the calculation process, the fast Fourier transform can be used to calculate the heat transfer rate. By using GPU to accelerate computation, it can effectively improve computation speed and reduce computation time when the object is simple and has fewer layers.

The complex amplitude distribution of the object light on the entire holographic surface can be expressed as

$$U_h(x_h, y_h) = \sum_{j=1:J} U_{hj}(x_h, y_h) = A_o(x_h, y_h)\exp[i\phi_o(x_h, y_h)] \tag{5.11}$$

(3) Holographic algorithm based on panel

A 3D object can be seen as a combination of a large number of triangular faces, with the vertex coordinates and normal vectors of each triangular face known. After a certain rotation and translation transformation, its angular spectrum is analytically known. Therefore, in holographic computing, the first step is to transform and

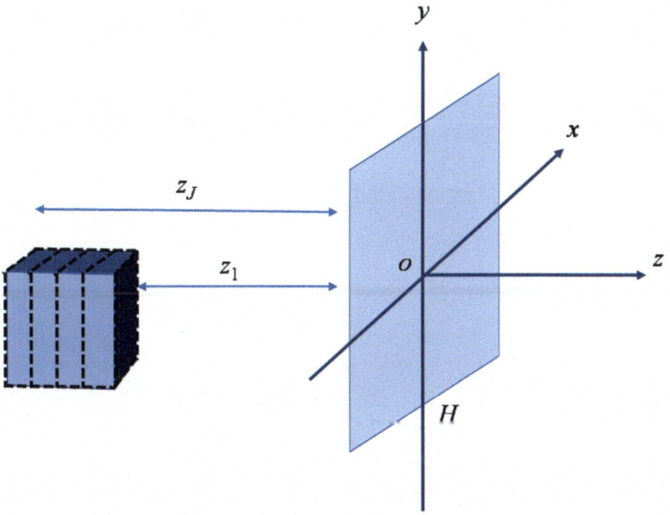

Fig. 5.3 Holographic computing model based on layer

decompose each triangular face of the three-dimensional object to obtain its angular spectrum. The total angular spectrum distribution is obtained by adding the angular spectra of all faces.

Figure 5.4 shows a schematic diagram of the holographic algorithm based on triangular facets. For a standard reference unit triangular surface, through two-dimensional fine transformation, triangular surface with specified orientation and arbitrary size in space can be obtained. Then 3D space rotation and translation can be carried out, and finally angular spectrum distribution can be analyzed.

5.4 Lensless Holographic Imaging

Lensless holographic imaging is an advanced and cutting-edge imaging technique that revolutionizes the way of visualizing and studying objects, from microscopic specimens to macroscopic structures. Lensless imaging system diagram is shown in Fig. 5.5. Unlike traditional imaging methods that rely on lenses and optical components, lensless holographic imaging [7] captures and reconstructs information about objects without the need for physical lenses.

At its core, lensless holographic imaging combines the principles of holography and computational algorithms to achieve high-resolution and non-invasive imaging. The process involves illuminating the object with a coherent light source, such as a laser, and recording the resulting interference pattern, known as a hologram. This

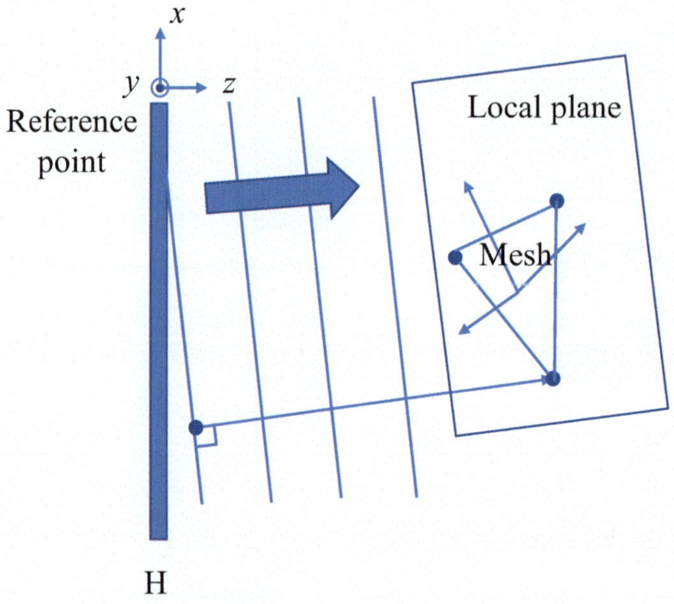

Fig. 5.4 Holographic computing model based on panel

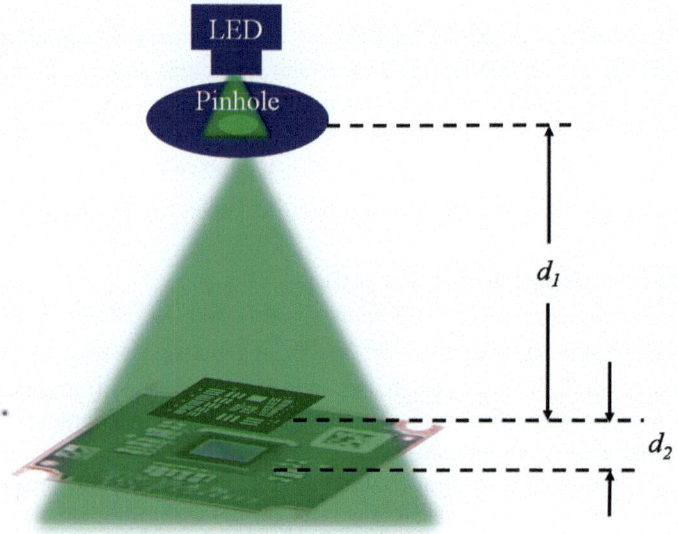

Fig. 5.5 Lensless imaging system diagram

hologram is then digitized and processed using sophisticated computational techniques, extracting phase and amplitude information. Through numerical reconstruction, the digitized hologram is transformed into a complex wavefront that carries a wealth of 3D information about the object's structure and depth. Researchers can then visualize this wavefront to obtain a detailed and accurate 3D representation of the imaged object.

The lensless holographic imaging technique offers numerous advantages over traditional imaging methods. First and foremost, it enables the capture of depth information, providing a realistic and comprehensive 3D view of the object being imaged [8]. Additionally, lensless holography eliminates the need for complex and bulky optical components, simplifying the imaging setup and reducing costs.

One of the most significant benefits of lensless holographic imaging is its applicability in various scientific and industrial fields. In microscopy, it allows researchers to study delicate and dynamic biological samples without the risk of damaging them, offering valuable insights into cellular processes and structures. Furthermore, lensless holographic imaging finds applications in materials science, nanotechnology, and quality control, where high-resolution imaging is crucial. The technique also excels in quantitative phase imaging, providing data about optical thickness and refractive index variations within the imaged object. This capability is particularly advantageous in studying transparent and translucent materials.

As computational power and algorithms continue to advance, lensless holographic imaging is continually improving, offering even higher resolutions and faster data processing. It holds great promise for future developments in imaging technologies and contributes significantly to our understanding of the microscopic world and beyond. With its non-invasive nature, versatility, and impressive capabilities, lensless holographic imaging stands at the forefront of modern imaging methods, shaping the future of scientific research and industrial applications.

5.4.1 Projective Lensless Imaging

Projective lensless imaging is a cutting-edge imaging technique that represents a paradigm shift in the way we capture and reconstruct visual information, as shown in Fig. 5.6. Departing from traditional lens-based imaging systems, this innovative approach leverages computational algorithms and advanced mathematical models to capture and process light data, enabling the creation of high-resolution images without the need for physical lenses.

In conventional imaging, lenses are essential components that focus light rays onto a sensor, allowing for the formation of sharp and detailed images. However, lenses come with limitations, such as size, weight, and optical aberrations, which can constrain imaging systems in various applications. Projective lensless imaging overcomes these limitations by reconstructing images directly from the captured raw light data, offering greater flexibility, compactness, and computational power.

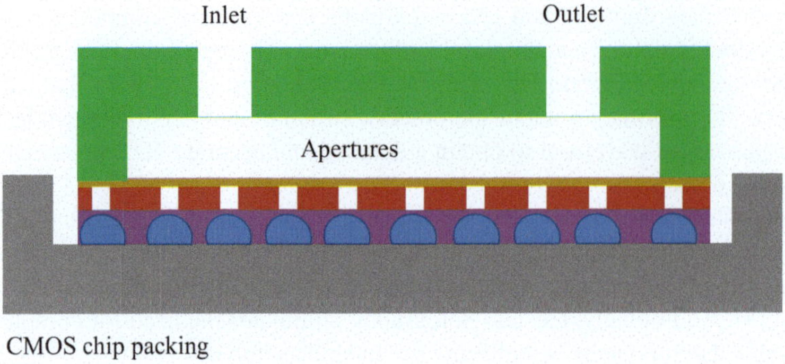

Fig. 5.6 Lensless projection imaging model

In the process of projective lensless imaging, the sample is placed directly on the surface of the sensor, and a light source with limited spatial coherence directly illuminates the phase object. The projection of the sample is directly collected by the image sensor. Additional post-processing steps may be applied to enhance the final image, such as noise reduction, image restoration, and resolution improvement. Projective lensless imaging allows for versatile imaging setups and configurations, enabling customizability and adaptability for various imaging tasks.

5.4.2 Phase Retrieval Algorithm

Phase retrieval algorithms are powerful computational methods used to reconstruct the phase information of a complex-valued signal from its amplitude measurements. In various scientific and engineering applications, signals are often acquired in the form of intensity measurements, but the phase information, which encodes valuable details about the signal's structure and coherence, may not be directly available.

Phase retrieval algorithms aim to recover the missing phase information, enabling the reconstruction of the complete complex-valued signal. These algorithms have significant implications in fields such as optics, astronomy, crystallography, electron microscopy, and diffraction imaging, where the accurate retrieval of phase information is crucial for image reconstruction and analysis.

(1) GS phase retrieval algorithm

The Gerchberg-Saxton (GS) algorithm is an iterative phase retrieval algorithm used to reconstruct the phase of a complex-valued signal from its amplitude measurements. It was proposed independently by Gerchberg and Saxton in the late 1970s and has since become a widely used method in various fields, such as optics, electron microscopy, and astronomy.

GS algorithm is particularly useful when phase information cannot be directly measured, but only the amplitude or intensity of the signal is available, which is a

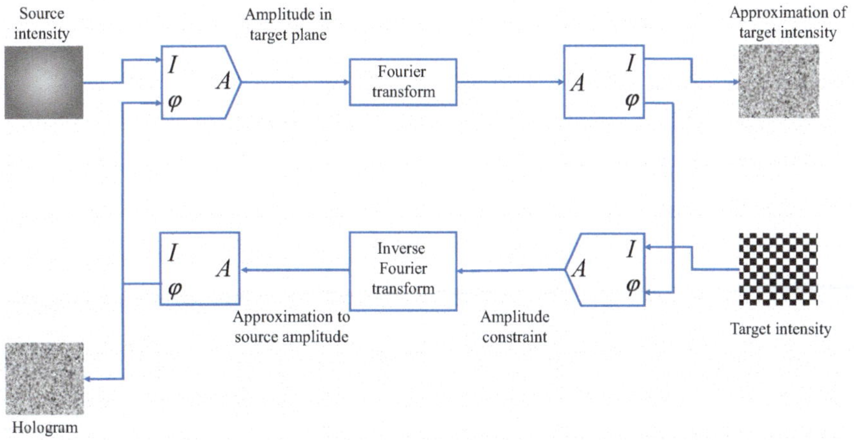

Fig. 5.7 Schematic diagram of GS algorithm

common scenario in many experimental setups. GS algorithm is shown in Fig. 5.7 and GS algorithm involves the following steps:

(1) Initialization: Start with an initial guess for the complex-valued signal's phase. This initial guess can be random or based on any prior knowledge or constraints.
(2) Fourier transform: Compute the Fourier transform of the complex-valued signal, obtaining its amplitude and phase information.
(3) Amplitude constraint: Replace the phase information of the Fourier-transformed signal with the known intensity measurements. This step ensures that the updated signal in the Fourier domain has the correct amplitudes.
(4) Inverse Fourier transform: Perform the inverse Fourier transform to obtain the updated signal in the original domain with the modified phase.
(5) Phase constraint: Replace the amplitude of the updated signal with the original amplitude measurements. This step ensures that the modified signal in the original domain has the correct amplitudes.
(6) Iteration: Repeat steps (2) to (5) for a predetermined number of iterations or until convergence is achieved. In each iteration, the phase information is alternately updated between the Fourier domain and the original domain while preserving the known amplitude constraints.

The GS algorithm continues to alternate between Fourier domain and original domain updates, iteratively refining the phase estimation until the phase converges to a stable solution. It's important to note that the GS algorithm is sensitive to noise and requires sufficient amplitude measurements to accurately reconstruct the phase. Depending on the specific application and the characteristics of the signal, additional constraints or regularization techniques may be incorporated to improve the accuracy and stability of the phase retrieval process.

Despite its simplicity, the GS algorithm has proven to be effective in a wide range of applications, and it serves as a foundational method in the field of phase retrieval and computational imaging. Researchers continue to explore variations and enhancements to this algorithm, contributing to its ongoing utility and relevance in modern scientific and engineering practices.

(2) Multi-height phase retrieval algorithm

The multi-height phase retrieval algorithm is an extension of the standard phase retrieval algorithms, designed to address the limitations of single-distance phase retrieval when dealing with complex and challenging samples. The multi-height phase retrieval algorithm overcomes the phase problem by utilizing multiple intensity measurements taken at different distances from the sample. By incorporating these multiple measurements, the algorithm can iteratively refine the estimation of the sample's phase, leading to more accurate reconstruction of the sample's structure. The multi-height phase retrieval algorithm includes the serial computing model and the parallel computing model.

The single-beam multiple-intensity phase reconstruction (SBMIR) is the serial computing model, as shown in Fig. 5.8. In the SBMIR calculation process, after completing the amplitude constraint of the current recording plane, the updated wavefront is transmitted to the next recording plane. Therefore, SBMIR can be regarded as a multi-distance scanning serial phase recovery algorithm.

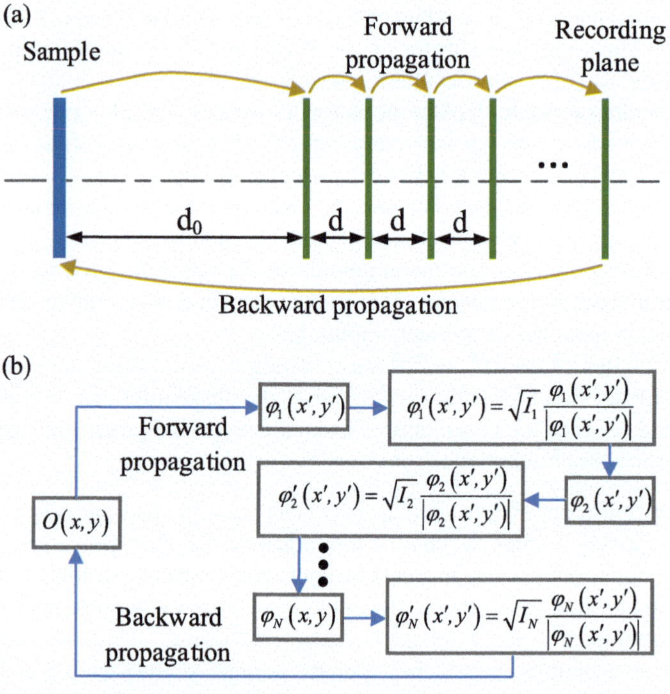

Fig. 5.8 Schematic diagram of SBMIR: **a** schematic diagram of optical system, **b** calculation process diagram

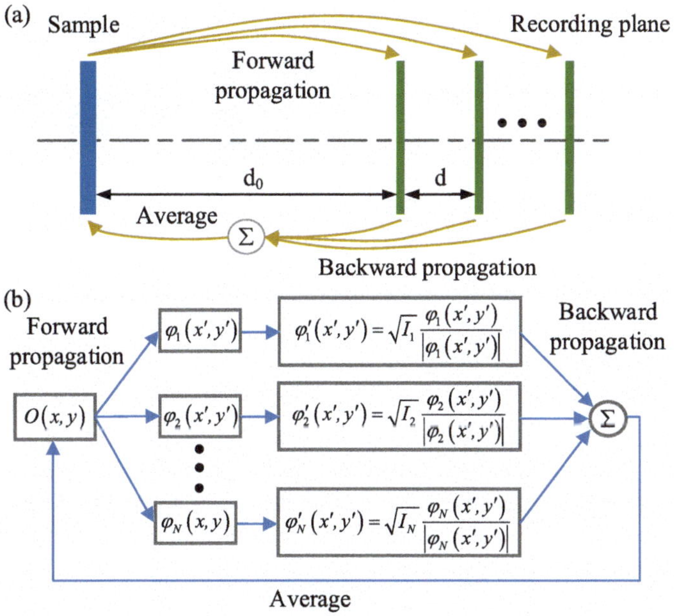

Fig. 5.9 Schematic diagram of APR: **a** schematic diagram of optical system, **b** calculation process diagram

Multi-height scanning phase recovery algorithm in parallel computing mode, amplitude phase retrieval algorithm (APR), as shown in Fig. 5.9. The APR algorithm transfers the updated images from different planes back to the sample plane and averages these images to obtain the final image, thus the algorithm has better robustness. Compared to off-axis holographic imaging, the advantage of multi distance scanning imaging is that it does not require any reference light and is computationally simple.

The key advantage of the multi-height phase retrieval algorithm is that it leverages the additional information from multiple intensity measurements to resolve ambiguities in phase retrieval, leading to enhanced accuracy in reconstructing the object's structure. However, it also demands a higher level of data acquisition and computational complexity, as multiple measurements need to be performed at various distances.

(3) Ptychography phase retrieval algorithm

Ptychography is an advanced imaging technique used for phase retrieval and high-resolution imaging of microscopic samples. The imaging principle of ptychography is shown in Fig. 5.10. It is particularly valuable in the fields of X-ray microscopy, electron microscopy, and coherent diffractive imaging. Ptychography overcomes the limitations of traditional imaging methods by providing a powerful way to reconstruct the phase and amplitude of an object's wavefront from its diffraction patterns.

The ptychography phase retrieval algorithm typically involves the following steps:

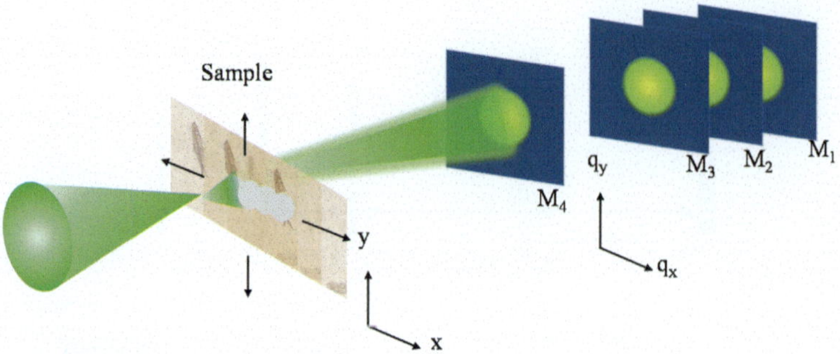

Fig. 5.10 Schematic diagram of ptychography

(1) Scanning: A coherent light source illuminates the sample. Instead of capturing a single diffraction pattern, a series of overlapping diffraction patterns is collected as the illumination is scanned over the sample in a grid-like fashion. These diffraction patterns contain information about the phase and amplitude of the sample's wavefront.
(2) Data acquisition: The recorded diffraction patterns are digitized and stored as intensity measurements. The positions and overlap of the illumination spots during the scanning process are also recorded.
(3) Initial guess: An initial estimate of the sample's complex wavefront (phase and amplitude) is made. This initial guess can be based on prior knowledge, simulations, or other techniques.
(4) Phase retrieval: The ptychography algorithm iteratively refines the estimate of the wavefront by comparing the simulated diffraction patterns, generated using the current estimate, with the measured diffraction patterns. The algorithm uses optimization techniques, such as least-squares fitting or maximum likelihood estimation, to minimize the discrepancy between the simulated and measured diffraction patterns.
(5) Propagation and update: The phase-retrieved wavefront is then propagated back to the sample plane using the known positions and overlaps of the illumination spots during the scanning process. This process updates the estimated sample wavefront.
(6) Iteration: Steps (4) and (5) are repeated iteratively until the phase retrieval converges, and the sample's wavefront is accurately reconstructed.

The ptychography phase retrieval algorithm's power lies in its ability to overcome limitations imposed by the "phase problem" encountered in traditional imaging techniques, where phase information is lost in the intensity-only measurements. By scanning and collecting overlapping diffraction patterns, Ptychography effectively captures the phase information and reconstructs high-resolution images of the sample.

Ptychography has significant applications in various fields, including materials science, biology, and nanotechnology, where it enables researchers to study complex and delicate samples with unprecedented detail and accuracy. Its ability to perform high-resolution imaging without the need for complex and expensive lenses or optics makes it a valuable tool in modern scientific research and imaging technologies.

5.4.3 *Pixel Super-Resolution Holographic Imaging*

Pixel super-resolution holographic imaging is an advanced technique used to enhance the resolution of holographic images beyond the physical limitations of the imaging system. Holography is a powerful imaging method that captures and reconstructs the three-dimensional information of an object using interference patterns of light waves. However, the resolution of holographic images is limited by the pixel size of the image sensor or the detector used to record the hologram.

Pixel super-resolution in holographic imaging addresses this limitation by employing computational techniques to improve the effective resolution of the reconstructed holographic image [9]. The principle of pixel super-resolution is shown in Fig. 5.11. It enhances fine details and allows for better visualization of structures at smaller scales than what the original sensor's pixel size could achieve.

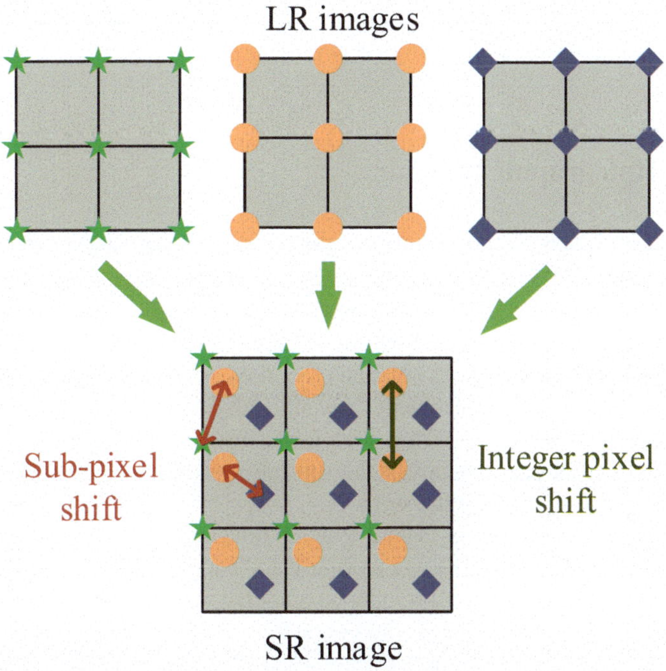

Fig. 5.11 Schematic diagram of sub pixel super-resolution

In pixel super-resolution holographic imaging, the holographic image of the object is captured using a sensor or detector. The sensor has a finite pixel size, which sets the initial resolution limit. To achieve super-resolution, multiple holograms of the same object are acquired with sub-pixel shifts. These sub-pixel shifts can be achieved by physically moving the object or by using digital techniques to shift the hologram digitally. Then, the acquired holograms, along with the sub-pixel shifts, are then processed using computational algorithms to perform phase retrieval. The phase retrieval algorithm aims to reconstruct the complex-valued wavefront of the object from the holographic data. Finally, the information from multiple sub-pixel shifted holograms is combined to reconstruct a super-resolved holographic image. This is achieved through interpolation or deconvolution techniques that effectively enhance the resolution beyond the initial pixel size of the sensor.

Pixel super-resolution holographic imaging has found applications in various fields, including microscopy, biological imaging, and materials science. It allows researchers to visualize and study fine structures and details that were previously beyond the resolution capabilities of the imaging system.

It's essential to note that super-resolution techniques in holographic imaging rely heavily on sophisticated computational algorithms and accurate sub-pixel registration. Moreover, the choice of the specific algorithm and the success of the super-resolution process depend on the quality and quantity of acquired holographic data. Ongoing research in this area continues to refine and develop new algorithms, making pixel super-resolution holographic imaging an exciting and promising frontier in advanced imaging technologies.

5.5 3D Holographic Imaging

In the realm of imaging technology, few advancements have captured the imagination as profoundly as 3D holographic imaging. Far beyond the confines of traditional photography and flat displays, holography has emerged as a revolutionary technique capable of recreating three-dimensional representations of objects with astonishing realism. By harnessing the intricate properties of light waves, 3D holographic imaging transcends the boundaries of conventional visual media, offering viewers an immersive. This additional dimension of data enables holograms to reconstruct the depth and spatial relationships of the subject. Thus the image can be obtained and defy the constraints of the flat surface.

The potential applications of 3D holographic imaging are vast and diverse. From art and entertainment, where holograms dazzle audiences with their visual splendor, to medical imaging, where they provide unparalleled insights into the human body, this technology continues to push the boundaries of innovation. As research and development continue to advance, the prospect of holographic telepresence and interactive data visualization promises to transform the way we communicate and comprehend complex information.

This introduction delves into the captivating world of 3D holographic imaging, exploring its underlying principles, current applications, and the exciting possibilities it holds for the future. As this groundbreaking technology evolves, it has the potential to reshape how we perceive and interact with the visual world, ushering us into an era of unprecedented visual experiences and unlocking new realms of creativity and scientific understanding.

5.5.1 3D Tomographic Holographic Imaging

In the field of medical imaging and scientific research, 3D tomography stands as a powerful and indispensable tool for visualizing the internal structures of objects in three-dimensional detail [10–12]. Tomography, derived from the Greek words "tomos" (meaning "slice") and "graphia" (meaning "describing" or "recording"), refers to the process of capturing cross-sectional images of an object to reconstruct a comprehensive 3D representation. This non-invasive imaging technique has revolutionized the way we diagnose medical conditions, study materials, and explore the complexities of various subjects.

3D tomography employs sophisticated imaging technologies, such as computed tomography (CT), magnetic resonance imaging (MRI), positron emission tomography (PET), and single-photon emission computed tomography (SPECT), to generate a series of 2D slices of the target object from different angles. These slices provide a detailed view of the object's internal structure at various depths, akin to peering through successive layers, like the pages of a book.

Through intricate computational algorithms, these 2D slices are then integrated and reassembled, meticulously stacking them together to construct a complete 3D representation. Figure 5.12 shows schematic diagram of three-dimensional tomography imaging. The result is a volumetric image that grants researchers, medical professionals, and scientists an unparalleled view into the intricate details of the subject of interest.

3D tomographic holographic imaging is an advanced imaging technique that combines holography and tomography to create three-dimensional reconstructions of complex objects or scenes. This method enables the visualization and analysis of the internal structures of objects in three dimensions with high accuracy and precision.

(1) Principles of 3D Tomography

A three-dimensional object can be seen as a set of object points, which can be divided into many layers along the depth direction z-axis. Chromatography is the process of layering along the depth direction at certain intervals, and then reconstructing the object by projecting it along the cross-section of each layer of the three-dimensional object. As shown in Fig. 5.13, the principle of chromatography is to first layer the object in the depth z direction, and then calculate the complex amplitude function of the object light field located on the holographic surface for each depth plane according to the different diffraction distance between each depth plane and the holographic

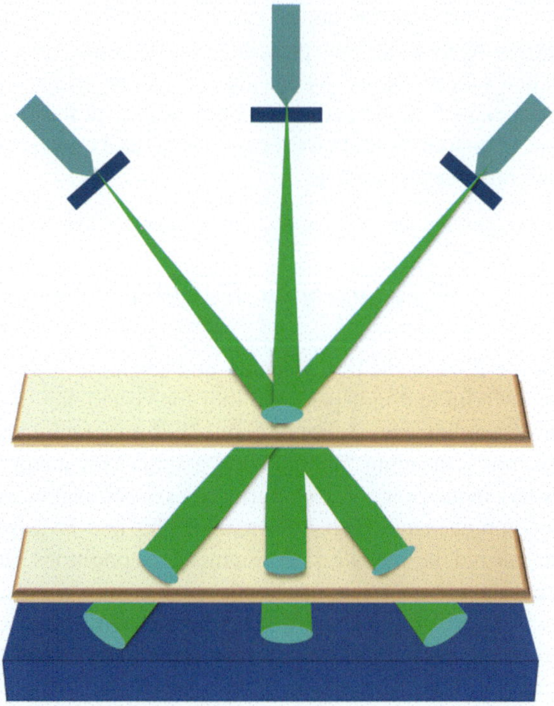

Fig. 5.12 Schematic diagram of three-dimensional tomography imaging

surface. After adding the complex amplitude function of the object light field to the reference light, a hologram is generated to achieve the calculation and reconstruction of three-dimensional objects.

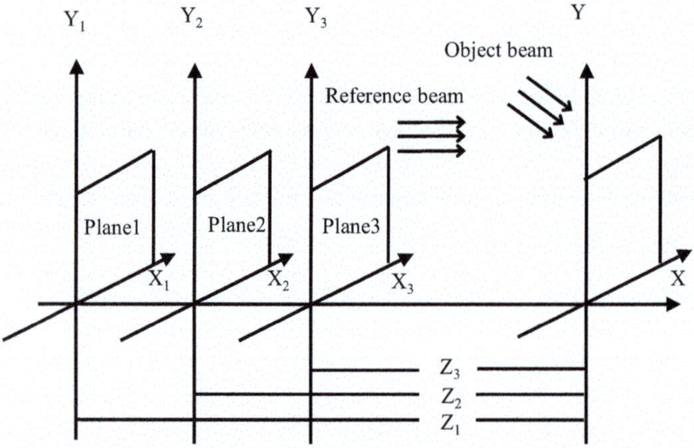

Fig. 5.13 Principles of three-dimensional tomography

Assuming the object is divided into n layers along the depth z direction. The complex amplitudes of all three-dimensional objects are represented as follows:

$$U(x, y, z) = \sum_{i}^{n} U_i(x, y) \tag{5.12}$$

Perform Fresnel diffraction on each layer of the object surface, and the complex amplitude function of the i_{th} layer of the object surface located on the holographic surface is:

$$\begin{aligned} E_i(x, y) = {} & \frac{\exp(\mathrm{i}kz_i)}{\mathrm{i}kz_i} \iint U_i(x_i, y_i) \exp[\mathrm{i}\phi_i(x_i, y_i)] \\ & \exp\left\{ \frac{\mathrm{i}k}{2z_i} \left[(x - x_i)^2 + (y - y_i)^2 \right] \right\} \mathrm{d}x_i \mathrm{d}y_i. \end{aligned} \tag{5.13}$$

The superposition of all layers and surfaces on the holographic surface results in a hologram, and its expression is:

$$E(x, y) = \sum_{i=1}^{n} E_i(x, y) = A_0(x, y) \exp[\mathrm{i}\phi_0(x, y)], \tag{5.14}$$

where $A_0(x, y)$ represents the amplitude of the holographic surface, and $\exp[i\phi_0(x, y)]$ represents the phase of the holographic surface.

(2) Diffraction holographic tomography

Diffraction holographic tomography is an advanced imaging technique that combines holography and tomography to achieve high-resolution, three-dimensional reconstructions of microscopic objects. By utilizing the principles of holography and tomography, this innovative method provides a comprehensive view of the internal structures of specimens, making it invaluable in various scientific and biomedical applications.

Traditional holography records both the intensity and phase of light waves scattered by an object, resulting in a hologram that encodes three-dimensional information about the scene. On the other hand, tomography is a tomographic imaging technique used to capture cross-sectional slices of objects, which are then combined to reconstruct a 3D representation of the subject. In diffraction holographic tomography, coherent light is used to record the diffraction patterns produced by an object from various angles, creating multiple holograms from different perspectives. The holographic data is then subjected to tomographic algorithms, allowing for the reconstruction of the object's 3D structure with exceptional detail and accuracy.

The object of interest is illuminated with coherent light from various angles. Each angle of illumination captures unique diffraction patterns resulting from the object's internal structures. Besides, holograms are also acquired for different orientations of the rotating sample. Then, a holographic recording medium, such as a holographic plate or a digital sensor, captures the diffraction patterns produced by the object at each illumination angle. Sophisticated computational algorithms process the recorded diffraction patterns, incorporating the principles of tomography to reconstruct the 3D internal structure of the object. By combining the data from multiple angles, a comprehensive and high-resolution 3D representation is generated.

As technology continues to evolve, diffraction holographic tomography holds great promise in pushing the boundaries of imaging capabilities. Its ability to visualize the inner workings of microscopic objects with exceptional resolution and non-invasive characteristics opens up new avenues for scientific exploration and discovery. Three-dimensional tomographic reconstructions based on multi-angle illumination lens-free microscope is an advanced imaging technique that allows for the reconstruction of three-dimensional (3D) structures of microscopic objects using a lens-free microscope with multi-angle illumination. This method enables researchers to obtain high-resolution 3D images of samples without the need for traditional lenses, providing a cost-effective and compact solution for various scientific and biomedical applications.

In the process of 3D tomography, the sample of interest is illuminated from multiple angles using coherent light sources, such as LEDs or lasers. Each illumination angle provides different perspectives of the sample, capturing its interactions with light from various directions. As the coherent light interacts with the sample, it undergoes scattering and interference, generating a holographic pattern. This pattern contains information about the sample's 3D structure encoded in its intensity and phase. Then, a digital sensor captures the holographic patterns formed by the multi-angle illumination. This sensor records the interference patterns of the scattered light waves. Advanced algorithms, often based on iterative methods, are applied to process the recorded holographic data and reconstruct the 3D structure of the sample. By combining the information from different angles, a high-resolution 3D image of the object is generated.

The key advantages of using a multi-angle illumination lens-free microscope for 3D tomographic reconstructions include: lens-free imaging, high resolution, wide field of view and label-free imaging. Overall, three-dimensional tomographic reconstructions based on multi-angle illumination lens-free microscopes offer an innovative approach to 3D imaging, enabling researchers to explore microscopic structures in a cost-effective, label-free, and high-resolution manner. The continuous development and integration of advanced computational algorithms further enhance the capabilities and potential applications of this imaging technology.

3D tomographic holographic imaging [12] finds applications in various fields, such as biomedical imaging, materials science, and non-destructive testing. In biomedical imaging, it can be used to study complex biological samples, such as cells, tissues, and microorganisms, providing valuable insights into their internal structures and dynamics. In materials science and non-destructive testing, 3D tomographic holographic imaging enables the examination of internal defects and structural properties of materials, improving our understanding of their characteristics and performance.

It's important to note that 3D tomographic holographic imaging is a computationally intensive process, as it involves handling a large amount of data and implementing sophisticated algorithms for phase retrieval and tomographic reconstruction. Advances in computational power and algorithms have significantly improved the efficiency and accuracy of this imaging technique, making it a valuable tool for non-invasive and high-resolution 3D imaging in various scientific and industrial applications.

5.5.2 Structured Light 3D Holographic Imaging

3D structured light refers to the structured light field with periodic distribution in both the transverse and axial directions. Unlike 2D Structured light, the axial periodicity of 3D structured light can modulate a certain depth of object information in the axial range to improve the axial resolution of the system. When imaging objects in 3D, it is expected to achieve rapid tomography of 3D objects by combining non-coherent digital holography.

Structured light 3D holography is a groundbreaking imaging technique that combines the principles of structured light projection and holography to create highly accurate and realistic three-dimensional representations of objects. By encoding depth information into the interference patterns of projected structured light, this innovative method offers precise visualization of complex 3D structures with remarkable detail and depth perception.

In traditional holography, coherent light is used to record both the intensity and phase information of light waves scattered by an object, resulting in a hologram that stores the object's 3D information. On the other hand, structured light involves projecting carefully designed patterns of light onto an object's surface and observing the deformation of these patterns caused by the object's contours. By combining these techniques, structured light 3D holography leverages their complementary strengths, resulting in a powerful imaging solution.

In the process of structured light 3D Holography, a distinct pattern of light, such as grids, stripes, or sinusoidal waves, is projected onto the object. As the structured light pattern interacts with the object's surface, it generates unique deformations and shadows that contain information about the object's 3D structure. When the structured light pattern interacts with the object, it scatters and creates interference patterns. These patterns carry valuable information about the object's depth and shape. A recording medium, such as a holographic plate or a digital sensor, captures the scattered light patterns. This medium records both the intensity and phase information of the light waves, forming a hologram that encodes the complete 3D representation of the object. Holographic Reconstruction: Using advanced computational algorithms, the recorded hologram is reconstructed. The algorithms decode the interference patterns to retrieve the phase information, enabling the accurate reproduction of the object's 3D structure. The result is a high-fidelity and immersive three-dimensional image.

Structured Light 3D Holography offers several key advantages. The structured light patterns provide precise information about the object's surface, resulting in accurate and detailed 3D reconstructions. And the imaging process is non-invasive and does not require physical contact with the object, making it ideal for fragile or sensitive subjects. The technique captures depth information, enabling viewers to perceive the object's spatial relationships more naturally, enhancing the realism of the 3D image. Structured light 3D holography can be applied to a wide range of objects and materials, from microscopic biological specimens to large-scale industrial components.

(1) Principle of Structured light tomography

In optical imaging systems, the image $U(x, y)$ generated by an object on the CCD acquisition plane can be divided into two parts: the defocused part and the in focus part.

$$U(x, y) = I(x, y) \cdot U_{in}(x, y) + U_{out}(x, y) \tag{5.15}$$

where $U_{out}(x, y)$ represents the defocused part $I(x, y)$ represents structured light. Set the phase-shifting interval to π/4 to carry out Analysis of algorithms of three-step phase-shifting tomographic reconstruction. The images of objects illuminated by Structured light with different initial phases can be expressed as

$$\begin{aligned} U_1(x, y) &= [1 + \cos(2\pi f_0 x + \varphi)] \cdot U_{in}(x, y) + U_{out}(x, y) \\ U_2(x, y) &= \left[1 + \cos\left(2\pi f_0 x + \varphi + \pi/4\right)\right] \cdot U_{in}(x, y) + U_{out}(x, y) \\ U_3(x, y) &= \left[1 + \cos\left(2\pi f_0 x + \varphi + \pi/2\right)\right] \cdot U_{in}(x, y) + U_{out}(x, y) \end{aligned} \tag{5.16}$$

The expression for tomography in focal images is

$$U_{in}(x, y) = \left| U_1(x, y) + U_2(x, y) \cdot \exp\left(i\frac{\pi}{4}\right) + U_3(x, y) \cdot \exp\left(i\frac{\pi}{2}\right) \right| \tag{5.17}$$

Structured light lighting technology has good tomographic ability and high compatibility with other technologies.

(2) Structured light illumination holographic imaging

The difference between 3D Structured light illumination incoherent digital holography and 2D Structured light illumination is that 3D objects contain Structured light cloth within a certain axial depth range. Therefore, when the three-dimensional object hologram is diffracted, propagated and reconstructed without axial scanning, Structured light stripes are included in object plane reconstruction at different depths within a certain depth range, as shown in Fig. 5.14. The tomographic reconstruction in the focal plane can be realized by combining the tomographic reconstruction algorithm. Compared with the two-dimensional Structured light illumination incoherent digital holography, the imaging speed is improved and the error caused by mechanical movement is reduced.

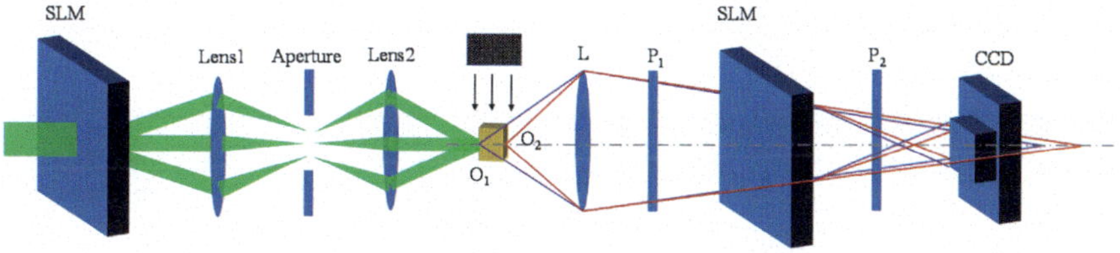

Fig. 5.14 Holographic imaging with 3D structured light illumination

For 3D objects illuminated by 3D Structured light, Structured light can be loaded within a certain axial depth range of the object. The hologram $H(x, y)$ of an object collected on the CCD plane can be expressed as

$$H(x, y) = O(x, y) \otimes PSF(x, y) \quad (5.18)$$

where $PSF(x, y)$ represents the point spread function of the system, and $O(x, y)$ represents the distribution of three-dimensional objects illuminated by structured light as

$$O(x, y) = S(x, y) \cdot I(x, y), \quad (5.19)$$

where $S(x, y)$ represents the distribution function of the object itself, and $I(x, y)$ represents the field distribution of three-dimensional structured light

$$I_{3Dstr}(x, y, z) = \sum_m I(z) J_m(x, y) \quad (5.20)$$

The distribution of holograms collected from the CCD acquisition plane is

$$H(x, y, z) = \left[S(x, y, z) \cdot \sum_m I_m(z) J_m(x, y) \right] \otimes PSF(x, y, z) \quad (5.21)$$

In order to eliminate the DC term and twin images, the phase-shifting technology is needed in the hologram recording process. During the phase-shifting process, the object function convolutions the point source hologram $PSF_i(x, y, z)$ generated when the mask with different phase values is loaded on the SLM, respectively, to obtain four holograms $H_i(x, y, z)$ (i = 1,2,3,4). After the four holograms are phase-shifting, a complex hologram $H_o(x, y, z)$ of a three-dimensional Structured light illuminating object is obtained

$$H_o(x, y, z) = H_1(x, y, z) - H_3(x, y, z) - \mathrm{i}[H_2(x, y, z) - H_4(x, y, z)] \quad (5.22)$$

3D structured light illumination can achieve tomographic imaging while obtaining lateral and axial resolution improvement of objects.

5.6 General Conclusions

Computer-generated holography is a revolutionary imaging technique that has transformed the way we create and visualize three-dimensional holographic content. By leveraging the computational power of computers, CGH enables the generation of highly realistic and immersive holographic images and videos, surpassing the limitations of traditional holography.

CGH relies on advanced algorithms to compute the complex interference patterns necessary for creating holographic representations of complex 3D scenes or objects. These algorithms take into account the intricate wavefronts of light and the inherent diffraction phenomena, allowing for precise and accurate reconstructions.

The benefits of computer-generated holography are vast. It opens up new possibilities for applications in various fields, such as entertainment, education, medical imaging, scientific visualization, and industrial design. With CGH, it is possible to recreate lifelike and interactive 3D visualizations that captivate audiences and enhance the understanding of complex structures and phenomena.

However, CGH also presents challenges, such as computational complexity and the need for high-performance hardware to achieve real-time holographic rendering. As technology advances, these challenges are gradually being overcome, and CGH is becoming more accessible and feasible for a broader range of applications.

The future of computer-generated holography holds great promise. Advancements in hardware and computational power will lead to even more realistic and interactive holographic displays. As the technology matures, it will likely find its way into everyday applications, transforming the way we perceive and interact with visual information.

In conclusion, computer-generated holography represents a significant step forward in holographic imaging, enabling us to create and experience 3D content in ways that were once considered science fiction. Its potential to revolutionize industries and reshape how we visualize the world makes CGH a fascinating and rapidly evolving field of study and innovation. As research and development continue, the impact of computer-generated holography is expected to grow, paving the way for exciting advancements in holographic technology and applications in the years to come.

Acknowledgements This work was supported by the National Natural Science Foundation of China (Nos. 11874132, 61975044, 12074094, 12374271, and 62375065); Interdisciplinary Research Foundation of HIT (No. IR2021237).

References

1. Ouyang W, Xu X, Lu W, Ni Z, Fei H, Shih-Chi C. Ultrafast 3D nanofabrication via digital holography. Nat Commun. 2023;14:1716.

2. Barsi C, Wan W, Fleischer J. Imaging through nonlinear media using digital holography. Nat Photon. 2009;3:211–5.
3. Blinder D, Chlipala M, Kozacki T, Schelkens P. Photorealistic computer generated holography with global illumination and path tracing. Opt Lett. 2021;46(9):2188–91.
4. Blinder D, Nishitsuji T, Schelkens P. Three-dimensional spline-based computer-generated holography. Opt Express. 2023;31(2):3072–82.
5. Blinder D, Nishitsuji T, Kakue T, Shimobaba T, Ito T, Schelkens P. Analytic computation of line-drawn objects in computer generated holography. Opt Express. 2020;28(21):31226–40.
6. Melde K, Mar AG, Qiu T, Fischer P. Holograms for acoustics. Nature. 2016;537(7621):518–22.
7. Boominathan V, Robinson JT, Waller L, Veeraraghavan A. Recent advances in lensless imaging. Optica. 2022; 9(1): 1–16
8. Tian F, Yang W. Learned lensless 3D camera. Opt Express. 2022;30(19):34479–96.
9. Defienne H, Cameron P, Ndagano B, Lyons A, Zhao MJ, Harvey AR, Charbon E, Fleischer JW, Faccio D. Pixel super-resolution with spatially entangled photons. Nat Commun. 2022;13(1):3566.
10. Choi W, Fang-Yen C, Badizadegan K, Oh S, Lue NN, Dasari RR, Feld MS. Tomographic phase microscopy. Nat Methods. 2007;4(9):717–9.
11. Charrière F, Pavillon N, Colomb T, Depeursinge C, Heger TJ, Mitchell EA, Marquet P, Rappaz B. Living specimen tomography by digital holographic microscopy: morphometry of testate amoeba. Opt Express. 2006;14(16):7005–13.
12. Sung Y, Choi W, Badizadegan CK, Dasari RR, Feld MS. Optical diffraction tomography for high resolution live cell imaging. Opt Express. 2009;17(1):266–77.

Chapter 6
Non-line-of-Sight Imaging

Yu Ji and Shutian Liu

Abstract Non-line-of-sight imaging is a technique used in computer vision and remote sensing to reconstruct the shape, appearance, or movement of objects or scenes that are not directly visible from a camera or sensor due to obstacles or occlusions.

6.1 Introduction

Non-line-of-sight imaging is a technique used in computer vision and remote sensing to reconstruct the shape, appearance, or movement of objects or scenes that are not directly visible from a camera or sensor due to obstacles or occlusions. In traditional imaging, the camera captures objects that are within its line of sight, but non-line-of sight imaging extends this capability to recover information about hidden objects or scenes. Non-line-of-sight imaging, also known as "imaging around corners" is shown in Fig. 6.1.

The key principle behind non-line-of-sight imaging is to indirectly capture and interpret the light that interacts with these hidden objects through reflection and scattering. Here is how non-line-of-sight imaging typically works:

1. **Light Propagation**: Light from a source (e.g., a laser or other light-emitting device) is emitted into the scene. This light propagates through the environment and interacts with various surfaces and objects.
2. **Scattering and Reflection**: As light encounters surfaces and objects, it undergoes scattering and reflection. Some of the light may bounce off these surfaces and objects and follow a path that eventually reaches a camera or sensor.
3. **Detection of Time-of-Flight**: In many non-line-of-sight imaging systems, the time-of-flight of the emitted light is measured. This means that the system records

Y. Ji (✉) · S. Liu
School of Physics, Harbin Institute of Technology, Harbin 150001, China
e-mail: JY_0603@163.com

S. Liu
e-mail: stliu@hit.edu.cn

Z. Liu et al. (eds.), *Computational Optical Imaging*, Advances in Optics and Optoelectronics, https://doi.org/10.1007/978-981-97-1455-1_6

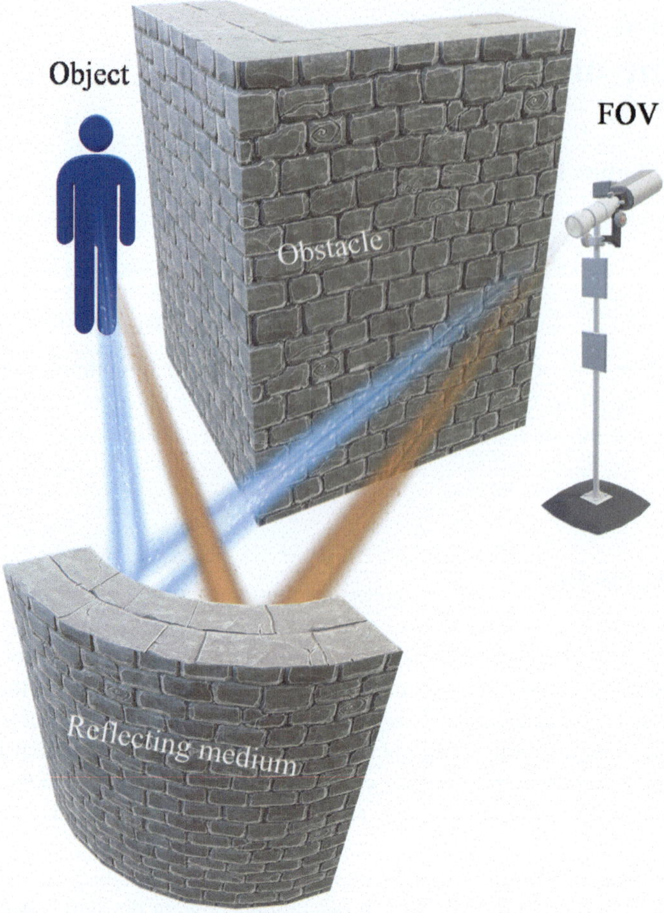

Fig. 6.1 Non-line-of-sight imaging technology

the time it takes for light pulses to travel from the source to the hidden object and back to a detector. The time delay provides information about the distance the light traveled.

4. **Data Collection**: The non-line-of-sight imaging system collects data about the arrival times of photons (individual particles of light) at the detector. This data includes information about the time delays and the number of photons detected.
5. **Computational Reconstruction**: Advanced computational algorithms are employed to process the collected data and reconstruct the hidden object or scene. This reconstruction involves estimating the 3D shape, appearance, or movement of the hidden objects based on the time-of-flight data, the known properties of light, and the geometry of the scene.

6. **Iterative Optimization**: Non-line-of-sight imaging often involves iterative optimization techniques. These algorithms refine the initial reconstruction by iteratively adjusting parameters until the estimated object or scene closely matches the observed data.
7. **Visualization**: The final output of non-line-of-sight imaging is typically a reconstructed image or 3D model of the hidden object or scene. This allows researchers or users to visualize and analyze the previously hidden information.

It's important to note that non-line-of-sight imaging is a challenging and computationally intensive task. Achieving high-quality reconstructions often requires sophisticated hardware and software solutions, as well as an understanding of the underlying physics of light propagation and scattering.

A rectangular coordinate system is established based on the intermediary wall. The coordinates on the intermediary wall are (since the z coordinates of all points on the intermediary wall are 0, so the z coordinates of the points on the intermediary wall are omitted for clearer writing), and the coordinates in the hidden space are (x, y, z). Suppose that the laser is irradiated at the $\boldsymbol{L}(\xi_L, \kappa_L)$ point of the intermediary wall, and is reflected by the intermediary wall and irradiated at the $\boldsymbol{O}(x_O, y_O, z_O)$ point on the surface of the object. The reflectivity of the object at the $\boldsymbol{O}$ point is $\alpha(x_O, y_O, z_O)$, and finally it is diffused by the object back to the detector field $\boldsymbol{D}(\xi_D, \kappa_D)$ on the intermediary wall. Since the distance between the non-line-of-sight imaging system and the intermediary wall is easily detected, the distance between the system and the intermediary wall will not be considered in the subsequent discussion. Assuming that the detected photon distribution is $s(\boldsymbol{D}, \boldsymbol{L}, t)$, the basic forward model of non-line-of-sight imaging is:

$$s(\boldsymbol{L}, \boldsymbol{D}, t) = \iiint_O \alpha(O) \frac{1}{2\pi r_1^2} \frac{1}{2\pi r_2^2} \delta(ct - r_1 - r_2) \boldsymbol{G}(\boldsymbol{L}, \boldsymbol{D}, \boldsymbol{O}) \mathrm{d}\boldsymbol{O}, \tag{6.1}$$

where $r_1 = |\boldsymbol{L} - \boldsymbol{O}| = \sqrt{(\xi_L - x_O)^2 + (\kappa_L - y_O)^2 + z_O^2}$ represents the distance from the laser point to the object. $r_2 = |\boldsymbol{D} - \boldsymbol{O}| = \sqrt{(\xi_D - x_O)^2 + (\kappa_D - y_O)^2 + z_O^2}$ indicates the distance from the object to the detection point. $\boldsymbol{G}(\boldsymbol{L}, \boldsymbol{D}, \boldsymbol{O})$ is a geometric factor [1], which contains the bidirectional reflectance distribution functions (BRDF) of Lambertian body reflection and diffuse reflection, the expression is

$$\boldsymbol{G}(\boldsymbol{L}, \boldsymbol{D}, \boldsymbol{O}) = \cos\langle \boldsymbol{O} - \boldsymbol{L}, \boldsymbol{n}_O \rangle \cos\langle \boldsymbol{O} - \boldsymbol{D}, \boldsymbol{n}_O \rangle, \tag{6.2}$$

where $\boldsymbol{n}_O$ is the normal at a point $\boldsymbol{O}$ on the surface of the object.

After discretization, Eq. (6.1) can be written as

$$S = A_{\mathrm{NLOS}} \boldsymbol{\alpha}, \tag{6.3}$$

where $S \in \mathbb{R}^{n_L n_D n_t}$ is the detection signal after discretization. n_L and n_D are the number of laser points and detection points, respectively. n_t is the number of time grid points after discretization. $\boldsymbol{A}_{\mathrm{NLOS}} \in \mathbb{R}^{n_L n_D n_t \times n_a n_b n_c}$ is the measurement matrix after discretization. n_a, n_b and n_c are the number of lattice points in the discretized hidden space. $\boldsymbol{\alpha} \in \mathbb{R}^{n_a n_b n_c}$ is the reflectance matrix after discretization.

There are several problems in solving Eq. (6.3) directly:

1. In non-line-of-sight imaging, the equation is usually ill-defined and often has multiple solutions, so filtering and regularization are usually required to solve the equation;
2. The measurement matrix $\boldsymbol{A}_{\mathrm{NLOS}}$ is a 5-dimensional matrix (2-dimensional detection points and 3-dimensional hidden space), so as the number of scanning points and discretized lattice points increases, the size of the measurement matrix will become larger and larger, which will lead to high time and space complexity of the calculation, thus reducing the efficiency and robustness of the calculation [2].

6.2 Classifications of Non-Line-Of-Sight Imaging

Non-line-of-sight imaging techniques can be classified into several categories based on their underlying principles and methodologies. Table 6.1 shows the advantages and disadvantages of different optical non-line-of-sight imaging.

6.2.1 Common Classifications

1. **Time-of-Flight (ToF) Imaging**

Single-Pulse ToF: This approach involves emitting a single light pulse and measuring the time it takes to return after interacting with surfaces and objects in the environment.

Multi-Pulse ToF: Multiple light pulses are emitted in sequence, and the system records the time-of-flight for each pulse. This information can be used to reconstruct hidden objects more accurately.

Frequency-Modulated Continuous Wave (FMCW): FMCW ToF systems emit continuous waveforms with varying frequencies. The reflected signals are analyzed to extract distance information.

2. **LIDAR-Based Imaging**

Flash LIDAR: Flash LIDAR systems use an array of detectors to capture the entire scene simultaneously. They are suitable for real-time non-line-of-sight imaging applications.

Table 6.1 Characteristics of different non-line-of-sight imaging methods

Method	Advantage	Disadvantage
Based on time-of-flight measurements	Three-dimensional information can be reconstructed	Information that depends on the intermediary wall
	High resolution	The third scattering signal is weak
	The cost is relatively low	The algorithm has high time complexity
	Less affected by ambient light	High algorithm space complexity
Strength-based measurement	Low cost	Greatly affected by ambient light
		Very dependent on the prior information of the scene
	Data is easy to measure	Low signal-to-noise ratio
		Lack of in-depth information
		Low resolution
Coherence-based measurement	High resolution	Small detection field of view
		Lack of in-depth information
		High device requirements
Inverse projection restoration	Simple algorithm	The recovery effect depends on the scanned area
	High robustness	
	High tolerance for noise and calibration inaccuracies	
Optimization algorithm	Good recovery effect	High computing resource usage
		Long recovery time
Optical cone transformation algorithm	Greatly simplified optical field transmission model	Limited by confocal imaging system
	Greatly improve imaging speed	

Scanning LIDAR: Scanning LIDAR systems employ a rotating or moving laser source and detector to build a 3D point cloud of the hidden scene.

3. **Photonics and Computational Techniques**

Computational Reconstruction: These methods rely heavily on computational algorithms to reconstruct hidden objects or scenes. They often involve optimization techniques and may use machine learning to improve accuracy.

Geometric Model-Based: Some non-line-of-sight imaging techniques are based on geometric models that describe the propagation of light in a scene. These models are used to infer the presence and location of hidden objects.

Matrix Factorization: Matrix factorization techniques can be used to recover hidden object shapes by decomposing the time-of-flight data into meaningful components.

4. **Multi-Bounce and Multi-Pass Imaging**

Multi-Bounce Imaging: This approach considers multiple reflections and bounces of light to estimate the geometry of hidden objects more accurately.

Multi-Pass Imaging: Multiple measurements are taken from different viewpoints or with variations in the lighting conditions to enhance the quality of non-line-of-sight reconstructions.

5. **Indirect Imaging Methods**

Indirect Imaging via Shadows: Some non-line-of-sight imaging techniques work by analyzing the shadows cast by known objects on hidden surfaces. The shape of these shadows can reveal information about the hidden scene.

Indirect Imaging via Acoustic Signals: In scenarios where light cannot penetrate, acoustic signals, such as sound or ultrasound, can be used to gather information and reconstruct hidden objects.

6. **Applications-Specific Approaches**

Non-line-of-sight imaging techniques can be tailored to specific applications, such as medical imaging (e.g., endoscopy), autonomous vehicles (e.g., obstacle detection), surveillance, and archaeology. These applications may have unique requirements and constraints that influence the choice of non-line-of-sight imaging method.

6.2.2 Active Non-Line-Of-Sight Imaging

There are many methods for transient optical transmission in non-line-of-sight field research. Most of the methods are based on creating a pair of virtual light sources and detection points on the visible wall. The pulsed light hits a virtual light source (VS) on the wall, while another transient detector focuses on another virtual detection point (VD) on the wall. Ultra-high temporal resolution light sensors such as fringe cameras can be used to obtain transient responses (statistics of photons in time) about virtual detection points. For VS and VD, transient optical transmission data is obtained by scanning on the wall surface. The figure shown is a general transient non-line-of-sight imaging setup. Once the scene data is obtained, the problem of reconstructing a hidden scene or object can be transformed into a de-multiplexing problem that can use a recovery algorithm that satisfies the constraints of the scene data. Further use of scene prior and regularization methods can improve the robustness of reconstruction.

Figure 6.2 shows the top view of transient non-line-of-sight imaging based on time resolution. The incident laser has a certain pulse frequency and pulse width (100 fs–100 ps), and the laser shines on the VS on the visible wall. The light is

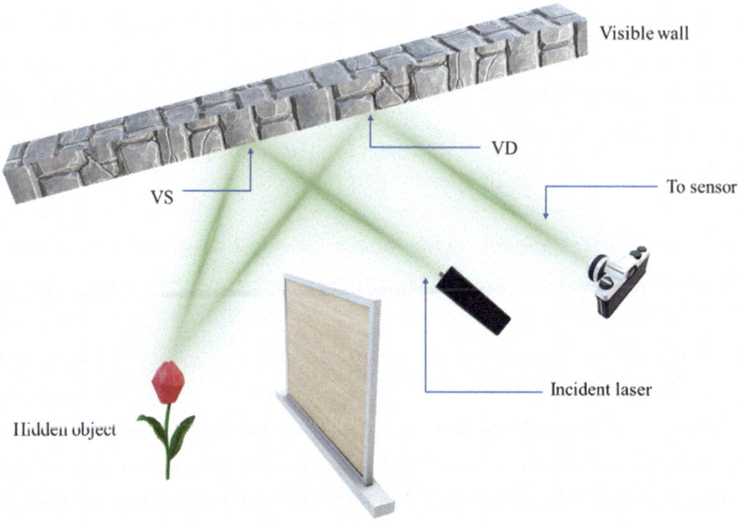

Fig. 6.2 Transient non-line-of-sight imaging setup diagram

scattered from the wall, some of the light is scattered to the hidden area and hits the hidden object, and some of the light on the hidden object will hit the VD on the visible wall again, and finally be received by the light sensor.

In the detection process of the confocal non-line-of-sight imaging system, it is assumed that the laser intensity emitted by the system is I_0. The distance between the system and the intermediary wall is R, and the distance between the intermediary wall and the hidden object is r. The reflectance of the medium wall and object is α_{wall} and α_{ob}, respectively. The area of the hidden object is S_{ob}, the field of view area of the detector on the medium wall is S_{FOV}, and the receiving aperture of the detector is S_{det}. The overall detection efficiency of the system is η, then the echo signal intensity I detected by the system is

$$I = I_0 \alpha_{wall} \frac{\alpha_{ob} S_{ob}}{2\pi r^2} \frac{\alpha_{wall} S_{FOV}}{2\pi r^2} \frac{\eta_{sys} S_{\text{det}}}{2\pi R^2}. \tag{6.4}$$

In the demonstration experiment in the laboratory, it is assumed that the reflectivity of the object and the intermediate wall are $\alpha_{\text{wall}} = \alpha_{\text{ob}} = 1$, the distance from the system to the wall and the distance from the wall to the object $R = r = 1$ m, the overall efficiency of the system is $\eta = 10\%$, the surface area of the object is $S_{\text{ob}} = 1\text{ m} \times 1\text{ m}$, and the diameter of the field of view of the detector on the wall is $d_{\text{FOV}} = 1$ cm, the sensor surface diameter is $d_{\text{det}} = 62.5\ \mu\text{m}$. According to Eq. (6.4), the attenuation of a non-line-of-sight imaging system is $-10 \times \log(I/I_0) = 148$ dB. At the same distance, the attenuation of the direct imaging system is 97 dB. Because non-line-of-sight imaging has more distance to the fourth power attenuation of hidden space than direct imaging, the attenuation of the detection signal in non-line-of-sight imaging is much larger than that in direct imaging.

On the other hand, the light signal directly scattered back by the intermediary wall is relatively strong, which will saturate the single photon detector, so if all the photons in this part are received by the detector, the single photon detector will reach saturation, which will cause the detector to fail to detect enough tertiary reflected photons.

Therefore, the first challenge facing the remote non-line-of-sight imaging system is how to improve the intensity of the target object back wave signal.

Most of the current confocal non-line-of-sight imaging systems use a beam splitter (BS) to separate the receiving light path from the transmitting light path, as shown in Fig. 6.3. In this system, the transmitting and receiving optical paths overlap for a long time. Because each end face inside the system, such as the optical fiber end face, the galvanometer end face, etc., is not a completely ideal mirror, so the end reflection will occur at the coincidence of these optical paths. Second. Since the emitting laser is also not an ideal Gaussian beam, the edge part of the emitting laser will shine on the device inside the optical path, such as the optical frame, resulting in the scattering of the laser. In this traditional non-line-of-sight imaging system using a beam splitter, the noise caused by the end reflection and internal scattering of the system is much stronger than the signal of the non-line-of-sight target. In addition, the advantages and disadvantages of different optical non-line-of-sight systems are given in Table 6.2.

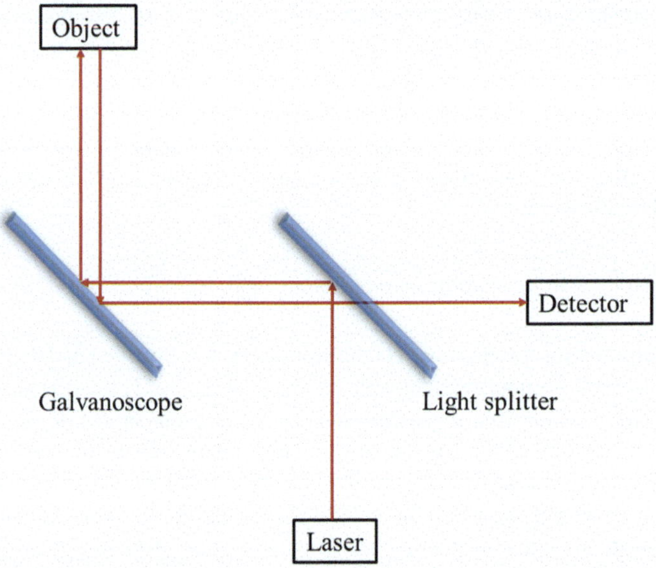

Fig. 6.3 Optical path diagram of a non-line-of-sight imaging system using beam splitter

Table 6.2 Characteristics of different non-line-of-sight imaging systems

Imaging system	Advantage	Disadvantage
Stripe camera as detector	Time resolution up to 2 ps	Large size
	High sensitivity	Expensive
		Low integration
The photonic mixing device acts as the detector	High integration	Low sensitivity
	Cheap	Low sensitivity
Gated enhanced charge-coupled device camera as detector	High sensitivity	Low time resolution
	Slightly cheaper	General ns magnitude
		Large volume
Geiger avalanche photodiode acts as a detector	High sensitivity (single photon level)	Abundant information collection
	High-time resolution (tens of ps)	
	Cheap	
	Small size	Complex algorithm
	Lightweight	
	Arrayable integration	

Due to the short duration of each pulse of the pulsed light, the echo signal collected each time can be received before the next pulse light transmission, which can simplify the overall optical path so that the VS and the VD on the wall are merged into the same point so that the transmit and return signals are in a time-sharing serial state [3].

Using confocal imaging Settings, a three-dimensional transient image $\tau(x', y', t)$ can be reconstructed into a three-dimensional discrete voxel $\rho(x, y, z)$, representing each point in the scene. In the continuous domain, the relationship between the 3D reconstructed discrete voxels and the 3D transient image $\tau(x', y', t)$ can be expressed by the following formula

$$\tau\left(x', y', t\right) = \iiint_{\Omega} \frac{1}{r^4}\rho(x, y, z)\delta\left(2\sqrt{(x'-x)^2+(y'-y)^2+z^2}-tc\right)\mathrm{d}x\mathrm{d}y\mathrm{d}z, \tag{6.5}$$

where c is the velocity of light. Equation (6.5) represents the transient measurement of the captured photon stream intensity value at the point (x', y') and time sampled $\tau(x', y', t)$, starting time calculated as the pulse laser begins scattering from the same point. Ω represents the three-dimensional half-space of the hidden scene located on one side of the wall ($z > 0$). The $x^2 + y^2 + z^2 - (tc/2)^2$ inside of the Dirac function represents a four-dimensional spacetime hypercone surface. The superconical surface models the propagation of light from the visible wall to the hidden scene and then to the visible wall. The distance function can be expressed in various forms $r = \sqrt{(x'-x)^2 + (y'-y)^2 + z^2} = tc/2$. Represents the distance the light travels at time t, so that only the $1/r^4$ terms that depend on time can be taken out of the integral expression. Similarly, the Eq. (6.5) can be rewritten using the equivalent substitution $z = \sqrt{u}$, $dz/du = 1/(2\sqrt{u})$ and $v = (tc/2)^2$ as

$$\underbrace{v^{\frac{3}{2}}\tau\left(x', y', 2\sqrt{v}/c\right)}_{R_t\{\tau\}(x',y',v)} = \iiint_{\Omega} \underbrace{\frac{1}{2\sqrt{u}}\rho(x, y, \sqrt{u})}_{R_z\{\rho\}(x,y,u)} \times \underbrace{\delta\left(\left(x' - x\right)^2 + \left(y' - y\right)^2 + z + u - v\right)}_{h(x'-x,y'-y,v-u)} \mathrm{d}x\mathrm{d}y\mathrm{d}u. \tag{6.6}$$

Equation (6.6) can be expressed as a three-dimensional convolution process $R_t\{\tau\} = h * R_z\{\rho\}$, where $*$ represents the three-dimensional convolution operator. h is a known three-dimensional point diffusion function (PSF), which describes the transient response of a single scattering point. $R_z\{\cdot\}$ is paired along the z-axis, resampled and multiplied by the attenuation coefficient $1/2\sqrt{u}$. $R_t\{\cdot\}$ resamples τ along the time axis and multiplies the coefficient $v^{3/2}$. Resampling transient images and reconstructed voxels can be expressed as $\tau_t = R_t\{\tau\}$ and $\rho_z = R_z\{\rho\}$, respectively. The non-line-of-sight imaging model in confocal Settings can be rewritten into a three-dimensional convolution operation:

$$\tau_t = h * \rho. \tag{6.7}$$

In discrete space can be rewritten into matrix form

$$\tau = R_t^{-1} F^{-1} \hat{H} F R_z \rho, \tag{6.8}$$

where τ and ρ are in vector form, the matrix F represents the three-dimensional discrete Fourier transform, and $\hat{H}$ is the diagonal matrix of the point diffusion function h in the frequency domain. Starting from the transient image τ, the 3D voxel ρ_* can be reconstructed through the 3D deconvolution (using the Wiener filter) process

$$\rho_* = R_z^{-1} F^{-1} \left[\frac{\hat{H}^*}{|\hat{H}|^2 + \frac{1}{\alpha}} \right] F R_t \tau, \tag{6.9}$$

where α represents the sensitivity parameter of the reverse point spread function to noise that can be manually adjusted.

The module diagram of the remote non-line-of-sight imaging system is shown in Fig. 6.4. The whole system is composed of a receiving and luminescence circuit, laser, detector, time-correlated single photon counting (TCSPC) module, synchronization signal generator and algorithm processing module.

The confocal imaging model will be further complicated by the introduction of the specular reflection term, and the imaging process of the whole non-visual field can be perfected by combining the ideal diffuse reflection of the intermediate

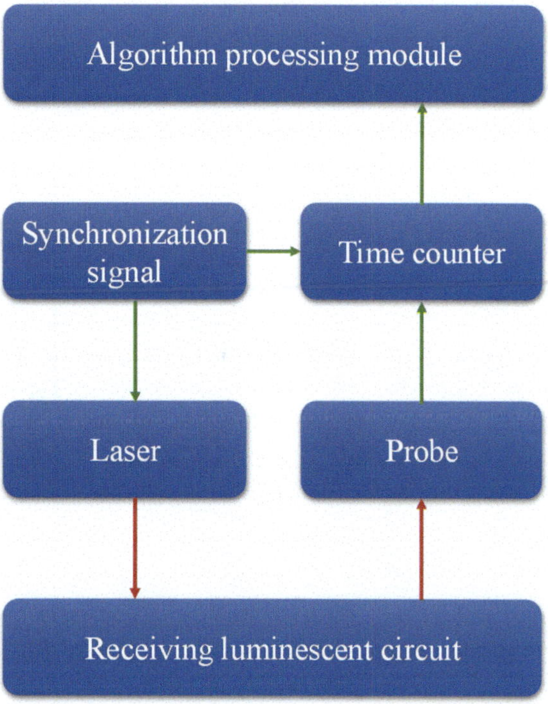

Fig. 6.4 Remote non-line-of-sight imaging system module diagram

reflector and specular reflection. One of the most commonly used methods to model the reflection characteristics of materials is the bidirectional reflection distribution function (BRDF). The definition diagram of its model is shown in Fig. 6.5. Its specific function definition [4] is as follows

$$f_r(\mathbf{x}, \boldsymbol{\omega}_i, \boldsymbol{\omega}_o) = \frac{dL_o(\mathbf{x}, \boldsymbol{\omega}_o)}{dE(\mathbf{x}, \boldsymbol{\omega}_i)} = \frac{dL_o(\mathbf{x}, \boldsymbol{\omega}_o)}{L_i(\mathbf{x}, \boldsymbol{\omega}_i)\cos\theta_i d\boldsymbol{\omega}_i}. \tag{6.10}$$

In Eq. (6.10), $E(\mathbf{x}, \boldsymbol{\omega}_i)$ represents the irradiance received at point $\mathbf{x}$, whose unit is W/m^2. $L_o(\mathbf{x}, \boldsymbol{\omega}_o)$ represents directional radiance along $\boldsymbol{\omega}_o$ at the illumination point $\mathbf{x}$, whose unit is $W/m^2 \cdot sr$.

The Lambert lighting model is suitable for modeling the reflection characteristics of all kinds of materials. The reflection intensity of the light is only related to the intensity of the incident light, the Angle between the incident light vector and the normal vector of the reflector and the surface properties of the reflector. The ambient light term is expressed as:

$$I_{\text{Ambdiff}} = K_a I_a, \tag{6.11}$$

where K_a, $(0 < K_a < 1)$ is the diffuse reflection coefficient of a given material against ambient light. I_a is ambient light intensity.

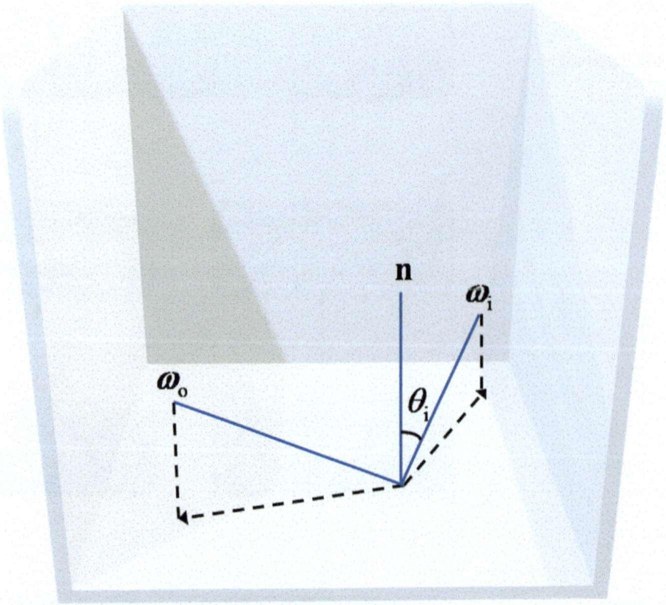

Fig. 6.5 BRDF model definition diagram

The specific calculation formula of diffuse reflection term is as follows

$$I_{\text{Diffuse}} = K_d \cdot I_d \cdot \cos\theta,$$

$$\cos\theta = N \cdot L, \tag{6.12}$$

where I_{Diffuse} represents the reflected light intensity value, K_d represents the attribute of the reflection intensity value of the surface of the diffuse reflection object, the unit normal vector of the incident light is L, I_d represents the intensity value of the incident light, and N represents the unit normal vector of the diffuse surface at the irradiation point of the incident light. $\cos\theta$ is the dot product of L and N, which is the cosine of the Angle between the two vectors. Using the Lambert model, you can only simulate the reflection effect of the ideal diffuse surface material, and you cannot deal with the reflection effect of the mirror highlight material. Therefore, the Lambert lighting model can be written as:

$$I = I_{\text{Ambdiff}} + I_{\text{Diffuse}} = K_a I_a + K_d \cdot I_d \cdot (N \cdot L). \tag{6.13}$$

Phong illumination model [5], as an empirical model in computational imaging, has perfected the defect that the Lambert model cannot deal with the specular highlight reflection effect. The illumination model consists of three structural components

$$I = I_{\text{Ambient}} + I_{\text{Diffuse}} + I_{\text{Specular}} \tag{6.14}$$

The global light quantity I is obtained by the accumulation of I_{Ambient} (ambient light component), I_{Diffuse} (diffuse reflection component) and I_{Specular} (specular reflection component). The diffuse reflection component can be equivalent to the diffuse reflection quantity in the Lambert model. The specular reflection component can be defined by the following calculation:

$$I_{\text{Specular}} = K_S I_d (R \cdot V)^{n_{Gloss}},$$

$$R = 2(L \cdot N)N - L, \tag{6.15}$$

where K_S represents the specular reflection coefficient of the material, I_d represents the intensity value of the incident light, L represents the unit normal vector of the incident light, N represents the unit normal vector of the diffuse surface at the irradiation point of the incident light, V represents the observation direction vector, R represents the specular reflection light vector of the incident light L at the incident point. $R \cdot V$ is the cosine value of the specular reflected light vector of incident light L and the cosine of the angle in the direction V. n_{Gloss} is a high concentrating index. The larger the n_{Gloss}, the higher the light concentration, and the smaller and brighter the light spots formed by the reflecting surface; conversely, the smaller the n_{Gloss}, the smaller the light concentration, indicating that the rougher the reflecting surface, the larger and darker the light spots formed by the reflecting surface. The global illumination of the Phong model can be expressed as

$$I = K_a I_a + K_d \cdot I_d \cdot (N \cdot L) + K_s I_d ((2(L \cdot N)N - L) \cdot V)^{n_{Gloss}}, \tag{6.16}$$

Using the improved Blinn-Phong lighting model [6], it is possible to reduce vector computation time and improve rendering speed, with components defined as follows

$$I_{\text{Specular}} = K_S I_d (N \cdot H)^{n_{Gloss}},$$

$$H = \frac{L + V}{|L + V|}. \tag{6.17}$$

When $N \cdot H$ is used to replace $R \cdot V$ in the Phong model, its advantage over the Phong model is that it takes much less time to compute H half angle vector than to compute R vector, but at the same time, H vector carries less spatial information than R vector.

The silicon-based ultrafast detector is an emerging technology that provides a more powerful solution for ultrafast time-resolved imaging and is the most promising device in the field of non-line-of-sight imaging applications. Single photon avalanche diodes (SPAD) are photodetectors that can detect and time-stamp a single photon event on a time scale of tens of picoseconds. Avalanche photodiodes operate in a similar principle to single-photon avalanche diodes and provide improved photon detection throughput, but at the expense of timing resolution and sensitivity. Recently, other techniques, such as silicon photomultiplier tubes or superconducting nanowire single-photon detectors, have begun to push the extremes of time-resolved photon counting and may prove useful for general imaging applications.

In the following, an image formation model for a single photon avalanche diode (adapted from O'Toole et al. [8]) is described. Suppose a single-photon avalanche diode is synchronized with an ultrafine lighting source such as a pulsed laser, and the light pulses are emitted into the scene. Then, some of the light is scattered back to the detector.

The distribution of photon g with time is used to describe the laser pulse. The photon flux r incident on the detector during the discrete time interval t is as follows

$$r[t] = (\tau * g)[t] + a[t], \tag{6.18}$$

where $*$ is a one-dimensional convolution operator. $a[t]$ is ambient photon flux. τ is the time impulse response of the scene. The impulse response simulates photon travel time, albedo, and any other optical transport effects. For scenarios that exhibit only direct light transmission, that is, when light bounces off an object exactly once before returning to the detector, τ is the discrete Dirac function. In the presence of global optical transmission, (such as caustic, mutual reflection, or subsurface scattering), τ simulates the response of the scene to infinitely short laser pulses. Note that the t in this symbol represents the relative time when the laser pulse is emitted.

An ideal photon counter that processes an incoming photon stream within a specific time window samples the rate function λ as

$$\lambda[t] = \eta(r * f)[t] + d, \tag{6.19}$$

where $\eta \in [0, 1]$ is the photon detection probability, which is composed of the quantum efficiency of a single photon avalanche diode and the avalanche probability. The dark count rate d [Hz] represents the number of false events detected in the absence of photons. Finally, f is the jitter of the detector, the uncertainty in the modeling timestamp mechanism. Jitter in the order of tens to hundreds of ps is common for modern single-photon avalanche diodes.

Photons hitting a single photon avalanche diode produce an avalanche of electrons time-stamped by a time digitizer. However, after an event is detected, the single-photon avalanche diode must be reset (or extinguished) before another event can be recorded. The dead zone time is usually in the order of 100 ns. Assuming that the length of the emitted laser pulse is significantly less than the dead time of the single-photon avalanche diode, at most one of the photons in that pulse can trigger the event. Note that this is not necessarily the first photon to arrive. Therefore, the presence or absence of an event detected within a short window is a Bernoulli test.

In general, the Bernoulli test is repeated N times by firing N laser pulses at a rate that leaves enough time for the single-photon avalanche diode to reset between pulses. For example, a low-level counting mechanism implemented by a field programmable gate array accumulates events for a certain "exposure time" and stores the number of detected events for each time bin in a histogram h. Shin et al. [9] recently showed that the probability of detecting a certain number of events in a histogram box can be modeled as a Poisson distribution

$$h[t] \sim \mathcal{P}(N\lambda[t]) \tag{6.20}$$

Note, however, that this is an approximation that models a single photon event as independent and assumes a low photon flux. In many cases, this low-flux assumption is explicitly enforced by using attenuated laser power capture measurements so that the number of detected photons does not exceed 5% of the number of emitted laser pulses. In the case of high-throughput imaging, the dependencies between the detection boxes can also be explicitly modeled and corrected [9].

6.2.3 *Polarized Non-Line-Of-Sight Imaging*

Because there is no active light source in the passive non-line-of-sight imaging algorithm, the information obtained by the imaging device simply depends on the self-luminous or reflected ambient light of the target scene. The reflection information obtained by imaging equipment will lead to a sharp decrease in signal-to-noise ratio due to the interference of ambient light, which brings great difficulties to imaging work. At present, the research of passive non-line-of-sight imaging is mostly based on the condition of steady-state optical transmission, and the information obtained by each imaging unit of the imaging equipment is equivalent to the integral in the time dimension and space dimension. Compared with the active method, the information in the time dimension is lost. In recent years, the research of passive non-line-of-sight imaging has begun to obtain corresponding results step by step. According to the different imaging principles, researchers have successfully restored the image of the target scene by using the intensity, coherence and polarization of the reflected light of the target scene, which lays a foundation for the follow-up passive imaging method. In addition, some researchers have proposed some new non-line-of-sight imaging methods, such as non-line-of-sight imaging (infrared imaging) and so on. According to the achievements of passive non-line-of-sight research in recent years, this paper introduces the current research status of imaging principles and imaging equipment, analyzes the difficulties encountered, and makes corresponding prospects.

The digital holography and statistical averaging methods are used to recover the three-dimensional object from the scattering field in Fig. 6.6. The shadowing of the object image is due to the loss of phase information caused by the scattering caused by the diffuser. Holography is a technique that can recover the phase information lost due to intensity recording in traditional photography. Only a continuous laser and a traditional camera are needed to perform not only remote observation, but also remote deformation measurement of objects hidden behind a strong diffuser or around a corner far from the recording position.

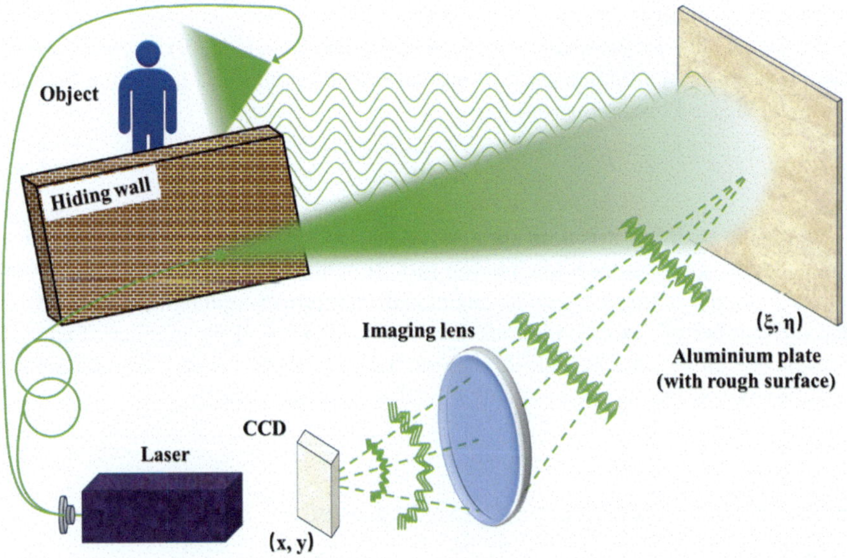

Fig. 6.6 Experimental schematic of 3D object retrieval from scattering field by digital holography and statistical averaging method

Non-line-of-sight imaging is performed in real-time using diffuse light in Fig. 6.7. Due to the "memory effect" of speckle correlation, high-resolution images of speckle light taken by a standard camera can encode sufficient information to achieve corner imaging with opaque layers and limited diffraction resolution. Using spatially incoherent light and a variety of samples (from white paint to dynamic biological samples), experimental demonstration of single imaging through scattering media and surrounding corners, without wavefront shaping, time-gating, or interferometric detection, requires only a common camera to image through an opaque scattering layer using speckle correlation.

A novel model of speckle formation and motion in which the sensor and light source observe the object only indirectly through a diffuse wall, the wall acting as a virtual bare mirror (no lens) sensor, and speckle is a high-frequency "noise-like" intensity pattern created when coherent light is reflected off an optically rough surface. Based on the statistical and geometric characteristics of speckle images, a novel clustering algorithm is developed, which enables the recognition of the motion paths of multiple independently moving non-visual scene objects. For the first time, off-the-shelf imaging components are used with extremely high precision (<10 microns) to track individual tracks of multiple objects around corners. Micronix VT-21L, a 2D electric moving platform with high precision (error less than 1 micron), was used to move the target object along a small and complex track. The imaging system was placed as shown in Fig. 6.8.

Passive non-field-of-view imaging based on the infrared band uses thermal imaging equipment to realize non-view imaging in an invisible light band, with

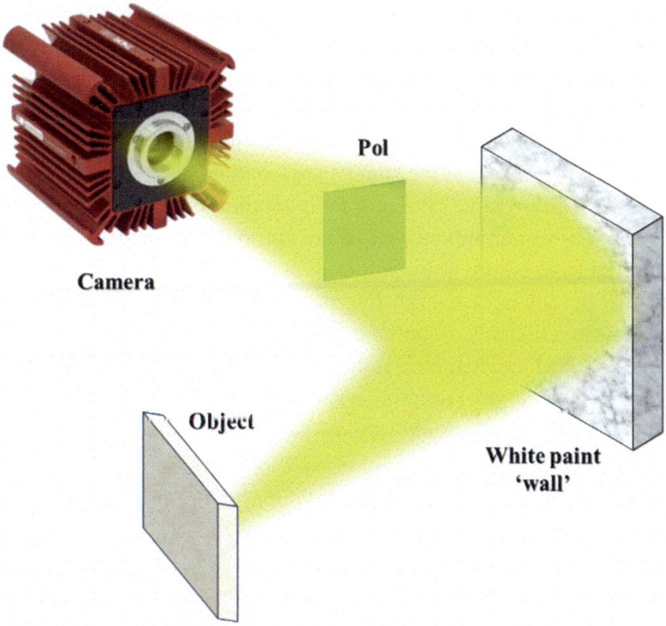

Fig. 6.7 A schematic diagram of a single imaging at the corner using backscattered light

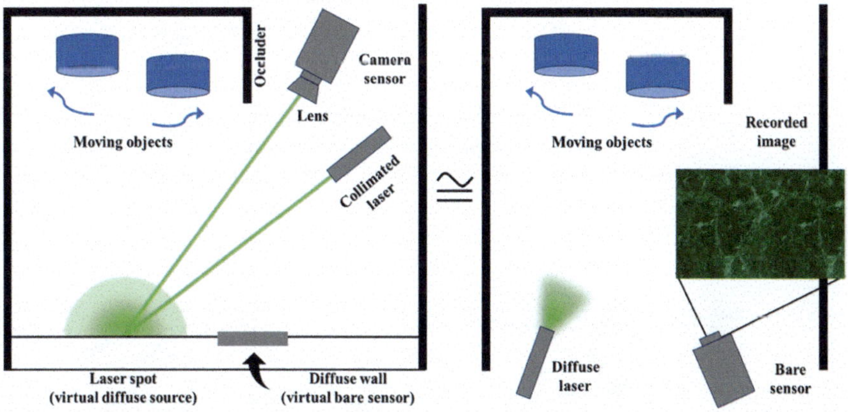

Fig. 6.8 Speckle imaging system

the help of the stronger specular reflectance in this band than in the visible band, and the opaque material of visible light can be penetrated in infrared band. As well as a series of properties such as blackbody radiation, non-view scene objects can be regarded as self-luminescence, passive non-view imaging is extended to invisible light band. It also expands a new direction for the development of non-view imaging. Divitt et al. realized speckle imaging of objects behind the occlusion and at corners using bispectrum and phase retrieval methods in the Mid-infrared Wave Infra band red Range (MWIR) [10]. MWIR imaging takes advantage of the ability of blackbody radiation self-luminescence above absolute zero, in which case, the object being imaged does not need an external illumination source. And it can be imaged at a higher resolution at the same aperture than at a longer wavelength. This paper also successfully realizes the non-field-of-view imaging under the occlusion of materials (such as semiconductors) that are completely opaque to visible light but can be penetrated by medium-wave infrared, as shown in Fig. 6.9.

Spatially incoherent light can be interpreted as broadband light emitted by an object, propagating a distance u, and then hitting a scattering surface. When light is scattered from the surface, it travels a distance v and is then absorbed by the array detector. After paraxial approximation, a speckle pattern is expected to be produced on the detector, where each speckle takes the shape of the object. Therefore, the

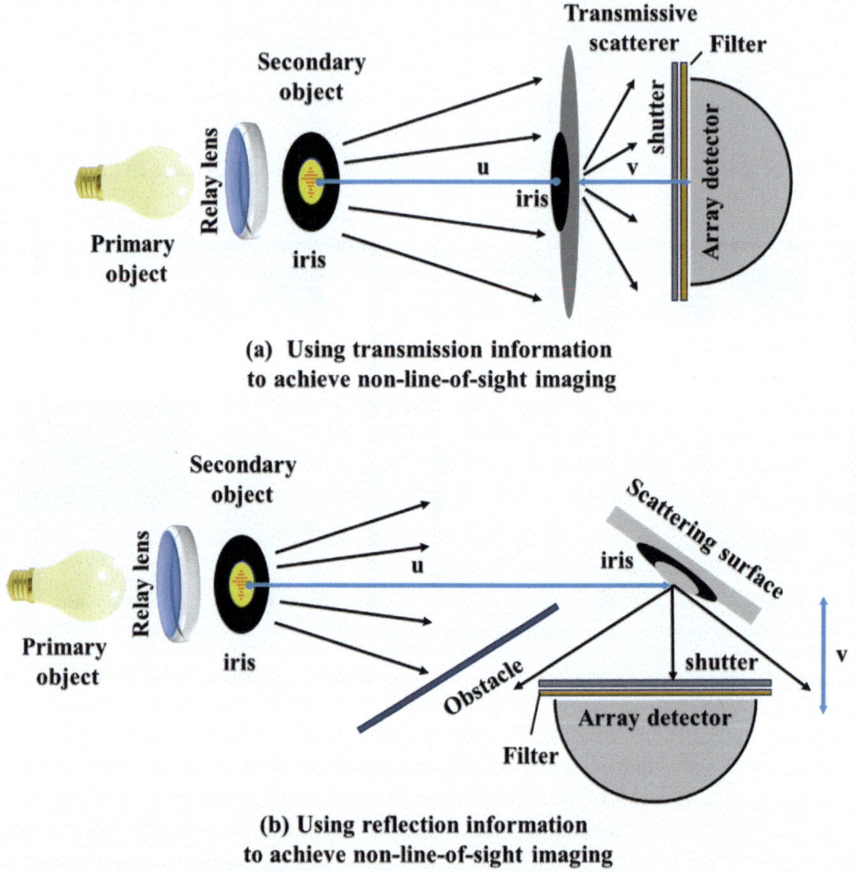

Fig. 6.9 MWIR non-line-of-sight imaging

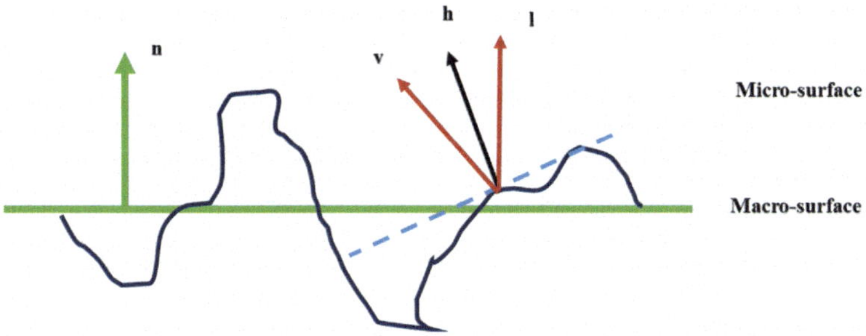

Fig. 6.10 Cook-torrance model of BRDF

convolution between the object intensity A and the intensity point diffusion function G is the image intensity at each wavelength λ. Then, zoom by the magnification factor $M = v/u$:

$$I(x, y; \lambda) = [A(\xi, \eta; \lambda) * G(\xi, \eta; \lambda)]|_{\xi=\frac{u}{v}x, \eta=\frac{u}{v}y} + B(x, y; \lambda), \tag{6.21}$$

where I is the 2-dimensional intensity distribution, $*$ indicates 2-dimensional spatial convolution, and B represents background intensity noise. The function G represents the image intensity that would be found in the image plane given a point-like object. An image I that approximately obeys Eq. (6.21) can be used to extract the object intensity A through a variety of methods including.

The bidirectional reflectance distribution function (BRDF) represents the reflection characteristics of the medium and reflects the energy relationship between the incident optical and the outgoing light in Fig. 6.10. Describe the Cook-Torrance model of BRDF:

$$f(l, \mathrm{v}, n) = \frac{D(h)F(\mathrm{v}, h)O(l, \mathrm{v}, h)}{4(n \cdot l)(n \cdot \mathrm{v})}, \tag{6.22}$$

where l represents the unit vector of the incident direction of light; v is the unit vector of the exiting ray; n is the reflection indicating macroscopic surface normal vector; h represents the normal vector of the spectator surface element, which is the unit vector of the Angle bisector between the incident direction and the exit direction of the light. D is the normal distribution function, indicating the distribution of micro-surface elements by reflection, and the normal vector probability density function describing micro-surface elements. F is the Fresnel effect, which describes the material of the reflective surface and the Fresnel reflectance that changes with the Angle of view. If the Fresnel reflectance is sparse, it is denoted F_0. O represents the geometric occlusion function, which describes the mutual occlusion of microscopic surface elements.

The reflection model is represented by Fig. 6.11, which can be expressed as

$$f(\theta_i, \theta_o) = a \cdot f_{\mathrm{BP}}(\theta_i, \theta_o) + b \cdot f_{\mathrm{TS}}(\theta_i, \theta_o). \tag{6.23}$$

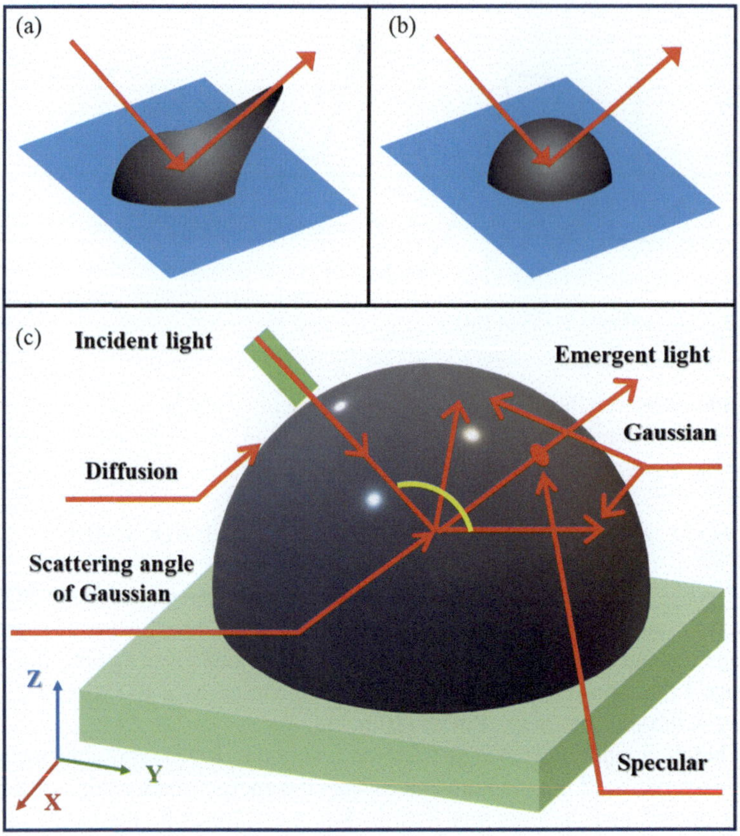

Fig. 6.11 Reflection model

This is a linear combination model. The corner BP represents the BP model, and the TS represents the TS model. The Blinn-Phong model can be described as:

$$f_{\mathrm{BP}}(\theta_i, \theta_o) = \frac{k_{\mathrm{L}}}{\pi} + k_{\mathrm{G}} \frac{8+s}{8\pi} z^s, \ z = \max(0, \vec{h} \cdot \vec{n}), \tag{6.24}$$

and the Torrance-Sparrow model can be mathematically expressed as

$$f_{\mathrm{TS}}(\vec{\theta}_i, \vec{\theta}_o) = \frac{k_{\mathrm{L}}}{\pi} + \frac{k_m}{4\pi(\vec{n} \cdot \vec{\theta}_i)} D \cdot F \cdot G, \tag{6.25}$$

where L is Lambertian reflection, m is specular reflection, k is reflectivity, h is angular bisection vector, n is normal vector, s is high light index (material surface smoothness), D is luminous contribution ratio, F is Fresnel coefficient, G is light shielding each other.

6.3 Applications of Non-Line-Of-Sight Imaging

Non-line-of-sight imaging has a wide range of applications in various fields, thanks to its ability to recover information about objects or scenes that are not directly visible to a camera or sensor due to obstacles or occlusions. Here are some notable applications of non-line-of-sight imaging:

1. **Search and Rescue Operations**: Non-line-of-sight imaging can be crucial in locating and rescuing individuals who are trapped or hidden in disaster scenarios, collapsed buildings, or other emergency situations. It helps responders identify the positions of survivors behind obstacles.
2. **Archaeology and Cultural Heritage**: Non-line-of-sight imaging can be used to explore archaeological sites and recover hidden structures, artifacts, or inscriptions without physically disturbing the site. It's valuable for preserving historical and cultural heritage.
3. **Autonomous Vehicles**: In the context of autonomous vehicles (self-driving cars and drones), non-line-of-sight imaging enhances perception capabilities. It helps vehicles detect pedestrians, vehicles, or obstacles that are hidden from their direct line of sight, improving safety and navigation.
4. **Surveillance and Security**: Non-line-of-sight imaging can aid in surveillance and security applications. It can detect and track individuals or objects that are obstructed by walls, buildings, or other obstacles, enhancing security measures in public spaces and buildings.
5. **Medical Imaging**: Non-line-of-sight imaging techniques have been explored for medical applications, such as endoscopy and minimally invasive surgeries. They allow physicians to visualize and navigate inside the body to diagnose and treat medical conditions.
6. **Industrial Inspection**: Non-line-of-sight imaging can be used in industrial settings for inspecting hidden or hard-to-reach areas, such as pipelines, machinery, and infrastructure. It helps identify defects and perform maintenance without disassembling equipment.
7. **Virtual Reality and Gaming**: In virtual reality (VR) and augmented reality (AR) applications, non-line-of-sight imaging can help improve the realism of virtual environments. It can simulate interactions with hidden or occluded objects, enhancing the user experience in gaming and simulations.
8. **Remote Sensing and Environmental Monitoring**: Non-line-of-sight imaging is valuable in remote sensing applications. It can provide insights into forested areas, underwater environments, and complex terrains by revealing hidden objects, vegetation, or geological features.
9. **Defense and Law Enforcement**: Military and law enforcement agencies can benefit from non-line-of-sight imaging for surveillance, reconnaissance, and situational awareness. It helps detect concealed threats and gather intelligence in challenging environments.

10. **Human–Computer Interaction**: Non-line-of-sight imaging can enable innovative human–computer interaction methods, such as touchless gesture control and tracking of user movements in three-dimensional space. It has potential applications in gaming, user interfaces, and interactive displays.
11. **Art and Entertainment**: In the entertainment industry, non-line-of-sight imaging can be used for special effects, scene reconstruction, and immersive experiences in theater productions, museums, and art installations.
12. **Scientific Research**: Researchers use non-line-of-sight imaging for scientific investigations, such as studying light propagation, understanding complex natural phenomena, and conducting experiments in optics and photonics.

Non-line-of-sight imaging is a rapidly evolving field, with ongoing research to improve its accuracy and applicability across various domains. Advances in hardware and computational techniques continue to expand its potential and make it more practical for real-world applications.

Acknowledgements This work was supported by the National Natural Science Foundation of China (Nos. 11874132, 61975044, 12074094, 12374271, and 62375065); Interdisciplinary Research Foundation of HIT (No. IR2021237).

References

1. Xu F, Shulkind G. Revealing hidden scenes by photon-efficient occlusion-based opportunistic active imaging. Opt Express. 2018;26(8):9945–62.
2. Zhu S, Sua Y, Rehain P, Huang Y. Single photon imaging and sensing of highly obscured objects around the corner. Opt Express. 2021;29(25):40865–77.
3. O'Toole M, Lindell D, Wetzstein G. Confocal non-line-of-sight imaging based on the light-cone transform. Nature. 2018;555(7696):338–41.
4. Nicodemus F. Directional reflectance and emissivity of an opaque surface. Appl Opt. 1965;4(7):767–75.
5. Phong B. Illumination for computer generated images. Commun ACM. 1973;18(6):311–7.
6. Blinn J. Models of light reflection for computer synthesized pictures. ACM Siggraph Comput Graph. 1977;11(2):192–8.
7. O'Toole M, Heide F, Lindell D, Zang K (2017) Reconstructing transient images from single-photon sensors. Proc CVPR : 1539–1547
8. Shin D, Xu F, Venkatraman D. Photon-efficient imaging with a single-photon camera. Nat Commun. 2016;7(1):12046.
9. Heide F, Diamond S, Lindell D, Wetzstein G. Sub-picosecond photon-efficient 3D imaging using single-photon sensors. Sci Rep. 2018;8(1):17726.
10. Divitt S, Gardner D, Watnik A. Imaging around corners in the mid-infrared using speckle correlations. Opt Express. 2020;28(8):11051–64.

Chapter 7
Coded Aperture Imaging

Yutong Li, Guodong Zhang, and Zhengjun Liu

Abstract Ordinary lenses cannot reflect, refract or focus X-rays, so coded aperture imaging arises at the historic moment. Modulation masks can change the distribution of beam in imaging systems. This chapter introduces the theoretical basis of coded aperture imaging technology and the basic structure of the computational imaging models. Then the method of optimal coded template and the estimation method of point spread function (PSF) are introduced. The spatial and spectral dimensional coded structures of coded aperture spectral imaging techniques are analyzed and compared. A multi-image encryption method based on compressed coded aperture imaging and pixel position scrambling operation is introduced. It focuses on the principles of self-interference coded aperture correlation holography and non-interference coded aperture correlation holography including diffraction model, reconstruction algorithm, noise source, development and application, etc.

This chapter introduces the theoretical basis of coded aperture imaging technology and the basic structure of the imaging model. Then the method of optimal coded template and the estimation method of point spread function are introduced. The spatial and spectral dimensional coded structures of coded aperture spectral imaging techniques are analyzed and compared. A multi-image encryption method based on compressed coded aperture imaging and pixel position scrambling operation is introduced. This paper focuses on the principles of self-interference coded aperture correlation holography and non-interference coded aperture correlation holography including diffraction model, reconstruction algorithm, noise source, development and application, etc.

Y. Li · Z. Liu (✉)
School of Physics, Harbin Institute of Technology, Harbin 150001, China
e-mail: zjliu@hit.edu.cn

G. Zhang
Middle School, Harbin Normal University, Harbin, China

Z. Liu et al. (eds.), *Computational Optical Imaging*, Advances in Optics and Optoelectronics, https://doi.org/10.1007/978-981-97-1455-1_7

7.1 Introduction

Coded aperture imaging is to put designed mask or aperture with a specific pattern of holes or opaque regions in front of the image sensor or detector. When capturing an image, instead of using a traditional lens to focus light onto the sensor, the coded aperture is used to adjust the transmittance of light. The pattern of holes or opaque regions in the mask allows only specific portions of the light to pass through and reach the sensor. Among them, the aperture shape and arrangement of the mask plate have certain rules, which are not chaotic. The coded mask can be equivalent to a spatial filter. Optimizing the reasonable coded shape and arrangement will have more high-frequency information in the spatial domain and improve the system resolution. Coded and decoded are two key steps in coded aperture imaging.

At the coded end, the original image is projected onto the coded board according to the optical principle, and the projected image will appear in the detector after the light passes through the coded board, and the projected images are superimposed on each other to form a two-dimensional coded image, that is, the convolution of the objective function and the coded aperture point spread function. Because pinholes are randomly distributed, it is necessary to build a unique binary array for each specific application. These patterns are selected based on factors like resolution requirements, light sensitivity, and computational complexity. After capturing the modulated image using the coded aperture, the decoded phase involves extracting the original image information from the captured data. The purpose of decoded is to improve the spatial resolution of image and the performance of imaging system. Coded aperture imaging has a large field of view, and a point target has an effect on the whole receiving plane after passing through the coded aperture. At the same time, coded aperture imaging technology is not very sensitive to noise interference, and high imaging signal-to-noise ratio and light collection efficiency can be obtained by using this technology. The schematic diagram of the principle of coded aperture imaging is shown in Fig. 7.1.

At present, there are mainly two kinds of coded apertures. One is multi-pinhole coded classified according to the spatial distribution of sub-apertures in the aperture plane, including random array, non-redundant array (NRA), uniform redundant array

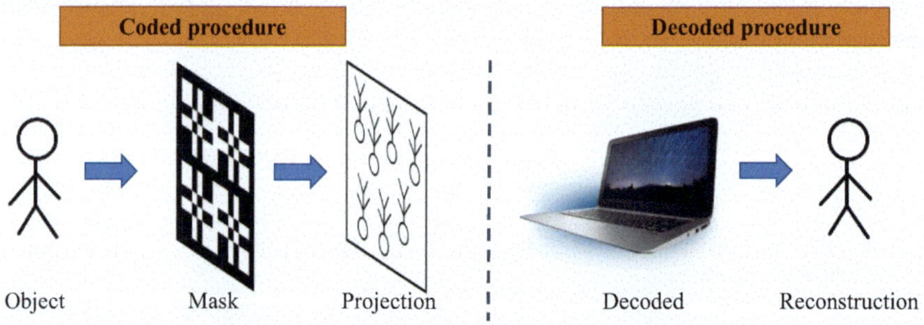

Fig. 7.1 Principle diagram of coded aperture imaging

(URA), modified uniform redundant array (MURA) and so on. The other is classified according to the shape of each aperture, mainly including single-ring hole, multi-ring hole and square hole.

Among them, the uniform redundant array coded mode is widely used, which increases the light collection efficiency by an order of magnitude on the premise of retaining the high resolution of a single pinhole, and can obtain the image of the target under weak X-ray. At the same time, the signal-to-noise ratio (SNR) is greatly improved. The theoretical SNR of multi-pinhole is $\sqrt{N/2}$ times that of single pinhole, where N is generally 1000–10,000. It can be seen that the improvement of SNR is considerable. In addition, tomographic images can be obtained, that is, the depth information of the target can be obtained.

In this chapter, we firstly focus on the basic principle of optical coded aperture imaging and the method of selecting the optimal coded aperture, and introduce the point spread function in the coded imaging system.

In the second part, the coded classification is introduced, including spatial and spectral dimensional coded. The spatial dimension coded is modulated first and then divided, and the spectral dimension coded is divided first and then modulated. Both of the two coded methods can reconstruct the complete target image and spectrum by pre-coded and post-decoded.

In the third part, a multi-image encryption method based on compressed coded aperture imaging and pixel position scrambling operation is introduced. In the process of encryption, each plaintext image is compressed and encoded by its corresponding coded mask, which is generated by code division multiple access (CDMA) modulation by a given auxiliary random matrix. Then, all the encoded images are recorded by charge coupled devices to generate superimposed images. The superimposed image is scrambled to get the ciphertext image. In the process of decryption, participants with the correct key group can successfully recover the corresponding plaintext image through location index key, auxiliary key, and compression sensing reconstruction algorithm.

In the final part, the coded aperture correlation holography (COACH) is introduced, and the system is divided into two categories in principle, namely self-interference COACH and interferenceless coded aperture correlation holography (I-COACH). The former adopts the self-interference technique and the latter uses the speckle correlation technique. This paper mainly introduces the principles of self-interference coded aperture correlation holography and non-interference coded aperture correlation holography, and analyzes the parameters related to I-COACH resolution, the source of noise, the method of noise suppression and so on. The development route of COACH is introduced, including 4D imaging, synthetic aperture far-field super resolution and quantitative phase measurement.

7.2 Principle of Coded Aperture Imaging

In Fig. 7.2, when the point light source is located at the front focal plane of the whole optical system and imaged by the optical system under ideal conditions, the image point obtained at the image plane should be the ideal image of the point light source. However, if it is in the out-of-focus position, the blurred spot is obtained on the image plane. Because the traditional imaging system will suppress the high spatial frequency information in out-of-focus images, if a physical mask with broadband frequency response is placed at the pupil position, the mask plate will extract different frequency information from the scene and transmit it to the detector. The detector can be divided into multiple regions, each region contains a part of frequency domain information of the whole scene, and all the frequency information on the detector can be superimposed to obtain the whole frequency domain information in the scene, that is, it contains low frequency, intermediate frequency and high frequency information. Using the superimposed frequency domain information to restore can improve the image resolution and save the high spatial frequency information of the restored image.

7.2.1 Coded Imaging Process

For lens imaging system, the input image can be decomposed into several elementary functions, and the light intensity distribution of the input image on the image plane is mathematically equal to the intensity superposition as follows

$$g(x, y) = \sum_{x_0, y_0} h(x, y, x_0, y_0) \cdot g_0(x_0, y_0) + n(x, y), \tag{7.1}$$

where $h(x, y, x_0, y_0)$ is the image spot produced by different point light sources in the field of view, and this function is only related to the spatial position of the input

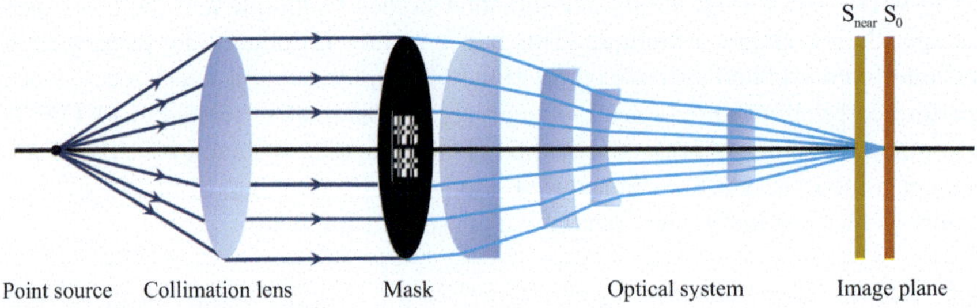

Fig. 7.2 Principle of lens imaging

point light source. The function $g_0(x_0, y_0)$ is the original input image and $g(x, y)$ is the degraded image. The $n(x, y)$ represents noise.

The system is further written in convolution integral form as follows

$$g(x, y) = h(x, y) * g_0(x_0, y_0) + n(x, y), \tag{7.2}$$

where $*$ represents convolution operator and the function $h(x, y)$ is point spread function. Then, perform Fourier transform on Eq. (7.2) to obtain

$$G(\mu, v) = H(\mu, v) \cdot G_0(\mu, v) + N(\mu, v), \tag{7.3}$$

where $H(\mu, v)$ is the amplitude-frequency characteristic of the lens imaging system.

The Fourier transform $G'(\mu, v)$ of the restored image obtained by filtering restoration is

$$G'(\mu, v) = G'(\mu, v)/H(\mu, v) = G_0(\mu, v) + N(\mu, v)/H(\mu, v), \tag{7.4}$$

where contains the desired original image $G_0(\mu, v)$, and adds a noise term. Since $H(\mu, v)$ appears in the denominator of the noise term, when H is small or equal to zero and noise $N(\mu, v)$ is not small in Fourier frequency domain, the noise is amplified. This is called ill-conditioned property, which means that small noise interference in degraded images will have a great impact on the restored images in those frequency spectrums where $H(\mu, v)$ is small. According to simple optical analysis, the frequency domain values of many transfer functions are very small, and the restoration of most images will encounter ill-conditioned characteristics of solution. An important consideration of image restoration method is how to control the influence of noise on restoration when it is morbid.

7.2.2 Image Reconstruction and Decoded Process

The convolution of the transfer function and the input signal can be equivalent to a kind of filtering of the input signal. The coded aperture transfer function can be regarded as a low-pass filter, and the high-frequency components of the input signal are suppressed or even lost. The process of image restoration and decoded is actually a deconvolution process, in which high-frequency noise will be amplified. Therefore, in order to get the real solution as much as possible, it is necessary to consider noise suppression in the deconvolution method and make an appropriate compromise between signal recovery and noise amplification.

(1) Pathological properties in image restoration

Deconvolution belongs to a kind of inverse problem in mathematical problems, and one of the common important attributes is their ill-conditioned characteristics. For an integrable function, then

$$\lim_{\omega \to \infty} \int_a^b h(t, \tau) \sin \omega\tau \mathrm{d}\tau = 0 \quad t \in [c, d]. \tag{7.5}$$

For functions $y(t)$ and $h(t, \tau)$, $x(t)$ is a solution of Eq. (7.5). When $\omega \to \infty$, $x_I(t) = x(t) + A \sin \omega t$ is also a solution of Eq. (7.5). For any given small value $\varepsilon_1 > 0$, there exists ω_1, so that when $\omega > \omega_1$, we can get

$$\int_a^b h(t, \tau) \sin \omega\tau \mathrm{d}\tau < \varepsilon_1. \tag{7.6}$$

If $\omega_2 \geq \omega_1$, then

$$A \int_a^b h(t, \tau) \sin \omega\tau \mathrm{d}\tau = \varepsilon_2 < A\varepsilon_1 \tag{7.7}$$

and

$$\int_a^b h(t, \tau)[x(\tau) + A \sin \omega_2\tau] \mathrm{d}\tau = y(t) + \varepsilon_2. \tag{7.8}$$

This means that a small deviation of observation $y(t)$ may cause a high-frequency parasitic component with large amplitude to be superimposed on the true solution.

(2) Optimal selection method of coded aperture

In order to recover the original image, it is necessary to estimate the point spread function (PSF) of the system first, and the estimation result of PSF function depends on the selected encoded board code type. We list four coded mask pattern as shown in Fig. 7.3. Figure 7.3a is mainly used for depth estimation of defocused images. This coded pattern has many distinguishable zero-crossing points in Fourier domain, but it is not suitable for image deblurring operation because the power spectrum of this coded pattern decays rapidly at high frequency and the captured image will lose a lot of information. The "Veeraraghavan" coded pattern in Fig. 7.3b can be applied to image deblurring operation, which has a wide passband and can meet the requirements of imaging system for large optical information capacity, with less loss of captured image information. The circle coded aperture in Fig. 7.3c is mainly used for focused imaging. When it is applied to deblurring operation, it will cause a lot of noise in the restored image, produce a lot of artifacts and lose high-frequency information of the image. The optimized coded aperture in Fig. 7.3d is an aperture pattern with broadband frequency response, so this coded pattern is mainly used for image deblurring, and its response at high frequency and low frequency is better than that of "Veeraraghavan" coded pattern. The selection of coded type can also be determined by Kullback-Leibler divergence method and modulation transfer function method.

Kullback-Leibler divergence measures the similarity of fuzzy patterns, which refer to aperture shapes and are located at different positions. Kullback-Leibler divergence can be used to describe the discrimination and sensitivity of aperture pattern to depth information. Kullback-Leibler divergence is an asymmetric measure of the difference between two probability distributions P and Q. P represents the theoretical distribution of accurate calculation, or the actual observed value, which represents the real distribution of data. Q represents the theoretical distribution, model distribution or understanding as a factor affecting the P distribution. The physical meaning of KL divergence shows that when the samples in P are encoded by Q, more bits are needed than those in P, where P refers to the existing data set and Q represents the theoretical result.

KL divergence is asymmetric. If it becomes symmetric, we introduce

$$D_{KL}(p,q) = \frac{1}{2} \times \left[D_{KL}(p,q) + D_{KL}(p,q)\right]. \tag{7.9}$$

The discrete definition and continuous definition of KL divergence are shown in the following formula

$$D_{KL}(P\|Q) = \sum_{i} P(i) \log \frac{P(i)}{Q(i)}, \tag{7.10}$$

$$D_{KL}(P\|Q) = \int_{-\infty}^{\infty} p(x) \log \frac{p(x)}{q(x)} \mathrm{d}x, \tag{7.11}$$

where $p(x)$ and $q(x)$ are the probability density distributions of random variables p and q, respectively. $D_{KL}(P\|Q)$ is a numerical value. KL divergence value must be non-negative, that is $D_{KL}(P\|Q) \geq 0$. Only when the two distributions are equal, the KL divergence is equal to 0.

In Eq. (7.1), the clear image in the original scene is set to g_0, whose derivative obeys Gaussian distribution and the average value is 0.

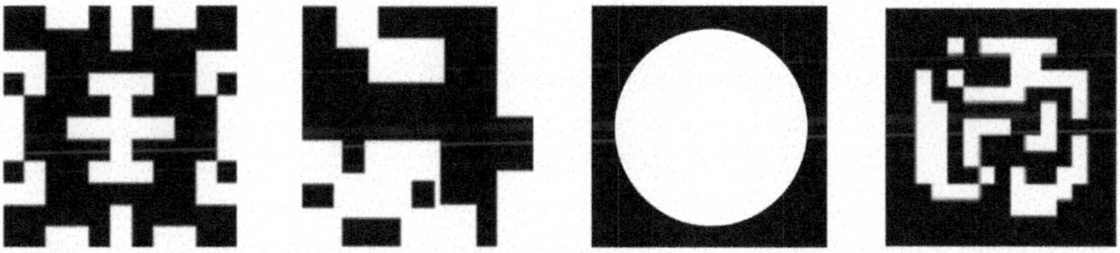

(a) Font coded aperture (b) Veeraraghavan coded aperture (c) Circle coded aperture (d) Optimized coded aperture

Fig. 7.3 Coded mask pattern

$$P(g_0) \propto \exp\left(-\frac{1}{2}\partial g_0^T \Theta^{-1} g_0\right) = N(0, \Theta),$$
$$\Theta = \left[\partial\left(C_1^T C_1 + C_2^T C_2\right)\right]^{-1}, \tag{7.12}$$

where C_1 and C_2 are derivative convolution matrices, which are equivalent to gradient filters $g_1 = [1 - 1]$ and $g_2 = [1 - 1]^{-1}\,[1 - 1]^{-1}$. Scalar ∂ represents variance and is a constant value.

Turning the image g into frequency domain, its distribution is diagonal, and the frequency distribution of this image can be obtained by Fourier transform

$$P_k(G) \propto \exp\left(-\frac{1}{2}G^T \sigma^{-1} G\right) \quad \sigma = H_k^T \overline{\Theta} H_k + \eta^2 I,$$
$$\overline{\Theta} = \left[\partial\left(F_1^T F_1 + F_2^T F_2\right)\right]^{-1}, \tag{7.13}$$

where $H_k(\mu, v)$ is the result of Fourier transform of the point spread function, F_1 and F_2 are Fourier transforms of C_1 and C_2, respectively.

According to the statistical model of the image, the calculation formula of KL divergence can be simplified as

$$D_{KL}(P_m, P_n) = \sum_{\mu,v}\left(\frac{\sigma_m(\mu, v)}{\sigma_n(\mu, v)} - \log\frac{\sigma_m(\mu, v)}{\sigma_n(\mu, v)}\right). \tag{7.14}$$

If the ratio of $\sigma(\mu, v)$ corresponding to aperture filters located at different depth positions is high, it means that the KL divergence value is large, and it also shows that the distribution at different depth positions will be obviously different. The greater the distribution difference between the blurred images $P_m(\mu, v)$ and $P_m(\mu, v)$ on the imaging plane, the easier it is to distinguish the position information of E_m and E_n on the object side.

For a space invariant system or a displacement invariant system. If the noise of the system is ignored, the formula Eq. (7.2) is Fourier transformed to obtain

$$G(\mu, v) = G_0(\mu, v) \cdot \tilde{H}(\mu, v), \tag{7.15}$$

where $\tilde{H}(\mu, v)$ is the result of Fourier transform of point spread function

$$\tilde{H}(\mu, v) = \iint h(\xi, \eta) \exp[-\mathrm{i}2\pi(\mu\xi + v\eta)][-\mathrm{i}2\pi(\mu\xi + v\eta)]\mathrm{d}\xi\,\mathrm{d}\eta$$
$$= H_c(\mu, v) - \mathrm{i}H_s(\mu, v). \tag{7.16}$$

Here $\tilde{H}(\mu, v)$ is the optical transfer function (OTF) of the system. The function consists of real part $H_c(\mu, v)$ and imaginary part $-H_s(\mu, v)$, , and its physical meaning shows the transmission ability of the system in frequency domain. The modulation transfer function (MTF) is obtained by finding the modulus of this function, and the angle between the imaginary part and the real part is called the phase transfer function (PTF).

When selecting the shape of coded aperture, the optimal aperture can be selected according to the modulation transfer function curve of each coded template. If the area enclosed by the corresponding MTF curve and the vertical and horizontal axes of a coded aperture is larger, it shows that the comprehensive optical quality of the coded template is better, which is characterized by strong contrast and high resolution.

The above two methods can realize the selection of coded aperture, but the coded aperture also needs to meet the physical limitations in the spatial domain:

(1) The coded template must be a binary template, because it is difficult to make a non-binary template accurately;
(2) The aperture ratio of coded template is generally 50%, which can improve the utilization rate and sensitivity of light energy and reduce the influence caused by noise;
(3) The coded template must be consistent with the diameter of the aperture stop, and the opening position of the template must be far away from the edge of the template to ensure the effective modulation of light waves by the mask plate;
(4) In order to reduce the influence of diffraction in the aperture area on the imaging quality, the aperture size of the coded unit should not be too close to the wavelength of light wave. The spatial resolution of the coded template should be low enough to avoid introducing strong diffraction effect.

(3) Recovery and decoded method

Early image restoration mainly adopts analog methods such as optical filtering. With the development of computer technology, digital methods have been used for image reconstruction in recent years. At present, many algorithms have been proposed for digital reconstruction of coded imaging.

(1) Cross-correlation method, that is, matched filtering method, is a restoration method that can maximize the utilization of signal-to-noise ratio. Cross-correlation of coded images with coded aperture is carried out to obtain the reconstruction of objects. In practical calculation, $G(\mu, v)$ is multiplied by a function M corresponding to the aperture. The choice of M is that MG can estimate $G'(\mu, v)$ and suppressing noise.

In the balanced cross-correlation method, the general choice is

$$M = (1 + p)H^T - p. \tag{7.17}$$

For a suitable p, restoration can suppress the noise better.
(2) Photon tracing method is similar to cross correlation method. For each photon detected, project it to the position where it may be emitted. If a photon touches an opaque aperture, it is counted in the background. If a photon hits a transparent aperture position, count this photon as the contribution of the source object.

After all photons are processed, the background is subtracted from the source to get the reconstructed image.

(3) Wiener filtering is to find a filter whose transfer function is $H_w(\mu, v)$, so as to minimize the mean square error between the recovered target estimate $f'(x, y)$ and the real $f(x, y)$, that is

$$E\left\{\left[f(x, y) - f'(x, y)\right]^2\right\}. \tag{7.18}$$

Then

$$H_w(u, v) = \frac{H^*(u, v)}{|H(u, v)|^2 + s_n(u, v) \big/ s_o(u, v)}, \tag{7.19}$$

where $H^*(u, v)$ is the complex conjugate of the transfer function of the coded aperture imaging system, $s_o(u, v)$ and $s_n(u, v)$ are the power spectral densities of the image and noise, respectively. In many cases, $s_o(u, v)$ and $s_n(u, v)$ are unknown. For the convenience of calculation, $s_n(u, v) \big/ s_o(u, v)$ can be set to an empirical value ε according to different processing objects, and then

$$H_w(u, v) = \frac{H^*(u, v)}{|H(u, v)|^2 + \varepsilon}. \tag{7.20}$$

(4) Max entropy model (MEM) is a restoration method of degraded data, which is widely used in many fields. In MEM, the image is regarded as a random distribution of photons, and the most likely distribution is the restored image. Photon distribution is constrained by energy. The objective function is

$$\max\left\{-f^T \ln f - n^T \ln n\right\}. \tag{7.21}$$

The constraint conditions are

$$Y = HF + N. \tag{7.22}$$

Although MEM can get good results, the huge amount of calculation limits the scale of application problems.

(5) Barrent introduced simulated annealing method into coded imaging to solve the problem that the nonlinearity of the problem may only get the local optimal solution under constraints. Simulated annealing can ensure the global optimal solution of the problem, but it also needs many iterations, and the calculation is huge.

In image restoration, starting from the initial image without photon distribution, a temperature kT is assumed. At the random pixel position, calculate the change ΔE of the objective function if a photon is added. If $\Delta E < 0$, we can add this photon; Otherwise, calculate $\exp(-\Delta E/\text{kT})$, and compare it with a random number ξ. When $\exp(-\Delta E/\mathrm{kT}) > \xi$, this photon can still be added. When $\exp(-\Delta E/\mathrm{kT}) < \xi$, the image will remain unchanged. Continue to add photons randomly until a balanced distribution is obtained, then reduce kT and repeat the above process, and finally get the required restored image.

7.3 Coded Aperture Spectral Imaging

The coded aperture spectral imager uses an area array coded template in the optical path for staring imaging. In the coded aperture spectral imaging system, the position of the coded mask determines the coded modulation mode of the system. Among them, spatial dimension coded is modulation and then splitting, and spectral dimension coded is splitting and then modulation. Both coded methods can reconstruct a complete target image and spectrum by coded and decoded.

7.3.1 Spatial Dimensional Coded Aperture Spectral Imaging System

The principle of spatial dimension coded aperture spectral imaging system is shown in Fig. 7.4. The scene light is a data cube structure composed of light of various bands and geometric shapes. When the staring imaging mode is adopted, the encoded device will encode the data set, and the data will be split by the dispersion device. The dispersion direction of the system is consistent with the line direction of the encoder, and there is no interference between lines of different bands. Therefore, the encoded data cube will be repeatedly superimposed on the detector to form the effect of map aliasing. The effective energy collected by the system is stronger, which effectively improves the image signal-to-noise ratio. Multi-channel multiplexing technology captures multi-band data at one time, reducing the number of encoded times. And the system is composed of a single coded device and a dispersive device. Similar to the traditional spectral imaging system, it is easier to realize the system by using less dispersive hardware and post-optical path.

In the space-dimensional coded aperture imaging system, the coded device encodes the scene light to construct a data cube, which requires that the data of each point can be accurately received by the detector, so the coded device and the detector need to meet the strict proportional relationship between the coded device unit and the detector pixel size. The imaging resolution of spatial coded aperture depends on the resolution of detector and the resolution of coded device. The pursuit of greater imaging resolution requires a larger detector array size, which leads to an increase in the amount of data, an increase in image reconstruction due to the aliasing of maps, a more complicated data cube dissociation, and difficult algorithm implementation.

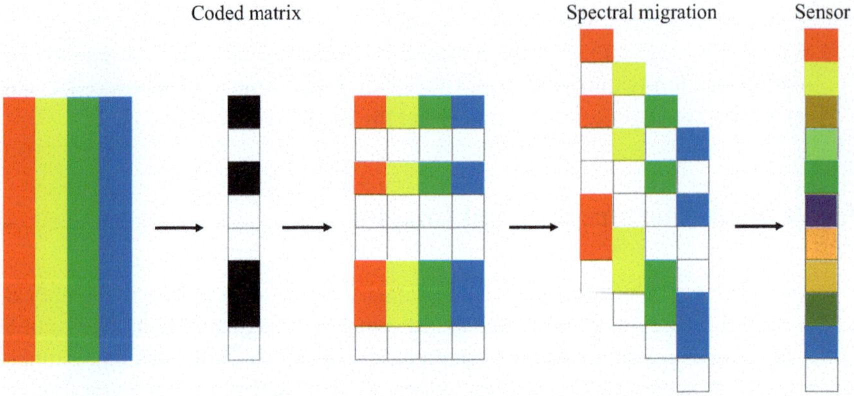

Fig. 7.4 Principle of spatial dimensional coded aperture spectral imaging system

7.3.2 Spectral Dimensional Coded Aperture Spectral Imaging System

The principle of spectral dimension coded imaging system is shown in Fig. 7.5. Based on the system structure of spatial dimension coded, dispersive devices and relay lens groups are added to spectral dimension coded. The scene data cube needs twice dispersion before it can be captured by the detector. When the first dispersion splitting is carried out, the data cube shifts for the first time along the band, and the deformed data cube is modulated by the coded device, and then the reverse dispersion is realized by the dispersion device, and the deformed data cube returns to the normal state to realize the light combination process. This working mode of splitting light first and then combining light can detect the coded data cube of spectral band under the condition that the physical object or scene space basis is unchanged. By viewing the detector results in different bands, the original scene images encoded in different bands can be observed.

The object or scene goes through the process of "splitting-coded-combining" optical system, and the detector can obtain some coded data cubes. Multiple cubes containing only partial spectral data can be obtained by changing the coded matrix for many times, and the complete image can be reconstructed when the number of data measurements is reached.

Spectral dimension coded aperture imaging has low requirements for coded devices and detector devices, and the spatial distribution of imaging is not needed in the light splitting and combining mode, and there is no direct relationship between the coded aperture unit and the pixel size of the detector, which greatly reduces the processing difficulty, selection difficulty and assembly difficulty of the optical system and is beneficial to the actual system construction. On the other hand, the spatial dimension data of the detector image plane do not overlap, and the data

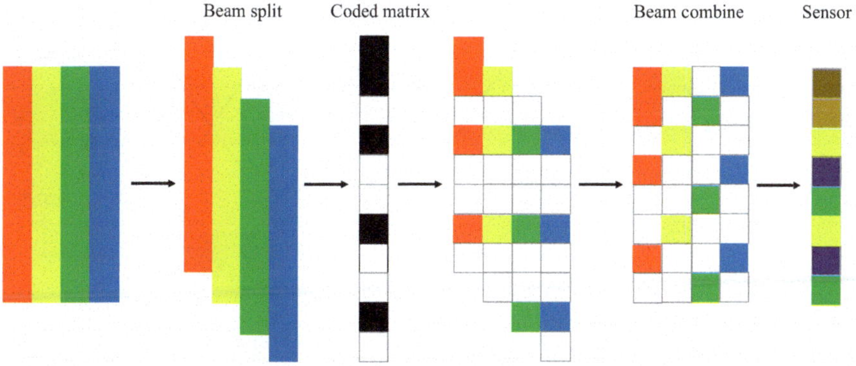

Fig. 7.5 Principle of spectral dimensional coded aperture spectral imaging system

correlation is small, so the processing algorithm is relatively easy. The imaging resolution of spectral dimension coded aperture system only depends on the resolution of detector pixel, not on the resolution of coded aperture device, which further enhances the development and application potential.

The optical path of spectral dimension spectral imaging system has added dispersive elements and post-optical system, which makes data demodulation relatively more complicated, system calibration as a whole more complicated, and the cost is further increased. Moreover, the whole process needs multiple acquisitions to obtain enough reconstruction information, which leads to the increase of coded times and the inability to realize real-time imaging.

7.4 Multi-image Encryption via Coded Aperture Imaging

7.4.1 Coded Aperture Compressive Temporal Imaging

In the previous chapters, we have introduced the principles of compressed sensing imaging and coded aperture imaging. Figure 7.6 is a schematic diagram of coded aperture compressive temporal imaging (CACTI) [1]. It is assumed that the size of the original image $I_t(x, y)(t = 1, 2, 3)$ and the projected coded mask $M_t(x, y)$ of each frame are 3×3 pixels. Due to the high refresh frequency of DMD, $I_t(x, y)$ and $M_t(x, y)$ are projected at the same time, and finally the total superimposed light intensity B in time and space within one exposure time of CCD can be obtained, which is expressed as:

$$B = \sum_{t=1}^{3} M_t \cdot I_t, \tag{7.23}$$

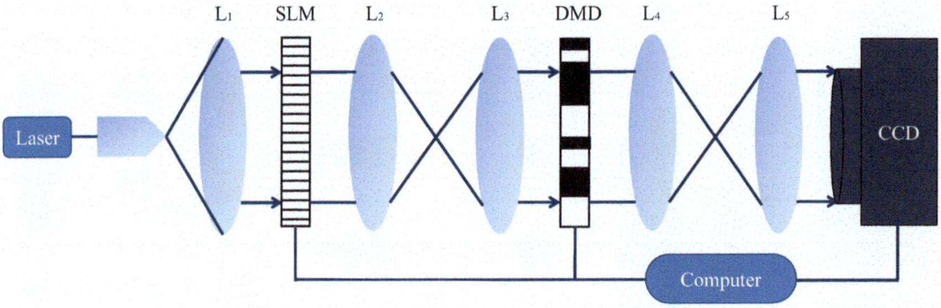

Fig. 7.6 Principle of coded aperture compressive temporal imaging

where "·" is expressed as Hadamard product.

In Fig. 7.7, the size of the matrix Φ is 9×27, which is regarded as a matrix combination of Ω_1, Ω_2, and Ω_3 with the size of 9×9. All elements of the coded mask M_1, M_2, and M_3 will be placed on the main diagonals of Ω_1, Ω_2, and Ω_3, respectively. At the same time, the original images I_1, I_2, and I_3 are rearranged into a column vector x with a size of 27×1, and the superimposed intensity m will be rearranged into a column vector y with a size of 9×1, so it can be regarded as a compressed sensing sampling process. The process of image decoded and reconstruction can also be realized by compressed sensing reconstruction, and the original multi-frame image can be recovered from one measurement.

Figure 7.8 shows the construction process of an encoded mask M_t taking a matrix with a size of 6×6 as an example. Wherein the highlighted part is two consecutive image pixels.

(1) Firstly, an auxiliary random matrix R with the same size as M_t is generated.
(2) A pseudo-random number generator is introduced to generate several row shift vectors p_t and several column shift vectors q_t, respectively.
(3) R is modulated by several generated row shift vectors p_t and column shift vectors q_t, and the required coded mask m can be obtained.

Each frame plane is modulated by a shift code, which can be regarded as code division multiple access (CDMA).

7.4.2 Multi-image Encryption Scheme

Assuming that the size of n images to be encrypted is $m \times m$ of $I_t(x, y)(t = 1, 2, \ldots, n)$, the encryption process is as follows:

(1) Each secret image m is sent to a spatial light modulator (SLM) for display, and then projected onto a digital micromirror device (DMD) by a $4f$ system.

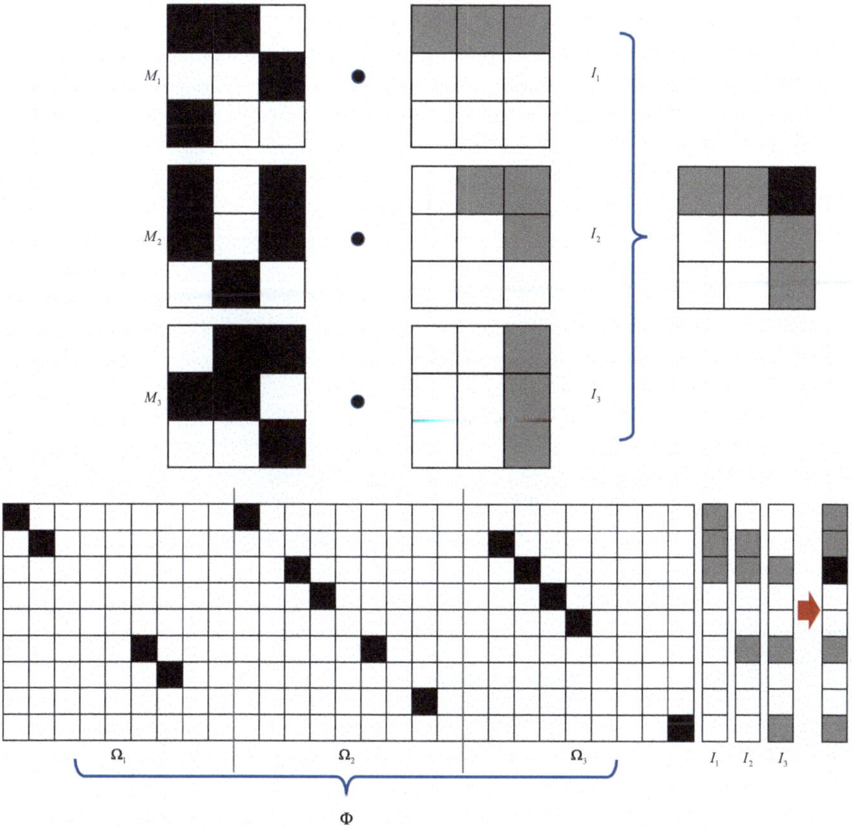

Fig. 7.7 Schematic diagram of sampling process of compression coded aperture

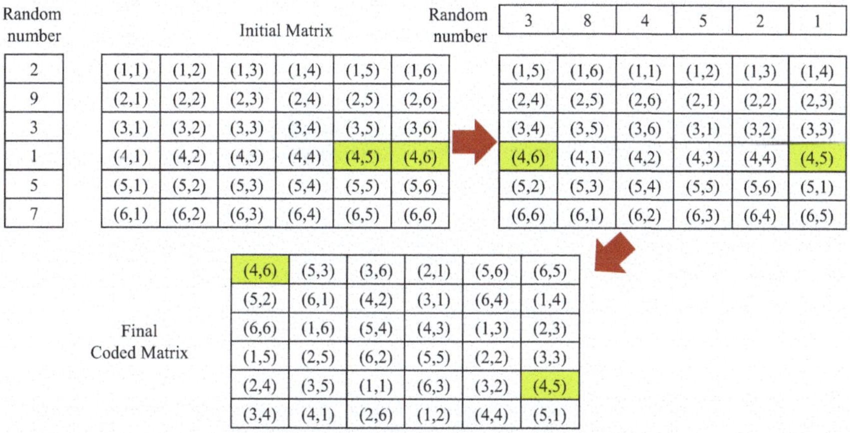

Fig. 7.8 Construction process of coded mask

(2) Input the coded mask M_t corresponding to each image I_t into DMD for coded. n coded masks M_t are obtained by code division multiple access (CDMA) modulation of auxiliary random matrix R with the size of $m \times m$. Among them, the random matrix R and the parameters a, c, m and x_0 used in the linear congruence method are auxiliary keys. □
(3) The encrypted total superimposed light intensity is recorded by CCD camera, and the superimposed image is stored in the computer.
(4) Scrambling and encrypting the superimposed image to obtain a ciphertext image, and storing the index key.

The corresponding decryption process is shown in Fig. 7.9, and the specific steps are as follows:

(1) Rearrange the pixel positions of the ciphertext image according to the correct index key to obtain the encrypted superimposed images of n secret images.
(2) The encoded mask M_t is reconstructed by collecting complete auxiliary keys (a, c, m, x_0 and random matrix R).
(3) Decrypt the original secret image by using compressed sensing reconstruction algorithm.

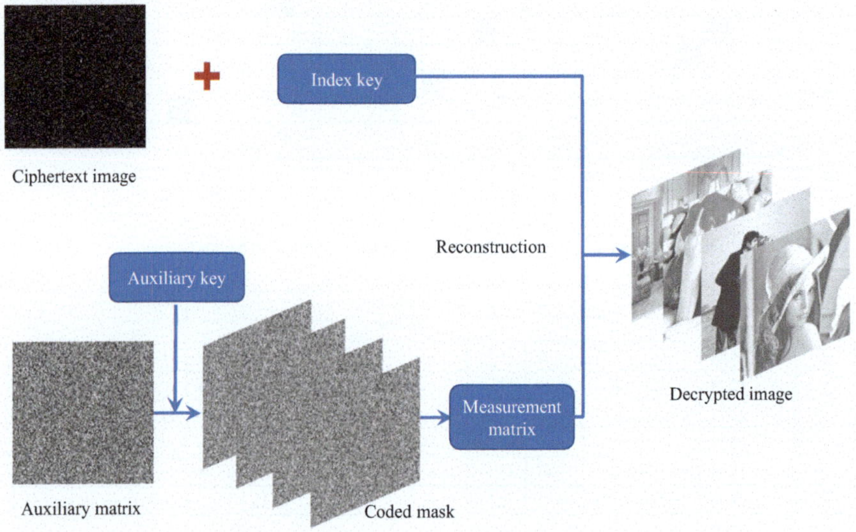

Fig. 7.9 Decryption process of multi-image encryption system based on time domain compressed aperture imaging

7.5 Coded Aperture Correlation Holography

Holography is a technology that can realize lensless imaging, which is not limited by lens aberration and can realize high-precision imaging on a relatively small scale. Holography encodes the complex amplitude information of an object into interference fringes through a highly coherent light source. Coded aperture correlation holography (COACH) uses incoherent light as light source [2], which is different from coherent holography in two aspects: first, it does not use coherent light, so there are no problems such as laser speckle noise. On the other hand, it is no longer limited by the coherent length of the light source, and it is easy to realize far-field imaging. Self-interference COACH system skillfully uses the partial coherence characteristics of light source to realize stereo imaging. At present, COACH system is divided into self-interference COACH and interferenceless coded aperture correla tion holography (I-COACH) [3, 4]. The former adopts self-interference technology, while the latter adopts speckle correlation technology.

7.5.1 The Classification of COACH System

(1) Scalar diffraction model of self-interference COACH system

Early image restoration mainly adopts analog methods such as optical filtering. With the development of computer technology, digital methods have been used for image reconstruction in recent years. At present, many algorithms have been proposed for digital reconstruction of coded imaging.

In the self-interference COACH system, the two beams of interference are all from objects, and the schematic diagram of the system is shown in Fig. 7.10. Set the horizontally polarized light to 0 and the counterclockwise direction to be positive. Wherein that polarization angle of the two polarizers are 45°, and the modulation polarization angle of the spatial light modulator is 0°. When the LED light source passes through the first polarizer, it can be converted into 45° linearly polarized light, which can be divided into the combination of 0° and 90° linearly polarized light. In which 0° linearly polarized light is modulated by SLM and regarded as modulated light. 90° linearly polarized light is not modulated and regarded as reference light. The polarization angles of the modulated light and the reference light are orthogonal and will not interfere. The modulated light and the reference light are analyzed by the polarizer again, and the polarization angles of the reference light and the modulated light are both 45°, resulting in self-interference.

The reference light of optical system is plane wave, and the complex amplitude distribution is A. The amplitude distribution of modulated light is $G(u, v)$, and (u, v) is the plane coordinate of the image sensor plane. The intensity image captured by the sensor is

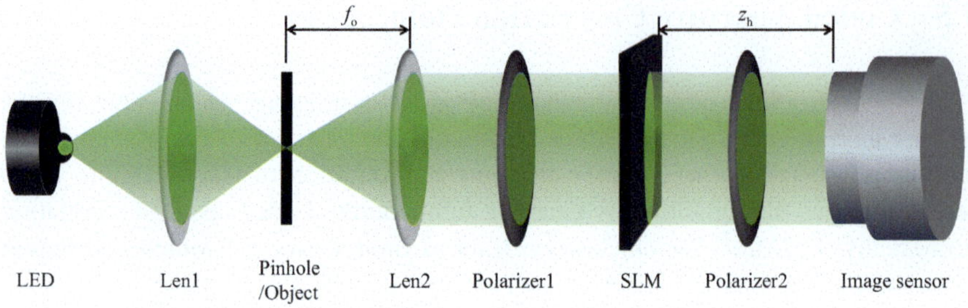

Fig. 7.10 Schematic diagram of self-interference COACH system

$$I_k(u, v) = |A + G(u, v)\exp(\mathrm{i}\theta_k)|^2, \quad \theta_k = \frac{2\pi}{3}(k-1), k = 1, 2, 3, \tag{7.24}$$

where $I_k(u, v)$ is a hologram with different phase shift. Three-step phase shift is used to eliminate twin images and conjugate terms. The phase-shifted hologram can be expressed as

$$\begin{aligned} A^*G(u, v) &= I_1\left[\exp(-\mathrm{i}\theta_3) - \exp(-\mathrm{i}\theta_2)\right] \\ &+ I_2\left[\exp(-\mathrm{i}\theta_1) - \exp(-\mathrm{i}\theta_3)\right] \\ &+ I_3\left[\exp(-\mathrm{i}\theta_2) - \exp(-\mathrm{i}\theta_1)\right], \end{aligned} \tag{7.25}$$

where A is a constant. The complex amplitude of an object can be expressed by a set of points as

$$t(x, y) = \sum_m a_m \delta(x - x_m, y - y_m). \tag{7.26}$$

The object is located in the front focal plane of the lens, and the outgoing light is two similar plane waves. One of them is the reference light $A_m \exp[\mathrm{i}2\pi(ux_m + vy_m)\lambda^{-1} f_0^{-1}]$ in the form of plane wave with inclination angle. The other is phase-shifted modulated light $\exp(\mathrm{i}\theta_k)B_m \exp[\mathrm{i}2\pi(ux_m + vy_m)\lambda^{-1} f_0^{-1}]$. The intensity image captured by the sensor can be expressed as

$$I_k'(u, v) = \sum_m \left| A_m \exp\left[\frac{\mathrm{i}2\pi(ux_m + vy_m)}{\lambda f_0}\right] + \exp(\mathrm{i}\theta_k) B_m \exp\left[\frac{\mathrm{i}2\pi(ux_m + vy_m)}{\lambda f_0}\right]\right|^2. \tag{7.27}$$

The phase-shifted hologram can be expressed as

$$\begin{aligned} H(u, v) &= I_1'\left[\exp(-\mathrm{i}\theta_3) - \exp(-\mathrm{i}\theta_2)\right] + I_2'\left[\exp(-\mathrm{i}\theta_1) - \exp(-\mathrm{i}\theta_3)\right] \\ &+ I_3'\left[\exp(-\mathrm{i}\theta_2) - \exp(-\mathrm{i}\theta_1)\right] \\ &= \sum_m A_m^* B_m G(u - u_m, v - v_m). \end{aligned} \tag{7.28}$$

The intensity image captured by the sensor can be expressed as the cross-correlation between $H(u, v)$ and $G(u, v)$ gives the reconstruction result, which is expressed as

$$\begin{aligned} H(u, v) &= \sum_m \left[A_m^* B_m G(u - u_m, v - v_m) \otimes G(u, v)\right] \\ &\approx \sum_m \left[A_m^* B_m \delta(u - u_m, v - v_m)\right] \approx t\left(\frac{u'}{M_T}, \frac{v'}{M_T}\right), \end{aligned} \tag{7.29}$$

where $M_T = z_h / f_0$ is the lateral amplification factor.

(2) Scalar diffraction model of I-COACH system

The optical path of I-COACH system is shown in Fig. 7.11. The modulation polarization direction of the spatial light modulator is consistent with that of the polarizer. Definitions $Q(a)$ and $L(\overline{r}/z)$ are defined as

$$Q(a) = \exp\left[\mathrm{i}\pi a\lambda^{-1}\left(x^2 + y^2\right)\right], \tag{7.30}$$

$$L(\overline{r}/z) = \exp\left[\frac{\mathrm{i}2\pi}{\lambda z}\left(r_x x + r_y y\right)\right]. \tag{7.31}$$

The amplitude of any point $r_s = (x_s, y_s)$ on the pinhole plane is $\sqrt{I_s}$, which can be expressed as $\sqrt{I_s}C_1 L(\overline{r}_s/z_s)Q(z_s^{-1})$ after passing through the lens, and the light field modulated by the spatial light modulator can be expressed as $\sqrt{I_s}C_1 L(\overline{r}_s/z_s)Q(z_s^{-1})Q(f_0^{-1})\exp[\mathrm{i}\Phi_k(r)]$. The intensity image captured by the final image sensor is

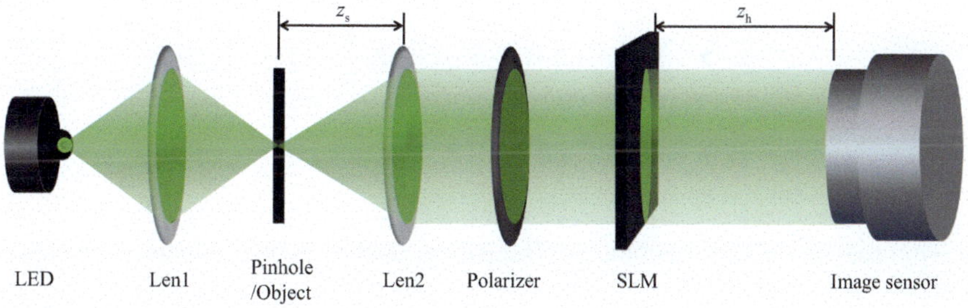

Fig. 7.11 Schematic diagram of I-COACH system

$$I_k(\overline{r}_0; \overline{r}_s, z_s) = \left| \sqrt{I_s} C_1 Q\left(\frac{1}{z_s}\right) L\left(\frac{\overline{r}_s}{z_s}\right) Q\left(-\frac{1}{f_0}\right) \exp[\mathrm{i}\Phi_k(\overline{r})] * Q\left(\frac{1}{z_h}\right) \right|^2. \tag{7.32}$$

Set $z_1 = z_s f_0 (f_0 - z_s)$ and Eq. (7.32) further simplified as

$$\begin{aligned} I_k(\overline{r}_0; \overline{r}_s, z_s) &= \left| \sqrt{I_s} C_1 L\left(\frac{\overline{r}_s}{z_s}\right) Q\left(-\frac{1}{z_s}\right) \exp[\mathrm{i}\Phi_k(\overline{r})] * Q\left(\frac{1}{z_h}\right) \right|^2 \\ &= I_k\left(\overline{r}_0 - \frac{z_h}{z_s}\overline{r}_s; 0, z_s\right). \end{aligned} \tag{7.33}$$

Object can be regarded as a collection of infinite points. Because the system is illuminated by incoherent light, the intensity distribution can be regarded as the superposition of PSF intensity. An object can be represented as

$$O(\overline{r}_s) = \sum_n^N a_n \delta(\overline{r} - \overline{r}_{s,n}). \tag{7.34}$$

The intensity value captured by the detector is

$$I_{OBJ,k}(\overline{r}_0; z_s) = \sum_n a_n I_k\left(\overline{r}_0 - \frac{z_h}{z_s}\overline{r}_s; 0, z_s\right). \tag{7.35}$$

In the actual optical path, multiple PSFs are needed to synthesize point spread hologram (PSH), which can be expressed as

$$H_{\mathrm{PSH}}(\overline{r}_0, z_s) = \sum_{k=1}^K [I_k(\overline{r}_0, z_s) \exp(\mathrm{i}\theta_k)]. \tag{7.36}$$

The synthetic hologram is

$$\begin{aligned} H_{\mathrm{OBJ}}(\overline{r}_0, z_s) &= \sum_{k=1}^K [I_{\mathrm{ORJ},k}(\overline{r}_0; z_s) \exp(\mathrm{i}\theta_k)] \\ &= \sum_{k=1}^K \sum_n \left[a_n I_k\left(\overline{r}_0 - \frac{z_h}{z_s}\overline{r_s}; 0, \overline{z}_s\right) \exp(\mathrm{i}\theta_k) \right] \\ &= \sum_n \left[a_n H_{\mathrm{PSH}}\left(\overline{r}_0 - \frac{z_h}{z_s}\overline{r_s}; 0, \overline{z}_s\right) \right]. \end{aligned} \tag{7.37}$$

The cross-correlation operation of $H_{\mathrm{OBJ}}(\overline{r}_0, z_s)$ and $H_{\mathrm{PSH}}(\overline{r}_0, z_s)$ in the reconstruction process is expressed as

$$\begin{aligned} R &= H_{\text{OBJ}}(\bar{r}_0, \bar{r}_s) \otimes H_{\text{PSH}}(\bar{r}_0, \bar{r}_s)n \\ &= \iint \sum_n \left[a_n H_{\text{PSH}}\left(\bar{r}_0 - \frac{z_h}{z_s}\bar{r}_{s,n}; z_s \right) H^*_{\text{PSH}}(\bar{r}_0 - \bar{r}_R, z_s) \right] \\ &= \sum_n \left[a_n \Lambda \left(\bar{r}_R - \frac{z_h}{z_s}\bar{r}_{s,n} \right) \right] \approx O\left(\frac{\bar{r}}{M_T} \right), \end{aligned} \tag{7.38}$$

where Λ is a δ-like function.

Lensless interferenceless coded aperture correlation holography (LI-COACH) imaging system is similar to I-COACH, but there is no Len 2, but it is represented by a spatial light modulator.

7.5.2 *The Development of COACH*

I-COACH and LI-COACH no longer need reference light, but the system still needs multi-point spread function (PSF) for reconstruction, which has some problems such as noise. According to the characteristics of low noise correlation, the reconstructed images of masks with different coded phases can be averaged to reduce the noise contrast. Because the coded phase mask (CPM) is a scattering-like medium, the background noise generated after reconstruction will vary with CPM. After averaging several different reconstruction results, the signal-to-noise ratio will also be improved. In addition, from the perspective of designing CPM, Rai and Rosen proposed the method of improving the signal-to-noise ratio by sparse CPM [5], and the PSF of sparse CPM can effectively weaken the noise speckle problem.

COACH system has introduced matched filter (MF), phase only filter (POF), non-linear Filter (NLF) and improved phase filter reconstruction to suppress noise [6, 7]. When multiple CPMs are used, multiple PSFs synthesize point spread hologram (PSH). And MF is cross-correlated with the hologram through PSH to get the reconstruction result. The smaller the full width at half maximum of the autocorrelation peak of PSH, the higher its resolution. The smaller the average value of the non-peak part of the autocorrelation, the smaller the correlation background noise. COAH introduces phase filtering, and the spectral phase of image signal contains most of the information of the signal. Moreover, the full width at half maximum of the cross-correlation peak between the signal with only the spectrum phase and the original signal is also smaller than the autocorrelation of the original signal. However, the first two methods need multiple PSFs, while the non-linear reconstruction (NLR) algorithm only needs a single PSF [6], but there are some problems such as long reconstruction time. Although the improved cross-correlation reconstruction method in AI-COACH is not as good as nonlinear reconstruction, it can be reconstructed by a single PSF, and its speed advantage is obvious.

COACH can realize stereoscopic color imaging of monochrome camera based on 3D imaging, and the CPM has different responses to different wavelengths. The device diagram is shown in Fig. 7.12. In the I-COACH system, CPM can be combined and spliced into a lens with larger diameter by synthetic aperture, so as to realize far-field super-resolution, as shown in Fig. 7.13 [8–11].

I-COACH system records PSF on the basis of CPM, and does not directly image the object on the image sensor, so the processing accuracy is much lower than that of super-large lens. In order to ensure the resolution and quality of I-COACH, two points need to be noted: firstly, the amplitude of the spectrum plane of CPM should be

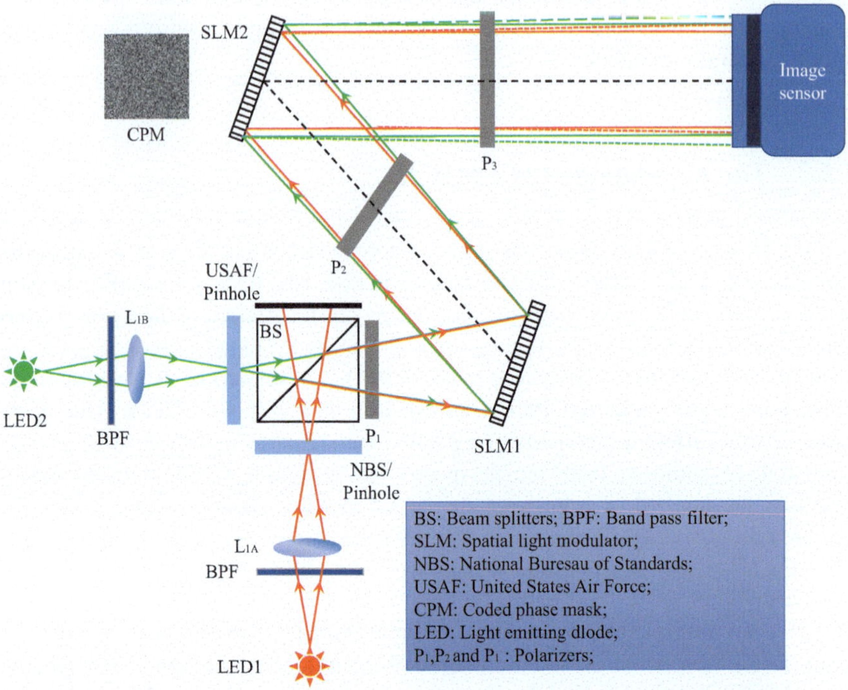

Fig. 7.12 Experimental setup for spectrum and space resolved 4D imaging by COACH

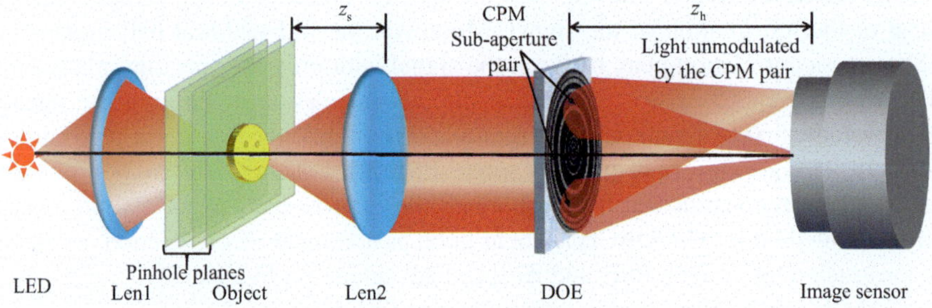

Fig. 7.13 Schematic of far-filed superresolution I-COACH

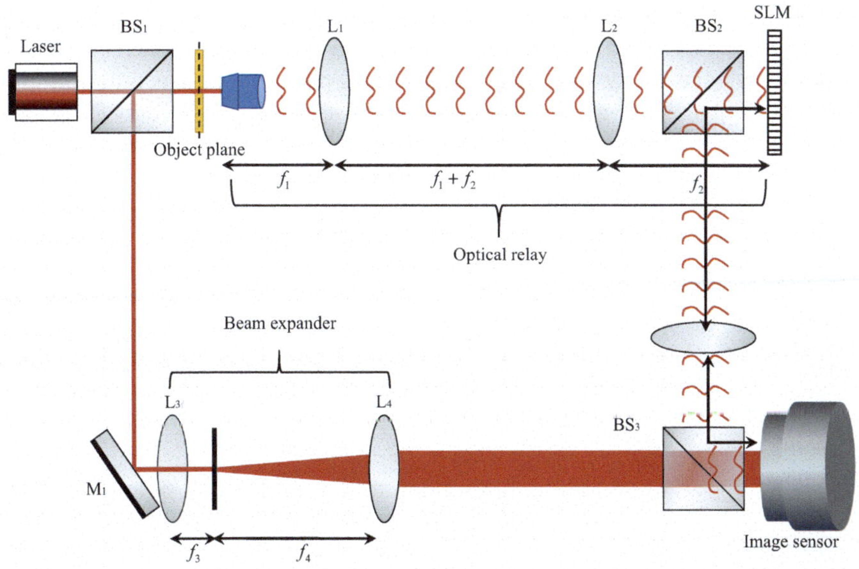

Fig. 7.14 Experimental setup for quantitative phase imaging by COACH

as uniform as possible, which can be achieved by Gerchberg-Saxon (G-S) algorithm. Secondly, CPM with high enough scattering degree is needed, and it will not limit the imaging resolution of the system. Self-interference COACH can achieve quantitative phase measurement [12], but the optical system needs to use laser and design special CPM to achieve quantitative phase measurement without speckle in large field of view, as shown in Fig. 7.14.

7.6 General Conclusions

Optical coded aperture imaging technology, in fact, is to add a mask at the pupil of the optical system, which will modulate the incident light, that is, after a point passes the coded aperture, its information will be scattered on the whole detection surface, so that the target information on the detection surface will be increased. Finally, the image on the detection surface is restored by using the image reconstruction algorithm.

In this chapter, firstly, the principle of optical coded aperture imaging is introduced, and the selection method of optimal coded template and the estimation method of point spread function in the system are discussed in detail.

Secondly, the related basis of coded aperture imaging technology is introduced, and the spatial dimension coded and spectral dimension coded structure of coded aperture spectral imaging technology are analyzed and compared in detail.

Thirdly, a multi-image encryption scheme based on time-domain compression coded aperture imaging is introduced. By using the coded mask generated by a given auxiliary random matrix to compress and encode the plaintext image in time domain, and scrambling the pixel position of the superimposed image captured by CCD, the original plaintext image is encrypted into a ciphertext image. Then the auxiliary key and the index key are transmitted to the receiver for decryption.

Finally, the principles of self-interference coded aperture correlation holography and non-interference coded aperture correlation holography system are introduced, including diffraction model, reconstruction algorithm, noise source, development and application, etc.

Acknowledgements This work was supported by the National Natural Science Foundation of China (Nos. 11874132, 61975044, 12074094, 12374271, and 62375065); Interdisciplinary Research Foundation of HIT (No. IR2021237).

References

1. Llull P, Liao X, Yuan X, Yang J, Kittle D, Carin L, Sapiro G, Brady JD. Coded aperture compressive temporal imaging. Opt Express. 2013;21(9):10526–45.
2. Vijayakumar A, Kashter Y, Kelner R, Rosen J. Coded aperture correlation holography–a new type of incoherent digital holograms. Opt Express. 2016;24(11):12430–41.
3. Vijayakumar A, Rosen J. Interferenceless coded aperture correlation holography–a new technique for recording incoherent digital holograms without two-wave interference. Opt Express. 2017;25(12):13883–96.
4. Kumar M, Vijayakumar A, Rosen J. Incoherent digital holograms acquired by interferenceless coded aperture correlation holography system without refractive lenses. Sci Rep. 2017;7(1):11555.
5. Rai MR, Rosen J. Noise suppression by controlling the sparsity of the point spread function in interferenceless coded aperture correlation holography (I-COACH). Opt Express. 2019;27(17):24311–23.
6. Rai MR, Vijayakumar A, Rosen J. Non-linear adaptive three-dimensional imaging with interferenceless coded aperture correlation holography (I-COACH). Opt Express. 2018;26(14):18143–54.
7. Wan Y, Liu C, Ma T, Qin Y, Lv S. Incoherent coded aperture correlation holographic imaging with fast adaptive and noise-suppressed reconstruction. Opt Express. 2021;29(6):8064–75.
8. Bulbul A, Vijayakumar A, Rosen J. Superresolution far-field imaging by coded phase reflectors distributed only along the boundary of synthetic apertures. Optica. 2018;5(12):1607–16.
9. Rai MR, Vijayakumar A, Rosen J. Superresolution beyond the diffraction limit using phase spatial light modulator between incoherently illuminated objects and the entrance of an imaging system. Opt Lett. 2019;44(7):1572–5.
10. Bulbul A, Rosen J. Super-resolution imaging by optical incoherent synthetic aperture with one channel at a time. Photon Res. 2021;9(7):1172–81.
11. Kashter Y, Vijayakumar A, Rosen J. Resolving images by blurring: superresolution method with a scattering mask between the observed objects and the hologram recorder. Optica. 2017;4(8):932–9.
12. Hai N, Rosen J. Coded aperture correlation holographic microscope for single-shot quantitative phase and amplitude imaging with extended field of view. Opt Express. 2020;28(19):27372–86.

Chapter 8
Phase Imaging

Ziyang Li, Xuyang Zhou, Yiran Wang, Shutian Liu, and Zhengjun Liu

8.1 Introduction

A monochromatic coherent light field can be represented by a two-dimensional complex amplitude, and the amplitude part of the complex amplitude is defined as the phase of the light field. Conventional photodetector devices can only detect the amplitude or intensity portion of the light field and cannot directly detect the phase information of the light field. Optical imaging systems based on intensity measurements, including cameras and the human eye, are sufficient for the observation of the macroscopic world in daily life. However, most biological cells and tissues are transparent or semi-transparent phase objects that absorb very little light, but the spatial distribution of their refractive indices or thicknesses is inhomogeneous, mainly in the phase portion of the light field, and thus conventional photodetector devices cannot satisfy the detection needs. In these cases, it is necessary to obtain the phase information of the target light field utilizing phase imaging [1–5].

Phase imaging techniques can be divided into two categories: qualitative phase imaging and quantitative phase imaging. Qualitative phase imaging is mainly used for microscopic observation, while quantitative phase imaging can digitally measure and analyze phase information. In this chapter, we first introduce phase contrast microscopy, a qualitative phase imaging method that converts the phase difference of a sample into an amplitude difference, allowing the observation of transparent phase objects that cannot be observed under a conventional microscope.

This chapter focuses on the quantitative phase imaging. Firstly, interferometric phase imaging and digital holographic imaging based on the principle of interference are introduced. Among the non-interferometric methods, wavefront sensor phase

Z. Li · X. Zhou · Y. Wang · S. Liu · Z. Liu (✉)
School of Physics, Harbin Institute of Technology, Harbin 150001, China
e-mail: zjliu@hit.edu.cn

S. Liu
e-mail: stliu@hit.edu.cn

Z. Liu et al. (eds.), *Computational Optical Imaging*, Advances in Optics and Optoelectronics, https://doi.org/10.1007/978-981-97-1455-1_8

imaging is introduced to solve the phase distribution of the light wave field by analyzing the wavefront gradient.

In the following part, the phase retrieval method by solving the mathematical inverse problem is introduced. The method requires only measuring the intensity distribution information of the optical wavefield, which greatly reduces the imaging difficulty.

Regarding single-frame phase retrieval algorithms, this chapter introduces the Gerchberg-Saxton (GS) algorithm and the hybrid input-output (HIO) algorithm. Regarding multi-frame phase retrieval algorithms, this chapter introduces the single-beam multiple-intensity phase reconstruction (SBMIR) and the amplitude-phase phase retrieval (APR) algorithms for coaxial multi-distance coherent diffraction imaging (MDCDI) system, the multi-wavelength illumination microscopy, and the ptychographic iterative engine (PIE) imaging system. The Fourier ptychographic microscopy (FPM) imaging system extends the methods of PIE from the spatial domain to the frequency domain, and realizes quantitative phase imaging through the method of Fourier spectral extension.

In the final part, the non-iterative phase imaging methods are introduced. The transport of intensity equation (TIE) computes the phase distribution in the sample plane by solving a partial differential equation, without the need for iterative solutions. Differential phase contrast (DPC) utilizes either a split light source or a split detector to reconstruct an intensity image that is linearly correlated with the phase gradient of the sample by comparing the differences in the intensity distributions collected by the detectors.

8.2 Phase Contrast Microscopy

Most biological cells and tissues are transparent for microscopic imaging and absorb very little light, so observation using conventional bright-field microscopy suffers from low imaging contrast and does not produce distinct diffraction patterns, making it difficult to observe cellular details. To overcome this shortcoming, researchers have devised various exogenous labeling strategies, among which fluorescent labeling means utilizing the affinity of different cell parts for specific dyes to emit distinctly differentiated fluorescence spectra to improve the observation contrast. However, the photobleaching properties and the phototoxicity of fluorescents may interfere with endogenous biological processes and pose a serious challenge for prolonged imaging of living cells [6]. Since then, new ideas of label-free imaging have been explored, by measuring the phase change of light waves before and after they pass through the sample to determine the internal structure of the sample.

The structural features of most colorless and transparent biological cells and the different refractive index distributions of the components can produce phase delays with different spatial distributions of the incident light waves. In 1942, Zernike et al. made it easier to observe the details of a target by introducing a phase difference into

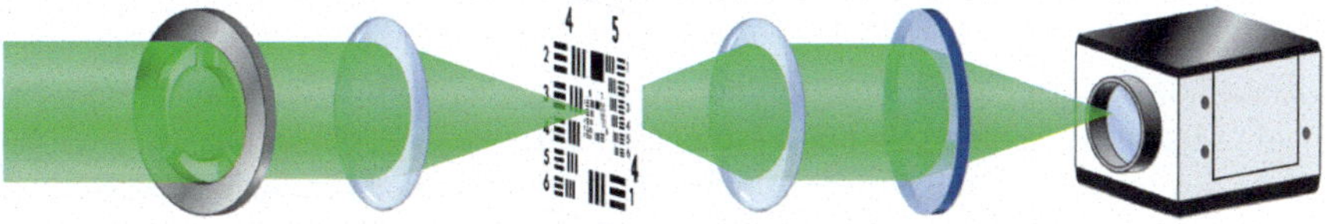

Fig. 8.1 The optical system of the phase contrast microscopy. CA: condense annulus. CL: condenser lens. OL: objective lens. PL: phase plate

the optical system, which enhances the contrast of a transparent object by altering the phase of the light [7].

The optical system of the phase contrast microscopy is shown in Fig. 8.1. The light from the ring light source is collimated and directed onto the sample plane. The light that passes directly through the sample is the background light, which converges to the focal plane through the objective lens. The diffracted light passing through the sample is dispersed on the focal plane. A phase plate with an additional phase delay is placed at the focal plane to change the phase relationship between the background light and other diffracted light. By changing the additional phase difference, the light intensity distribution in the image plane can be changed, and the background light can also be weakened to improve the imaging contrast.

Phase contrast microscopy is based on the principle of spatial filtering to change the spectral phase of the object light wave, and successfully convert the phase difference into amplitude difference, thus greatly improving the distinguishability of transparent objects under the optical microscope. Zernike won the 1953 Nobel Prize in Physics for his work, which provided a new method of observing transparent objects in the microscope and was of great significance to research in the fields of biology, medicine, and materials science.

8.3 Interferometric Phase Imaging

Although phase contrast microscopy enables in vivo observation of transparent biological cells, the relationship between the measured intensity and the sample phase is not strictly linear, thus the visualization of phase information is only at a qualitative level. With the advent of digital image sensors and the improvement of information optics theory, quantitative phase imaging has become a major important direction in the field of imaging technology.

Interferometric phase imaging is the classical method of quantitative phase imaging. Utilizing the interference effect of coherent light, the invisible phase information is converted into visible interference fringes by spatial superposition of two beams of coherent light, which in turn can be recorded by a photodetector.

After obtaining the interference fringes, the phase can be demodulated from the interferogram by the fringe analysis technique to realize the phase acquisition.

Based on the principle of interference effect of coherent light, quantitative phase imaging technology based on holography interferometry is epochal progress in the optical discipline, and the ability of holograms to reproduce the three-dimensional image of an object is incomparable to other technologies. However, the recording of holograms usually involves a series of cumbersome processes such as exposure, development, and fixation, which makes it difficult to achieve real-time recording and reproduction. In addition, the image reconstruction process still requires reference beam irradiation, which can only observe the virtual image of the object from the naked eye, and cannot be recorded and quantitatively analyzed.

In 1967, Goodman and Lawrence first proposed the idea of digital holography, which is a typical computational imaging system [8]. Its recording optical path and ordinary holography are basically the same, the difference is that the use of charge coupled device (CCD) and other light-sensitive electronic components instead of ordinary photographic dry plate to shoot holograms, and will be recorded digital holograms stored in the computer, and then use digital computing methods for digital reproduction of the hologram, as shown in Fig. 8.2. The coherent laser is divided into object light and reference light by a beam splitter, and the object light is reflected by the sample object and interferes with the reference light in the camera plane, and the reference light reflection angle is adjusted to realize the three-dimensional reconstruction of the sample object.

Despite the rapid development of interferometric phase imaging represented by digital holography, the basic principle is still interferometry: by introducing additional coherent reference light, the invisible phase information is converted into intensity signals such as interferometric fringes, so that it can be captured by conventional imaging devices, and through a series of fringe analysis algorithms, the phase can then be demodulated from the interferogram. However, the interference-based phase retrieval technique requires extremely stringent measurement conditions. First of

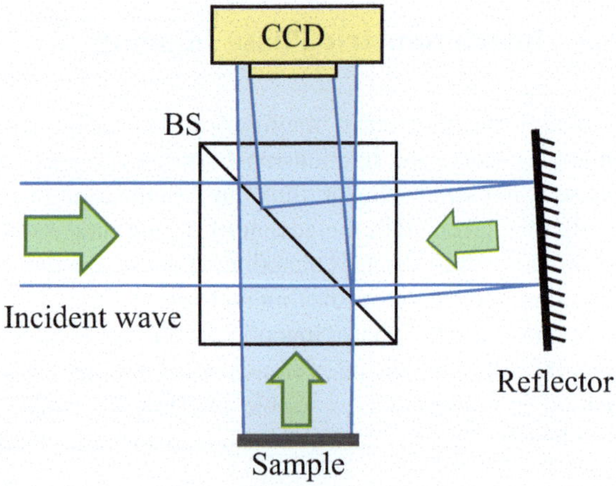

Fig. 8.2 Digital holographic phase measurement device. BS: beam splitter

all, in order to guarantee measurement accuracy, the imaging system must require a highly temporally and spatially coherent light source; at the same time, the introduction of additional reference light paths leads to more stringent requirements for airflow disturbances and environmental vibrations.

8.4 Wavefront Sensor Phase Imaging

Wavefront sensing phase imaging is a non-interferometric phase imaging technique. Considering the case that the incident light wave is a plane wave, its wavefront should be a flat surface, however, when the plane wave passes through the diffracted sample, the wavefront is no longer flat but shows different gradients locally, as shown in Fig. 8.3. Phase imaging of the sample can be realized by comparing the wavefront aberration of the beam passing through the sample, which is defined as the deviation between the ideal wavefront and the actual wavefront.

A Hartmann screen is a special type of diaphragm that controls the transmission of light at different locations utilizing holes arranged in a certain regular pattern, which are ultimately incident on the receiving screen with different gradients. When the gradient at each point is determined, the shape of the beam wavefront can be inverted. The Hartmann screen has low energy utilization for light waves, so Shack et al. improved the conventional Hartmann screen by using a microlens array as a replacement [9]. The optical wavefield to be measured was focused into a series of lattice arrays utilizing a microlens array. By analyzing the displacement of the actual wavefront relative to the reference ideal aberration-free case of the regular spacing of the lattice array, the wavefront gradient of the wavefield to be measured can be obtained, and its integration can be obtained from the distribution of the wavefront phase. The principle is shown in Fig. 8.4. The wavefront sensing phase

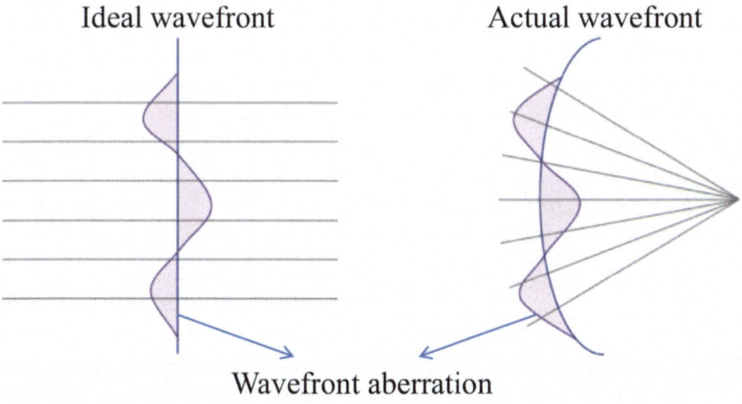

Fig. 8.3 The schematic diagram of wavefront aberration

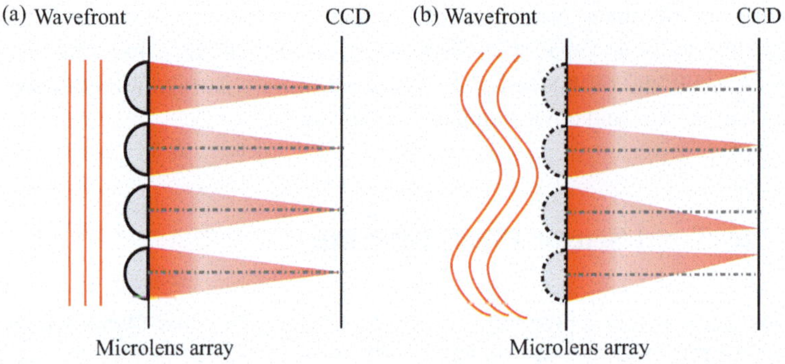

Fig. 8.4 The principle of wavefront sensor phase imaging: **a** the reference ideal aberration-free case, **b** the actual wavefront and its lattice distribution

imaging system has no special requirements for optical hardware, low environmental requirements, no need for calibration, a compact optical path, and good measurement accuracy. However, due to the limitation of the physical size of the Hartmann screen and microlens array, the detected signals do not fully utilize all the effective pixels of the imaging device, resulting in constraints between spatial resolution and phase.

8.5 Single-Frame Phase Retrieval Algorithm

Since it is extremely difficult to directly measure the phase distribution of a light wave field, it is easier to measure the information of the amplitude or intensity distribution of the light wave field. Therefore, the process of reconstructing the phase from the amplitude or intensity distribution can be considered as a mathematical inverse problem, and the algorithms that solve this type of inverse phase problem from the light intensity inversion are collectively referred to as phase retrieval algorithms, and the algorithm that requires only one measurement image is known as single-frame phase retrieval algorithm.

8.5.1 GS Algorithm

In 1972, Gerchberg and Saxton first proposed a single-frame phase retrieval algorithm [10], called the GS algorithm, which reconstructs the phase information by using the intensity image of the spatial and the Fourier domain. When the Fourier intensity of the samples in the spatial and frequency domain is known, the phase distribution can be reconstructed through iteration.

First, define the amplitude A_1, A_2 in the spatial and frequency domains and initialize the spatial domain phase φ with random data or constants. Based on the initialization input, combining the complex amplitude $f = A_1 \cdot \exp(i\varphi)$ in the spatial domain. Assuming that the transformation relation is between the spatial and Fourier domains, the Fourier transform of f is calculated and $F = |F| \cdot \exp(i\theta)$ is obtained. Apply Fourier domain constraints and replace the existing amplitude with the known frequency domain amplitude A_2 to synthesize the frequency domain complex amplitude $F' = A_2 \cdot \exp(i\theta)$. Finally, the inverse Fourier transform is computed and spatial domain constraints are imposed to synthesize the spatial domain complex amplitude and proceed to the next iteration. When the specified number of iterations is reached or the evaluation criteria of image quality are satisfied, the iteration is stopped, and the spatial domain and frequency domain phase distribution φ and θ are obtained. The algorithm process is shown in Fig. 8.5.

The total energy of the spatial and frequency domain amplitudes needs to be balanced before calculations can be performed. For some non-dissipative systems, such as Fresnel diffraction and Fourier transforms, the energy integrals in the two planes need to be equal. Without energy matching, unstable results will occur when calculating the amplitude error.

Here are the simulation results of the GS algorithm. Define Fig. 8.6a as a spatial domain constraint and Fig. 8.6b as a frequency domain constraint. After 100 iterations, the reconstructed phase of spatial domain and frequency domain phase are shown in Fig. 8.6c and d.

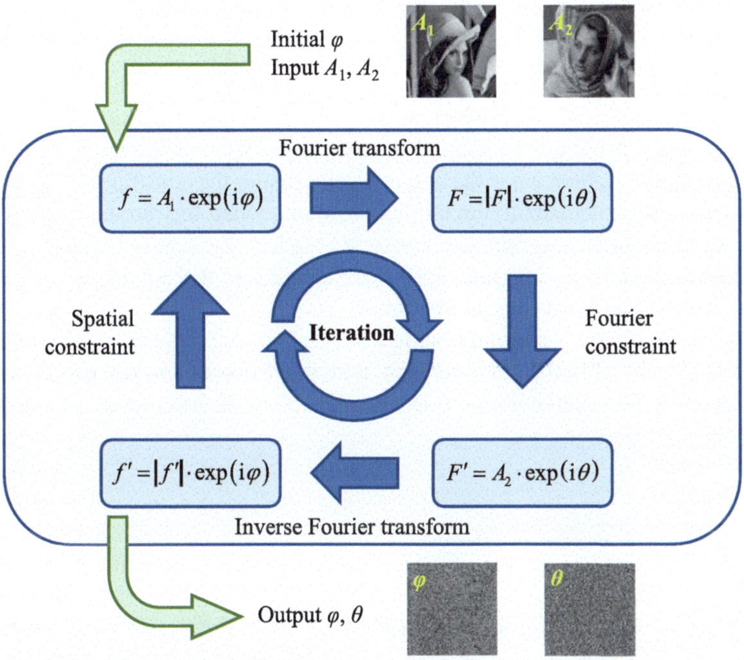

Fig. 8.5 The process of GS algorithm

Fig. 8.6 The simulation result of GS algorithm: **a** the amplitude of the spatial domain, **b** the amplitude of the frequency domain, **c** the reconstructed phase of the spatial domain, **d** the reconstructed phase of the frequency domain

The GS algorithm, as a classical phase retrieval algorithm, also has several limitations. Firstly, the GS algorithm is sensitive to the initial value, and different initial value settings will lead to different output results. Secondly, the GS algorithm lacks a global search mechanism and is prone to convergence stagnation during the iteration process. Finally, when the data contains large noise, it will not be possible to reconstruct the desired results.

8.5.2 HIO Algorithm

In 1982, Fienup introduced the negative feedback mechanism into the phase retrieval algorithm and proposed the HIO algorithm [11]. The output result of the current iteration is not considered the optimal solution and is used as the input for the next round of iteration. Rather, when the output result satisfies the constraints of the spatial domain, no change is made to the output. However, when the output results do not satisfy the constraints, appropriate changes are made to the output to make it satisfy the constraints. Thus, by making appropriate changes to the inputs, the output can be pushed in the desired general direction.

The process of HIO algorithm is shown in Fig. 8.7. The computational procedure is roughly equivalent to the GS algorithm, with the difference being the updating of the phase in the new spatial domain. The updated phase function $\varphi'(x, y)$ is given as

$$\varphi'(x, y)=\begin{cases} \theta(x, y), & (x, y) \in \gamma, \\ \varphi(x, y) - \beta \cdot \theta(x, y), & (x, y) \notin \gamma, \end{cases} \tag{8.1}$$

where β is a relaxation factor selected based on specific circumstances, usually defined as a constant. γ is the region of interest.

Here are the simulation results of the HIO algorithm. Figure 8.8a is selected as the reference image and the center is selected as the region of interest. The relaxation factor β is 0.8, iterated 1000 times. Figure 8.8b shows the reconstruction results of the HIO phase retrieval algorithm. In order to evaluate the convergence of phase retrieval,

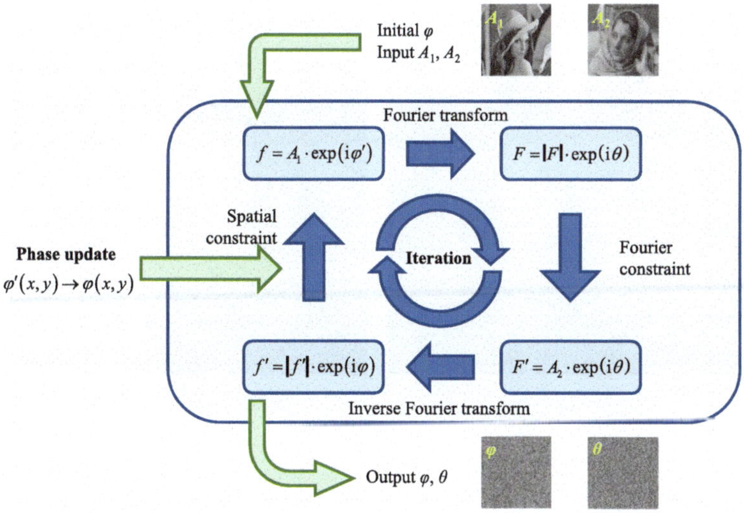

Fig. 8.7 The process of HIO algorithm

the logarithm of mean square error (LMSE) of the reference and reconstructed images for different numbers of iterations can be calculated. The calculation formula for LMSE is defined as:

$$\text{LMSE} = \log_{10}\left\{\sum_{x,y} [f(x, y) - f_0(x, y)]^2\right\}, \tag{8.2}$$

where $f_0(x, y)$ is the function of the reference image. The LMSE for different numbers of iterations is shown in Fig. 8.8c. The reconstruction results in good convergence.

The HIO algorithm has better convergence and noise robustness compared to the GS algorithm. However, the a priori information of the samples provided by the single-frame phase retrieval algorithm is still limited, which makes it difficult to reconstruct high-resolution images stably in practice.

8.6 Multi-frame Phase Retrieval Algorithm

In the process of phase retrieval, in order to ensure the existence and uniqueness of the solution and to avoid problems such as convergence hysteresis due to the iteration being trapped in local minima, the redundancy of the information can be increased by adding more measurements so that the constraints of more measurements can be utilized to improve the convergence and reliability of the algorithm.

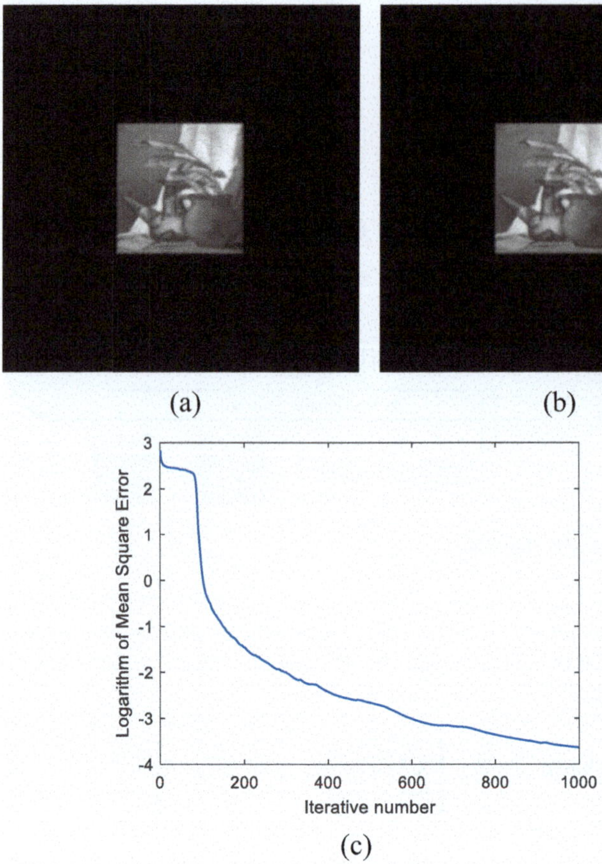

Fig. 8.8 The simulation result of HIO algorithm: **a** the reference image, **b** the reconstruction image after 1000 times of iterations, **c** the LMSE for different numbers of iterations

8.6.1 Single-Beam Multiple-intensity Phase Reconstruction Algorithm

Multi-frame phase retrieval applied to coaxial MDCDI is defined as multi-distance phase retrieval algorithms, and the imaging system is shown in Fig. 8.9. A fiber-coupled laser is condensed and directed, and the detector collects intensity of the diffraction images $I_1, I_2, \ldots I_N$ at N different distances, which is determined by the complex amplitude $E_n(x, y; z_n)$ on the nth measurement plane $(n = 1, 2, \ldots, N)$. The complex amplitude and the collected intensity by CCD can be represented as [12]:

$$E_n(x, y; z_n) = \mathcal{F}^{-1}\left\{\mathcal{F}[f(x, y)] \cdot \exp\left[\mathrm{i}\frac{2\pi z_n}{\lambda}\sqrt{1 - (\lambda f_x)^2 - \left(\lambda f_y\right)^2}\right]\right\}, \quad (8.3)$$

$$I_n(x, y; z_n) = E_n(x, y; z_n) \cdot E_n^*(x, y; z_n), \tag{8.4}$$

where $E_n^*(x, y; z_n)$ denotes the complex conjugate of $E_n(x, y; z_n)$. $\mathcal{F}$ and $\mathcal{F}^{-1}$ represent the Fourier transform and inverse Fourier transform. $f(x, y)$ represents the transmittance function of the sample, λ is the wavelength of the incident wave, f_x and f_y are spatial frequency coordinates determined by the image size and pixel dimensions. After obtaining the diffraction intensity information, we reconstruct the complex hologram of the sample by using a multi-distance phase retrieval algorithm.

Multi-distance phase retrieval algorithms can be divided into two main categories: serial transmission and parallel weighting. The representative algorithm for serial-transfer phase retrieval is the SBMIR algorithm [13], and the principle is shown in Fig. 8.10.

Firstly, define the light field of the first detection plane:

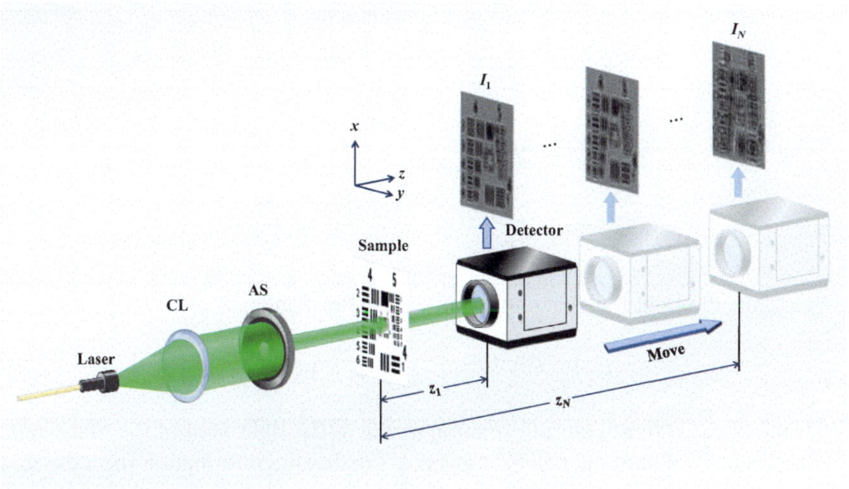

Fig. 8.9 The imaging system of coaxial MDCDI system. CL: condenser lens; AS: aperture stop

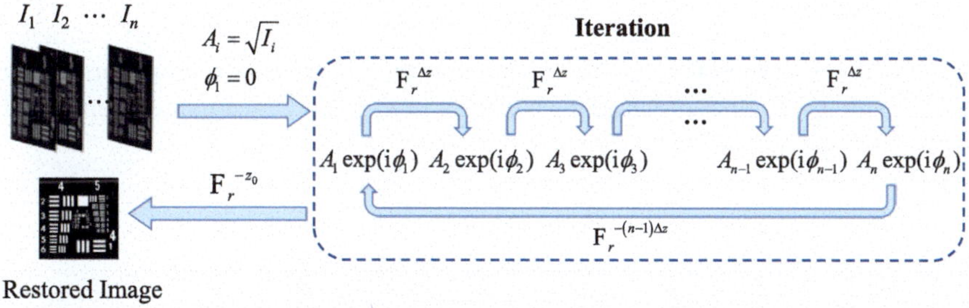

Fig. 8.10 The principle of SBMIR algorithm

$$E_1 = A_1 \exp(i\phi_1), \tag{8.5}$$

in which $A_1 = \sqrt{I_1}$, and ϕ_1 is a zero or random matrix. Then propagate the light field information of the first plane to the next detection plane:

$$E_2' = \mathcal{F}_r^{\Delta z}(E_1) = A_2' \exp(i\phi_2), \tag{8.6}$$

in which $\mathcal{F}_r^{\Delta z}$ represents the propagation of Fresnel diffraction, with a propagation distance of Δz. After that, the calculated light field amplitude is replaced with the a priori information recorded by the detector:

$$E_2 = \frac{E_2'}{|E_2'|}\sqrt{I_2} = A_2 \exp(i\phi_2). \tag{8.7}$$

Repeat the propagation process until it reaches the last detector plane, where the complex amplitude of the light field is represented as:

$$E_n' = \mathcal{F}_r^{\Delta z}(E_{n-1}) = A_{n-1}' \exp(i\phi_{n-1}). \tag{8.8}$$

Then update the light field amplitude:

$$E_n = \frac{E_n'}{|E_n'|}\sqrt{I_n} = A_n \exp(i\phi_n). \tag{8.9}$$

Propagate the light field back to the first detector plane:

$$E_1 = \mathcal{F}_r^{(n-1)\Delta z}(E_n) = A_1' \exp(i\phi_1). \tag{8.10}$$

Set the updated light field information as the initial light field of the first plane, repeat the iteration until the results converge. Finally, the information from the first detector plane is propagated to the sample plane to obtain the reconstructed image.

Here are the simulation results of the SBMIR algorithm. The center wavelength of the laser is 532 nm, the pixel size is 1.67 μm, and the sample is shown in Fig. 8.11a. The initial diffraction distance is 1 mm, and 7 diffraction images with the spacing is 0.1 mm are taken, and the number of iterations is 100 times. The sample reconstructed by SBMIR is shown in Fig. 8.11b. The relationship of normalized cross-correlation (NCC) with the number of iterations is shown in Fig. 8.11c, and the relationship of LMSE with the number of iterations is shown in Fig. 8.11d.

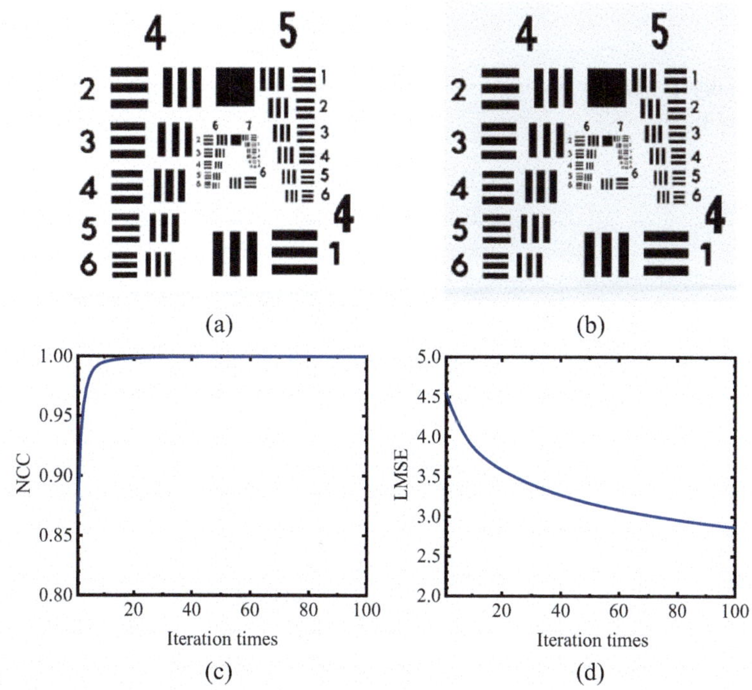

Fig. 8.11 The simulation result of SBMIR algorithm: **a** the original sample, **b** the SBMIR reconstruction sample, **c** the NCC for different numbers of iterations, **d** the LMSE for different numbers of iterations

8.6.2 *Amplitude-Phase Retrieval Algorithm*

The representative algorithm for parallel-transmission phase retrieval is the APR algorithm [14], and the principle is shown in Fig. 8.12.

Firstly, define the light field of the first detection plane:

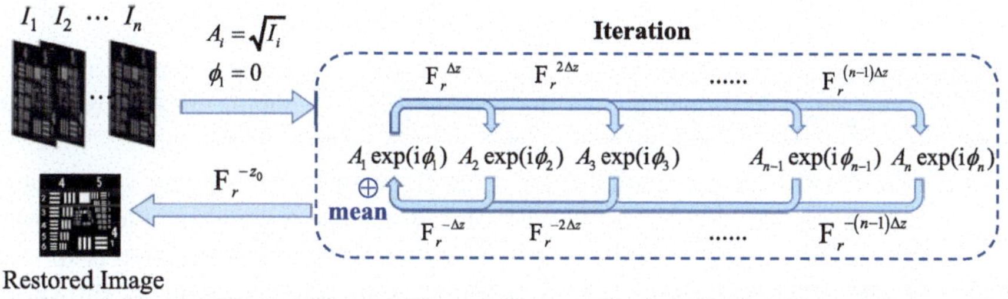

Fig. 8.12 The principle of APR algorithm

$$E_1 = A_1 \exp(i\phi_1). \tag{8.11}$$

Secondly, propagate the light field information from the first plane to the 2nd to nth detection planes:

$$\begin{aligned} E'_2 &= \mathcal{F}_r^{\Delta z}(E_1) = A'_2 \exp(i\phi_2), \\ E'_3 &= \mathcal{F}_r^{2\Delta z}(E_1) = A'_3 \exp(i\phi_3), \\ &\cdots \\ E'_n &= \mathcal{F}_r^{(n-1)\Delta z}(E_1) = A'_n \exp(i\phi_n). \end{aligned} \tag{8.12}$$

Then, replace the amplitude of the light field in different planes with the intensity recorded by the detector respectively:

$$\begin{aligned} E_2 &= \sqrt{I_2} \exp(i\phi_2) = A_2 \exp(i\phi_2), \\ E_3 &= \sqrt{I_3} \exp(i\phi_2) = A_3 \exp(i\phi_3), \\ &\cdots \\ E_n &= \sqrt{I_n} \exp(i\phi_n) = A_n \exp(i\phi_n). \end{aligned} \tag{8.13}$$

Propagate the updated light field information back to the first camera plane:

$$\begin{aligned} E_{2,1} &= \mathcal{F}_r^{-\Delta z}(E_2) = A_{2,1} \exp(i\phi_{2,1}), \\ E_{3,1} &= \mathcal{F}_r^{-2\Delta z}(E_2) = A_{3,1} \exp(i\phi_{3,1}), \\ &\cdots \\ E_{n,1} &= \mathcal{F}_r^{-(n-1)\Delta z}(E_n) = A_{n,1} \exp(i\phi_{n,1}). \end{aligned} \tag{8.14}$$

Take the average of the light field propagated back to the first plane, and set the updated light field information as the initial light field of the first plane.

$$E_1 = \frac{1}{n} \sum_{i=1}^{n} E_{i,1}. \tag{8.15}$$

Finally, repeat the iteration until the results converge. The information from the first detector plane is propagated to the sample plane to obtain the reconstructed image.

Here are the simulation results of the APR algorithm. The imaging system and parameter settings are the same as SBMIR algorithm. The original sample is shown in Fig. 8.13a. The sample constructed by APR after 100 times of iteration is shown in Fig. 8.13b. The relationship of NCC with the number of iterations is shown in Fig. 8.13c, and the relationship of LMSE with the number of iterations is shown in Fig. 8.13d.

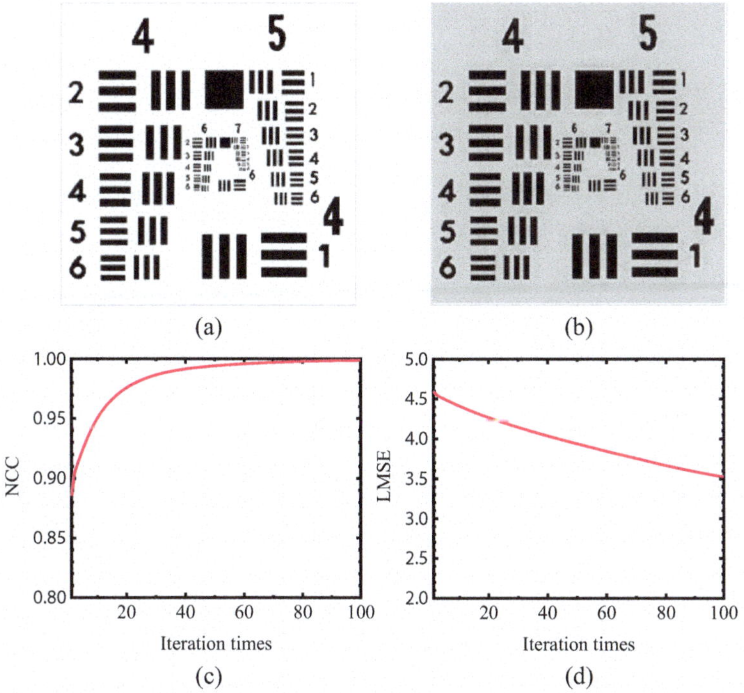

Fig. 8.13 The simulation result of APR algorithm: **a** the original sample, **b** the APR reconstruction sample, **c** the NCC for different numbers of iterations, **d** the LMSE for different numbers of iterations

Next, we compare the noise robustness of the SBMIR and APR algorithms. When the samples are added with Gaussian noise with variance equal to 0.1, the results after 100 iterations of the SBMIR and APR algorithms are shown in Fig. 8.14a, b, respectively, The NCC comparison results of SBMIR and APR are shown in Fig. 8.14c, and the LMSE comparison results are shown in Fig. 8.14d. Compared to SBMIR, APR has better noise robustness and can reduce the impact of environmental noise, light source coherence, and other factors on reconstruction quality.

8.6.3 Multiwavelength Illumination Microscopy

The MDCDI system requires moving the detector along the axis for scanning, however, in some imaging scenarios, the detector is spatially constrained and cannot move freely. Considering that a change in wavelength is equivalent to a change in the optical path, the light source is replaced with a multi-wavelength light source, and changing the wavelength of the incident light is physically equivalent to changing the distance of the detector from the sample.

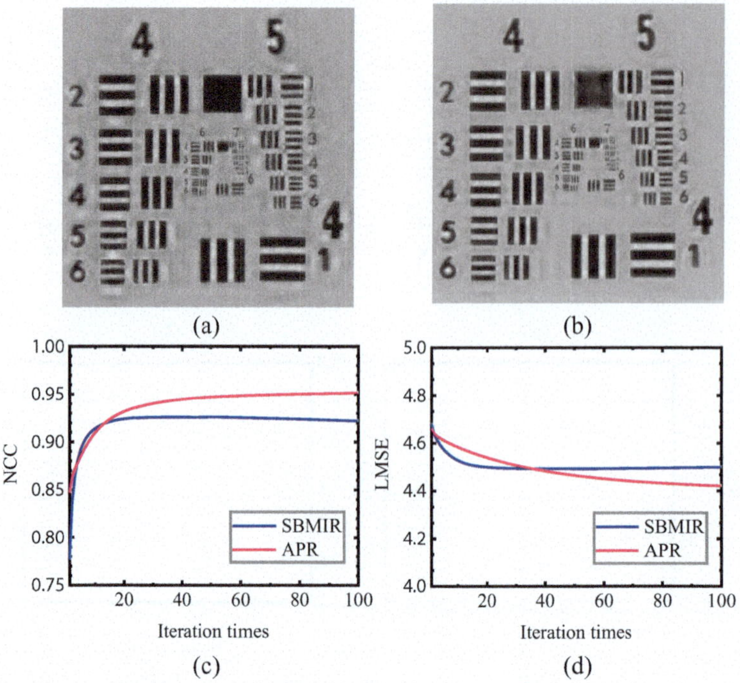

Fig. 8.14 The noise robustness comparison of the SBMIR and APR algorithms: **a** SBMIR reconstructed sample, **b** APR reconstructed sample, **c** NCC comparison results, **d** LMSE comparison results

The multiwavelength illumination microscopy imaging system is shown in Fig. 8.15 [15]. The multi-wavelength light source emits a coherent beam of light to illuminate the sample, and the detector records the diffraction intensity distribution of the sample. The light wave at the plane between the sample and the detector should be approximately a plane wave. Therefore, the distance z_0 between the light source and the sample must be large enough, and the distance z_1 between the sample and the detector must be small enough, usually less than 1 mm [16].

The initial monochromatic coherent light wave field is diffracted through the sample and the detector receives the intensity information. After matching with the phase distribution and transmitted to the sample plane by inverse diffraction, the amplitude and phase of the light wave field are updated. Then the light wave field at the next wavelength in the sample plane and the detector plane is calculated. The above process is iterated until the error is less than the set threshold, and the amplitude and phase distribution of the light wave field in the sample plane is output.

Exactly because the change to the wavelength is equivalent to the change of the optical range, when the light source emits different wavelengths, the light intensity distributions received by the detector are equivalent to the light intensity distributions received by the detector of the MDCDI experimental system at different distances.

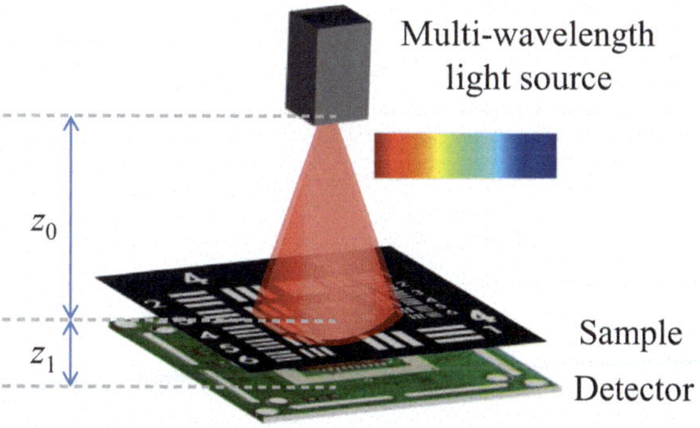

Fig. 8.15 The imaging system of the multiwavelength illumination microscopy imaging system

The light intensity distribution is used as an input to the SBMIR or APR phase retrieval algorithm, and the corresponding diffraction distance is adjusted according to the incident wavelength, and the reconstruction results of the samples can be obtained after the multi-frame phase retrieval algorithm.

8.6.4 Ptychographic Iterative Engine

Since conventional coherent diffraction imaging requires isolated imaged samples, the field-of-view is greatly limited. In 2004, Rodenburg et al. proposed a lensless PIE imaging system by combining ptychographic scanning with phase retrieval [17]. The PIE moves an illumination point confined to a small aperture relative to the sample, scans and records diffraction points at different locations, and finally reconstructs the sample by phase retrieval. This method requires overlap between neighboring regions, and it is the high degree of redundancy in the information of the large amount of light intensity data acquired by ptychographic imaging that allows the non-convexity of the phase retrieval problem to be greatly mitigated, resulting in improved spatial resolution and a wider field-of-view.

The imaging system of the PIE imaging system is shown in Fig. 8.16. The illumination beam $p(r)$ limited by the pupil is incident on the sample $O(r)$, and the pupil or sample is fixed on a two-dimensional displacement device for array scanning. The PIE process is as follows:

Assuming the sample is thin enough, the aperture function at the nth position is:

$$p_n(r) = p(r - r_n), \tag{8.16}$$

in which $p(r)$ can be presented as:

$$p(r) = \begin{cases} 1, & \text{if } r \leq d/2, \\ 0, & \text{if } r > d/2, \end{cases} \tag{8.17}$$

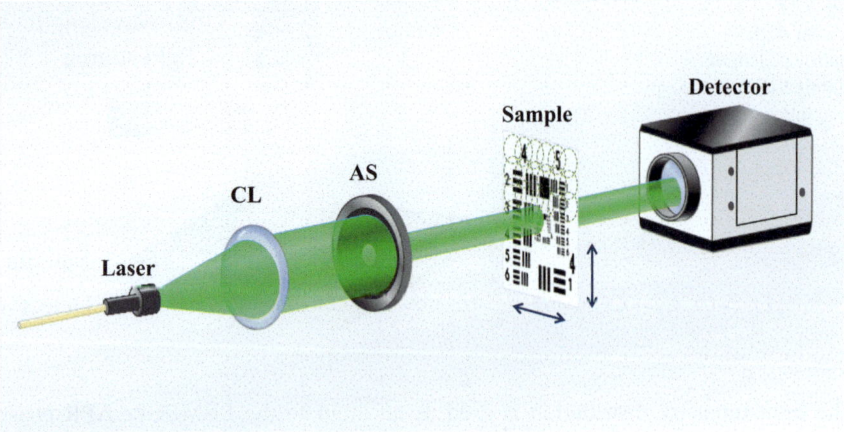

Fig. 8.16 The imaging system of the PIE imaging system

where d is the diameter of the aperture. The output function of the sample is:

$$\varphi_n(r) = O_n(r)p(r - r_n), \tag{8.18}$$

where $O_n(r)$ is the light field at the sample plane in the nth iteration. When $n = 1$, $O_n(r)$ is a constant or random matrix. The light field in the detector plane is:

$$\varphi_n\left(r'\right) = \mathrm{A}[\varphi_n(r)], \tag{8.19}$$

where A is the angular spectrum propagation operator. Replacing the amplitude of the calculated light field with the amplitude of the light field recorded by the detector:

$$\varphi_n'\left(r'\right) = \frac{\varphi_n\left(r'\right)}{|\varphi_n(r')|}\sqrt{I_n}. \tag{8.20}$$

Propagate the updated light field to the sample plane:

$$\varphi_n'(r) = \mathrm{A}^{-1}\left[\varphi_n'\left(r'\right)\right], \tag{8.21}$$

where A^{-1} is the angular spectrum inverse propagation operator. The sample function is then updated:

$$O_{n+1}(r) = O_n(r) + \frac{p_n(r)}{|p_n(r)|_{\max}} \frac{p_n^*(r)}{\left(|p_n(r)|^2 + \alpha\right)}\beta\left[\varphi_n'(r) - \varphi_n(r)\right]. \tag{8.22}$$

Calculate the light field propagated from the next aperture of the sample to the camera plane through angular spectrum diffraction:

$$\varphi_{n+1}(r) = O_{n+1}(r)p(r - r_{n+1}),$$
$$\varphi_{n+1}(r') = \mathrm{A}[\varphi_{n+1}(r)]. \tag{8.23}$$

Two adjacent apertures need to ensure an overlap rate of no less than 60%. Iterate the above process until the error threshold is satisfied.

The PIE algorithm requires the spatial distribution of illumination light to be accurately determined, but under actual experimental conditions, the distribution of illumination light is easily affected by factors such as position disturbances of optical components and mechanical errors of displacement tables. Extended layered scanning imaging technology (ePIE) can simultaneously update the sample function and beam probe function during the iteration process, improving the reconstruction accuracy and convergence speed in the complex amplitude reconstruction process of objects [18].

In the ePIE algorithm, the sample function update formula is:

$$O_{n+1}(r) = O_n(r) + \alpha \frac{p_n^*(r)}{|p_n(r)|_{\max}^2}[\varphi_n'(r) - \varphi_n(r)]. \tag{8.24}$$

And the beam probe update equation is:

$$p_{n+1}(r) = p_n(r) + \beta \frac{O_n^*(r)}{|O_n(r)|_{\max}^2}[\varphi_n'(r) - \varphi_n(r)]. \tag{8.25}$$

The regularized ptychographical engine (rPIE) introduces the concept of low-order mixed states into the solution process of psychographic imaging by adding a regularization term to the equation for solving the object function, where the regularization term is not a constant but a spatially varying coefficient whose magnitude depends on the intensity distribution of the illumination spot [19]. The regularization term is not a constant but a spatially varying factor whose size depends on the intensity distribution of the illumination spot. By matching the appropriate regularization coefficients, the method has a faster convergence rate and less stringent initialization requirements for the illumination spot. In the rPIE algorithm, the sample function update formula is:

$$O_{n+1}(r) = O_n(r) + \frac{p_n^*(r)[\varphi_n'(r) - \varphi_n(r)]}{(1-\alpha)|p_n(r)|^2 + \alpha |p_n(r)|_{\max}^2}. \tag{8.26}$$

8.6.5 *Fourier Ptychographic Microscopy*

In 2013, Zheng et al. further extended ptychographic imaging from the spatial domain to the frequency domain and proposed the FPM [20]. The FPM imaging system includes a conventional microscope with a low numerical aperture and an LED array light source, as shown in Fig. 8.17. Using the LED array light source, images were acquired using a low-magnification objective lens in different illumination directions. According to the Fourier frequency shift property, the spectrum is shifted by the phase shift in the null domain, and the high-frequency information beyond the cutoff frequency becomes measurable. Compared with conventional optical microscopy, FPM can achieve high spatial resolution with a large field-of-view.

The main process of FPM is as follows. Firstly, initialize the high-resolution image:

$$f_h = \sqrt{I_h}\exp(\mathrm{i}\varphi_h), \tag{8.27}$$

where I_h and φ_h can be defined as zero matrices. Correspondingly, initialize the Fourier spectrum as:

$$F_h = \mathrm{F}\,(f_h).$$

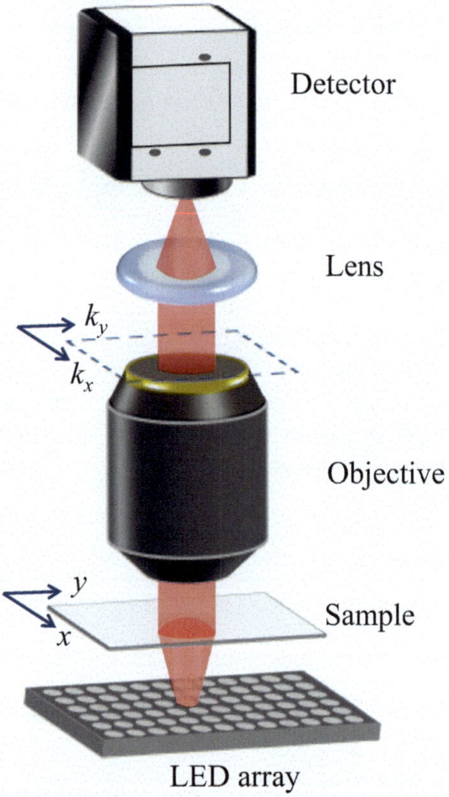

Fig. 8.17 The imaging system of the FPM imaging system

$$F_h = \mathcal{F}(f_h). \tag{8.28}$$

A small subregion is selected in the Fourier domain of the initial high-resolution image, the objective lens is equivalent to a low-pass filter, the filter function of the objective lens in Fourier space H_l is a circular region of radius NA $\cdot$ k_0, in which NA is the numerical aperture, $k_0{=}2\pi / \lambda$ is the wavenumber in vacuum, and λ is the wavelength of the incident light as

$$F_l = F_h \cdot H_l,\, H_l = \text{circ}(\text{NA} \cdot k_0). \tag{8.29}$$

The inverse Fourier transform is applied to generate a new low-resolution target image:

$$f_l = \mathcal{F}^{-1}(F_l) = \sqrt{I_l}\exp(\mathrm{i}\varphi_l). \tag{8.30}$$

Update the amplitude information with the intensity information captured by the detector and update the high-resolution image of the corresponding region in the Fourier domain:

$$\begin{aligned} f_{lm}^i &= \sqrt{I_m}\exp(\mathrm{i}\varphi_l),\\ F_h^i\left(k_x^i, k_y^i\right) &= \mathcal{F}\left(f_{lm}^i\right), \end{aligned} \tag{8.31}$$

in which I_m is the measured intensity, k_x^i and k_y^i are Fourier domain coordinates.

Change the orientation of the LED array illumination such that each shifted subregion corresponds to a unique low-pass filter position, and ensure that neighboring subregions overlap to achieve convergence. Repeat the above steps until the entire high-resolution image in Fourier space has been modified by updating all low-resolution intensity measurements.

$$F_h\left(k_x^N, k_x^N\right) = F_h^N\left(k_x^N, k_y^N\right), \tag{8.32}$$

in which N is the total number of LED lighting. It has been a complete iteration cycle. Repeat the iteration until the Fourier spectrum converges. Transform Fourier domain spectrum into spatial domain to restore high-resolution field.

$$f_h = \mathcal{F}^{-1}(F_h). \tag{8.32}$$

Differently from PIE, FPM replaces probe movement with angular diversity of illumination light, and the ptychographic in FPM occurs in the spectral domain. Instead of a diffraction pattern, a low-resolution focused image of the sample is detected directly. The spectra from different illumination angles are partially overlapped in the Fourier domain. The finite aperture of the imaging system becomes a frequency-domain sub-aperture in FPM, i.e., a ptychographic in the frequency domain.

8.7 Non-iterative Phase Imaging

8.7.1 Transport of Intensity Equation

One class of non-interferometric phase retrieval method corresponding to the iterative method is called the TIE method. In 1982, Teague proposed that under the paraxial approximation, a second-order elliptic partial differential equation can be derived using the Helmholtz equation for describing the quantitative relationship between the intensity of an optical field propagating along the direction of the optical axis and the phase perpendicular to the optical axis plane. In 1983, Teague proposed a method for solving the equation based on the Green's function [21].

Unlike iterative phase retrieval methods, this TIE does not use scalar diffraction formulas to invert the phase. Consider a monochromatic coherent beam propagating along the z-axis direction, TIE can be expressed as:

$$-k\frac{\partial I(x,y,z)}{\partial z}=\nabla\cdot[I(x,y,z)\nabla\varphi(x,y,z)], \tag{8.34}$$

in which $k{=}2\pi/\lambda$ is the wave number of the incident coherent beam, ∇ is the Hamiltonian operator, $I(x,y,z)$ is the intensity distribution on a plane with a propagation distance of z, and $\varphi(x,y,z)$ is the phase distribution to be solved.

The TIE phase retrieval algorithm requires the capture of intensity images in multiple out-of-focus planes. To capture these intensity images, a $4f$ system is usually required. In a $4f$ system, the sample object plane and the probe image plane are strictly conjugate, so that the intensity information on the out-of-focus planes can be obtained by moving the probe plane. The derivative of axial light intensity can be obtained by the biplane finite difference method. Take the intensity pattern of the sample in the axial direction in the two planes before and after the image plane, and the two images have equal distances from the center, as shown in Fig. 8.18. The axial light intensity can be approximated as:

$$\frac{\partial I(x,y,z)}{\partial z}\approx\frac{I(x,y,z{+}\Delta z)-I(x,y,z-\Delta z)}{2\Delta z}. \tag{8.35}$$

A representative TIE solution method is presented below. Define $I_{-1}{=}I(x,y,z{+}\Delta z)$, $I_0{=}I(x,y,z)$, and $I_1{=}I(x,y,z{+}\Delta z)$. Equation (8.34) can be simplified as:

$$\nabla\cdot[I_0\nabla\varphi]{=}-\frac{2\pi}{\lambda}\frac{I_1-I_{-1}}{2\Delta z}. \tag{8.36}$$

Define auxiliary function $\nabla\psi{=}I_0\nabla\varphi_0$ to replace the left side of Eq. (8.35):

$$\nabla^2\psi{=}\frac{I_1-I_{-1}}{2\Delta z}. \tag{8.37}$$

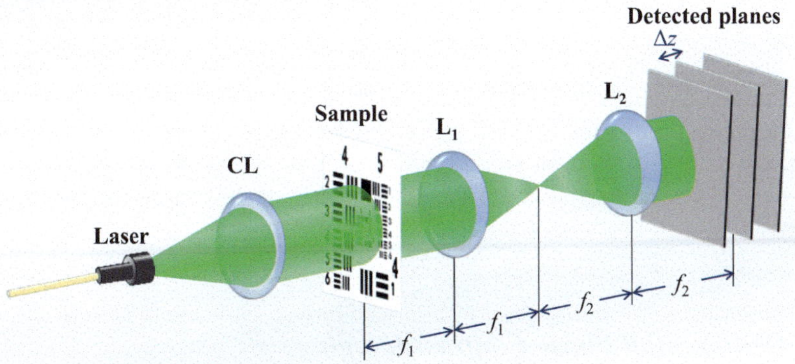

Fig. 8.18 4*f* system-based TIE imaging system

Utilizing the Fourier transform property of $\mathcal{F}(\nabla^2\psi) = -(k_x^2 + k_y^2)\mathcal{F}(\psi)$, perform Fourier transform on both sides of Eq. (8.37):

$$-(k_x^2 + k_y^2)\mathcal{F}(\psi) = -\frac{2\pi}{\lambda}\mathcal{F}\left(\frac{I_1 - I_{-1}}{2\Delta z}\right). \tag{8.38}$$

Simplifying the equation Eq. (8.38), we get:

$$\psi = \mathcal{F}^{-1}\left[\frac{2\pi}{\lambda}\frac{1}{k_x^2 + k_y^2}\mathcal{F}\left(\frac{I_1 - I_{-1}}{2\Delta z}\right)\right]. \tag{8.39}$$

Finally, using the Frankot-Chellappa algorithm to integrate the phase gradient, the phase distribution can be obtained as [22, 23]:

$$\varphi_0(x, y) = \mathcal{F}^{-1}\left[\frac{-ik_x\mathcal{F}(\varphi_x) - ik_y\mathcal{F}(\varphi_y)}{2\pi(k_x^2 + k_y^2)}\right], \tag{8.40}$$

where φ_x and φ_y are reconstructed phase derivative, and they can be represented as:

$$\begin{cases} \varphi_x = \frac{1}{I_0}\frac{\partial\psi}{\partial x}, \\ \varphi_y = \frac{1}{I_0}\frac{\partial\psi}{\partial y}. \end{cases} \tag{8.41}$$

The TIE method can reproduce the phase distribution of the sample plane light field by recording only three intensity distributions, which has the advantage of fast phase acquisition. However, the above method solves the derivative of the phase function and then integrates it to obtain the phase distribution of the measured optical field. Therefore, this method can only be used for samples with continuously varying phases.

The distance Δz between neighboring light field planes also affects the phase retrieval results. When Δz is too small, the ambient noise in the experiment greatly affects the phase retrieval results. However, when Δz is too large, higher-order differential errors lead to significant phase-blurring effects. Therefore, the optimal Δz should depend on the maximum differential distance of the object and the noise intensity.

8.7.2 Differential Phase Contrast Imaging

Another class of non-interferometric direct phase retrieval methods is called DPC imaging. In 1984, Hamilton et al. proposed DPC by combining a scanning microimaging system with a separation sensor [24]. This method successfully realizes the qualitative observation of the information of the sample by measuring the intensity difference generated by the phase of the sample. The scanning DPC imaging system is shown in Fig. 8.19. The laser is converged by the objective lens to the sample. The laser is converged on the sample through the objective lens, then deflected by the modulation of the sample phase, and finally received by two semicircular split detectors. The phase gradient in the DPC system has a symmetric response, so the phase gradient of the sample obtained by the two detectors is symmetrically distributed. By scanning the phase gradient variations caused by differences in sample thickness, an intensity image linearly related to the sample phase gradient is obtained.

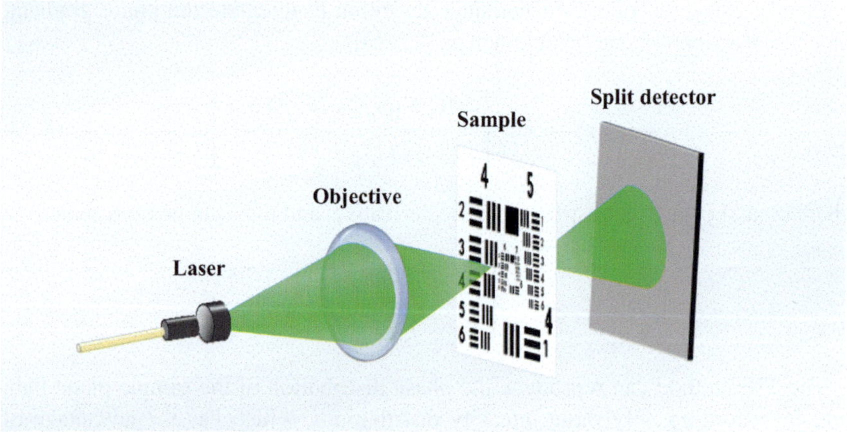

Fig. 8.19 Scanning DPC imaging system

Fig. 8.20 Full-field DPC imaging system

Since the scanning DPC imaging system requires a laser to scan the sample point by point, the imaging speed is slow and the image quality is susceptible to laser interference. Therefore, full-field-of-view DPC imaging has been developed. In 2009, Mehta et al. [25] first proposed the basic principle of full-field differential phase contrast imaging, in which asymmetric illumination and aperture modulation are used to obtain the phase gradient of the sample, as shown in Fig. 8.20. The system uses an opaque semicircular paper to modulate the light source to generate an asymmetric illumination pattern to irradiate the sample, and the light transmitted through the sample is converged on the detector by the objective lens to form a full-field-of-view image of the sample, and the phase gradient of the sample can be obtained from only two images.

The DPC system has the characteristics of using partially coherent imaging, simple structure, and easy construction and transformation. Compared with other quantitative phase methods, it has high imaging efficiency and can achieve fast or even real-time imaging. Currently, it has been successfully applied in multiple fields such as optical microscopy, biomedicine, and life sciences.

8.8 General Conclusions

In this chapter, a variety of phase imaging techniques are presented. Firstly, phase contrast microscopy, a qualitative phase imaging technique, is introduced. This method utilizes the principle of spatial filtering to change the phase of the spectral plane of an object light wave through the difference in refractive index distribution of a transparent phase-type sample, effectively converting the phase difference into an intensity contrast and improving the contrast of the observation. However, phase contrast microscopy can only be used for qualitative observations.

Interferometric phase imaging is a classical phase imaging method that converts invisible phase information into visible interference fringes by spatially superimposing two beams of coherent light. Digital holography is an important application of interferometric phase imaging and can be used for three-dimensional reconstruction of the sample object.

Wavefront sensing imaging is a non-interferometric phase imaging technique. Using a Hartmann screen or microlens array, the displacement of the actual wavefront relative to the ideal aberration-free case is analyzed, and the gradient of the wavefront is integrated over the gradient of the wavefront surface to obtain the phase distribution of the wavefront.

The phase retrieval algorithms do not require direct measurement of the phase distribution of the optical wave field, but consider the reconstruction process of the phase as a mathematical inverse problem. The single-frame phase retrieval algorithms only need one measurement image, and reconstruct the amplitude and phase of the sample image by iterative means. This chapter focuses on two representative algorithms, GS and HIO. Simulation results demonstrate that the HIO algorithm has better convergence and noise robustness compared to the GS algorithm.

The single-frame phase retrieval algorithm is extended to a multi-frame phase recovery algorithm by adding more measurements to increase the a priori information and utilizing the constraints of more measurements to improve the reliability and convergence. In this chapter, multi-distance phase imaging, multi-wavelength illumination microscopy, PIE and FPM are introduced. Multi-distance phase imaging is usually combined with the MDCDI system, which takes the intensity information collected by the detector at different distances from the sample as the input to the algorithm, and its imaging principles can be divided into two categories: serial transmission and parallel weighting, represented by the SBMIR and APR algorithms, respectively. Simulation results show that when the situation is ideal, both SBMIR and APR can reconstruct the samples more satisfactorily, and SBMIR exhibits better NCC and LMSE than APR. However, when the samples are added with noise, APR exhibits better noise robustness than SBMIR. Multiwavelength illumination microscopy converts the distance in the coaxial CDI system to the wavelength of the incident light. The PIE imaging system uses monochromatic coherent light illumination, where the illumination beam is constrained by a probe. The probe scans in the sample plane, and the neighboring probes need to ensure a sufficient overlap. To correct the distribution of the illuminated probes and improve the imaging results of PIE, ePIE and rPIE have been proposed. FPM extends the scanning of PIE from the spatial domain to the Fourier frequency domain and replaces the movement of the probes in PIE with the diversity of illumination angles from the LED array light source to the sample. FPM enables higher spatial resolution and wider field-of-view than conventional optical microscopy.

Finally, non-iterative phase imaging is introduced in this chapter. TIE directly calculates the phase distribution of light wavelengths in the sample plane by solving the second-order elliptic partial differential equations. To capture these intensity images, a $4f$ system is usually required. A representative TIE solution method in this chapter is derived in this chapter. DPC imaging is another non-iterative phase imaging technique. Scanning DPC reconstructs an intensity image linearly correlated with the phase gradient of the sample through the difference in intensity received by semicircular split detectors. The full-field DPC imaging system uses asymmetric illumination and aperture modulation to obtain the phase gradient of the sample, which enlarges the imaging field-of-view and improves imaging efficiency.

Acknowledgements This work was supported by the National Natural Science Foundation of China (Nos. 11874132, 61975044, 12074094, 12374271, and 62375065); Fundamental Research Funds for the Central Universities (No. HIT.DIJJ.2023175); Interdisciplinary Research Foundation of HIT (No. IR2021237).

References

1. Sobieranski A, Inci F, Tekin H, Yuksekkaya M, Comunello E, Cobra D, Wangenheim A, Demirci U. Portable lensless wide-field microscopy imaging platform based on digital inline holography and multi-frame pixel super-resolution. Light Sci Appl. 2015;4: e346.
2. Huang Z, Memmolo P, Ferraro P, Cao L. Dual-plane coupled phase retrieval for non-prior holographic imaging. PhotoniX. 2022;3(1):3.
3. Rivenson Y, Zhang Y, Günaydın H, Teng D, Ozcan A. Phase recovery and holographic image reconstruction using deep learning in neural networks. Light Sci Appl. 2018;7:17141.
4. Zuo C, Li J, Sun J, Fan Y, Zhang J. Transport of intensity equation: a tutorial. Opt Lasers Eng. 2020;135: 106187.
5. Wu J, Zhang H, Zhang W, Jin G, Cao L. Single-shot lensless imaging with Fresnel zone aperture and incoherent illumination. Light Sci Appl. 2020;9:53.
6. Zuo J, Vartanyants I, Gao M, Zhang R, Nagahara L. Atomic resolution imaging of a carbon nanotube from diffraction intensities. Science. 2003;300(5624):1419–21.
7. Zernike F. Phase contrast, a new method for the microscopic observation of transparent objects part II. Physica. 1942;9(10):974–86.
8. Goodman J, Lawrence R. Digital image formation from electronically detected holograms. Appl Phys Lett. 1967;11(3):77–9.
9. Shack R, Platt B. Production and use of a lenticular Hartmann screen. J Opt Soc Am. 1971;61:656–61.
10. Gerchberg RW, Saxton WO. A practical algorithm for the determination of phase from image and diffraction plane pictures. Optik. 1972;35:237–46.
11. Fienup J. Phase retrieval algorithms: a comparison. Appl Opt. 1982;21(15):2758–69.
12. Kuzmenko A, Butok O. Phase retrieval for arbitrary complex-valued objects using structured illumination. Opt Express. 2023;31(15):24505–15.
13. Pedrini G, Osten W, Zhang Y. Wave-front reconstruction from a sequence of interferograms recorded at different planes. Opt Lett. 2005;30(8):833–5.
14. Liu Z, Guo C, Tan J, Wu Q, Pan L, Liu S. Iterative phase-amplitude retrieval from multiple images in gyrator transforms. J Opt. 2015;17: 025701.
15. Bao P, Zhang F, Pedrini G, Osten W. Phase retrieval using multiple illumination wavelengths. Opt Lett. 2008;33(4):309–11.
16. Wu X, Sun J, Zhang J, Lu L, Chen R, Chen Q, Zuo C. Wavelength-scanning lensfree on-chip microscopy for wide-field pixel-super-resolved quantitative phase imaging. Opt Lett. 2021;46(9):2023–6.
17. Rodenburg J, Faulkner H. A phase retrieval algorithm for shifting illumination. Appl Phys Lett. 2004;85(20):4795–7.
18. Maiden A, Rodenburg J. An improved ptychographical phase retrieval algorithm for diffractive imaging. Ultramicroscopy. 2009;109(10):1256–62.
19. Maiden A, Johnson D, Li P. Further improvements to the ptychographical iterative engine. Optica. 2017;4(7):736–45.
20. Zheng G, Horstmeyer R, Yang C. Wide-field, high-resolution Fourier ptychographic microscopy. Nat Photon. 2013;7(9):739–45.
21. Teague M. Deterministic phase retrieval: a Green's function solution. J Opt Soc Am. 1983;73(11):1434–41.

22. Frankot R, Chellappa R. A method for enforcing integrability in shape from shading algorithms. IEEE Trans Pattern Anal Mach Intell. 1988;10(4):439–51.
23. Gao P, Pedrini G, Zuo C, Osten W. Phase retrieval using spatially modulated illumination. Opt Lett. 2014;39(12):3615–8.
24. Hamilton D, Sheppard C, Wilson T. Improved imaging of phase gradients in scanning optical microscopy. J Microsc. 1984;135(3):275–86.
25. Mehta S, Sheppard C. Quantitative phase-gradient imaging at high resolution with asymmetric illumination based differential phase contrast. Opt Lett. 2009;34(13):1924–6.

Chapter 9
Light Field Camera

Pengtao Zhao and Shutian Liu

9.1 Introduction

In the traditional two-dimensional imaging framework, there are limitations in solving many problems such as computing scene depth and acquiring the geometric shape of objects. The fundamental reason for these limitations lies in the fact that traditional imaging methods can only record the positional information of the light rays, losing the highly coupled information such as the angle of the light rays with respect to scene depth, object geometry, and scene occlusion relationships, in other words, the three-dimensional (3D) information. The theory of light field imaging, precisely, addresses the fundamental problem of collecting information about the angles of light rays. Therefore, computational photography based on light field imaging theory has become a hot topic in research both domestically and internationally in recent years.

The light field camera, based on a micro-lens array (hereinafter referred to as the light field camera), is currently the mainstream device for capturing light fields. By utilizing a light field camera, it is possible to simultaneously capture information about the angles and positions of light rays. Employing corresponding computational photography techniques enables the development of innovative image generation methods that overturn traditional imaging approaches, such as digital refocusing, synthetic aperture imaging, light field microscopy, panoramic depth image synthesis, and scene depth map acquisition. This provides more comprehensive and effective solutions for computer vision problems, including 3D reconstruction, panoramic stitching, viewpoint synthesis, target identification, and tracking.

P. Zhao (✉) · S. Liu
School of Physics, Harbin Institute of Technology, Harbin 150001, China
e-mail: zhaopengtao0627@163.com

S. Liu
e-mail: stliu@hit.edu.cn

Z. Liu et al. (eds.), *Computational Optical Imaging*, Advances in Optics and Optoelectronics, https://doi.org/10.1007/978-981-97-1455-1_9

However, compared to traditional imaging techniques, light field imaging technology also has certain limitations. For a single light field camera, given a fixed number of sensor pixels, the process of capturing light field camera signals involves a compromise between angle and position resolution since it needs to simultaneously acquire information about the position and angle of light rays. Constrained by imaging principles and sensor fabrication processes, a single-view light field camera exhibits poor sampling completeness for the target light field signal, resulting in limitations such as a finite field of view, lower depth recovery accuracy, and vulnerability to occlusion. Single-view light field data has limitations, and different light field camera imaging systems also exhibit variations in physical parameters and imaging models. These issues have become important bottlenecks that restrict the widespread application of light field cameras.

Just as multi-view geometry theory resolved previously considered unsolvable problems in computer vision, extending the light field computational theory from a single viewpoint to multiple viewpoints can effectively address the challenges faced by single-view light field systems. Therefore, researching universal light field camera imaging models and precise decoding methods for light field data, exploring parameter calibration, data correction, and unified parameterization methods that align with the characteristics of light fields, serve as the theoretical foundation for optimizing light field computation and expanding its applications. Currently, several research institutions have undertaken studies on light field camera imaging models and parameter calibration methods. These studies largely focus on the dual-plane light field model, addressing key issues such as dual-plane parallelism calibration, distortion model calibration, and spatial pose estimation for light field cameras.

This chapter will center its discourse on the methodologies pertaining to light field acquisition and representation. It aims to expound upon the imaging model of light field cameras and delineate their sampling patterns concerning the acquisition of pertinent light field information. Furthermore, the chapter seeks to encapsulate a comprehensive overview of techniques employed for the representation of light field data. Through a meticulous examination of light field decoding models, the theoretical underpinnings of light field parameterization, and methodologies for delineating errors inherent in light field data, the chapter endeavors to elucidate the fundamental principles underlying the application of light field cameras in domains such as wavefront detection, image depth estimation, and the principles governing digital refocusing.

9.2 Light Field

9.2.1 Concept of Light Field

The concept of a light field refers to the distribution and properties of light in space, encompassing parameters such as intensity, phase, and polarization. In the fields of physics and optics, the light field is typically described in the form of electromagnetic waves. It can be mathematically characterized and analyzed using methods such as wave equations or Maxwell's equations. In computer graphics and virtual reality, the term light field refers to a mathematical model describing the lighting conditions in a scene, applicable for real-time rendering, ray tracing, and other applications. The study of light fields is crucial for understanding the propagation and interaction of light, as well as for generating high quality images.

In 1936, Gershun, in his exploration of artificial light sources, was the first to define the light field as the spatial distribution of light radiation in all directions at different positions. Later, Adelson and Bergen applied the light field theory to computer vision, introducing the concept of the plenoptic function. Assuming that at time t, the human eye is positioned in space coordinates (x, y, z) and receives light from the direction (θ, φ) with a wavelength λ, this light ray can be represented as a seven-dimensional function

$$L = L(x, y, z, \theta, \varphi, \lambda, t). \tag{9.1}$$

In theory, Eq. (9.1) can be used to construct the light field function. However, in practice, it is impossible to measure the complete and continuous light field function of an object. Only an approximation of the light field function, obtained through discretization and dimensionality reduction, can be detected. The reduction of dimensions in the light field function is achieved through reasonable assumptions. Assuming the light field function is statistical and remains constant over time, the temporal dimension can be eliminated. Additionally, when the wavelength of the collected light is narrowed down to a specific range, the wavelength dimension can be ignored. Consequently, the reduced-dimensional light field can be represented as

$$L = L(x, y, z, \theta, \varphi). \tag{9.2}$$

When light propagates in free space, its radiance remains constant. Building upon this, researchers Levoy and Hanrahan from Stanford University further refined the light field theory. They eliminated redundant parameters from the five-dimensional light field function through the process of light field parameterization, resulting in a simplified four-dimensional light field function as

$$L = L(x, y, \theta, \varphi). \tag{9.3}$$

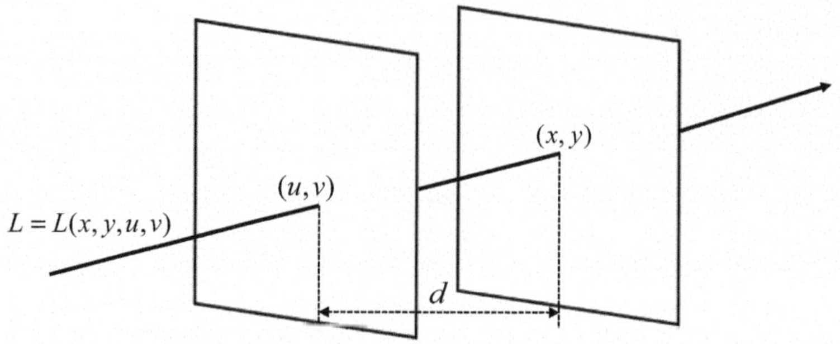

Fig. 9.1 Double-plane parameterization of the light field function

The above equation represents the light field in free space. When analyzing the transmission of the light field in a camera, representing the light field using angles can be inconvenient. Typically, angles are converted to spatial coordinates, as illustrated in Fig. 9.1. θ and φ can be expressed as

$$\theta = \tan^{-1}\frac{x-u}{d},\ \phi = \tan^{-1}\frac{y-v}{d}, \tag{9.4}$$

therefore, Eq. (9.3) can be transformed into

$$L = L(x, y, u, v). \tag{9.5}$$

9.2.2 Acquisition Methods for Light Field Images

In contrast to traditional 2D optical imaging methods, capturing a light field image requires a more intricate and sophisticated optical imaging system. Light field imaging technology is categorized based on the principles and configurations of optical systems. It mainly includes multi-sensor array camera imaging, represented by camera arrays, time-sequential multiplexing light field imaging, and single-sensor multi-channel multiplexing light field imaging. Multi-sensor array imaging involves the arrangement of multiple imaging devices in a planar or spherical configuration to capture multi-perspective information about the scene. A prominent example is the camera array proposed by the Stanford University Light Field Imaging Lab, depicted in Fig. 9.2a. The light field imaging method utilized by the Tsinghua University Information Science and Technology Laboratory, employing a spectral sensor array, typically involves a complex structure and a sizable volume. To facilitate light field acquisition for dynamic scenes, additional equipment, including a high-speed synchronous triggering device, is required. Furthermore, achieving the correct encoding and acquisition of four-dimensional light fields mandates that all cameras

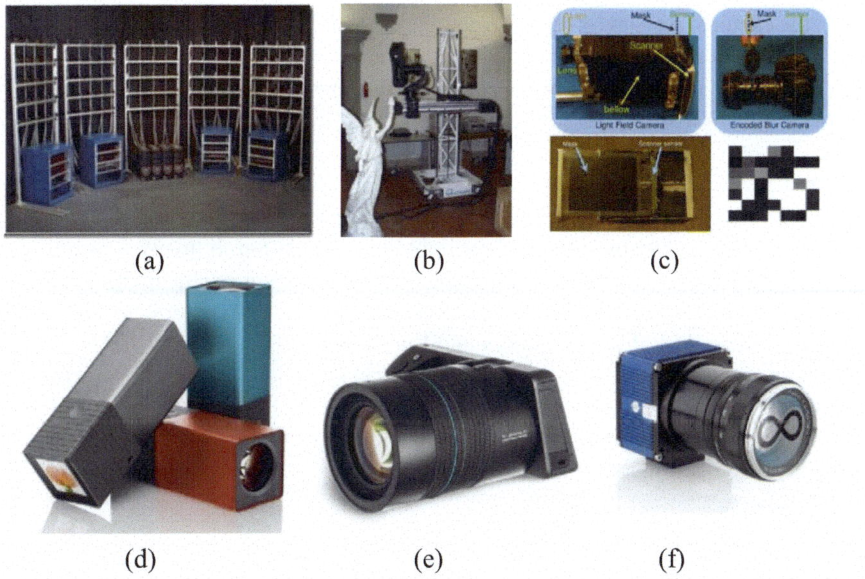

Fig. 9.2 Typical light field capture devices. **a** Stanford University camera array, **b** Stanford University Gantry system, **c** mask camera, **d** Lytro light field camera, **e** illumination light field camera, **f** Raytrix light field camera

be configured or corrected to identical parameters, thereby introducing additional complexity to the system.

Time-sequential multiplexing light field imaging technology involves sampling the light field by mechanically displacing a single sensor. Unlike camera arrays, time-sequential systems lack dynamic acquisition capabilities. A notable example is the Gantry system proposed by Stanford University, as depicted in Fig. 9.2b. Cameras achieve multiple captures of the scene from different angles by moving along a fixed track. While this setup enables perspective acquisition with a free baseline compared to a fixed array, it imposes stringent requirements on the positioning precision of the mechanical system. Single-sensor multiplexing light field imaging is classified into frequency domain and spatial domain multiplexing based on the multiplexing principle. This approach involves mapping the four-dimensional light field onto a 2D sensor through specialized optical structures. These imaging devices have a simple structure, high integration, and enable single-exposure capture. However, achieving a balance between spatial and angular resolution requires a trade-off, necessitating the sacrifice of some spatial resolution to meet the demands of angle sampling, especially within the constraints of a specific sensor resolution.

Frequency domain multiplexing light field imaging involves adding attenuation and scattering elements inside a traditional camera. These elements project components of different spatial and angular frequencies captured by the camera onto different regions of the image sensor, aiming to record the light field. Through corresponding decoding algorithms, the four-dimensional light field can be reconstructed from the captured images. A typical representative of frequency domain multiplexing light field cameras is the mask camera proposed by Veeraraghavan [1], as shown in Fig. 9.2c. The mask camera avoids introducing aberrations and alignment issues associated with adding new lens components. However, the mask element absorbs approximately 50% of the collected light, limiting the use cases of mask light field cameras.

Spatial domain multiplexing light field imaging is currently the most extensively researched and widely used method for light field acquisition. It involves inserting a microlens array between the main lens of a traditional camera and the image sensor to achieve light field recording. The most representative light field cameras are Ng's non-focusing handheld light field camera and Lumsdaine's focusing light field camera, also known as standard light field camera 1.0 and standard light field camera 2.0. These two structures, commercialized by Lytro and Raytrix respectively, are tailored for the consumer photography market and industrial measurement market. Their structures are illustrated in Fig. 9.2d–f.

9.3 Imaging Principle of Light Field Camera

9.3.1 Light Field Data Acquisition

The image acquisition in a traditional camera involves the passage of light through the main lens, projecting onto the CCD. Each pixel on the CCD represents the energy integration of all the light rays reaching that position, as depicted in Fig. 9.3a. The geometric representation of all the light rays reaching the gray pixel corresponds to a gray strip in the light ray-space diagram, as shown in Fig. 9.3b. The reason these rays form a vertical strip in the light ray-space diagram is that their u-coordinates are all different, covering all positions of the main lens. Therefore, it can be understood that the geometric representation of the light received by each pixel on the CCD will form a strip in the light ray-space diagram with the same x-coordinate as that pixel.

After each strip of light rays reaches the corresponding pixel on the CCD, it is recorded as a pixel value. This pixel value is directly proportional to the received radiance of the incident light. Therefore, the radiance value received by any pixel in a traditional camera (in the two-dimensional case) is

$$E_d(x) = \frac{1}{d}\int L_d(u, x)\cos^2\theta du, \tag{9.6}$$

where $L_d(u, x)$ represents the light field in the space from the main lens plane to the CCD plane, d is the distance between the main lens plane and the CCD plane, θ is the angle between the light ray (u, x) and the normal to the imaging plane. In a traditional camera, only the positional information of the light field x is recorded, and the information in the u-direction of the light field is integrated out.

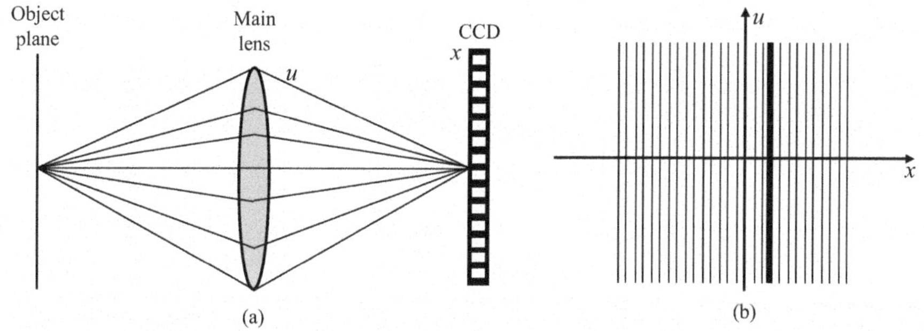

Fig. 9.3 The imaging principle. **a** The imaging principle of a traditional camera, **b** space-light ray diagram

Unlike traditional cameras, a light field camera needs to convert the four-dimensional information of the light field (position and direction information) into two-dimensional plane information and record it through a CCD. The optical structure of a light field camera mainly consists of the main lens, microlens array, and CCD. The microlens array is located at the imaging plane of the object, as shown in Fig. 9.4a. In reality, the microlens array is much smaller compared to the main lens for illustrative purposes, and it has been magnified in the figure. The rays emitted by the main lens pass through the gray microlens and form sub-images on the recording plane of the CCD. The geometry of these rays corresponds to the gray strip in the light ray-space diagram shown in Fig. 9.4b. This is because these rays share the same x coordinate, which is the position coordinate of the gray microlens. However, the u coordinates of these rays are all different, covering all positions of the main lens, as shown in Fig. 9.4a. A sub-image is sampled and recorded by multiple pixels on the CCD, and the u coordinates of the set of rays recorded by each pixel correspond to a sub-region of the main lens (as outlined by the dashed box in the figure). The u coordinates of these rays vary within the sub-region, so each set of rays recorded by each pixel corresponds to a small square in the gray strip area of the light ray-space diagram.

As shown in Fig. 9.5, the sets of rays received by different sub-images' two gray pixels on the CCD correspond to a single gray small square in the light ray-space diagram. The x-coordinate of the small square represents the position coordinate of the lens through which the light passes, and the u-coordinate represents the position coordinate of the sub-region of the main lens through which the light passes (as outlined by the dashed box in the figure).

The column width (in the x-direction) of the small square equals the diameter of the microlens, and the number of squares divided in each column (in the u-direction) equals the sampling pixels of the sub-image. From this, it can be inferred that the set represented by a small square in the light ray-space diagram in Fig. 9.5b will be received by one pixel on the CCD and recorded as a pixel value. All rays (or the light field) in the dual-plane space of the main lens and microlens array are discretely sampled in the light ray-space diagram by these small squares. Different sets of rays

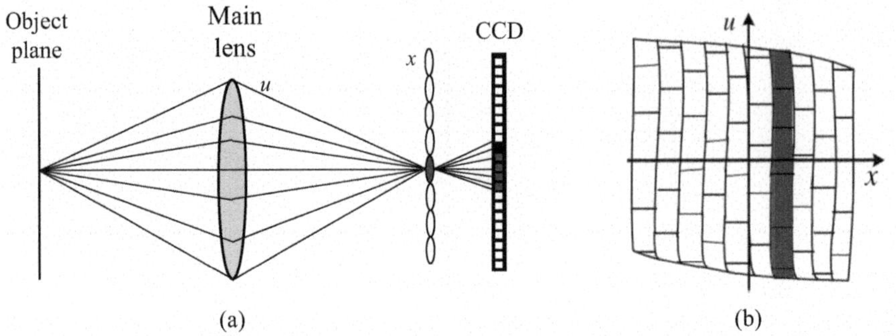

Fig. 9.4 The imaging principle of a light field camera. **a** Light field camera imaging, **b** space-light ray diagram

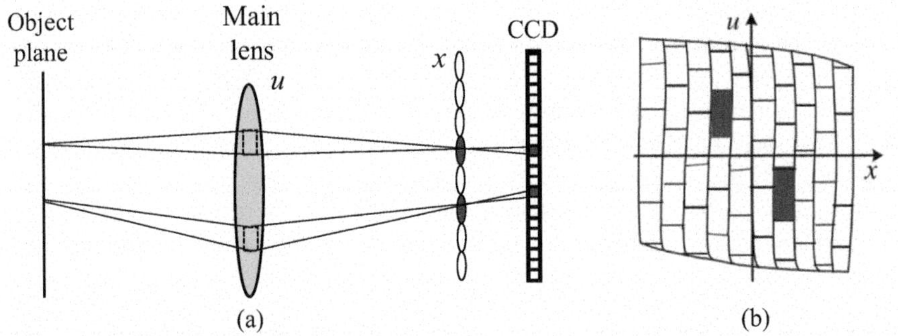

Fig. 9.5 The imaging principle of a light field camera. **a** Light field sampling in a light field camera, **b** space-light ray diagram

in different small squares are recorded by different pixels on the CCD. Thus, the position and directional information of the light field, under discrete sampling, are recorded by the light field camera in the form of a two-dimensional image.

9.3.2 Imaging Relationship of Light Field Camera

The two-dimensional image recorded by the light field camera imaging can be referred to as a light field image. The schematic diagram of the light field camera imaging an object is shown in Fig. 9.6, where the microlens array is located at the imaging plane of the object, and the CCD is positioned at the focal plane of the microlenses. As the plane of the object and the plane of the microlens array are in a conjugate relationship, the image of the object is formed on the front surface of the microlens array. If we consider the plane of the microlens array as the imaging plane and each microlens as a pixel, then the object image is sampled and recorded by

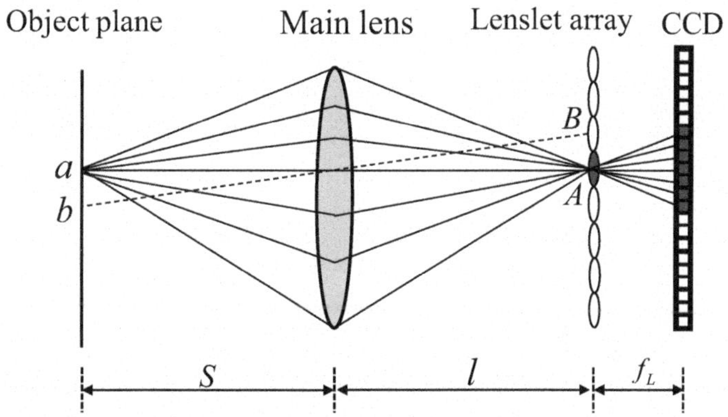

Fig. 9.6 Diagram of light field camera imaging

these microlens pixels. As the entrance pupil is circular, each entrance pupil image is distributed in a circular pattern. When the light field camera captures images of a target, the f-number of the microlens and the main lens parameters L/D (where D is the main lens diameter) need to satisfy certain matching requirements. Since the microlens focal length f_L is much smaller than the distance L from the microlens to the main lens, the main lens plane and the CCD plane satisfy the conjugate relationship.

When $f_L < l/\mathrm{D}$, the pupil images are distributed smaller and farther apart on the CCD. In this case, the information contained in each pupil image does not overlap, but the effective working area of the CCD is not fully utilized, resulting in a waste of the CCD pixel sampling rate for the pupil images; When $f_L > l/\mathrm{D}$, the pupil images formed by individual microlenses overlap on the CCD, leading to information overlap between the pupil images. This ultimately affects the extraction of light field information; When $f_L = l/\mathrm{D}$, the size of the pupil image is essentially the same as that of the microlens, maximizing the utilization of CCD pixels without overlapping between pupil images. At this point, the light field camera reaches its optimal working state.

When imaging the target, the microlens's f and the main lens parameter (l/D) of the light field camera need to meet certain matching requirements, as illustrated in Fig. 9.6. Since the focal length of the microlens is much smaller than the distance from the microlens to the main lens (l), the main lens plane is in conjugate relation with the CCD plane. According to the imaging relationship, each microlens forms a pupil image for the entrance pupil. When the f of the microlens is respectively greater than, equal to, and less than l/D, the acquired light field images by the light field camera are illustrated in Fig. 9.7. It is evident that when $f_L > l/\mathrm{D}$, the pupil images formed by each microlens overlap on the CCD, leading to the mixing of information between the pupil images, ultimately affecting the extraction of light field information. When $f_L < l/\mathrm{D}$, the pupil images are distributed smaller and farther apart on the CCD, in this case, the information in each pupil image does not

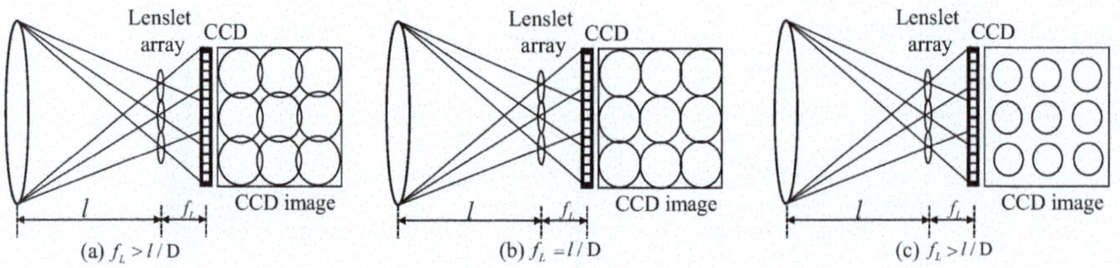

Fig. 9.7 The imaging diagrams of the light field camera with microlenses at different f-values

overlap, but the effective working area of the CCD is not fully utilized, wasting the sampling rate of CCD pixels for the pupil images. When $f_L = l/\text{D}$, the pupil images are roughly the same size as the microlens, maximizing the utilization of CCD pixels without overlap between pupil images. At this point, the light field camera achieves its optimal operating state.

9.4 Wavefront Sensing with Light Field Cameras

9.4.1 Wavefront Sensing Principle

Contrary to the light field imaging of the light field camera, in wavefront sensing, the microlens array of the light field camera is directly placed at the focal plane of the main lens, where $L = f_L$ in Fig. 9.8. When defining the local slopes falling onto microlenses 1, 2, and 3 as negative, zero, and positive, respectively, and considering a flat wavefront as the incident wave, the local slopes at points A, B, and C on the wavefront are all zero.

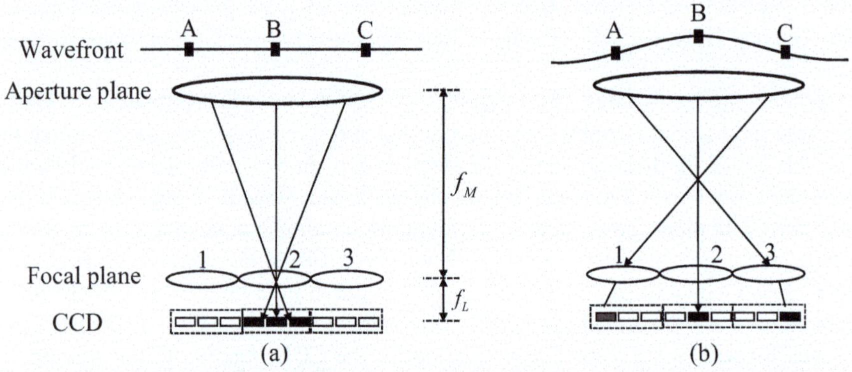

Fig. 9.8 Schematic diagram of the wavefront sensing principle of a light field camera. **a** Planar incident wave, **b** distorted incident wave

Light rays emanating from points A, B, and C, after passing through the main lens, converge onto microlens 2. Microlens 2 measures an incoming light slope of 0, aligning with the actual wavefront slope. When the incident wavefront is distorted, as shown in the illustration, the local wavefront slopes at points A, B, and C become positive, zero, and negative, respectively. Light rays from these points, after passing through the main lens, fall onto microlenses 3, 2, and 1, respectively. Consequently, the measured incoming light slopes are positive, zero, and negative, aligning with the actual wavefront slopes. This indicates that the local slopes of the incident wavefront at different positions are recorded in the light field camera's light field image. By analyzing the light field image, one can obtain the local slopes of different parts of the incident wavefront.

As shown in Fig. 9.9, a ray passing through a point A on the front surface of the main lens and reaching a point B on the front surface of the microlens array can be divided into two processes: first, refraction through the main lens, and then transmission through free space over a distance f_M (where f_M is the focal length of the main lens). The transmission matrices T_F for the main lens and T_l for free space are given by

$$T_F = \begin{bmatrix} 1 & 0 \\ -1/f_M & 1 \end{bmatrix}$$
$$T_l = \begin{bmatrix} 1 & f_M \\ 0 & 1 \end{bmatrix}. \tag{9.7}$$

Assuming the positions and directions of A are represented by τ and α, and those of B are represented by x and β, according to the principles of geometrical optics, there is the following relationship between the two points

$$\begin{bmatrix} x \\ \beta \end{bmatrix} = \begin{bmatrix} 1 & f_M \\ 0 & 1 \end{bmatrix} \begin{bmatrix} 1 & 0 \\ -1/f_M & 1 \end{bmatrix} \begin{bmatrix} \tau \\ \alpha \end{bmatrix}, \tag{9.8}$$

where f_M is the focal length of the main lens. Thus, we have

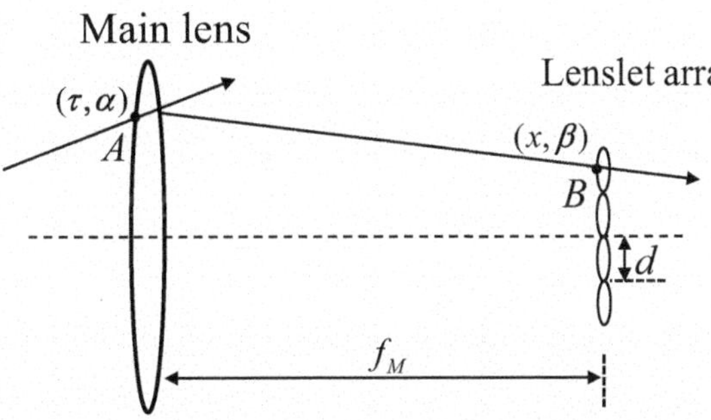

Fig. 9.9 The transmission of rays in the light field camera

$$\begin{bmatrix} x \\ \beta \end{bmatrix} = \begin{bmatrix} f_M\alpha \\ -\tau/f_M + \alpha \end{bmatrix}, \tag{9.9}$$

therefore, the slope of point A can be expressed as

$$\alpha = \frac{x}{f_M}. \tag{9.10}$$

Assuming point B is located at the nth microlens, and the total number of microlenses is 2N (in one-dimensional direction), the range of x is $[(n-1)d, nd]$. If the center position of the nth microlens is approximated as the position of point B, then

$$\alpha \approx \frac{(n-0.5)d}{f_M}, \tag{9.11}$$

where d is the diameter of the microlens. From the above equation, it can be seen that the incident direction of the light ray can be estimated by the ratio of the illuminated microlens position to the focal length of the main lens.

Equation (9.11) indicates the relationship between the light ray passing through point A and the light field image. However, the wavefront slope at point A is actually related to all the light rays passing through point A. As shown in Fig. 9.10, since each microlens images the entrance pupil, all the rays passing through point A converge to the corresponding position of the entrance pupil image. This position corresponds one-to-one with point A. Assuming that the nth microlens (in one-dimensional direction) receives the number of photoelectrons from k rays denoted by $I_n(\tau)$, and these k rays pass through the center position of the nth microlens for approximation, it can be known from Eq. (9.11) that the wavefront slope at point A can be expressed as

$$S_\tau(\tau) = \frac{1}{f_M} \frac{\sum_{n=-N+1}^{N} I_n(\tau)\cdot(n-0.5)d}{\sum_{n=-N+1}^{N} I_n(\tau)}. \tag{9.12}$$

Expanding (9.12) to the two-dimensional case, assuming the micro-lens array is 2M × 2N and the corresponding light pupil image for the (m, n) microlens is $I_{mn}(\tau, \nu)$, the relationship between the incident wavefront slope and the light field image is given by

$$S_\tau(\tau, \nu) = \frac{1}{f_M} \cdot \frac{\sum_{m=-M+1}^{M}\sum_{n=-N+1}^{N} I_{mn}(\tau, \nu)(n-0.5)d}{\sum_{m=-M+1}^{M}\sum_{n=-N+1}^{N} I_{mn}(\tau, \nu)},$$

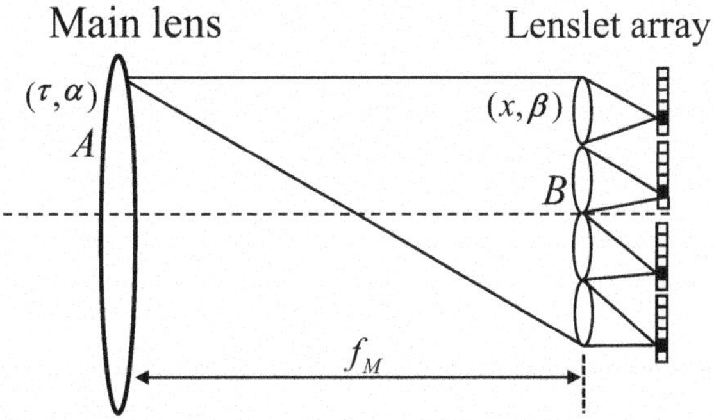

Fig. 9.10 The trajectory of the light passing through point A in the light field camera

$$S_{\eta}(\tau, \nu) = \frac{1}{f_M} \cdot \frac{\sum_{m=-M+1}^{M} \sum_{n=-N+1}^{N} I_{mn}(\tau, \nu)(m \quad 0.5)d}{\sum_{m=-M+1}^{M} \sum_{n=-N+1}^{N} I_{mn}(\tau, \nu)}. \tag{9.13}$$

Extracting the slope of the incident wavefront from the light field image can also be achieved through a recombination process, as illustrated in Fig. 9.11. Taking the 1st pixel in each light pupil image as an example, the pixels located at the corresponding positions of these micro-lenses in the micro-lens array are recombined, resulting in a recombined sub-image called the reconstructed group sub-image. This recombined sub-image essentially represents the image on the focal plane of all the light rays passing through the 1st sub-region of the light pupil, and in a physical sense, it is equivalent to the sub-aperture image of a Shack-Hartmann sensor. Arranging all the virtual sub-aperture images based on their sub-region positions forms the recombined image. Therefore, if the light field image is denoted as $I_{mn}(\tau, \nu)$, the virtual sub-aperture images can be represented as $I_{\tau\nu}(m, n)$.

After obtaining the virtual sub-aperture images, the centroid of each sub-aperture image is estimated using the centroid algorithm as

$$\begin{aligned} S_c(\tau, \nu) &= \frac{1}{f_M} \cdot \frac{\sum_{m=-M+1}^{M} \sum_{n=-N+1}^{N} I_{\tau\nu}(m, n)(n - 0.5)d}{\sum_{m=-M+1}^{M} \sum_{n=-N+1}^{N} I_{\tau\nu}(m, n)}, \\ S_c(\tau, \nu) &= \frac{1}{f_M} \cdot \frac{\sum_{m=-M+1}^{M} \sum_{n=-N+1}^{N} I_{\tau\nu}(m, n)(m - 0.5)d}{\sum_{m=-M+1}^{M} \sum_{n=-N+1}^{N} I_{\tau\nu}(m, n)}. \end{aligned} \tag{9.14}$$

The relationship between the slope of the incident wavefront and the centroid can be expressed as

$$S_{\tau} = \tau_c / f_M, \quad S_{\nu} = \nu_c / f_M. \tag{9.15}$$

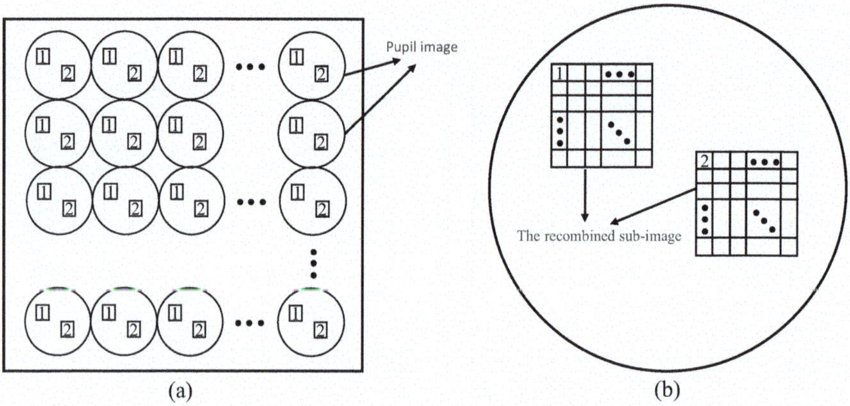

Fig. 9.11 Schematic diagram of light field image reconstruction. **a** Light field image, **b** reconstructed image

9.4.2 Error Analysis of Wavefront Sensing

The wavefront detection process of a light field camera mainly consists of three steps: The first step involves obtaining the light field image, where system errors arise from registration errors between the microlens array and CCD in the light field camera. The second step entails reconstructing the image based on the light field image and calculating the centroids of various sub-images, introducing centroid measurement errors, along with inherent principle errors in the light field camera. The third step involves reconstructing the wavefront based on slope data, with error sources stemming from the accuracy of the reconstruction algorithm.

In a light field camera, the function of the microlens is analogous to a slope filter. When the incident wave slope varies within the bandwidth of the slope filter, the output signal of the light field camera remains constant. However, in this scenario, the output signal of the light field camera fails to accurately respond to changes in the incident wave slope, leading to detection errors. This type of error is determined by the optical structure of the light field camera and is therefore referred to as principle error.

As illustrated in Fig. 9.12, when a plane wave passes through the main lens of the light field camera, it generates an Airy disk on the front surface of the microlens array. The radius of the Airy disk is denoted as $1.22\lambda f_M/\mathrm{D}$, where f_M is the focal length of the main lens, λ is the central wavelength of the sensor, and D is the diameter of the main lens.

When the slope of the incident plane wave is 0 (i.e., the wavefront is perpendicular to the optical axis), the center of the Airy disk is located at the junction of the central two microlenses, and its light is evenly distributed and received by these central two microlenses. As the slope of the incident plane wave gradually increases from 0 to $1.22\lambda/\mathrm{D}$, the center of the Airy disk gradually moves towards the first microlens

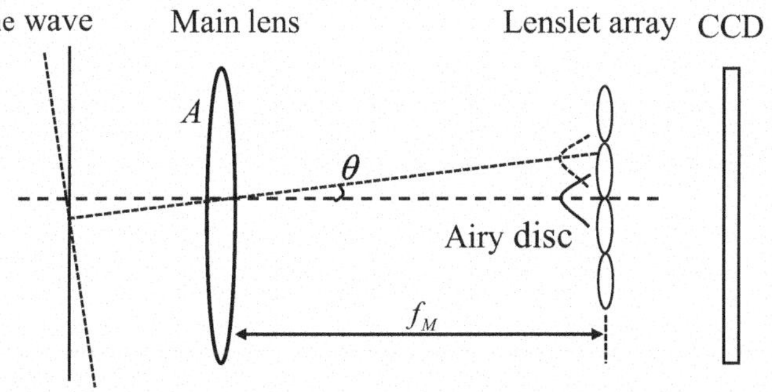

Fig. 9.12 Plane waves are incident upon the light field camera

above the optical axis, as shown in Fig. 9.13a. As the slope of the incident plane wave gradually increases from 1.22λ/D to $(d/f_M - 1.22\lambda/\mathrm{D})$, the center of the Airy disk completely enters the first microlens above the optical axis, and the optical energy of the Airy disk is fully received by this microlens, as depicted in Fig. 9.13b. As the slope of the incident plane wave gradually increases from $(d/f_M - 1.22\lambda/\mathrm{D})$ to d/f_M, the optical energy of the Airy disk is jointly received by the first and second microlenses above the optical axis, as shown in Fig. 9.13c.

Summarizing the analysis above, the relationship between the light field camera slope signal S_1 and the incident plane wave signal θ can be represented by Fig. 9.14. It is evident that the maximum slope error caused by principle errors is:

$$\delta_\theta = \frac{d}{2f_M} - 1.22\frac{\lambda}{\mathrm{D}}, \tag{9.16}$$

where d is the microlens diameter, f_M is the main lens focal length, λ is the central wavelength of the sensor, and D is the main lens diameter. It can be inferred that as d

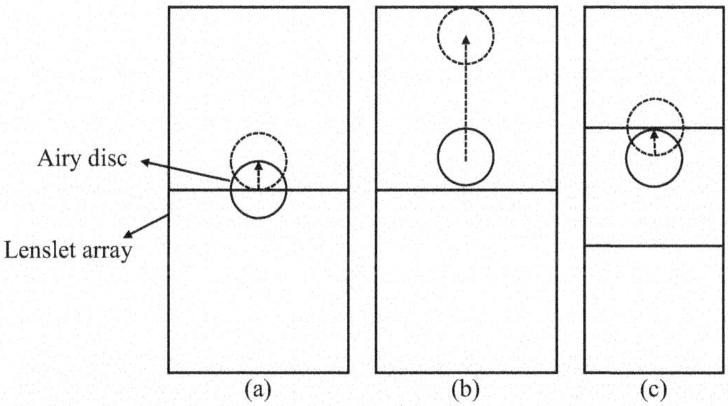

Fig. 9.13 Schematic representation of the Airy disk's position on the front surface of the microlens array

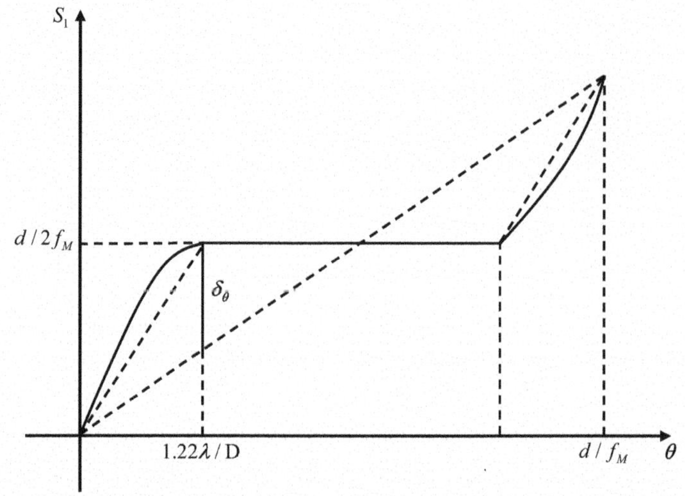

Fig. 9.14 Graph illustrating the correlation between the output signal of a light field camera and the incident wave slope signal

decreases and f_M increases, the principle error decreases. When the microlens diameter satisfies condition $d = 1.22\lambda f_M/\mathrm{D}$, the principle error disappears. However, as the microlens size decreases, the spatial resolution of the light field camera decreases accordingly.

The primary sources of centroid measurement error in a light field camera include discrete sampling errors of pixels, photon noise errors, and readout noise errors [2–4]. Discrete sampling error of pixels refers to the CCD's inherent discreteness when capturing a light spot, causing the continuous grayscale image to be sampled and segmented into a stepped, discrete grayscale image. The degree of discrete sampling error increases with lower CCD resolution. Photon noise error arises from the stochastic nature of photon noise in both time and space, leading to minute variations in the coordinates of the optical wave centroid. Photon noise follows a Poisson distribution. Readout noise originates from fluctuations in dark current, charge storage, and transfer within the photoelectric components. This results in random outputs from CCD pixels even in the absence of a signal input, and readout noise follows a Gaussian distribution.

For a point target, utilizing the centroid equation allows the derivation of centroid measurement error induced by photon noise as follows [5]

$$\sigma_{cp}^2 = \sigma_A^2/V, \tag{9.17}$$

where σ_A is the equivalent Gaussian width of the spot, and V represents the total number of photoelectrons received by the sampling window during the reassembly of the sub-image. It is evident that the centroid measurement error induced by photon noise is influenced by both the equivalent Gaussian spot size and the number of photons within the sampling window. The wavefront measurement error introduced by photon noise in the light field camera can be estimated as follows: the diffraction-limited spot radius of a point target, imaged through the main lens onto the front surface of the microlens array, is

$$s = 1.22\lambda f^{\#}, \tag{9.18}$$

where $f^{\#}$ is the f-number of the main lens. When the number of pixels corresponding to a microlens is $N_D \times N_D$, the virtual sub-aperture count is $N_D \times N_D$, and the virtual sub-aperture count typically requires $N_D \geq 4$. Consequently, $s < d$ is obtained, indicating that under weak turbulence conditions, photons emitted from a point source (received by the main lens) are primarily concentrated on the four central microlenses of the microlens array. Thus, the equivalent Gaussian width of the spot satisfies the following relationship

$$\sigma_A < 2d. \tag{9.19}$$

From the relationship between centroid offset and wavefront slope, it can be determined that the root mean square value of the slope measurement error induced by photon noise is

$$\sigma_\theta = \frac{\sigma_{cp}}{f_M}, \tag{9.20}$$

substituting Eqs. (9.17) and (9.19) into Eq. (9.20), it can be inferred

$$\sigma_\theta < \frac{2d}{f_M \cdot \sqrt{V}}. \tag{9.21}$$

Thus, the root mean square value of the wavefront measurement error induced by photon noise can be obtained

$$\sigma_0 = \sigma_\theta \mathrm{D}/N_D < \frac{2\mathrm{D}d}{N_D f_M \sqrt{V}}, \tag{9.22}$$

substituting the size of the microlens $d = 0.5N\mathrm{D}\lambda f^{\#}$ into Eq. (9.22), we obtain

$$\sigma_0 < \frac{\lambda}{\sqrt{V}}. \tag{9.23}$$

The wavefront measurement error introduced by readout noise to the light field camera can be estimated as follows. Research indicates that the overall tilt of the turbulent wavefront follows a Gaussian distribution, with a mean square value of

$$\sigma_\alpha = 0.18\lambda^2 b^{-1/3} r_0^{-5/3}, \tag{9.24}$$

where $b = D_{\text{sup}}$ is the length of the subregion, according to the principles of geometric optics, there is a relationship between the centroid X_c and α

$$\langle X_c^2 \rangle = \langle \sigma_\alpha^2 f_M^2 \rangle = 0.18\lambda^2 b^{-1/3} r_0^{-5/3} f_M^2. \tag{9.25}$$

Assuming the length of the subregion is $b = r_0$, Eq. (9.25) can be simplified to

$$\langle X_c^2 \rangle = \langle \sigma_\alpha^2 f_M^2 \rangle = 0.18\lambda^2 b^{-2} f_M^2. \tag{9.26}$$

From the relationship between centroid offset and wavefront slope, it can be determined that the root mean square value of the slope measurement error induced by readout noise is

$$\sigma_\theta = \frac{\sigma_{cr}}{f_M}. \tag{9.27}$$

The root mean square value of the wavefront measurement error induced by readout noise can be obtained as σ_0

$$\begin{aligned} \sigma_0 &= \sigma_\theta D/N_D = \sigma_\theta b \\ &= \frac{b}{f_M}\sqrt{\frac{\sigma_r^2}{V^2} \cdot ML\left(\frac{L^2-1}{12}d^2 + 0.18\lambda^2 b^{-2} f_M^2\right)} \\ &= \frac{\sigma_r}{V}\sqrt{ML\left(\frac{L^2-1}{12}d^2 \cdot \frac{b^2}{f_M^2} + 0.18\lambda^2\right)}, \end{aligned} \tag{9.28}$$

substituting the size of the microlens $d = 0.5ND\lambda f^{\#}$ into Eq. (9.28), we obtain

$$\sigma_0 = \frac{\lambda\sigma_r}{V} \cdot \sqrt{ML\left[0.021 \cdot \left(L^2 - 1\right) + 0.18\right]}. \tag{9.29}$$

According to Eq. (9.29), the number of sampled pixels for the signal spot is typically 2×2, while the actual number of pixels in a light field camera is $M \times L$. These surplus pixels do not contain signal information; instead, they introduce additional noise errors in centroid calculation. If the influence of these ineffective pixels can be eliminated, the accuracy of centroid measurement will be effectively improved. The threshold method is effective in achieving this. The implementation process involves analyzing the noise in the sub-aperture image, setting a threshold based on its noise level, and subtracting this threshold from all pixels in the sub-aperture image. Values greater than zero after subtracting the threshold are retained, while values less than zero are set to zero. Through the threshold method, the effective number of pixels in the sampling window in Eq. (9.29) will be retained, i.e., M and L will decrease. Consequently, the wavefront measurement error induced by readout noise will also decrease.

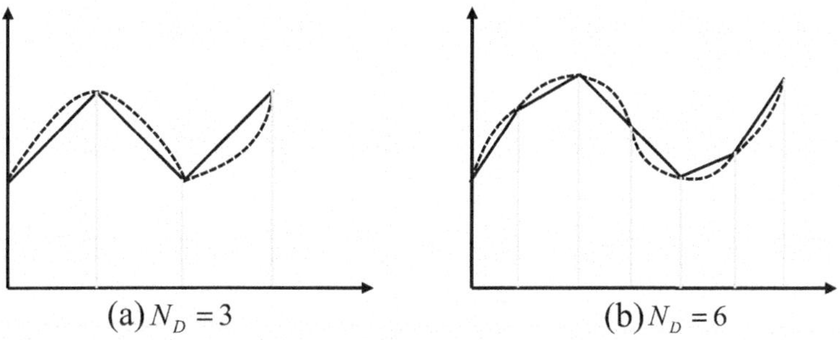

Fig. 9.15 Different wavefront sampling densities

The main factors affecting the accuracy of the reconstruction algorithm are the shape of the wavefront to be reconstructed and the slope measurement error. When the shape of the wavefront to be reconstructed is relatively simple (such as defocus aberration), the reconstruction algorithm can effectively reconstruct the wavefront. However, for wavefronts with high-order aberrations, the reconstruction algorithm may not perfectly reconstruct the wavefront due to insufficient wavefront sampling density. Figure 9.15 illustrates the concept of different wavefront sampling densities. Each solid segment represents a slope data, and the dashed line represents the reconstructed original wavefront. As the wavefront sampling density increases, the reconstructed wavefront becomes closer to the original wavefront, and the accuracy of the reconstruction algorithm improves. According to the imaging principles of a light field camera, if the number of pixels corresponding to each entrance pupil image is $N_D \times N_D$, it is equivalent to dividing the entrance pupil into $N_D \times N_D$ subregions. The wavefront of the incident wave is sampled by these subregions. Therefore, the higher the number of pixels $N_D \times N_D$, the higher the reconstruction accuracy.

The wavefront reconstruction error induced by slope measurement errors is contingent on the error transmission situation. Assuming phase error and slope error are denoted as ε and n respectively, and the true phase and true slope data are φ_0 and g_0 respectively, the relationship can be expressed as

$$\begin{cases} \varphi = \varphi_0 + \varepsilon, \\ g = g_0 + n, \end{cases} \tag{9.30}$$

assuming the relationship matrix between phase and slope data is denoted as M, the relationship can be expressed as follows

$$\varphi = \varphi_0 + \varepsilon = M \cdot g = M \cdot (g_0 + n), \tag{9.31}$$

thus, the relationship between phase error and slope error can be written as

$$\varepsilon_i = \sum_k M_{ik} n_k. \tag{9.32}$$

Assuming n is a zero-mean Gaussian white noise, the relation can be given as follows

$$\langle n_k n_l \rangle = \sigma_t^2 \delta_{kl}, \tag{9.33}$$

where σ_t^2 represents the variance of the slope error, the relationship can be calculated as follows

$$\begin{aligned}\langle \varepsilon_i \rangle &= \sum_k \sum_l M_{ik} M_{il} \langle n_k n_l \rangle \\ &= \sigma_t^2 \sum_k M_{ik}^2.\end{aligned} \tag{9.34}$$

Hence, the mean square error of the phase error induced by slope measurement errors in the reconstruction error is equal to

$$\begin{aligned}\sigma_\phi^2 &= \frac{1}{N_D^2} \sum_i \langle \varepsilon_i^2 \rangle \\ &= \frac{\sigma_t^2}{N_D^2} \sum_i \sum_k M_{ik}^2.\end{aligned} \tag{9.35}$$

Thus, it can be inferred that the wavefront reconstruction error is directly proportional to the slope measurement error σ_t^2 and inversely proportional to the number of pixels N_D^2.

9.5 Light Field Image Depth Estimation Algorithm

In contrast to traditional cameras that require multiple shots and are susceptible to mechanical shake and changes in viewpoint for depth estimation, a light field camera can achieve depth estimation with a single shot, offering convenience. A four-dimensional light field comprises spatial and angular information in four dimensions, while a regular image contains only spatial information in two dimensions, making it challenging to directly describe the four-dimensional light field in two dimensions. One approach is to fix two dimensions of the four-dimensional light field and project it into a two-dimensional image for observation. This process is known as light field visualization. Depending on the type of fixed dimensions, three different forms of visualization can be obtained: macro-pixel image [6], multi-view image [7], and epipolar plane image [8].

Based on these three visualizations, depth maps can be directly obtained using defocus-based methods, multi-view stereo matching algorithms, and EPI slope algorithms. In recent years, with the rise of deep learning technologies, there has been a proliferation of algorithms that use convolutional neural networks to obtain depth maps from light field images. Currently, depth acquisition techniques based on light field images can be categorized into three types: defocus-based depth map acquisition, depth map acquisition based on multi-view stereo matching algorithms, and depth map acquisition based on EPI slope. The following will introduce the principles of these three algorithms for calculating depth maps.

9.5.1 Wavefront Sensing Principle

The light field camera captures both the position and direction information of light rays. The recording of direction information can be conceptualized as the mapping of different directions of light rays passing through the (x, y) coordinates on the main lens plane to distinct coordinates on the imaging plane. By keeping spatial coordinates fixed and amalgamating pixels from different cameras capturing the same location, it becomes possible to represent the angle information of light rays passing through the point (x, y).

Representation of the 4D light field using macro-pixels offers a more intuitive and organized structure compared to the original arrangement in the light field image. This enhanced organization proves beneficial for subsequent data processing. The visualization of the 4D light field based on macro-pixel representation is remapped into a remap image. Leveraging digital refocusing techniques, multiple refocused images are generated, enabling the acquisition of refocused images at varying depths. The defocusing method makes depth measurement convenient by assessing the "blurriness" of pixels at different focus stacks. The defocusing method follows a general procedure: after calibrating the micro-lens centers on the original light field image captured by the light field camera to obtain redundant-free 4D light field data, multiple refocused images are obtained through the refocusing principle. Subsequently, the defocus depth is determined by evaluating the defocus depth using the maximum Laplacian gradient value at different focus positions. This process involves assessing the defocus depth using the maximum variance of the Laplacian gradient within an angular region. Finally, a Markov fusion is applied to merge and optimize the defocus depth map and correlation, resulting in the final depth map.

Following micro-lens center calibration on the light field camera, the 4D light field $L_0(x_0, y_0, u_0, v_0)$ can be decoded from the original light field image. Based on the light field dual-plane representation model and digital refocusing theory, altering the position of the imaging plane of the light field camera establishes a relationship between the coordinates of the new light field $L_n(x, y, u, v, \alpha_n)$, recorded at the new imaging depth, and the original light field $L_0(x_0, y_0, u_0, v_0)$ as follows

$$[x, y, u, v] = [x_0, y_0, u_0, v_0]A, \tag{9.36a}$$

$$A = \begin{bmatrix} 1\,0 & \alpha_n & 0 \\ 0\,1 & 0 & \alpha_n \\ 0\,0\,1-\alpha_n & 0 \\ 0\,0 & 0 & 1-\alpha_n \end{bmatrix}, \tag{9.36b}$$

where, α_n represents the sparsity ratio of the movement distance of the image plane, and α_n represents the index corresponding to different α_n values. Integrating $L_n(x, y, u, v, \alpha_n)$ along the u and v directions separately results in refocused images corresponding to different α_n values

$$I_n(x, y, \alpha_n) = \frac{1}{N_u \times N_v} \sum_{u=1}^{N_v} \sum_{v=1}^{N_v} L_n(x, y, u, v, \alpha_n), \tag{9.37}$$

where, N_u and N_v represent the size of the angular resolution of the 4D light field.

The defocus response $R_1(x, y, d, \alpha_n)$ of the 4D light field $L_n(x, y, u, v, \alpha_n)$ at different focusing depths is obtained by calculating the Laplacian values of the refocused images $I_n(x, y, \alpha_n)$ at different focusing depths

$$R_1(x, y) = \frac{1}{|w_D|} \sum_{x,y \in w_D} \left|\nabla^2 I_n(x, y, d, \alpha_n)\right|, \tag{9.38}$$

where, ∇^2 represents the Laplacian operator, and to enhance the robustness of the algorithm, the Laplacian mean for the current pixel is calculated with a window size w_D.

After obtaining multiple refocused images through digital refocusing, the correlation response value is measured for each refocused image using the angular variance of spatial pixels, expressed by the following equation. This provides a measure of the correlation response as

$$\sigma_{\alpha_n}(x, y)^2 = \frac{1}{N_v N_u} \sum_{v'} \sum_{u'} \left[I_n(x, y, u', v', \alpha_n) - I_n(x, y, \alpha_n)\right]. \tag{9.39}$$

The correlation response $R_1(x, y, c, \alpha_n)$ for different focus depths in the 4D light field $I_n(x, y, s, t, \alpha_n)$ is achieved by computing the angular variance of the 4D light field as

$$R_1(x, y, c, \alpha_n) = \frac{1}{|w_c|} \sum_{x,y \in w_D} \sigma_n(x, y, \alpha_n), \tag{9.40}$$

where w_c represents the window size for the current pixel to enhance the robustness of the algorithm. For each refocused image, there are now correlation response values for each pixel.

Form the refocusing depth map $D_1(x, y)$ by taking the index value n for which the defocus response $R_1(x, y, d, \alpha_n)$ is maximized, and form the correlation depth map $D_2(x, y)$ by taking the index value n for which the correlation response $R_1(x, y, c, \alpha_n)$ is minimized. The specific expressions are as follows

$$\begin{aligned} D_1(x, y) &= \arg\max_n [R_1(x, y, d, \alpha_n)] \\ C_1(x, y) &= \arg\max_n [R_1(x, y, c, \alpha_n)]. \end{aligned} \tag{9.41}$$

However, for some special scenes, it may be challenging to find the optimal response value. Therefore, confidence estimation is performed using the peak-to-sidelobe ratio equation. First, calculate the confidence for different depth maps

$$\begin{aligned} D_{\text{conf}}(x, y) &= D_1(x, y)/D'_1(x, y) \\ C_{\text{conf}}(x, y) &= C_1(x, y)/C'_1(x, y), \end{aligned} \tag{9.42}$$

where $D'_1(x, y)$ represents the depth map obtained when the defocus response $R_1(x, y, d, \alpha_n)$ is at its suboptimal maximum, and $C'_1(x, y)$ represents the depth map obtained when the correlation response $R_1(x, y, c, \alpha_n)$ is at its suboptimal minimum.

The reason for selecting defocus and correlation as two cues for depth fusion is that, in occluded regions, the depth values obtained from the defocus cue are highly sensitive, leading to severe artifacts, while the correlation cue exhibits strong robustness. In sensitive regions, the defocus cue fails to capture depth variations effectively, whereas the correlation cue can reveal depth discontinuities. In terms of noise, the correlation cue is sensitive, while the defocus cue demonstrates good robustness. Therefore, the defocus and correlation cues complement each other effectively.

The depth map corresponding to the extremum response value and the confidence in determining whether it is optimal are represented by the following equation

$$\{Z_1, Z_2\} = \{D_1, C_1\}, \ \{W_1, W_2\} = \{D_{\text{conf}}, C_{\text{conf}}\}. \tag{9.43}$$

9.5.2 Stereo Matching Method

If the two coordinates $u = u_0$ and $v = v_0$ of the light field camera plane are fixed, the four-dimensional light field can be projected onto the coordinates (u_0, v_0) of the sub-aperture images captured by the cameras. Consider each sub-image captured by each viewpoint in the light field camera as a whole and arrange the sub-images in the order of the coordinates of the camera plane. Based on the 4D light field, it is possible to represent it using "macro-pixel elements" in the form of a 2D visualized remap image. Each 'macro-pixel element' contains a number of pixels corresponding to angular information, representing the directional details of the 4D data. Extracting identical position pixels within each macro-pixel element results in a sub-image. This step corresponds to the sub-images captured by the camera array coordinates. By extracting all pixels within the macro-pixel element, multiple sub-images with distinct perspectives can be obtained. Integrating and arranging all these sub-images leads to the expression of the 4D light field in the form of sub-aperture images.

In the process of acquiring multi-view images with an array camera, the camera array positions are fixed. Each obtained sub-aperture image captures only the spatial information of the four-dimensional light field. Each image corresponds to different sampling points on the camera plane, with slight variations in perspective. These perspective differences manifest as positional disparities between the same objects in different sub-images. Therefore, all sub-aperture images collectively reflect the angular information in the light field. At this point, the depth estimation problem transforms into a multi-view stereo matching problem.

According to the Fourier transform theorem, the temporal displacement in the time domain transforms into the product of the original signal's Fourier expression and an exponential term in the frequency domain. Applying this transformation to an image, if the image I is shifted by Δx, the corresponding transformation in the Fourier domain is given by the following equation

$$\mathcal{F}\{I(x+\Delta x)\} = \mathcal{F}\{I(x)\}e^{-2\pi \mathrm{i} \Delta x}. \tag{9.44}$$

In order to facilitate the matching of light field images to sub-aperture images, two different matching cost measures were constructed: absolute error and gradient error sum. The final matching measure is obtained through a weighted combination, and it is a function of the image position x and the loss index l. The specific form is expressed by the following equation

$$C(x,l) = \alpha C_A(x,l) + (1-\alpha)C_G(x,l), \tag{9.45}$$

whereas, $\alpha \in (0, 1)$, represents the weight between the cost functions of absolute error C_A and the gradient error sum C_G, where C_A is defined by the following equation

$$C_A(x,l) = \sum_{u \in v} \sum_{x \in R_x} \min(|I(u_c, x) - I(u, x + \Delta x(u,l))|, \tau_1), \tag{9.46}$$

where R_x denotes the rectangular region centered at point x, and τ_1 is the truncation value for the matching cost measure; V represents all perspectives except the central view u_c.

By comparing the central sub-aperture image $I(u_c, x)$ with sub-aperture images $I(u, x)$ from other perspectives, a loss measure is constructed. This involves continuously moving around a point x on a sub-aperture image $I(u_i, x)$ from a certain perspective and taking the difference with the central sub-aperture image. This process utilizes phase-shift theory to obtain the pixel intensity after the movement, thereby obtaining the disparity between sub-aperture images from different perspectives and the central image. The displacement Δx in Eq. (9.46) is defined as follows:

$$\Delta x(u, l) = lk(u - u_c), \tag{9.47}$$

where k represents the depth unit in pixels, Δx linearly increases with the distance generated by the sub-aperture image u from other perspectives and the central sub-aperture image u_c. Therefore, the gradient error sum and cost function are constructed as the following equations

$$\begin{aligned} C_G = \sum_{u \in v} \sum_{x \in R_x} & \beta(u) \min(\text{diff}_x(u_c, u, x, l), \tau_2) \\ & + [1 - \beta(u)] \min(\text{diff}_y(u_c, u, x, l), \tau_2), \end{aligned} \tag{9.48}$$

where $\text{diff}(u_c, u, x, l) = |I_x(u_c, x) - I_x(u, x + \Delta x(u, l))|$ is the gradient difference in the x-direction between the central sub-aperture image and sub-aperture images from other perspectives. Similarly, diff_y represents the gradient difference in the y-direction between the matching image and the central image, where τ_2 is a constant used to enhance the algorithm's robustness. $\beta(u)$ is a weight controlling the gradient differences between the x and y directions, determined by the relative distance between the coordinates (u, v) of sub-aperture images from any perspective and the coordinates (u_c, v_c) of the central sub-aperture image

$$\beta(u) = \frac{|u - u_c|}{|u - u_c| + |v - v_c|}. \tag{9.49}$$

9.5.3 *Epipolar Plane Image Slope Method*

The first two visualization methods for the 4D light field involve fixing points on the camera plane to acquire multi-view images or fixing points on the imaging plane to obtain macro-pixel elements. The former emphasizes the reflection of spatial ray-angular information, while the latter focuses on the representation of angular information in spatial rays. If a specific coordinate on the camera plane (generally assumed as $u = u_0$) is fixed with a coordinate on the imaging plane (generally assumed as $x = x_0$), an epipolar plane image (EPI) can be obtained for a specific direction.

After simplifying the light field through a dual-plane representation, any point $P(X, Y, Z)$ in space can be expressed in terms of the imaging plane (x, y) and the camera plane (u, v). Therefore, the light field can be represented as $L(x, y, u, v)$. If we fix (y^*, v^*) and only vary (x, u), as shown in Eq. (9.50), we can obtain a two-dimensional plane $L_1(x, u)$. We refer to this as the EPI

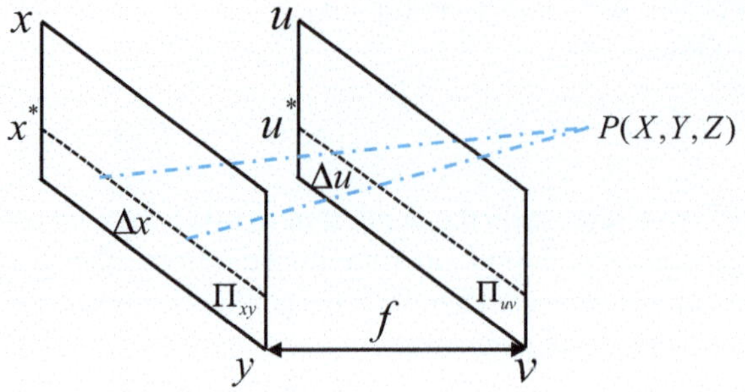

Fig. 9.16 Schematic diagram of depth estimation from EPI slopes

$$L_1(x, u) = L(x, y^*, u, v^*). \tag{9.50}$$

Depending on the imaging position, imaging a point in the real world at different locations, as shown in Fig. 9.16, when u changes, the coordinate x of point P will change. That is, the direction of the line in the EPI reflects depth information. Specifically, if the direction corresponding to the line is defined as θ, the actual depth Z can be expressed by

$$Z = f\frac{\Delta u}{\Delta x} = \frac{f}{\tan\theta}, \tag{9.51}$$

where f is the distance between the two parallel planes, angle θ is used to reflect the relationship between Δu and Δx, displaying the movement of scene points between views. Depth estimation essentially corresponds to estimating the slope of the lines in the EPI, that is, accurately determining θ.

Compared to regular images, EPI exhibits boundaries, representing the imaging pixels of the same point from different perspectives. EPIs should contain a set of linear structures, including disparity information among pixel points. Therefore, specific coordinates for a line can be determined using the points around that line. In previous methods, circles were created for each pixel to obtain the minimum tangent slope, thereby acquiring the depth value for that pixel. The rotated parallelogram involves creating a parallelogram operator for slope θ to obtain depth information. A parallelogram is created with the pixel to be estimated as the center, and the line passing through this center point divides the parallelogram into two differently sized parts. The line with the maximum histogram distance between the two parts corresponds to the desired disparity information for that pixel.

Firstly, a weighted function is utilized to create the size of the parallelogram window. The weights within the window are used to represent the distance weight of each pixel to the desired pixel disparity. The weighted function $w_\theta(i, j)$ is created using the derivative of the Gaussian function. Specifically, for each point in the EPI denoted as $L_{yv}(x, u)$ and its relationship with the reference point $p(x, u)$, the corresponding weighted pixel is defined by the following equation

$$w_\theta(i,j) = c \cdot d_\theta(i,j) \cdot \exp\left[\frac{-d_\theta(i,j)}{2\alpha^2}\right], \tag{9.52}$$

where $d_\theta(i,j) = i - [x + (j-u)\cdot\tan\theta]$ is the distance along the same row in the parallelogram from any point to the reference point P, $[x + (j-u)\cdot\tan\theta]$ assumes the matching point is in the j row, α is a scale parameter determined by the complexity of the data, and c is a normalizing function.

The obtained parallelogram is uniformly divided into equal parts with the matching point P as the center. The histogram distance between the two parts is measured using the distance measure L_2, expressed by the following equation

$$L^2 = \sum_i \frac{g_\theta^2(i) + h_\theta^2(i)}{g_\theta(i) + h_\theta(i)}, \tag{9.53}$$

where g_θ and h_θ are histograms obtained by dividing the parallelogram into two equal parts with the matching point P as the center. A larger value of $L^2(g_\theta, h_\theta)$ implies the presence of a line with slope θ corresponding to the depth values in the desired EPI.

The distance of the histogram in the horizontal EPI image $L_2(x,u)$ is denoted as $d_2(x,u,\theta)$. The value of θ corresponding to the maximum distance $d_2(x,u,\theta)$ is considered as the depth value of the horizontal EPI image $L_2(x,u)$, as shown in the equation

$$\theta_2(x,u) = \arg\max_\theta d_2(x,u,\theta). \tag{9.54}$$

After obtaining the optimal depth for the horizontal EPI image, the corresponding optimal depth for the vertical EPI image can also be determined. Following the acquisition of the initial depth maps for both horizontal and vertical EPI images, optimization and integration are necessary. For pixels in ambiguous regions, particularly those affected by occlusion, it is essential to assign accurate depth values. To measure the reliability of depth estimation, the confidence metric c is defined using the following equation

$$c = \exp\left(-\frac{d/d_{\max}}{2\sigma^2}\right). \tag{9.55}$$

After obtaining the confidence maps for the depth maps of the horizontal and vertical EPI images, the confidence value for each point in the entire image can be expressed by the following equation

$$d_{u,v}(x,y,\theta) = c_{y,v^*}(x,u^*)d_{y,v^*}(x,u^*,\theta) + c_{x,u^*}(y,u^*)d_{x,u^*}(y,v^*,\theta). \tag{9.56}$$

After obtaining the confidence values for each pixel, the integration of depth maps involves applying guided filtering with a smoothing filter for depth optimization. Finally, a global depth map is obtained through the combination of disparity optimization and weighted median filtering.

9.6 Light Field Camera Achieves Digital Refocusing

In simple terms, digital refocusing involves altering the image plane to achieve corresponding refocused effects at different focal points. This is achieved by projecting the light rays captured by the light field device onto a new imaging plane. Digital refocusing techniques can be categorized into two methods: spatial domain and frequency domain. This includes spatial domain digital integral refocusing and frequency domain Fourier transform refocusing techniques.

After capturing image information with a light field camera, digital refocusing techniques can be applied to sequentially exposed images. This constitutes a crucial application of light field imaging technology. Based on this principle, it is possible to deconvolve defocused images and reconstruct target images with accurate focal distances, thereby addressing challenges in scenarios involving automatic focusing.

Light field can be considered as a function that describes the quantity of light rays passing through any point and direction in space. The radiance emitted from any point in space is referred to as the light field of that point and can be represented by an energy function. Additionally, when extracting light rays from optical imaging devices, energy loss during transmission is typically not taken into account. However, in reality, light undergoes some degree of energy loss during propagation. As a result, the captured image information does not accurately represent the true colors in the scene. Therefore, it is necessary to perform color correction on the acquired images and, based on this correction, use refocusing techniques to obtain refocused images at different focal points.

According to the principles of light field imaging and the energy function of light rays, it can be derived that focusing the image at different focal planes corresponds to changing the distance between micro-lenses and the sensor. This leads to different slices in the spatial ray trajectory. Assuming the image is focused at a new imaging plane denoted as F', according to the refocusing principle in Fig. 9.17, it can be inferred from similar triangles that the coordinates $(u + (x - u)/\alpha, v + (y - v)/\alpha)$ represent the intersection of the plane F and the light rays. Therefore, the light ray $L_{\alpha F}$ can be expressed as

$$\begin{aligned} L_{\alpha F}(x, y, u, v) &= L_F(u + \frac{x-u}{\alpha}, v + \frac{y-v}{\alpha}, u, v) \\ &= L_F\left[u(1 - \frac{1}{\alpha}) + \frac{x}{\alpha}, v(1 - \frac{1}{\alpha}) + \frac{y}{\alpha}, u, v\right], \end{aligned} \tag{9.57}$$

where $\alpha = F'/F$, and let B_α be a 4×4 matrix, it can be obtained

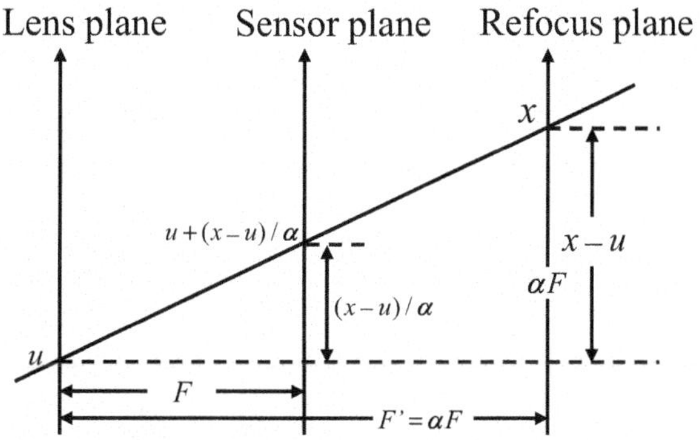

Fig. 9.17 Digital refocusing principle diagram

$$(x, y, u, v) \cdot B_\alpha = \left[u(1-\frac{1}{\alpha})+\frac{x}{\alpha}, v(1-\frac{1}{\alpha})+\frac{y}{\alpha}, u, v\right], \tag{9.58a}$$

$$B_\alpha = \begin{pmatrix} \frac{1}{\alpha} & 0 & 1-\frac{1}{\alpha} & 0 \\ 0 & \frac{1}{\alpha} & 0 & 1-\frac{1}{\alpha} \\ 0 & 0 & 1 & 0 \\ 0 & 0 & 0 & 1 \end{pmatrix}. \tag{9.58b}$$

Obtained from Eqs. (9.57) and (9.58a, 9.58b)

$$L_{\alpha F}(x, y, u, v) = L_F[(x, y, u, v) \cdot B_\alpha]. \tag{9.59}$$

Derived from Eq. (9.59), the energy function of the refocused light rays can be expressed as

$$E_{\alpha F} = \iint L_{\alpha F}(x, y, u, v)\mathrm{d}u\mathrm{d}v = \iint L_F \cdot B_\alpha \mathrm{d}u\mathrm{d}v, \tag{9.60}$$

where $E_{\alpha F}$ is the refocused energy obtained from the 4D light field.

Spatial domain integral refocusing is the process of determining the corresponding α values for different planes at the refocusing position based on the collected 4D light field data. The α values, in turn, determine the pixel positions in the image and their corresponding micro-lenses in Fig. 9.18.

Defining the operation factor transforming the depth F of the sensor to αF in the light field as P_α, then $P_\alpha(L_F)$ can be represented as the application of P_α in the light field L_F. It can be used to compute photos under different depth focus settings within the camera, representing the projection of four-dimensional slices onto a two-dimensional plane as

$$P_\alpha(L_F)(x, y) = \frac{1}{\alpha^2 F^2} \iint \left[u(1-\frac{1}{\alpha})+\frac{x}{\alpha}, v(1-\frac{1}{\alpha}), u, v\right]\mathrm{d}u\mathrm{d}v, \tag{9.61}$$

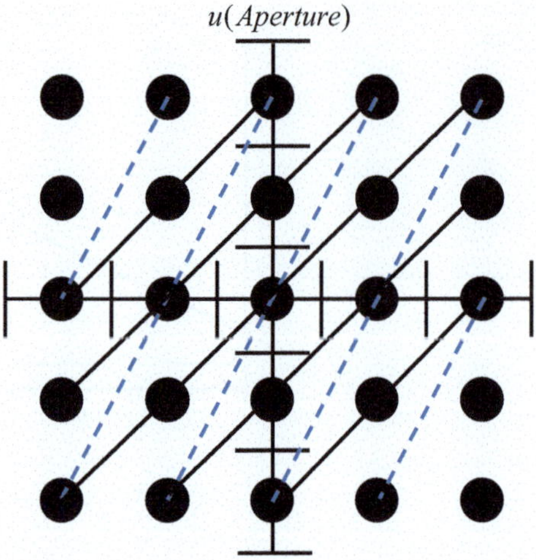

Fig. 9.18 The two parameter values α in the photographic transformation

according to the Fourier slice theorem, the above expression can be defined as

$$P_\alpha(L_F) = \frac{1}{\alpha^2 F^2} I_2^4 \circ B_\alpha[L_F], \tag{9.62}$$

where, I_2^4 represents the integration factor, and the image is represented using a two-dimensional function. The symbol $\circ$ represents the mapping relationship between variables, that is, the coordinate conversion operation in Eq. (9.61).

Therefore, integration is applied to the four-dimensional function. B_α is a four-dimensional basis transformation, which can be defined as

$$B_\alpha = \begin{pmatrix} 1 & 0 & -\alpha & 0 \\ 0 & 1 & 0 & \alpha \\ 0 & 0 & 1 & 0 \\ 0 & 0 & 0 & 1 \end{pmatrix}. \tag{9.63}$$

In utilizing the four-dimensional Fourier transform for refocusing, define S_2^4 as the slicing factor, meaning filling the last two dimensions of the function with zeros, and define F^n as the N-dimensional Fourier transform. Substituting $F^{-2} \circ S_2^4 \circ (B_\alpha^{-T}/|B_\alpha^{-T}|) \circ F^4$ for $I_2^4 \circ B_\alpha$, the expression becomes

$$P_\alpha = \frac{1}{\alpha^2 F^2} F^{-2} \circ S_2^4 \circ \frac{B_\alpha^{-T}}{|B_\alpha^{-T}|} \circ F^4, \tag{9.64}$$

substituting $\beta_\alpha = (1/(\alpha^2 F^2) S_2^4 \circ B_\alpha^{-T})$ into Eq. (9.64) yields

$$P_\alpha = F^{-2} \circ \beta_\alpha \circ F^4, \tag{9.65}$$

the value of β_α can be calculated using Eq. (9.65)

$$\beta_\alpha[G](x, y) = \frac{1}{\alpha^2 F^2} G(\alpha \cdot k_x, \alpha \cdot k_y, (1-\alpha) \cdot k_x, (1-\alpha) \cdot k_y). \tag{9.66}$$

Light passing through the microlens edge can undergo refraction and diffraction, leading to the blurring of the spectral frequencies of noise light. This, in turn, results in a decrease in the accuracy of the recorded signal.

9.7 General Conclusions

This chapter starts by exploring the fundamental concepts of light fields and the methods for acquiring information from light fields. The concept and parameterized representation of light fields are examined based on these studies. Building upon this research, the chapter delves into light field imaging techniques using a microlens array, describing the light field in a parameterized form and providing a thorough derivation of imaging formulas. Furthermore, the chapter discusses the principles of wavefront sensing in light field cameras, deriving the underlying technology and addressing various sources of errors in wavefront detection. An analysis of the impact of different factors on measurement accuracy is conducted, and approximate formulas for calculating various errors are derived. The chapter also explores the principles of depth map acquisition using defocusing methods, multi-view stereo matching algorithms, and EPI slope algorithms. Additionally, it discusses the principles behind achieving digital refocusing using light field cameras.

Acknowledgements This work was supported by the National Natural Science Foundation of China (Nos. 11874132, 61975044, 12074094, 12374271, and 62375065); Interdisciplinary Research Foundation of HIT (No. IR2021237).

References

1. Veeraraghavan A, Raskar A, Agrawal A, Mohan A, Tumblin J. Dappled photography: mask enhanced cameras for heterodyned light fields and coded aperture refocusing. ACM Trans Graph. 2007;26(3):69.
2. Cao G, Yu X. Accuracy analysis of a Hartmann-shack wavefront sensor operated with a faint object. Opt Eng. 1994;33(7):2331–5.
3. Thomas S, Fusco T, Tokovinin T. Comparisonof centroid computation algorithms in a Shack-Hartmann sensor. Mon Not R Astron Soc. 2006;371:323–36.
4. Lardiere O, Conan R, Clare R. Performance comparison of centroiding algorithms for laser guide star wavefront sensing with extremely large telescopes. Appl Opt. 2010;49(31):g78–94.
5. Jiang W, Xian H, Shen F. Detecting error of Shack-Hartmann wavefront sensor. Proc SPIE. 1997;3126:534–44.
6. Tao MW, Hadap S, Malik J, Ramamoorthi R (2013) Depth from combining defocus and correspondence using Light-Field cameras. IEEE Int Conf Comput Vis 673–682

7. Jeon HG, Park J, Choe G, Park J, Bok Y, Tai YW, Kweon IS (2015) Accurate depth map estimation from a lenslet light field camera. Int Conf Comput Vis Pattern Recognit 1547–1555
8. Zhang S, Sheng H, Li C, Zhang J, Xiong Z. Robust depth estimation for light field via spinning parallelogram operator. Comput Vis Image Underst. 2016;145:148–59.

Chapter 10
Adaptive Optical Imaging

Shurui Yang and Zhengjun Liu

10.1 Introduction

Babkock proposed the use of a wavefront sensor to detect wavefront aberration information and then utilize deformable optical elements to generate controllable optical phase shifts for compensating wavefront aberrations, giving rise to the fundamental concept of adaptive optics. Leighton developed a first-order active optical system for compensating image motion in astronomical telescopes, achieving the highest image quality at that time. Techniques for adaptive control of electromagnetic wavefronts emerged in the field of microwaves. Skolnik and King introduced the principles of phase conjugation and wavefront compensation. In the same year, Adams proposed the principle of maximizing the clarity of transmitted wave high-frequency vibrations and received wave imaging. These four principles became the fundamental principles of traditional correction-based adaptive optical systems [1].

Over the past six decades, adaptive optics technology has seen vigorous development and has been applied in various fields such as astronomy, space optics, lasers, and biomedical sciences [2]. In astronomy, it is used to overcome dynamic wavefront disturbances caused by atmospheric turbulence, improving the resolution and signal-to-noise ratio of optical instruments. In the field of space optics, it is used for remote sensing imaging, strategic defense systems, and to mitigate errors related to design, manufacturing, assembly, thermal effects, and structural deformation. In the laser domain, it is applied to correct issues like thermal deformation inside laser cavities, errors in optical component fabrication and assembly, non-uniform gain media, resonator misalignment, atmospheric turbulence disturbances, and thermal blooming effects, with the goal of enhancing the energy density delivered to target areas or addressing coherence degradation and reduced usability in laser communication. In the biomedical field, adaptive optics correct various types of static and dynamic

S. Yang · Z. Liu (✉)
School of Physics, Harbin Institute of Technology, Harbin 150001, China
e-mail: zjliu@hit.edu.cn

Z. Liu et al. (eds.), *Computational Optical Imaging*, Advances in Optics and Optoelectronics, https://doi.org/10.1007/978-981-97-1455-1_10

aberrations in systems to obtain clear images of biological tissues. Adaptive optics integrates knowledge from multiple disciplines, including optics, mechanics, electronics, thermodynamics, computer science, and control theory, making it a multidisciplinary field that corrects wavefront errors in real time based on actual wavefront conditions. Compared to traditional optical technology, it offers unique advantages in correcting dynamic errors in optical systems. Depending on the spatiotemporal characteristics of wavefront error sources, adaptive optics systems can employ suitable and diverse wavefront sensing and correction methods, utilizing a wide range of correction devices.

10.2 Fundamental Theory of Adaptive Optics

In principle, the light field from an ideal point source, without any phase disturbance, still exhibits an ideal point imaging in the far field. This is because, on an infinitely extended transmission wavefront, the light field maintains good spatial coherence. Optical focusing systems can observe the far-field distribution of the light source over a finite transmission distance, but at the same time, they truncate the transmission wavefront during propagation. The far-field distribution of the original point source is no longer ideal but is compressed within the diffraction limit imposed by the aperture of the optical system. This constitutes the diffraction limit of an optical system. In practical applications, the spatial coherence of the transmission wavefront on the optical system's entrance pupil is disrupted by factors such as the transmission medium and environmental disturbances, leading to the segmentation of the originally intact optical entrance pupil into sub-apertures with a coherent length scale. It is shown in Fig. 10.1.

The essence of adaptive optics is to restore the spatial coherence of the disrupted light transmission wavefront, recombine the light fields from various sub-apertures coherently in the far field, compress the far-field spot, and thereby approach the diffraction limit of the optical system, enhancing the resolution of the optical system.

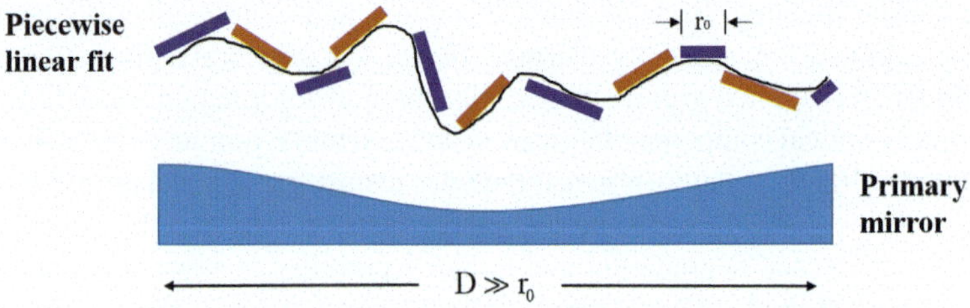

Fig. 10.1 The transmission medium disrupts the coherence of the target light field

10.2.1 Principle of Phase Conjugation

The core principle of adaptive optics is the real-time correction of optical beam wavefront aberrations to enhance the quality of beam transmission or the imaging performance of optical systems. The fundamental principle is phase conjugation, where an optical field with phase errors can be represented as

$$W_1 = |A|e^{i\varphi}. \tag{10.1}$$

In the equation, φ represents the phase fluctuations of the optical field caused by disturbances. The role of an adaptive optics system is to generate modulation that is conjugate to the incident optical field within the system.

$$W_2 = |A|e^{-i\varphi}. \tag{10.2}$$

As a result, the superposition of the two optical fields mentioned above allows for the compensation of phase errors, yielding an output nearly planar optical field. Adaptive optics typically corrects phase errors only, with no impact on the original optical field's amplitude. In situations with significant amplitude errors, the correction effectiveness may be affected, but for most applications, correcting phase errors alone is sufficient to meet practical requirements.

10.2.2 Correcting Adaptive Optics System

A correcting adaptive optics system utilizes a wavefront sensor to measure the phase of incident light in real time. It generates controllable optical phase shifts using optical elements that can be deformed as desired, thus providing real-time compensation for the wavefront aberrations of the incident light. This enables the incident light to output as a planar wave or spherical wave after passing through the wavefront corrector. A typical correcting adaptive optics system is shown in Fig. 10.2.

Traditional correcting adaptive optics systems are primarily composed of three main components: a wavefront sensor (WFS), a wavefront corrector (DM), and a control unit (CS). The wavefront sensor is used to measure wavefront errors, and the control unit drives the wavefront corrector to perform corrections based on the wavefront error information.

10.2.3 Wavefront and Its Representation

(1) **Zernike Orthogonal Polynomials**

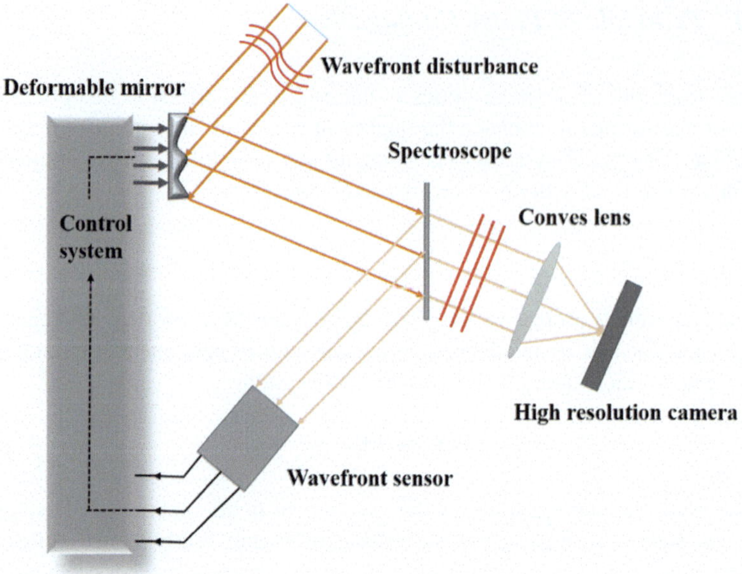

Fig. 10.2 Schematic diagram of a typical corrective adaptive optics system

Within a circular domain and using polar coordinates, wavefront aberrations can be represented as a linear combination of Zernike polynomials, as shown in the following equation:

$$\psi(r,\theta) = \sum_{i=0}^{N} a_i Z_i(r,\theta). \tag{10.3}$$

In this equation, r represents the radius normalized to the range [0,1], it is the polar angle, Zi denotes the coefficient of the i-th order Zernike polynomial, and N is the number of Zernike basis functions used in the expansion. In polar coordinates, Zernike polynomials are products of radial and azimuthal terms. Without a specific ordering scheme, defining a particular order of a Zernike polynomial requires two parameters: azimuthal order m and radial order n Zernike polynomials can be expressed as follows

$$\begin{aligned} Z_{even}(r,\theta) &= \sqrt{n+1}R_n^m(r)\cos(m\theta),\ m \neq 0, \\ Z_{odd}(r,\theta) &= \sqrt{n+1}R_n^m(r)\sin(m\theta),\ m \neq 0, \\ Z_i(r,\theta) &= R_n^0(r),\ m = 0. \end{aligned} \tag{10.4}$$

The function $R_n^m(r)$ is referred to as the radial function and is defined as follows

$$R_n^m(r) = \sum_{s=0}^{(n-m)/2} \frac{(-1)^s (n-s)!}{s![(n+m)/2-s]![(n-m)/2-s]!} r^{n-2s}, \tag{10.5}$$

And m and n must satisfy the following relationship:

$$\begin{cases} m \leq n, \\ n - m = 2p, (p = 0, 1, 2, \ldots), \end{cases} \tag{10.6}$$

The conversion relationship between the single index i and m, n for Zernike polynomials is as follows

$$i = \frac{n(n+2)+m}{2}, \tag{10.7}$$

The two-dimensional shapes corresponding to the first 35 orders of Zernike polynomials (excluding the piston term Z_0) are shown in Fig. 10.3. It can be observed that as the order increases, the corresponding spatial frequency of the surface shape gradually increases. Under the condition of the same azimuthal order, Zernike modes near the central axis of the pyramid have lower azimuthal spatial frequencies, while Zernike modes closer to the edges of the pyramid have higher azimuthal frequencies, resulting in complex wavefront edge shapes.

(2) **Legendre polynomials**

Legendre polynomials are a set of orthogonal polynomials commonly used to address problems involving spherical symmetry. They find extensive applications in physics, engineering, and various mathematical fields.

There exists a close relationship between Legendre polynomials and spherical harmonics. Spherical harmonics are functions used to describe problems with spherical symmetry, and Legendre polynomials constitute a part of them. Specifically, spherical harmonics can be expressed as a linear combination of Legendre polynomials. Legendre polynomials have widespread applications in physics, particularly in describing spherically symmetric potential fields, electric fields, magnetic fields, etc. They also appear in classical mechanics, quantum mechanics, electrodynamics, and other fields. Legendre polynomials play a crucial role in signal processing and numerical analysis. They can be utilized for data fitting, function approximation, and as basis functions in certain numerical algorithms. Legendre polynomials are a highly valuable mathematical tool, and their orthogonality property and recurrence relation make them an ideal choice for addressing problems related to spherical symmetry and solving specific mathematical and physical problems.

For certain specific regions, you can choose different orthogonal polynomials that are suitable for that region. For example, when applying adaptive optics to stripe laser systems and reconstructing the far-field wavefront, it is common to use Legendre polynomials that are orthogonal on the interval [−1, 1] to fit the wavefront. Legendre polynomials are the solutions to the Legendre equation with the natural boundary

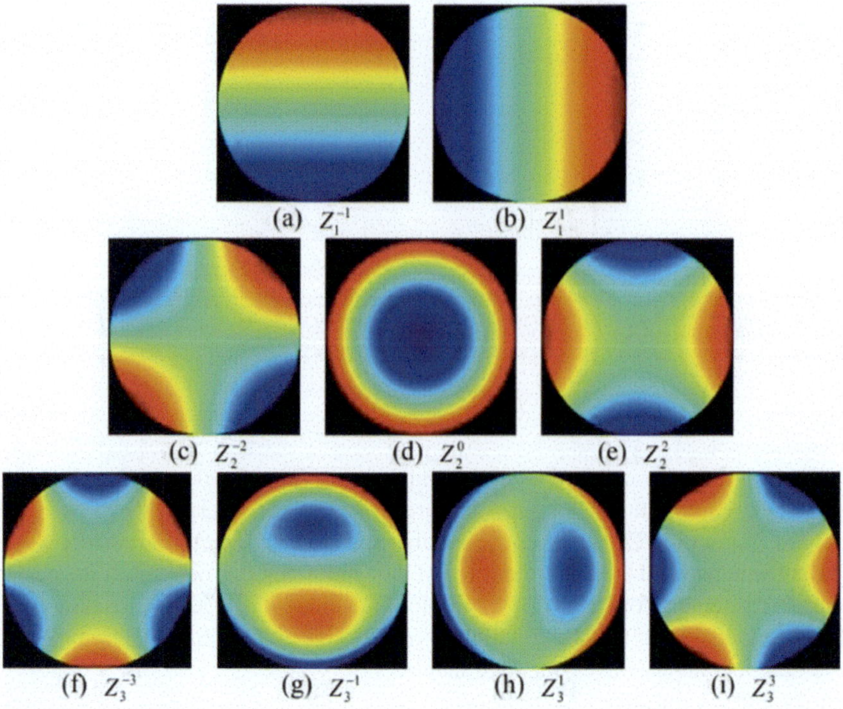

Fig. 10.3 Aberration wavefront maps under 1st to 3rd order Zernike mode

condition solutions remaining finite at $x = \pm 1$. Their expression is as follows

$$P_l(x) = \sum_{k=0}^{[l/2]} (-1)^k \frac{(2l-2k)!}{2^l k!(l-k)!(l-2k)!} x^{l-2k}. \tag{10.8}$$

In the equation, $[l/2]$ represents the greatest integer not exceeding $l/2$. The Legendre polynomials for the first few orders are as follows

$$\begin{aligned}
P_0(x) &= 1, \\
P_1(x) &= x, \\
P_2(x) &= \frac{1}{2}(3x^2 - 1), \\
P_3(x) &= \frac{1}{2}(5x^3 - 3x), \\
P_4(x) &= \frac{1}{8}(35x^4 - 30x^2 + 3), \\
P_5(x) &= \frac{1}{8}(63x^5 - 70x^3 + 15x),
\end{aligned}$$

$$P_6(x) = \frac{1}{16}(213x^6 - 315x^4 + 105x^2 - 5). \tag{10.9}$$

The integral results of Legendre polynomials of different orders within the interval $[-1,\ 1]$ are as follows

$$\int_{-1}^{+1} P_m(x)P_n(x)dx = \begin{cases} 0, n \neq m, \\ \frac{2}{2n+1}, n = m. \end{cases} \tag{10.10}$$

Since wavefronts are distributed in two-dimensional space, and Legendre polynomials are one-dimensional functions, when performing wavefront reconstruction, it is necessary to extend and define Legendre polynomials in two-dimensional space. The extension is done by multiplying Legendre polynomials that are mutually independent in the x and y directions, as follows

$$P_{m,n}(x, y) = P_m(x)P_n(y). \tag{10.11}$$

Describing the distribution of the wavefront space using Legendre product polynomials, the first 10 terms of the two-dimensional Legendre polynomials can be obtained, as shown in Table 10.1.

Unlike Zernike polynomials, each term of the Legendre polynomials does not have a direct relationship with traditional Seidel aberrations. Therefore, it is not possible, as with Zernike polynomials, to analyze the components and magnitudes of various aberrations present in the distorted wavefront through the magnitude of each order's coefficient.

Table 10.1 Expression of the first 10 terms of Legendre polynomials

terms	$P_{m,n}$	m	n	$P_{m,n}(x, y)$
1	$P_{0,0}$	0	0	1
2	$P_{1,0}$	1	0	x
3	$P_{0,1}$	0	1	y
4	$P_{2,0}$	2	0	$(3x^2 - 1)/2$
5	$P_{1,1}$	1	1	xy
6	$P_{0,2}$	0	2	$(3x^2 - 1)/2$
7	$P_{3,0}$	3	0	$(5x^3 - 3x)/2$
8	$P_{2,1}$	2	1	$(3x^2 - 1)y/2$
9	$P_{1,2}$	1	2	$(3y^2 - 1)x/2$
10	$P_{0,3}$	0	3	$(5y^3 - 3y)/2$

10.2.4 Quality Evaluation Standards for Adaptive Optics Systems

An adaptive optics system is a device that compensates for and corrects optical wavefront distortions to improve image quality. Its goal is to correct the wavefront phase at all positions within the optical receiving aperture to the same value. In the case of point source imaging, when an undistorted plane wave passes through the optical system and is imaged, the phase at various positions within the optical receiving aperture is identical. At this point, the image formed on the focal plane exhibits the highest energy concentration and maximum intensity, representing the optical diffraction limit of the imaging system. The corresponding image spot is called an "Airy disk." Conversely, if the wavefront experiences distortion, there are phase differences among points within the optical receiving aperture. In such cases, the energy concentration of the image spot decreases, the intensity at the center of the spot decreases, and the spot radius increases.

Therefore, the quality of correction for point source imaging in an adaptive optics system can be evaluated from three aspects: describing the phase differences among points within the optical receiving aperture, the radius of the imaging spot, and the intensity at the center of the spot. The corresponding metrics are Root Mean Square (RMS) wavefront phase error, optical diffraction limit factor, and Strehl ratio (SR).

The RMS wavefront phase error describes the extent of phase fluctuations of points within the optical receiving aperture, and it is calculated as follows

$$\mathrm{RMS} = \sqrt{\frac{1}{N}\sum_{i=1}^{N}[\varphi(x, y) - \langle\varphi(x, y)\rangle]^2}. \tag{10.12}$$

The optical diffraction limit factor is the multiple by which the radius of the imaged spot, formed by distorted incident light waves, relative to the optical diffraction limit spot radius. If the intensity distribution of the imaged spot is denoted as $I(x, y)$, then the corresponding spot radius R can be calculated using the following expression

$$R = \frac{\iint \sqrt{(x - \overline{x})^2 + (y - \overline{y})^2} I(x, y)\mathrm{d}x\mathrm{d}y}{\iint I(x, y)\mathrm{d}x\mathrm{d}y}. \tag{10.13}$$

The Strehl ratio is defined as the ratio of the maximum intensity of the actual imaged spot to the maximum intensity of the ideal imaged spot, as follows

$$SR = \frac{\max[I(x, y)]}{\max[I_0(x, y)]}. \tag{10.14}$$

In the equation, $I(x, y)$ is the intensity distribution of the actual imaged spot with wavefront aberrations, and $I_0(x, y)$ is the intensity distribution of the ideal imaged spot without wavefront aberrations.

In the case where the aberration wavefront phase $\varphi(\rho, \theta)$ is known, you can also calculate the Strehl ratio using the following expression.

When the Strehl ratio of the spot is greater than 0.8, it can be considered that the quality of the imaged spot is high, and the wavefront aberrations have been well corrected.

$$\begin{aligned} SR &= \frac{1}{\pi^2}\left|\int_0^{2\pi}\int_0^1 \exp[\mathrm{i}\varphi(\rho, \theta)]\rho \mathrm{d}\rho \mathrm{d}\theta\right|^2 \\ &= \frac{1}{\pi^2}\left|\int_0^{2\pi}\int_0^1 \sum_{n=0}^{\infty}\frac{[\mathrm{i}\varphi(\rho, \theta)]^n}{n!}\rho \mathrm{d}\rho \mathrm{d}\theta\right|^2. \end{aligned} \tag{10.15}$$

If the imaging target is a larger extended object, you need to consider the intensity distribution image of the imaged spot of a point source as the point spread function (PSF) of the imaging system. By convolving the original scene with the PSF, you obtain the corresponding imaging results of the extended target. In other words

$$f(x, y) = I(x, y) * h(x, y). \tag{10.16}$$

Then, the imaging results corresponding to the Airy disk are used as a reference image for evaluating the quality of the obtained imaging results. Common evaluation methods include Peak Signal-to-Noise Ratio (PSNR) and Structural Similarity (SSIM). The calculation method for PSNR is as follows

$$\mathrm{PSNR} = 10\log_{10}\left\{\frac{(2^k - 1)^2}{\mathrm{MSE}[f(x, y), I(x, y)]}\right\}. \tag{10.17}$$

In the equation, k represents the bit depth of the images, and $\mathrm{MSE}[f(x, y), I(x, y)]$ is the Mean Squared Error (MSE) between the imaging frame $f(x, y)$ and the reference image $I(x, y)$. If the size of the two images is the same, then the calculation method for the Mean Squared Error is as follows

$$\mathrm{MSE}[f(x, y), I(x, y)] = \frac{1}{N_x N_y}\sum_{x=0}^{N_x-1}\sum_{y=0}^{N_y-1}[f(x, y) - I(x, y)]^2. \tag{10.18}$$

The calculation method for structural similarity (SSIM) is as follows

$$\mathrm{SSIM}[f(x, y), I(x, y)] = \frac{(2\mu_f\mu_I + c_1)(2\sigma_{fI} + c_2)}{(\mu_f^2 + \mu_I^2 + c_1)(\sigma_f^2 + \sigma_I^2 + c_2)}. \tag{10.19}$$

In the equation, μ_f and μ_f represent the means of the imaging frame and the reference image, respectively. σ_f^2 and σ_I^2 represent the variances of the imaging frame and the reference image, respectively. σ_{fI} is the covariance between the two images, and the values of parameters c_1 and c_2 correspond to $(k_1L)^2$ and $(k_2L)^2$, respectively. Here $k_1 = 0.01$, $k_2 = 0.03$, L denotes the pixel value range of the images.

10.2.5 Adaptive Optical System Error Sources

(1) **Environmental errors**

When light waves propagate through the atmosphere, they are affected by various sources of errors, including molecular absorption and various types of scattering (atmospheric gas scattering, aerosol scattering, Rayleigh scattering, and resonance scattering). These effects lead to energy attenuation of the light signal and can cause depolarization of the light waves. Additionally, atmospheric turbulence results in random variations in the refractive index of the air, leading to random fluctuations in the wavefront amplitude and phase of the light waves. This results in phenomena such as light intensity scintillation, wavefront distortion, angle-of-arrival fluctuations, and beam wandering.

When high-energy lasers propagate through the atmosphere, the strong interaction of light with the atmosphere causes beam expansion and distortion, resulting in a reduction in the average light intensity on the target. As laser beams travel through the atmosphere, they heat the atmosphere due to the absorption of laser energy by atmospheric gas molecules and aerosol particles. This local heating causes the local air temperature to rise, leading to expansion and reduced density, which in turn results in a decrease in the local refractive index. When the laser beam continuously passes through this region, it will diverge. The phenomena caused by the nonlinear interaction between the atmosphere and the laser beam, such as beam spreading and distortion, are known as thermal blooming. Thermal distortion is one of the primary error sources affecting optical systems, especially space remote sensing systems.

In the case of space remote sensing systems, the thermal environment is primarily determined by the thermal radiation from the Sun and other celestial bodies, as well as the thermal disturbances generated by the satellite platform and other instruments inside the satellite. These thermal effects cause temperature changes and temperature gradients within the telescope, resulting in thermal stress and, ultimately, thermal distortion.

(2) **Adaptive optical system errors**

Typically, adaptive optical systems can reduce the system's wavefront aberrations by several orders of magnitude, allowing systems that would otherwise produce images nearly at the diffraction limit to approach this limit. However, adaptive optical systems themselves introduce new errors that can impact the overall imaging quality of the system. These are primarily manifested as follows:

(1) Optical Component Errors: The optical components in adaptive optical systems are subject to manufacturing, assembly, and alignment errors, as well as errors caused by external factors such as temperature, gravity, and vibrations. These errors can affect the performance of the adaptive optics system just as they do in conventional imaging optical systems.
(2) Wavefront Sensing Errors: Errors in wavefront sensing can result from inaccuracies in measuring the wavefront aberrations. These errors can affect the system's ability to correct aberrations accurately.
(3) Control System Errors: Errors in the control system can lead to inaccuracies in the correction process. These errors may result from the control algorithms, sensor noise, or other factors in the feedback loop that adjust the deformable mirrors or other optical elements.
(4) Actuator Errors: Actuators are responsible for driving the deformable mirrors or other optical components to their final positions to correct the aberrations. The precision and accuracy of these actuators can introduce errors into the correction process.

The intrinsic errors of the adaptive optical system itself cannot be corrected and, when combined with the residual errors after optical system correction, constitute the total system error, which ultimately determines the system's imaging quality.

10.3 Wavefront Sensor

A wavefront sensor is one of the fundamental components in adaptive optical systems. It needs to measure wavefront aberrations with sufficient spatial resolution and speed so that the Adaptive Optics (AO) system can compensate for real-time distorted wavefronts. Wavefront sensors do not directly measure the wavefront phase but instead measure the spatial gradient or Laplacian curve of the wavefront. The spatial gradient is commonly referred to as wavefront slope, and the Laplacian curve is known as wavefront curvature. These measurements can be used to estimate the wavefront through phase reconstruction algorithms [3].

Commonly used wavefront sensors include Hartmann-Shack wavefront sensors, curvature wavefront sensors, interferometric wavefront sensors, and pyramid wavefront sensors. Throughout the rest of this article, when referring to a wavefront sensor, we will be specifically talking about the Hartmann-Shack wavefront sensor. The Hartmann-Shack wavefront sensor employs a pupil plane measurement method to measure the local wavefront slope (the first derivative of the wavefront) within subapertures created by a microlens array. The position of the centroid of the light spot in each subaperture on the detector is directly proportional to the slope of the wavefront within that subaperture. The principle of the Hartmann-Shack sensor is shown in Fig. 10.4.

The lens array is placed in a plane that is conjugate to the pupil plane to sample the incoming wavefront. Initially, the lens array is calibrated with a reference plane

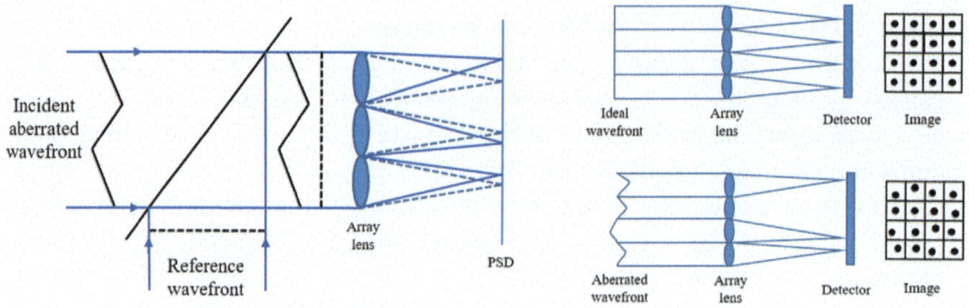

Fig. 10.4 Principle of the Hartmann-Shack wavefront sensor

wave, and theoretically, each lens forms a point source image at its focal point. When a perturbed wavefront is incident, each lens receives an approximately tilted wavefront and forms an off-axis point source image on its focal plane. By measuring the position of the spots, the angle at which the light wave arrives at each microlens can be directly estimated. This allows the calculation of the average wavefront slope over the range of each microlens using simple geometric relationships.

Finally, the entire wavefront is reconstructed based on the relationship between wavefront slope and wavefront phase. The relationship between the centroid offset of the light spot within each subaperture and the corresponding subaperture slope is as follows

$$\begin{cases} c_{xi} = \dfrac{\sum\limits_{i,j} x_{i,j} I_{i,j}}{\sum\limits_{i,j} I_{i,j}} = \alpha_{xi} f = \dfrac{\lambda f}{2\pi S_i} \iint\limits_{S_i} \dfrac{\partial \varphi}{\partial x} \mathrm{d}x\mathrm{d}y = \dfrac{\lambda f}{2\pi} G_{xi}, \\ c_{yi} = \dfrac{\sum\limits_{i,j} y_{i,j} I_{i,j}}{\sum\limits_{i,j} I_{i,j}} = \alpha_{yi} f = \dfrac{\lambda f}{2\pi S_i} \iint\limits_{S_i} \dfrac{\partial \varphi}{\partial y} \mathrm{d}y\mathrm{d}y = \dfrac{\lambda f}{2\pi} G_{yi}. \end{cases} \tag{10.20}$$

In this context, (c_{xi}, c_{yi}) represents the centroid coordinates of the i-th subaperture, $I_{i,j}$ and $(x_{i,j}, y_{i,j})$ are the coordinate values corresponding to the signal at location (i, j) in the CCD, f is the focal length of the microlens, α_{xi} and α_{yi} represent the angles of arrival of the wavefront in the x and y directions for the subaperture, λ is the wavelength of the incident light, S_i is the area of the *i-th* subaperture, φ is the wavefront under test, and G_{xi} and G_{yi} represent the average slopes of the wavefront in the x and y directions for the i-th subaperture.

Using a quadrant detector, it is possible to measure the centroid coordinates of the imaged spot formed by the distorted sub-wavefront.

$$\begin{cases} S_x = K\dfrac{(A+B)-(C+D)}{(A+B+C+D)}, \\ S_y = K\dfrac{(A+D)-(C+B)}{(A+B+C+D)}. \end{cases} \tag{10.21}$$

10.3.1 Wavefront Corrector

A wavefront corrector is a device used in adaptive optical systems to perform wavefront correction. It is a core component within adaptive optical systems, and in principle, this device can be any equipment capable of altering the shape of the perturbed wavefront. Depending on the method of wavefront correction, it can be categorized as either transmissive or reflective.

Transmissive wavefront correctors achieve phase modulation of light within a medium by changing the refractive index using a controller. An example of this is a liquid crystal spatial light modulator. Reflective wavefront correctors, on the other hand, change the optical surface shape of a mirror to alter the optical path length. Reflective wavefront correctors can further be divided into tilt-tip mirrors (TTM) and deformable mirrors (DM), depending on the type of aberrations they correct. In this context, the wavefront correctors used are all reflective.

TTMs are used to correct overall tilt in the wavefront. Since tilt correction requires a larger range of motion, the control loops for tilt mirrors and deformable mirrors are kept separate to reduce the correction pressure on the deformable mirror and simplify system complexity. Tilt mirrors find applications not only in adaptive optics but also in fields like precision optical tracking.

Deformable mirrors (DM) are the most critical component in adaptive optics systems. They play a pivotal role in transforming adaptive optics technology from theory to practical implementation. Typically, they use piezoelectric ceramics to drive a flexible reflective mirror. Schematic diagram of the control principle of the DM as shown in Fig. 10.5. The surface shape of this mirror can be rapidly modified in response to external voltage signals. Multiple individual actuators control the shape of the mirror's surface. These actuators push and pull the mirror surface, deforming it in a manner that approximates the turbulence-induced aberrations to compensate for incoming wavefront distortions.

10.4 Wavefront Controller

An adaptive optics system can be defined as a multivariable servo control system, similar to traditional servo systems, comprising a wavefront sensor, a control device, a real-time computer (RTC), and a deformable mirror. The goal of this servo system is to compensate for wavefront aberrations caused by atmospheric turbulence. Its design objective is to minimize the residual phase variance along the imaging path, thereby improving the system's point spread function. The input and output of the servo system are respectively the wavefront phase disturbance and the corrected residual phase. Figure 10.6 provides a schematic diagram of the control principle of an adaptive optics system.

The system is represented using a typical block diagram, which includes a controller (CC), a digital-to-analog converter (DAC), and a high-voltage amplifier

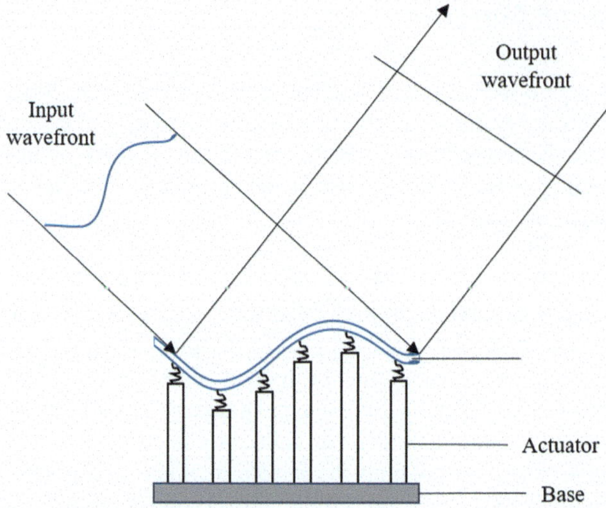

Fig. 10.5 Schematic diagram of the control principle of the deformable mirror

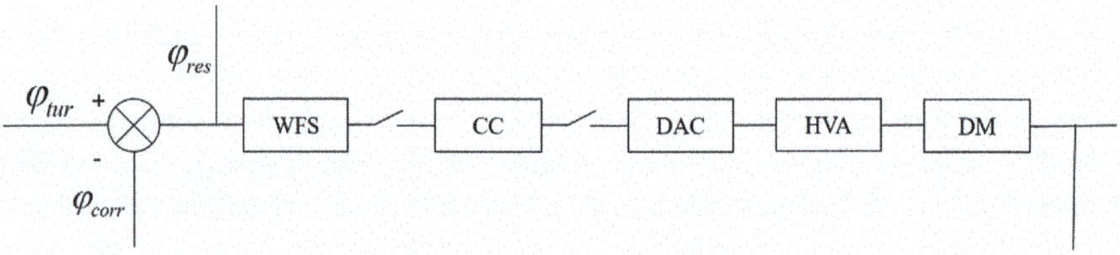

Fig. 10.6 Control block-diagram representation of an AO system

(HVA). The input of the servo system is the uncompensated wavefront, denoted as $\varphi_{tur}(x, y, t)$ which represents its value at point (x, y) and time t The wavefront sensor measures the wavefront correction residual phase values $\varphi_{res}(x, y, t)$ with noise. These values undergo image processing, slope calculation, wavefront reconstruction, and control operations within the controller (CC). Image processing involves denoising and other treatments of the acquired spot images from the wavefront sensor to enhance the accuracy of slope calculations. Wavefront reconstruction transforms the subaperture wavefront slopes into.wavefront phase. Common wavefront reconstruction algorithms include the zone method, modal method, and direct slope method. Wavefront reconstruction is a crucial computational step within the wavefront controller.

After wavefront reconstruction, corresponding compensation voltages are computed through control operations. These compensation voltages are then applied to the deformable mirror (DM) after passing through the digital-to-analog converter (DAC) and the high-voltage amplifier (HVA). The phase introduced by the DM is denoted as $\varphi_{corr}(x, y, t)$ and satisfies the following relationship:

$$\varphi_{res}(x, y, t) = \varphi_{tur}(x, y, t) - \varphi_{corr}(x, y, t). \tag{10.22}$$

In order to ensure real-time correction of wavefront distortions, an adaptive optical system must be a high-sampling-frequency multi-channel parallel control system. The wavefront controller needs to extract tens to over a thousand wavefront slope signals from the spot images output by the wavefront sensor and generate tens to over a thousand driver control signals after wavefront reconstruction. This places high demands on the computational capability and real-time performance of the wavefront controller.In the early stages of adaptive optics development, wavefront controllers often used large-scale analog circuits. With the advancement of digital electronics, the commonly used wavefront controllers today come in the following forms: DSP-based wavefront controllers, FPGA-based processing platform wavefront controllers, DSP + FPGA platform wavefront controllers, and wavefront controllers based on large-scale general-purpose CPUs and GPU computing platforms.

10.4.1 Conjugate Adaptive Optics System and Its Control Method

The conjugate adaptive optics system is one of the earliest adaptive optics systems proposed, and it is still widely used in many astronomical telescopes due to its simplicity and effectiveness. Examples of telescopes using this adaptive optics system include the LBT telescope in the United States, the VLT telescope at the European Southern Observatory in Chile, and several smaller-aperture telescopes worldwide. The conjugate adaptive optics system is a traditional linear adaptive optics system consisting mainly of three subsystems: a wavefront sensor for detecting aberrations in the wavefront, a tilt mirror or deformable mirror for correcting the aberrations in the wavefront, and a control computer for monitoring and decoding sensor information while generating control signals. Figure 10.7 illustrates the conjugate adaptive optics system used for astronomical observations. One of the most distinctive features of these systems is the inclusion of a wavefront sensor, making wavefront reconstruction and control operations a critical part of their functionality.

10.4.2 Commonly Used Wavefront Control Techniques

The closed-loop control process of the proportion-integration-differentiation (PID) controller is shown in Fig. 10.8. The components of the PID control algorithm mainly include proportional, integral, and derivative sections. This algorithm detects deviation signals through feedback to perform closed-loop control.

Commonly used wavefront control techniques include the PID algorithm, which is one of the most classical algorithms in the control field and best exemplifies

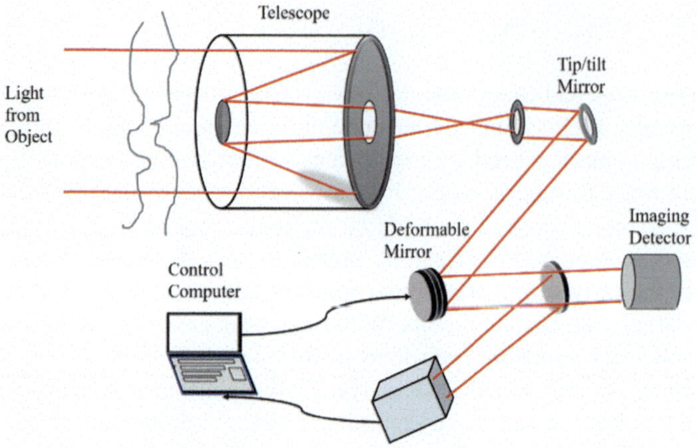

Fig. 10.7 Phase-conjugation-type adaptive optical system

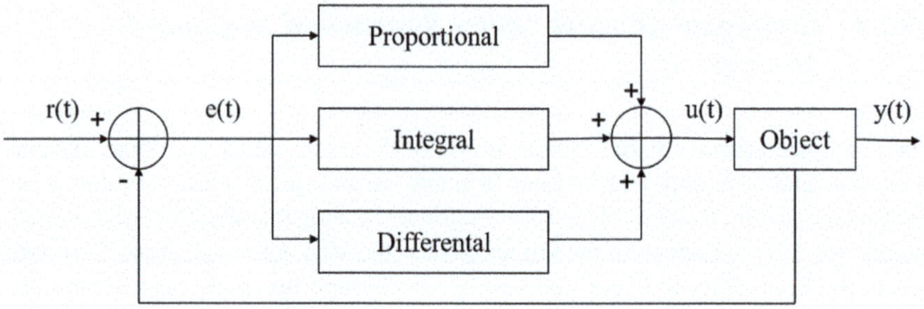

Fig. 10.8 PID control diagram

the concept of feedback in closed-loop control. In this algorithm, the proportional term reflects the deviation of the response system; as long as there is deviation, the proportional term comes into play. The integral term is primarily used to eliminate static errors, referring to the difference between the input and output quantities of a stable system. The integral term compensates for static errors by accumulating deviations. The differential term not only reflects changes in deviation but also allows for advanced system control based on the trend of these changes, improving response speed. The differential parameter exists only in a few controls, such as temperature control, while in other cases, this parameter is zero. Therefore, in Adaptive Optics (AO) systems, an incremental proportional-integral algorithm is mainly employed to correct distorted wavefronts.

10.5 Adaptive Optics Without Wavefront Sensor

Adaptive optics systems typically employ dedicated wavefront sensors (such as Hartmann-Shack sensors) to measure wavefront errors. Subsequently, based on the measured wavefront data, they control wavefront correctors (such as deformable mirrors) to apply conjugate compensation for wavefront errors. Traditional adaptive optics has found extensive applications in fields such as astronomical observations, high-energy lasers, fundus imaging, and more. However, due to the inherent complexity of the system structure and its working principles, its applicability is subject to certain limitations.

(1) The structure of adaptive optics systems is relatively complex, making it challenging to apply in situations with strict requirements for system reliability or cost. For instance, space-based optical remote sensing payloads on satellite platforms have stringent limitations on volume, weight, and power consumption. Therefore, when implementing adaptive optics for space-based optical remote sensing, simplification and specialized designs are necessary.

(2) Adaptive optics systems typically compensate for phase errors in the wavefront based on the phase conjugate principle but do not compensate for changes in wavefront amplitude. When there is significant amplitude variation in the wavefront, as in cases of strong scintillation, adaptive optics systems based on the phase conjugate principle may not meet the requirements. For example, applications involving laser transmission through near-horizontal atmospheric conditions, such as free-space optical communication and directed energy weapons, must consider the effects of scintillation.

(3) In some practical applications, obtaining suitable beacons is challenging. For instance, high-energy laser beams that focus on non-cooperative extended targets after atmospheric transmission require pre-compensation of wavefront errors caused by atmospheric turbulence to enhance the laser power density on the target. In such cases, there are typically no readily available beacon sources, and the wavefront control method may need to operate in a "target-in-the-loop" mode.

(4) Using traditional wavefront sensors to measure errors can be difficult in certain applications. For example, when using a Shack-Hartmann sensor to measure wavefront errors caused by biological tissues in confocal microscopy, fluorescence is generated throughout the entire illumination cone. Consequently, it is challenging to distinguish the source of the light waves in the sensor, leading to measurement ambiguity.

(5) When the brightness of the observed target is low, there is a need to balance the requirements for signal-to-noise ratio in both the wavefront measurement and imaging paths. In some cases, it may not be feasible to achieve both simultaneously.

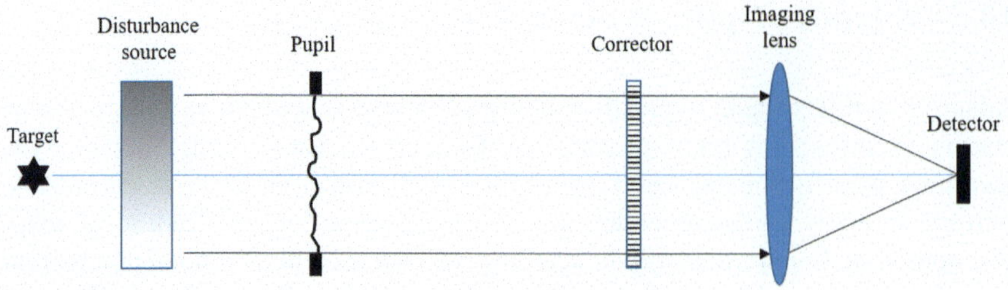

Fig. 10.9 A schematic diagram of the wavefront sensorless adaptive optics system

(6) Non-common path errors exist between the measurement path and the imaging path in adaptive optics systems. High precision wavefront correction requires calibration to remove these errors, and the system's correction accuracy is influenced by the calibration precision.

Due to the aforementioned issues with traditional adaptive optics, a new technology known as Wavefront Sensorless Adaptive Optics (WSLAO) has gained attention and widespread application. The concept of wavefront sensorless adaptive optics originated from the image sharpening technique proposed by Muller et al. in 1974. However, early image sharpening methods were not practically viable due to poor real-time correction capabilities. In the late 1990s, with the innovation of hardware technology and the introduction of new optimization control algorithms, wavefront sensorless adaptive optics technology began to flourish.

The structure of a wavefront sensorless adaptive optics system is illustrated in Fig. 10.9. Light emitted by the observed target (either point or extended) experiences distortion at the entrance pupil, caused by disturbances that can originate externally or within the optical system itself. The distorted wavefront is then converged onto a focal plane detector by an imaging lens after passing through the wavefront corrector. The fundamental idea behind wavefront sensorless adaptive optics is to initially establish a merit function based on the intensity information from the detector. Subsequently, specific optimization algorithms are employed to control the wavefront corrector in order to find the extremum of the merit function. When the merit function reaches its extremum, it is considered that the wavefront error has been corrected.

10.5.1 Objective Function

Wavefront sensorless adaptive optics systems often use an objective function to evaluate the quality of correction, which is also referred to as the system's performance evaluation function. The choice of the objective function varies depending on the specific optical system. Different evaluation functions typically lead the system to different final states. Additionally, an appropriate objective function can facilitate algorithm convergence. Therefore, selecting the right objective function is crucial in accordance with the system requirements. Currently, commonly used objective functions include the following

$$
\begin{aligned}
J_1 &= \iint I^2(x, y)\mathrm{d}x\mathrm{d}y, \\
J_2 &= \frac{I(x_0, y_0)}{I_0(x_0, y_0)}, \\
J_3 &= \iint I_0(x, y)I(x, y)\mathrm{d}x\mathrm{d}y, \\
J_4 &= \iint M(x, y)I(x, y)\mathrm{d}x\mathrm{d}y.
\end{aligned}
\tag{10.23}
$$

In the equations provided, $I(x, y)$ represents the intensity distribution of the distorted wavefront in the far field; $I(x_0, y_0)$ denotes the peak intensity of the distorted wavefront in the far field; $I_0(x, y)$ stands for the intensity distribution of the ideal wavefront in the far field; $I_0(x_0, y_0)$ represents the peak intensity of the ideal wavefront in the far field; and $M(x, y)$ represents the transmission function of the aperture. J_1 corresponds to the sum of the squared intensities at various points in the field. J_2 (Strehl ratio, SR) is the ratio of the peak intensity of the actual received field to the peak intensity of the ideal plane wave in the far field. J_3 represents the sum of the products of intensities at corresponding points between the actual received field and the ideal field. J_4 represents the energy enclosed within an annular region.

From the four commonly listed objective functions mentioned above, it is evident that each of these functions has a different emphasis. However, when an algorithm converges, each objective function will exhibit an extremum, which serves as the solution sought by the algorithm and the ultimate goal of system correction. Therefore, selecting an appropriate objective function is crucial in accordance with the requirements of the adaptive optics system. In this paper, we have chosen the Strehl ratio (SR) as the performance evaluation function. A higher SR indicates a smaller residual wavefront distortion after correction.

10.5.2 Objective Function

Signal-to-noise ratio (SNR) refers to the ratio of the power of the output signal to the power of noise at the receiving end. It reflects the fidelity of the signal after transmission and its ability to withstand noise. When a light beam propagates through the atmosphere, it is affected by atmospheric turbulence, leading to a significant degradation in the quality of the light beam and reducing the reliability of communication systems. Adaptive optics technology improves the quality of light beam transmission through turbulence, and the enhancement of communication system performance can be measured by the ratio of the system signal power to the noise power introduced by turbulence.

Selecting the Strehl ratio as the performance evaluation function, assuming that the noise introduced by atmospheric turbulence remains constant before and after the correction by the adaptive optics system, the ratio of the system's signal-to-noise ratio after correction to the signal-to-noise ratio before correction is used to assess the system's performance as

$$\frac{SNR'}{SNR} = \frac{I'}{I}. \tag{10.24}$$

where *SNR* is the signal-to-noise ratio of the system before correction, *SNR'* is the signal-to-noise ratio of the system after correction, and according to Eqs. (10.2–10.4), *I'/I* represents the ratio of the Strehl ratio after correction to the Strehl ratio before correction, which can be obtained.

$$\frac{SNR'}{SNR} = \frac{SR'}{SR}. \tag{10.25}$$

10.5.3 Optimization Algorithms in Adaptive Optics

The optimization algorithms commonly used in adaptive optics systems are typically iterative algorithms. The general process involves selecting a parameter type as a control signal and generating its initial value. The parameter is then iterated in a specific way to generate a series of result values that converge in the desired direction. The convergent extreme value that meets the objective conditions is the target solution.

Currently, the commonly used algorithms in adaptive optics systems include: Simulated Annealing (SA), Stochastic Parallel Gradient Descent (SPGD), Genetic Algorithm (GA), and the Gerchberg-Saxton (GS) algorithm. Simulated Annealing emulates the process of solid-state annealing. It begins by heating a solid to a high temperature, allowing the internal particles to reach a disordered state due to the increased temperature. Then, during the cooling process, the particles gradually return to the ground state. The algorithm searches in the parameter space in the direction of decreasing energy until it finds the optimal solution. Genetic Algorithm is inspired by the principles of natural selection and genetic evolution in biology. It conducts a global search by selecting, crossing over, and mutating individuals in the population to enhance their adaptability. It continues until individuals that meet the search criteria appear. In solving the wavefront correction problem, control information is first encoded into the chromosome. Through generations of genetic mutation operations, the chromosome evolves adaptively. When the objective function converges to an extremum, the control information contained in the chromosome is decoded to find the optimal solution. The Stochastic Parallel Gradient Descent (SPGD) algorithm estimates control parameters based on changes in the measured value of the objective function and changes in control parameters. It performs an iterative search of control parameters in the direction of gradient descent. The GS algorithm is based on the principle that the wavefront fields at two focal planes of a lens satisfy a Fourier transform relationship. It is an algorithm for wavefront reconstruction through iterative Fourier transforms.

10.6 Applications of Adaptive Optics

10.6.1 Applications of Adaptive Optics in Astronomical Observation

One of the earliest applications of adaptive optics technology is military telescopes used for observing artificial satellites [4]. Currently, the largest satellite imaging telescopes in the world are the 3.5 m telescope at the Starfire Optical Range in the United States and the 3.67 m Advanced Electro-Optical System at AMOS. Both of these telescopes use 941-element deformable mirrors as wavefront correctors, and the number of subapertures in the wavefront sensor is 1024. The interference of atmospheric turbulence in astronomical observations was the origin of adaptive optics technology research [5]. Compared to military telescopes, astronomical telescopes track targets with smaller angular velocities, and they typically operate in longer-wavelength infrared bands, resulting in relatively lower performance requirements for adaptive optics systems. However, when the wavefront sensor operates, there must be sufficiently bright reference light sources near the imaging target to provide wavefront distortion information. In actual observation areas, suitable reference light sources may not be readily available. Therefore, astronomical telescopes adaptive optics systems rely on laser guide stars to function properly. The principle of laser guide stars involves focusing laser light into a specific altitude in the atmosphere, utilizing the interaction of laser light with atmospheric substances to generate backscattered light signals. These signals serve as reference light sources for the adaptive optics system as illustrated in Fig. 10.10.

In order to achieve higher resolution in astronomical observations, researchers have begun to develop larger ground-based telescopes. The Thirty Meter Telescope (TMT) project, jointly developed by the United States, Japan, China, Canada, India, and other countries, is scheduled for completion in 2027. To correct optical wavefront distortions over a larger field of view, the telescope generates laser guide stars using six laser emission devices for its adaptive optics system. Combined with measurements from natural guide stars, the total dimension of the wavefront distortion measurement information reaches 35,000 dimensions. The wavefront corrector includes 3,000 driving units and operates at a deformable frequency of 800 times per second. The European Southern Observatory (ESO) is also developing the European Extremely Large Telescope (E-ELT) with a 39-m aperture. Its laser guide star system comprises 8 lasers [6], and the wavefront corrector utilizes a 2.5 m deformable mirror

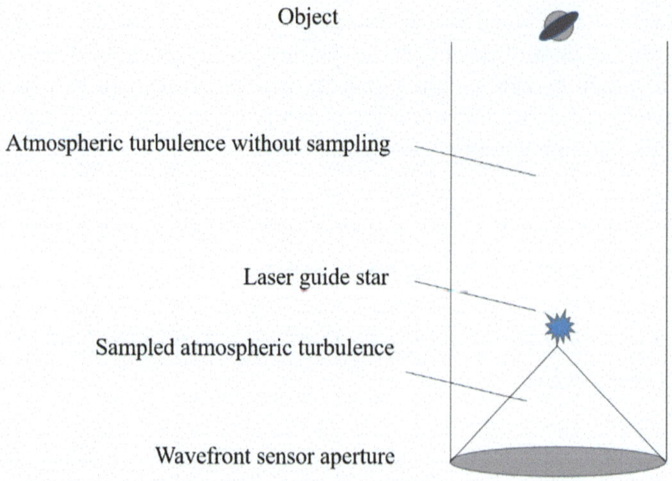

Fig. 10.10 Focusing non isoplanatic error of laser guide star

driven by 5,000 voice coil motors, operating at a wavefront correction frequency of 1 kHz.

In addition to ground-based telescopes, space telescopes are also equipped with adaptive optics systems [7]. For example, the 2.4-m aperture Hubble Space Telescope uses an adaptive optics system to correct for aberrations caused by factors such as gravity and temperature variations in space. As the next-generation space telescope, the 6.5-m aperture James Webb Space Telescope, planned for launch in 2021, is constructed from 18 mirror segments with a diameter of 1.3 m each. To minimize aberrations resulting from mirror segment alignment, Lockheed Martin has designed a 2,000-unit adaptive optics system for it. The telescope also uses a near-infrared camera as a wavefront sensor [8], achieving an imaging diffraction limit of 2 μm.

Adaptive optics technology is also applied in solar observations. Compared to nighttime astronomical telescopes, solar telescopes face more severe atmospheric turbulence due to the high temperature introduced by sunlight. In Fig. 10.11 traditional adaptive optics systems have a limited field of view, typically on the order of arcseconds, which may not encompass the entire Sun. To expand the system's correction field, multi-conjugate adaptive optics technology is employed. The idea behind this approach is to measure wavefront aberrations in multiple directions, then reconstruct wavefront errors induced by different layers of atmospheric turbulence. Finally, multiple wavefront correctors are used to correct phase aberrations caused by different turbulence layers.

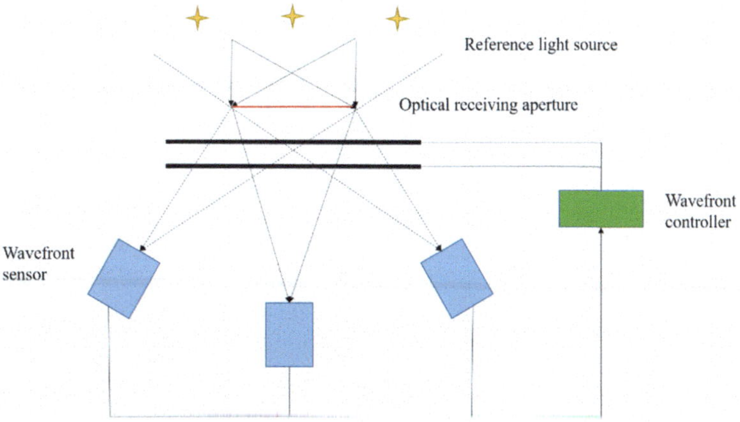

Fig. 10.11 Multi-conjugate adaptive optics (MCAO) system

10.6.2 Application of Adaptive Optics in Biology

The principle of the adaptive optics system for the human eye is as shown in Fig. 10.12. Similar to typical adaptive optics systems, this system mainly consists of a wavefront sensor, a wavefront corrector, and a wavefront controller. The most common adaptive optics system for the human eye consists of a Hartmann wavefront sensor, a deformable mirror, and control software [9]. The wavefront sensor continuously measures the optical aberrations of the human eye in real time, and the control system uses this aberration data to guide and control the deformable mirror to make corresponding adjustments to correct the eye's aberrations. Based on this correction, high-resolution imaging of the retina is achieved through a fundus illumination system and a high-sensitivity imaging system. A visual display device and a vision testing system are used to study the impact of optical aberrations on visual function. For high-resolution imaging of the retina, full correction of both low-order and high-order aberrations of the human eye is required to achieve near-diffraction-limited imaging resolution [10, 11]. However, for vision function research, the optical aberrations of the eye can be fully corrected, selectively corrected, or specific aberrations can be added to study the impact of different aberration components on visual function. This is why we refer to it as optical aberration control rather than correction. The light emitted by a semiconductor laser passes through spatial filtering and collimation before entering the eye.

After converging in the eye, it forms a spot on the retina. This spot can be considered as a point light source on the retina. The backscattered light carries information about the eye's optical aberrations and exits through the pupil. This scattered light enters the Hartmann wavefront sensor after passing through the deformable mirror and aperture-matching system. The Hartmann wavefront sensor segments and samples the incoming wavefront, and the aberrations of the human eye are obtained through wavefront reconstruction. The control system uses this aberration

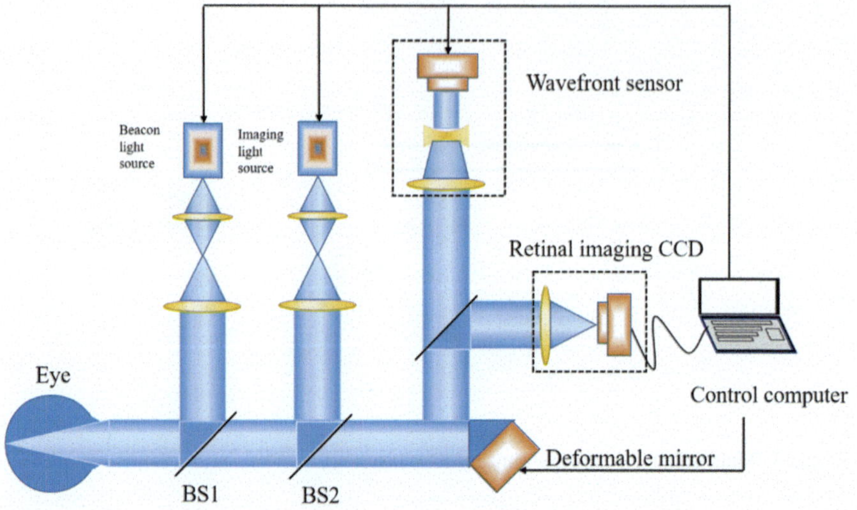

Fig. 10.12 Principle diagram of adaptive optics retinal high-resolution imaging

data to guide and control the deformable mirror to correct the optical aberrations of the eye. Once the aberration correction is stable, the subject observes the grating patterns displayed on the visual target presentation device and makes judgments. The subject's judgments are collected by external equipment and sent to a computer for analysis. The system automatically adjusts the spatial frequency or contrast of the grating based on the subject's judgments, following a predefined program mode to complete vision function testing or perceptual learning training.

10.6.3 Application of Adaptive Optics in High-Energy Lasers

In high-energy laser systems, the quality of the laser beam is constrained by various factors, such as uneven refractive index distribution of the laser working material, alignment errors in laser processing and assembly, and thermal effects of the working material. These factors lead to distortion in the laser output wavefront, reducing the quality of the output beam. Balancing high average laser output power and good beam quality has become a hot research topic. To simultaneously achieve high average laser output power and good beam quality, passive beam quality control technology can be employed to improve the system from two aspects. First, optimizing the laser design to enhance the performance of the device itself. This includes improving thermal management techniques, considering solid gain media with structures such as slabs, disks, or fibers to reduce thermal effects, and enhancing the processing and alignment precision of optical components to minimize wavefront distortion. Second, using an additional optical system to improve the output beam through techniques such as negative lens compensation and resonator cavity design, employing static compensation methods to eliminate aberrations in the laser system. In addition to passive beam quality control technology, active beam quality control technology can be used to compensate for wavefront distortions [10, 11].

Active compensation includes both nonlinear optics technology and adaptive optics technology. Nonlinear optics technology is based on the nonlinear interaction between the laser and certain media, generating phase conjugation of the incident wave for phase conjugate correction. However, nonlinear optics technology faces challenges in the high-energy laser field due to its low damage threshold. Adaptive optics-based beam purification technology, which actively controls the wavefront of the laser output beam, is a primary means for achieving beam purification in high-energy laser applications.

To overcome the limitations of traditional adaptive optics systems, TIL (Target-In-the-Loop) adaptive optics technology was initially proposed by TMeara at the Hughes Laboratory in the United States. Subsequently, detailed theoretical analysis of the TIL control concept was conducted by researchers, including Vorontsov from the U.S. Army Laboratory. TIL technology is applicable to non-cooperative targets, providing a new solution to address the energy concentration issue on the surfaces of extended non-cooperative targets.

The principle of the TIL adaptive optics system is illustrated in Fig. 10.13. The main laser emitted by a high-energy laser system is transmitted to the target surface after passing through atmospheric turbulence. After scattering on the target surface, a portion of the light is reflected back, and the wavefront sensor of the adaptive optics system senses the scattered light from the target surface. It obtains information about the laser intensity distribution on the target surface and generates control signals through the adaptive optics wavefront control system. These signals drive the corrector in real-time to correct the wavefront of the emitted laser, thereby compensating for the dynamic wavefront distortions caused by atmospheric turbulence and improving the energy concentration on the target surface. In a TIL-based adaptive optics system, a significant feature is the establishment of an optical feedback loop between the target and the adaptive optics system. This loop continuously detects the impact of introduced phase distortions by the corrector on the morphology of the light spot on the target surface, enabling the adjustment of the corrector's control signals. This control does not require prior knowledge of the model of the controlled system and relies directly on optimization algorithms.

The performance of the system depends on the characteristics of the emitted laser wave, the transmission properties of the medium, and the characteristics of the target. It is influenced by factors such as the wavelength, coherence, aperture size, intensity, phase of the emitted laser beam, non-uniformity in the refractive index of the transmission medium, and target factors such as velocity, shape, surface roughness, reflection, and scattering coefficients. In many practical engineering applications based on the TIL structure, the size of the target may exceed the characteristic size of the illuminated area on its surface (target thermal spot). This scenario illustrates the application of TIL systems to extended non-cooperative targets. In such cases, the

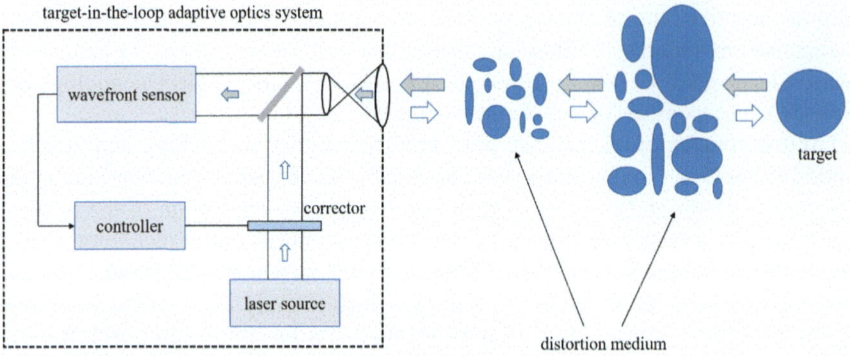

Fig. 10.13 TIL adaptive optics system principle diagram

characteristics of the scattered echo depend on the spatial distribution of the optical field on the target surface and the intrinsic properties of the target.

10.7 General Conclusions

Adaptive optics is a remarkable optical technology aimed at enhancing the performance of optical systems by real-time correction of wavefront aberrations, thereby improving system capabilities and image quality. This technology is based on the measurement of wavefronts and the real-time adjustment of optical elements through a feedback system, allowing it to address challenges such as atmospheric turbulence and manufacturing errors in optical components. The following is an in-depth discussion of the fundamental principles, system evaluation criteria, error sources, key components, and applications of adaptive optics.

The fundamental principle of adaptive optics lies in real-time monitoring of the wavefronts propagating through the optical system and adjusting optical elements in real-time through a feedback system to correct wavefront aberrations. This process involves wavefront sensing, error computation, wavefront correction, and control. The key components include the wavefront sensor, responsible for real-time measurement of incoming light wavefront aberrations, the wavefront corrector adjusting optical elements based on sensor information, and the wavefront controller coordinates this process to ensure stability and efficient system operation.

System evaluation criteria primarily include the Strehl ratio and the optical transfer function (OTF). The Strehl ratio measures the similarity between actual and ideal imaging, while the OTF describes the system's response to wavefront aberrations at different frequencies. These evaluation criteria directly reflect the performance and image quality of adaptive optics systems.

In adaptive optics, error sources mainly consist of atmospheric turbulence and manufacturing errors in optical components. Atmospheric turbulence arises from air movements in the atmosphere, causing wavefront distortions in the optical system. Manufacturing errors in optical components include factors such as the shape of lenses and surface smoothness, resulting in deviations under less-than-ideal manufacturing conditions.

Key components are integral to the adaptive optics system, comprising the wavefront sensor, wavefront corrector, and wavefront controller. The wavefront sensor monitors wavefront aberrations in real time, the wavefront corrector adjusts optical elements based on sensor data, and the wavefront controller coordinates these processes to ensure stable and efficient system operation.

In addition to traditional wavefront sensors, some non-wavefront sensor adaptive optics methods have garnered attention. Model-based approaches are one such method, utilizing pre-established system models for predicting and correcting wavefront aberrations, providing a solution for situations where direct wavefront measurement is challenging.

Adaptive optics finds widespread applications across various fields. In astronomical observations, it enhances telescope resolution and reduces the impact of atmospheric turbulence on images, allowing for more precise observations. In laser communication, adaptive optics technology is employed to overcome transmission obstacles caused by atmospheric turbulence, thereby improving communication quality. In the field of laser medical applications, it is utilized for real-time correction of wavefront aberrations in patients' eyes, enhancing the precision and safety of laser surgeries.

In summary, the continuous development of adaptive optics technology provides a robust tool for optimizing optical systems. By correcting wavefront aberrations in real-time, adaptive optics systems demonstrate excellent performance in complex environments, offering higher capabilities and reliability across various applications. Ongoing research and innovation will further propel the advancement of adaptive optics technology, expanding its potential in diverse application scenarios.

Acknowledgements This work was supported by the National Natural Science Foundation of China (Nos. 11874132, 61975044, 12074094, 12374271, and 62375065); Interdisciplinary Research Foundation of HIT (No. IR2021237).

References

1. Sichelstiel BA, Waters WM, Wild TA. Self-focusing array research model. IEEE Trans Antenn Propag. 1964;12(2):150–4.
2. Hardy JW. Active optics: a new technology for the control of light. Proc IEEE. 1978;66(6):651–97.
3. Verinaud C. On the nature of the measurements provided by a pyramid wavefront sensor. Opt Commun. 2004;233:27–38.
4. Daukantas P. Ground-based telescope for the 21th century. Opt Photon News. 2006;18(9):28–34.

5. Mugnier LM, Fusco T, Conan J. MISTRAL: a myopic edge-preserving image restoration method, with application to astronomical adaptive-optics-corrected long-exposure images. J Opt Soc Am A. 2004;21(10):1841–54.
6. Cao ZL, Mu QQ, Hu LF, Li DY, Peng ZH, Liu YG, Xuan L. Preliminary use of nematic liquid crystal adaptive optics with a 2.16-meter reflecting telescope. Opt Express. (2009);17(4):2530–37
7. Zhong L, Tian Y, Rao C. Speckle image reconstruction of the adaptive optics solar images. Opt express. 2014;22(23):29249–59.
8. Ragazzoni R. Pupil plane wavefront sensing with an oscillation prism. J Mod Opt. 1996;43(2):289–93.
9. Fernandez EJ, Iglesias I, Artal P. Closed loop adaptive optics in the human eye. Opt Lett. 2001;26(10):746–8.
10. Gao S, Gruev V. Gradient-based interpolation method for division-of-focal-plane polarimeters. Opt Express. 2013;21(1):1137–51.
11. Vorontsov M, Weyrauch T, Lachinova S. Speckle metric optimization based adaptive optics for laser beam projection and coherent beam combining. Opt Lett. 2012;37(14):2802–4.

Chapter 11
Fluorescence Imaging

Gangshan Liu and Zhengjun Liu

Abstract In this chapter, fluorescence imaging and microscopy are introduced in detail. It is the fundamental mechanisms, different imaging modalities and complex techniques facing the frontiers of biological and medical research. A comprehensive overview of fluorescence imaging and microscopy is given, including advanced techniques as well. From understanding fundamental mechanisms to exploring cutting-edge methods such as two-photon microscopy and fluorescence lifetime imaging, researchers are equipped with powerful tools to unravel the complexity of biological systems at the molecular level. Fluorescence imaging plays a key role in shaping our understanding of cellular dynamics, with great potential for future discoveries in computational imaging and life sciences.

11.1 Introduction

Stokes discovered the phenomenon of minerals such as fluorite emitting blue light when exposed to ultraviolet radiation, which he named fluorescence emission. The discovery of fluorescence emission marked the beginning of the era of fluorescence imaging. Fluorescence was commonly used to observe endogenous fluorescent substances such as minerals, crystals, resins, medicinal herbs, butter, chlorophyll, and vitamins. The invention of fluorescence microscopy expanded the application of fluorescence imaging techniques to biological research, enabling the observation of organisms by exciting endogenous fluorescent molecules in cells and tissues. Important endogenous fluorescent molecules in biology include amino acids, nicotinamide adenine dinucleotide (NADH), and flavin adenine dinucleotide (FAD), among others. Table 11.1 presents the excitation and emission wavelengths of commonly encountered endogenous fluorescent molecules in biology.

Fluorescence imaging techniques are used in the field of flow field diagnostics, with the most commonly used technique being planar laser-induced fluorescence

G. Liu · Z. Liu (✉)
School of Physics, Harbin Institute of Technology, Harbin 150001, China
e-mail: zjliu@hit.edu.cn

Z. Liu et al. (eds.), *Computational Optical Imaging*, Advances in Optics and Optoelectronics, https://doi.org/10.1007/978-981-97-1455-1_11

Table 11.1 Endogenous fluorescent molecules in organisms

Endogenous fluorescent molecules	Excitation wavelength (nm)	Emission wavelength (nm)
NADH	340	450
FAD	380–490	520–560
Tyrosine	270	305
Tryptophan	295	353
Melanin	340–400	360–560
Chlorophyll	465/665	673/726
Collagen	270–370	305–450

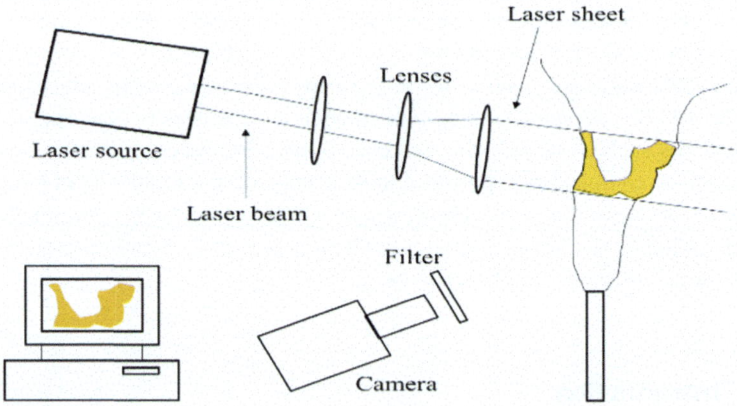

Fig. 11.1 Illustration of a planar laser induced fluorescence (PLIF) imaging system

(PLIF). PLIF, which is essentially a wide-field fluorescence intensity imaging technique. PLIF enables non-intrusive visualization of flow fields and diagnostics of flow parameters such as velocity, temperature, concentration, and pressure. A typical PLIF system is illustrated in Fig. 11.1, where the flow field is illuminated by a sheet of light to induce fluorescence from the target species, and a camera captures the intensity image of the fluorescence signal. For different types of flow fields, PLIF can detect intermediate species or tracer fluorophores: liquid PLIF [1] commonly employs soluble fluorophores such as fluorescein, rhodamine 6G, and rhodamine B; gas flow field PLIF commonly employs tracers such as NO, acetone (CH_3COCH_3), biacetyl ($(CH_3CO)_2$), toluene (C_7H_8), and pentan-3-one ($C_5H_{10}O$); intermediate species in flame and other reactive flows include CH, OH, HCO, CH_2O, NO, CN, etc. [2]. Table 11.2 lists the excitation and emission wavelengths of commonly used tracers and intermediate species in flow field diagnostics using PLIF. Intermediate species are often small molecules compared to large molecular tracers, and their excitation spectra are narrower, requiring precise tuning of the excitation light.

To achieve clearer, faster, and more comprehensive imaging, fluorescence imaging techniques have undergone significant developments. These include:

Table. 11.2 Excitation and emission wavelengths of common tracers and species for flow diagnosis with PLIF

Components/Tracers	Excitation wavelength (nm)	Emission wavelength (nm)
Fluorescein (Water)	400–520	490–600
Rhodamine 6G	450–550	500–600
Rhodamine B	500–600	550–650
Acetone	225–320	350–600
Toluene	240–270	~300
Pentan-3-one	250–320	250–320
OH	~282	~309
CH	~387	~430
HCO	~259	280–350
CH_2O	~355	400–500
CN	~356.5	~389
NO	~226	~250

(1) Super-resolution imaging techniques: such as stimulated emission depletion (STED) microscopy (~5.8 nm lateral spatial resolution), single molecule localization microscopy (SMLM) (~20 nm lateral spatial resolution), and structured illumination microscopy (SIM) (~115 nm lateral spatial resolution).
(2) Three-dimensional imaging: techniques such as laser scanning confocal microscopy, multiphoton fluorescence microscopy, and light sheet fluorescence microscopy.
(3) High-speed imaging: techniques such as frequency-domain reconstruction for multiple exposures (FRAME) technology [3].

These advancements have significantly pushed the boundaries of fluorescence imaging, allowing for imaging with improved spatial resolution, three-dimensional visualization, and high temporal resolution.

Additionally, there are fluorescence spectral imaging techniques, such as multispectral imaging, hyperspectral fluorescence imaging, and fluorescence lifetime imaging (FLIM), including techniques like time-correlated single photon counting (TCSPC), streak cameras, gated cameras, and frequency-domain FLIM. These fluorescence imaging techniques can be combined and integrated to enhance imaging capabilities.

11.2 Fluorescence Imaging

11.2.1 Mechanisms of Fluorescence Generation

Fluorescence is a phenomenon of light emission exhibited by certain atoms and molecules when they absorb light of specific wavelengths and transition from higher electronic energy states to lower energy states by photon emission. The generation of fluorescence involves three main processes: excitation of the molecule by absorbed photons, relaxation of excited-state electrons through vibrational relaxation between energy levels, and the return of the molecule to the ground state by emitting photons and releasing energy. These processes occur on a very short timescale, typically ranging from nanoseconds to femtoseconds. Although the entire process from excitation to fluorescence emission lasts only a billionth of a second for most molecules, it represents a fundamental interaction between light and matter. The study of the underlying mechanisms of fluorescence has become the basis for various fields such as time-resolved fluorescence, spectroscopy, and optical microscopy. On the other hand, fluorescence exhibits highly sensitive emission spectra, spatial resolution, and specificity, making fluorescence-related techniques important research tools in fields such as genetics, cell biology, and oncology.

For any molecule, depending on the overall electronic energy states and the symmetry of various electronic spin states, there exist several different electronic states (S_0, S_1, S_2). Each electronic state is further divided into different vibrational and rotational energy levels. At room temperature, apart from the lowest vibrational energy level of the ground state, very few molecules possess enough internal energy to exist in any other state. Therefore, the excitation process typically originates from this energy level. It is through the action of fluorescent probes and fluorescent dyes that molecules undergo electronic transitions and eventually generate fluorescence. Fluorophores that covalently bind to macromolecules such as nucleic acids, lipids, or proteins are referred to as fluorescent molecular groups.

The various processes involved in the absorption and emission of fluorescent molecular groups are typically represented by the Jablonski diagram shown in Fig. 11.2. The Jablonski diagram depicts the vertical stacking of energy levels, including the ground state (S_0), and the first (S_1) and second (S_2) excited singlet states. The thicker lines represent electronic energy levels, while the thinner lines represent various vibrational energy states, neglecting rotational states. Transitions between states are indicated by straight-line arrows, representing the transient nature of the excitation process associated with the absorption or emission of photons. Curved arrows represent internal transitions or nonradiative relaxation processes occurring on longer timescales.

Under the irradiation of incident light at the appropriate wavelength, molecules in the ground state (S_0) are excited to higher electronic and vibrational states, either S_1 or S_2. The initial excited state depends solely on the excitation frequency, with higher frequencies leading to higher excited states. Molecules in the condensed phase rapidly relax to the lowest vibrational level band of S_1, known as internal conversion

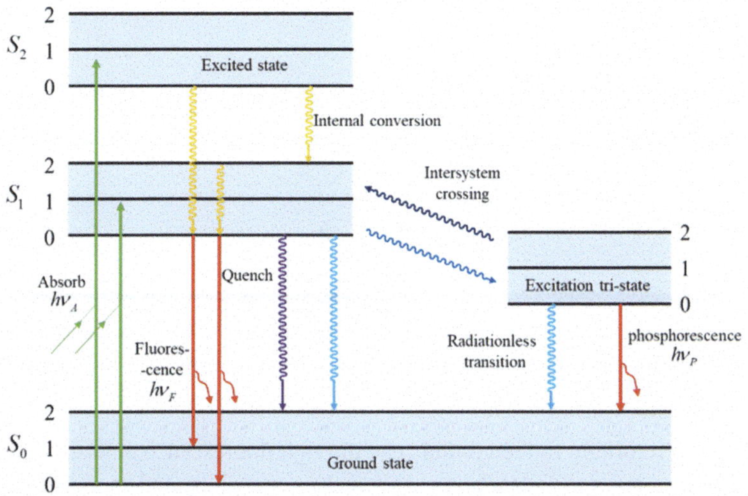

Fig. 11.2 Jablonski diagram energy level diagram

(transition to a lower energy state without emitting photons), typically occurring on the picosecond or smaller timescale. Since fluorescence lifetimes are usually on the order of 10^{-8} s, internal conversion is typically completed before emission. Internal conversion is often attributed to changes in bond lengths and dissipation of vibrational energy (thermal transfer). As a result, fluorescence emission usually arises from the thermally equilibrated excited state, specifically the lowest energy vibrational state of S_1.

Before ultimately returning to the ground state, excited-state molecules reside in the lowest excited singlet state S_1, for a duration on the nanosecond timescale. If this long-lived transition is accompanied by photon emission, it is referred to as fluorescence. Over the course of an excitation cycle, the number of molecules remaining in the excited state diminishes with time, resulting in a decay in the overall fluorescence intensity over time. The time it takes for the fluorescence intensity to decay to 1/e of its initial maximum value after excitation is referred to as the fluorescence lifetime. The process of fluorescence decay is described by exponential decay, following an exponential decay law

$$I(t) = I_0 e^{-kt}, \tag{11.1}$$

where I_0 represents the maximum fluorescence intensity during excitation, and k is the decay constant. Assuming that at time τ, the measured $I(t)$ is 1/e of I_0, then τ is defined as the fluorescence lifetime

$$\tau = k^{-1}, \tag{11.2}$$

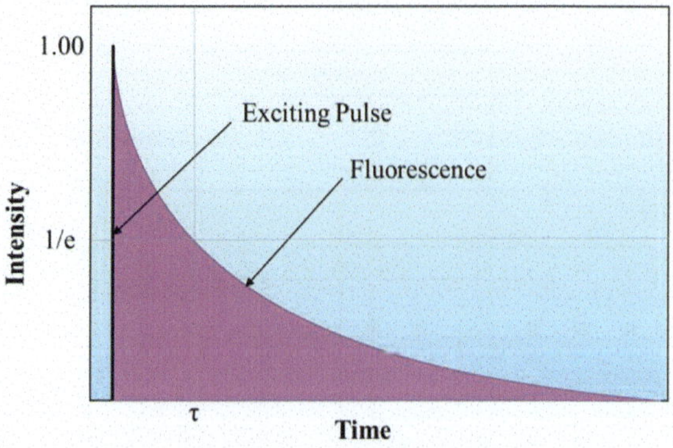

Fig. 11.3 Fluorescence decay curve

in addition to returning to the ground state through fluorescence emission, excited-state molecules can also undergo processes such as quenching and energy transfer, as indicated by the purple dashed line in the diagram, to return to the ground state. The fluorescence lifetime τ is typically related to the rate constants of these processes, and the decay constant k represents the sum of the rate constants for various processes

$$k = k_F + \sum k_i, \tag{11.3}$$

k_F represents the decay constant for the fluorescence emission rate, while k_i represents the decay rate constant for other non-radiative processes see Fig. 11.3.

When the excited-state molecules are coupled with normal thermal motion, the closely spaced vibrational energy levels of the ground state result in a broad range of photon energies emitted during the emission process. As a result, the observed fluorescence often exhibits an intensity curve with a certain width in the wavelength domain, rather than a single wavelength line. Under a defined excitation wavelength and excitation intensity, the relationship between emission wavelength and fluorescence intensity is represented by the fluorescence emission spectrum, also known as the fluorescence spectrum. Since fluorescence is primarily emitted from the lowest vibrational level of the first excited singlet state (S_1) back to the ground state, emission spectra obtained under different excitation wavelengths exhibit the same shape. Corresponding to the fluorescence emission spectrum is the fluorescence excitation spectrum, which measures the variation of fluorescence intensity with excitation wavelength and reflects the relative efficiency of fluorescence generation under different excitation wavelengths. Generally, the fluorescence emission spectrum and excitation spectrum exhibit a certain mirror symmetry relationship.

Before highly reactive excited-state molecules are bleached due to excitation light, most fluorophores can be excited and emit fluorescence hundreds to thousands of times. Several other transition pathways with varying probabilities compete with the fluorescence emission process. The energy of the excited state can be dissipated non-radiatively as heat, and excited fluorophores can also collide with another molecule to transfer energy via a second type of non-radiative process or intersystem crossing to the lowest triplet excited state. The latter case has a low probability, and photons that undergo triplet state crossing eventually return to the ground state through phosphorescence emission or transition back to the excited singlet state, resulting in delayed fluorescence.

11.2.2 Fluorescence Imaging Way

Fluorescence imaging has gained widespread application in the field of life science microscopy due to its prominent features such as high sensitivity and specificity. Apart from using fluorescent molecular labeling in samples, fluorescence imaging can be considered completely non-invasive and non-destructive, making it suitable for in vivo imaging. The development of fluorescence imaging has greatly benefited from the introduction of multidimensional microscopy techniques. In addition to acquiring three-dimensional spatial information, fluorescence imaging can also integrate information from other dimensions such as spectra, polarization of light, and fluorescence lifetime.

(1) Intensity-Resolved Imaging

The simplest way to measure fluorescence is to obtain its steady-state intensity, where the fluorescence intensity of a sample can provide relevant localization information or quantitative information such as fluorophore concentration and quantum efficiency. Under the condition where the excitation radiation is not depleted by the sample and the quantum efficiency remains constant, the fluorescence intensity measured from a specific volume of the sample is proportional to the fluorophore concentration. However, due to the influence of factors such as absorption, scattering, and sample properties, achieving absolute quantification of intensity measurements is challenging. In practical fluorescence intensity measurement techniques, variables such as illumination intensity, collection efficiency, spatial–temporal factors are often combined to obtain relative quantification of fluorescence intensity values.

(2) Spectral-Resolved Imaging

In fluorescence imaging, using fluorescent probes of different colors selectively labels structures within tissues and studies their interactions, while aiming to effectively capture all the information from these fluorophores during a single recording session. Different fluorophores have different fluorescence emission spectra, and by controlling the excitation wavelength, specific fluorophores can be selectively excited in a multi-label system or specific endogenous molecules can be targeted. During signal acquisition at specific wavelengths, different fluorophores can be separated into individual spectral channels or optimized signals compared to background fluorescence. For example, when autofluorescence interferes with exogenous fluorescence, selective suppression of autofluorescence can be performed to improve the signal-to-noise ratio of the image. Spectral imaging typically uses dispersion elements (prisms, gratings, etc.) to extend and sample the full spectrum along the spatial dimension, combined with point or imaging detection instruments (such as spectrometers) to record spectra of individual points or lines. Each pixel in a hyperspectral image contains complete information of the entire fluorescence spectrum, and by spectral separation and image integration, imaging information on the sample at each narrow continuous wavelength band can be obtained. The fused information from each spectral channel forms a hyperspectral image, which reveals the distribution of different spectral information in the sample. The high spectral resolution capability of hyperspectral imaging allows for a wide range of applications in disease diagnosis, clinical surgery guidance, and combined with imaging techniques such as confocal microscopy, it can enable detection of embryonic development and tracking of specific cells, among other applications in the field of biomedical imaging.

(3) Time-Resolved Fluorescence Imaging

While measuring steady-state fluorescence is straightforward, it has been demonstrated that a significant amount of information is lost in the process of time averaging. For example, the anisotropic decay of fluorescence in large molecules is often more complex than single exponential decay, and the precise shape of anisotropic decay contains information about the shape and flexibility of the large molecule. However, such shape information is lost during the time-averaging process of anisotropic decay. Large molecules can exist in multiple conformations, and the decay time of their bound probes can depend on the conformation. Multiple decay times can be obtained from intensity decay information, allowing analysis of the conformational states of fluorophores. On the other hand, steady-state intensity displays the average intensity resulting from the weighted average of two or more decay times. Time-resolved fluorescence measurements also provide valuable information for various reasons. Many molecular information in fluorescence can only be obtained through time-resolved measurements. For example, in the presence of energy transfer, measurements of the distribution of acceptors in space around donors and the rate of energy transfer are possible. Different fluorophores can be distinguished based on their lifetimes, even if their emission spectra overlap. Furthermore, when the physical and biochemical environment surrounding the fluorescent molecules changes and affects fluorescence, analyzing the changes in the expected lifetime of fluorescent molecules can provide quantitative information about ion concentration, interactions with pH and temperature in the environment, and other aspects. In contrast, steady-state intensity fluorescence often provides information limited to the structural aspect of the sample.

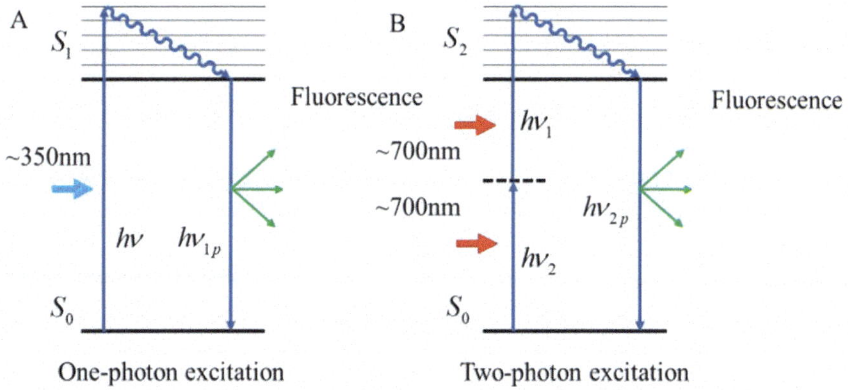

Fig. 11.4 Energy level diagrams for single-photon excitation and two-photon excitation

11.3 Fluorescence Microscopy Imaging Techniques

11.3.1 Principle of Fluorescence Microscopy Imaging

The generation of fluorescence involves several steps: Firstly, specific wavelengths of external light sources excite the molecules or atoms of the sample, causing changes in the energy of the absorbed photons and resulting in a transition effect. This leads to the molecules or atoms being in an excited state. Secondly, after a brief relaxation period, the excited photons return to the excited state and emit photons, exhibiting a blue-shifted characteristic. This is the process of fluorescence generation.

The phenomenon of two-photon fluorescence was first discovered by Maria, but it was not experimentally verified due to limited technological conditions at the time. It was not until the emergence of laser technology that two-photon excitation fluorescence techniques were validated. The principles of single-photon fluorescence and two-photon fluorescence techniques are illustrated in Fig. 11.4. In single-photon techniques, fluorescent molecules in the ground state absorb a photon and transition to higher energy levels, followed by spontaneous emission to return to the ground state and emit fluorescence. In contrast, in two-photon techniques, fluorescent molecules in the ground state simultaneously absorb two photons and transition to higher energy levels, subsequently emitting fluorescence in the form of spontaneous emission.

11.3.2 Fluorescence Spectrum and Quantum Yield

Fluorescence spectrum refers to the distribution of fluorescence intensity with respect to wavelength. It expresses the relative intensity of emitted wavelengths for a molecule under different excitation wavelengths. It consists of an excitation spectrum and an emission spectrum. Due to energy loss during the relaxation process following photon absorption by the molecule, a blue shift phenomenon known as Stokes shift occurs in fluorescence emission. The excitation spectrum characterizes the distribution of fluorescence intensity with respect to wavelength under different excitation wavelengths. On the other hand, the emission spectrum represents the distribution of fluorescence intensity with respect to wavelength for a sample molecule excited at a specific wavelength, indicating the fluorescence intensity distribution at a single excitation wavelength.

Fluorescence quantum yield can characterize the emission efficiency of a fluorophore and is defined as the ratio of the number of emitted photons to the number of absorbed photons. Therefore, the formula for fluorescence quantum yield is as follows

$$Q = \frac{r}{r+k}, \tag{11.4}$$

in the equation, r represents the emission rate of the fluorescent molecule, and k represents the rate of non-radiative deactivation. When the non-radiative deactivation rate k is much smaller than the emission rate r, the fluorescence quantum yield is approximately equal to 1. The fluorescence quantum yield is independent of the excitation wavelength, which means that it remains the same across the entire range of the absorption spectrum, including the center and the edges.

11.3.3 Two-Photon Fluorescence Microscopy

The research group led by Denk first developed laser scanning two-photon fluorescence microscopy and applied it to the field of biomedical research, marking the beginning of the significant role of two-photon fluorescence technology in the biological field. Subsequently, with the advancement of femtosecond laser technology, two-photon fluorescence technology has become increasingly important in biomedical diagnostics. In conventional single-photon fluorescence microscopy commonly used in biological detection, energy level transitions are induced by the absorption of single photons. In contrast, in two-photon microscopy, simultaneous absorption of two photons of the same or different wavelengths is required to induce transitions to higher energy levels in electrons. After relaxation, the electrons return to the ground state and emit fluorescence signals. Therefore, the excitation process in two-photon fluorescence as

$$E_{ex} - E_0 = h_c\left(\frac{1}{\lambda_1} + \frac{1}{\lambda_2}\right). \tag{11.5}$$

(1) Two-Photon Fluorescence Intensity

The fluorescence absorption cross-section parameter δ represents the ability of molecular probes in two-photon fluorescence technology to absorb photons. Currently, measurements of this parameter can be performed using techniques such as nonlinear transmission method, Z-scan method, two-photon transient absorption spectroscopy, and two-photon induced fluorescence method. It has been observed through these measurements that the fluorescence absorption cross-section in two-photon fluorescence is smaller compared to single-photon fluorescence. In two-photon fluorescence, two photons need to be simultaneously absorbed within a smaller fluorescence absorption cross-section. Therefore, the excitation efficiency of two-photon fluorescence is proportional to the square of the excitation light energy, as expressed by the following equation

$$R = \frac{\sigma I^2}{\lambda hc}, \tag{11.6}$$

where σ represents the fluorescence absorption cross-section, I represents the incident light intensity, λ represents the wavelength of the incident light, h represents Planck's constant, and c represents the speed of light in vacuum. By analyzing the above equation, it can be concluded that the two-photon fluorescence effect can only be achieved under specific conditions of high-energy excitation light and high numerical aperture objective focusing, relative to single-photon fluorescence technology.

The intensity of two-photon fluorescence can be expressed as

$$I = \mathrm{C}Na(T, w, \Delta w, \Delta w_1)I^2\left[\frac{\mathrm{A}}{\mathrm{A}+Q}\right], \tag{11.7}$$

where C is a constant related to the experiment, N is the material density for two-photon fluorescence generation, a is the two-photon absorption coefficient, which is a function of temperature T, excitation light frequency ω, two-photon linewidth Δw, and laser linewidth Δw_1. I_0 represents the laser intensity, A is the effective Einstein coefficient, and Q is the de-excitation rate of the excited state. Assuming the incident light is Gaussian distributed, that is

$$I_0 = I_1\frac{W_0}{W(Z)}\exp\left[-\frac{2r^2}{w(z)^2}\right], \tag{11.8}$$

substituting Eq. (11.8) into Eq. (11.7), we obtain the following expression

$$I = CNaI_1^2\frac{w_0}{w(Z)}\exp\left\{\left[-\frac{2r^2}{w(z)^2}\right]\times\left[\frac{\mathrm{A}}{\mathrm{A}+Q}\right]\right\}. \tag{11.9}$$

Two-photon fluorescence is a third-order nonlinear effect, and it can only be excited by photons at the focal point under tightly focused conditions. The intensity of two-photon fluorescence is proportional to the square of the incident light intensity, as indicated by Eq. (11.4). For a Gaussian beam as the incident light, the two-photon fluorescence exhibits a more confined spatial distribution, which is advantageous for improving the spatial resolution of laser scanning confocal microscopy.

(2) Two-photon fluorescence spatial resolution

An optical imaging system refers to the illumination of the sample in front of the objective lens, followed by the collection of light through a series of optical components. The spatial distribution information of the sample is expressed as a similar intensity distribution of light fields at another location. As the sample is treated as a collection of points, it produces a coherent superposition of Airy patterns at multiple distinct ideal image points in the image plane of the optical imaging system.

The point spread function (PSF) is the impulse response function of an optical system, which convolves the real object $f(x, y, z)$ with the PSF to obtain the point spread function of the imaging system. Different PSFs can characterize the spatial resolution characteristics of different imaging systems.

In two-photon microscopy, when the objective lens is of the circular achromatic type, the point spread function [4] of the microscope system is given by

$$PSF(v, u) = \left| 2\int_{0}^{1} P(\rho)J_0(V\rho)\exp\left(\frac{iu\rho^2}{2}\right)\rho d\rho \right|, \tag{11.10}$$

where $P(\rho)$ represents the pupil function, J_0 represents the zeroth-order Bessel function v and u represent the axial and radial coordinates, respectively.

$$v = kr(NA), \tag{11.11}$$

$$u = kz(NA)^2, \tag{11.12}$$

by substituting Eqs. (11.11) and (11.12) into Eq. (11.10), we obtain the point spread functions for confocal fluorescence microscopy and two-photon fluorescence microscopy

$$PSF_{con} = PSF_i \times PSF_{DET} = PSF^2(u, v), \tag{11.13}$$

$$PSF_{2p} = (PSF_{ex}) \approx PSF^2\left(\frac{v}{2}, \frac{u}{2}\right), \tag{11.14}$$

by simplifying the above equation in the focal plane of the objective lens, we obtain

$$PSF_{ef} = \left[\frac{2J_1(v)}{v}\right]^2, \tag{11.15}$$

$$PSF_{con} = \left[\frac{2J_1(v)}{v}\right]^4, \tag{11.16}$$

$$PSF_{2p} = \left[\frac{2J_1(v/2)}{(v/2)}\right]^4. \tag{11.17}$$

From the above equation, it can be observed that the point spread function of confocal microscopy is the square of that in conventional fluorescence microscopy. Due to the pinhole filtering effect in confocal microscopy, stray light present in the optical system is effectively eliminated, resulting in improved spatial resolution of confocal microscopy systems. Additionally, it can be inferred that the excitation frequency in two-photon fluorescence microscopy is half of that in single-photon microscopy, indicating that the wavelength of the excitation light is twice that in single-photon techniques. Therefore, in an ideal environment, the resolution of two-photon fluorescence microscopy systems is slightly lower compared to confocal fluorescence microscopy systems equipped with a pinhole.

(3) Two-Photon Fluorescence Penetration Depth

When a Gaussian beam propagates through a scattering medium, the variation of light intensity in the spatial coordinate system at the axial position $z = z_s$ in a thin scattering medium can be described by the following equation

$$I(\rho, z) = I_0\left[\frac{w_0}{w(z)}\right]^2 \exp\left[-\frac{2\rho^2}{w^2(z)}\right]\exp[-u_s(z + z_s)], \tag{11.18}$$

where w_0 is the waist of the Gaussian beam, ρ is the coordinate position of the beam, I_0 is the peak power of the beam, and us is the scattering coefficient. The expressions for w_0 and $w(z)$ are given by

$$w_0 = \left(\frac{\lambda z_0}{\pi}\right)^{1/2}, \tag{11.19}$$

$$w(z) = w_0\left[1 + \frac{z^2}{z_0^2}\right]^{1/2}, \tag{11.20}$$

as the laser beam propagates deeper into the scattering medium, the expression for the change in the beam's energy with propagation distance z is given by

$$I_{TPF}(0, z) \propto I^2(\rho = 0, z) = I_0^2\left[1 + \left(\frac{z}{z_0}\right)^2\right]^{-2} \times \exp[-2u_s(z + z_s)], \tag{11.21}$$

through the formula, we can observe that the energy of light decreases exponentially with the propagation distance in the medium.

11.3.4 *Application of Two-Photon Fluorescence Microscopy in Biology*

(1) Monitoring the movement of Ca^{2+} ions

Ca^{2+} plays a crucial regulatory role in cells by modulating enzyme activity and protein phosphorylation. Additionally, Ca^{2+} serves as a vital indicator of cell permeability regulation, making it essential to observe the transport and transmission processes of Ca^{2+} and changes in Ca^{2+} concentration in neurons.

SM Block and other researchers utilized two-photon laser scanning microscopy to observe changes in concentration in live mouse brains following stimulation, providing evidence of the relationship between Ca^{2+} concentration and organism stimulation. Ca^{2+} concentration also plays a significant role in biological processes such as cell division.

(2) Imaging of deep brain cortical tissue

Due to the low absorption of near-infrared laser by components such as water and tissue cells in vivo, two-photon microscopy using near-infrared light as the light source can achieve greater penetration depth. It is precisely because of this capability that two-photon fluorescence microscopy has been widely applied in the field of biological brain imaging.

11.4 Fluorescence Lifetime Microscopy

11.4.1 *Quantitative Theory of Fluorescence Lifetime Imaging*

The three important properties of fluorescent molecules are extinction coefficient, quantum yield, and fluorescence lifetime. The extinction coefficient represents the molar absorptivity of the target solvent at a specific excitation wavelength with a concentration of one mole per liter and a light path of 1 cm. The extinction coefficient is inversely proportional to the fluorescence lifetime. Quantum yield is the ratio of emitted photons to absorbed photons, representing the probability of a molecule in the excited state producing a fluorescence photon. Quantum yield is greatly influenced by environmental factors such as pH, concentration, and solvent polarity. Fluorescence lifetime is the time a molecule stays in the excited state before returning to the ground state. The time-resolved fluorescence signal can often be described using a single exponential decay model: $I(t) = I_0 \exp(-t/\tau)$, where $I(t)$ represents the fluorescence intensity at time t, I_0 is the initial intensity, and τ is the fluorescence lifetime. τ corresponds to the time it takes for the fluorescence intensity to decay to 1/e of its initial value.

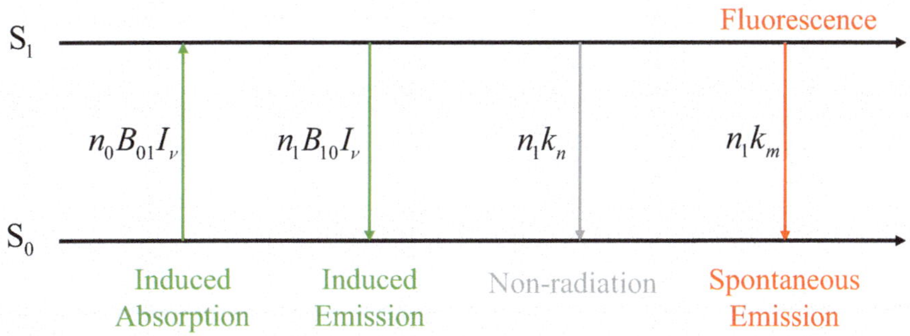

Fig. 11.5 Diagram of a two-level system

The fluorescence signal decay model can be derived from the rate equations of a two-level system. Figure 11.5 illustrates the main transition processes involved in a two-level system. n_0 and n_1 represent the particle concentrations in energy levels S_0 and S_1, respectively. B_{01}, B_{01}, k_m, and k_n are the rates of stimulated absorption, stimulated emission, spontaneous fluorescence emission, and non-radiative relaxation, respectively. I_ν is the integral intensity of the excitation light within the molecular absorption line.

The rate equation for the particle concentration on energy level S_1, n_1, can be expressed as follows

$$\frac{dn_1(t)}{dt} = n_0(t)(I_\nu B_{01}) - n_1(t)(I_\nu B_{01} + \sum_i k_i), \tag{11.22}$$

where $i = a$ and n_a. When the excitation light is in the form of ultrashort pulses, it can be approximated as a delta function. Therefore, after excitation is completed, $I_\nu = 0$ Solving Eq. (11.22), we obtain

$$n_1(t) = n_1^{AE} \exp\left[-\sum_i k_i t\right], \tag{11.23}$$

where n_1 is the particle concentration on state S_1 after excitation is completed. As fluorescence arises from spontaneous emission from state S_1, the time-resolved fluorescence signal can be expressed as

$$f(t) = \frac{\Omega}{4\pi} V k_f n_1(t) = A \exp(-t/\tau),$$
$$A = \frac{\Omega}{4\pi} V k_f n_1^{AE}, \tau = (\sum_i k_i)^{-1}, \tag{11.24}$$

Ω represents the solid angle of signal light collection, and V is the excitation volume. It can be observed that the fluorescence lifetime depends on all the rates of processes returning from the excited state to the ground state.

In traditional fluorescence intensity imaging, the time integral of the fluorescence signal strength $f(t)$ (Eq. (11.24) is taken as

$$S_a = \int f(t)\mathrm{d}t = A\tau = \frac{\Omega}{4\pi}\mathrm{V}n_1^{AE}QE, \tag{11.25}$$

$$QE = \frac{k_x}{\sum_i k_i} = \frac{\tau}{\tau_0}, \tag{11.26}$$

where $\tau_0 = 1/k_x$ represents the fluorescence lifetime when only spontaneous emission process exists, and QE stands for quantum efficiency. It can be observed that without knowing the value of QE when S_a is known, it is not possible to quantitate the concentration of excited state particles n_1^{AE}. Therefore, in applications requiring quantification of fluorophore concentration, the measurement of fluorescence lifetime τ is essential. By substituting τ into Eq. (11.26), QE can be obtained, and then n_1^{AE} can be estimated using Eq. (11.25), with the relationship between n_1^{AE} and fluorophore concentration as follows

$$n_1^{AE} = Xm_{v,J}\int I_v\mathrm{d}t \times B_{01}, \tag{11.27}$$

in which, X represents the concentration of the target component, $m_{v,J}$ are the proportions of particles in the initial states v and J, respectively, and I_v is integrated over time the total energy of the excitation light at the absorption wavelength. It can be seen that, under the condition of known fluorophore properties, environmental parameters, and excitation light parameters, the fluorescence lifetime can be measured to quantify the fluorophore concentration X. The above is the theoretical basis for concentration quantification measurement using FLIM.

In addition to quantifying fluorophore concentration, fluorescence lifetime can be used to determine the non-fluorescent radiative rate k_n, thereby obtaining environmental information related to non-fluorescent radiation processes. Any process that causes fluorescence intensity decay can be called fluorescence quenching, including static quenching, collisional quenching, and Förster resonance energy transfer (FRET). Static quenching is the process where the quencher combines with the fluorophore in the ground state to form a non-fluorescent compound, reducing the number of excitable fluorophores. Collisional quenching is the process where fluorophores in the excited state collide with another molecule and return to the ground state through non-fluorescent radiation. FRET is the phenomenon where a fluorophore (donor) in the excited state resonates with an acceptor, leading to the return of the donor to the ground state and the excitation of the acceptor. It can be seen that static quenching does not change the fluorescence lifetime. Therefore, the next section will introduce the physical mechanisms of collisional quenching and FRET-induced lifetime changes, as well as applications derived from these two mechanisms.

11.4.2 Fluorescence Lifetime Change

(1) Collisional Quenching

Collisional quenching mechanisms [5] include intersystem crossing, Dexter electron exchange, and photoinduced electron transfer (PET). Figure 11.6a illustrates the Dexter electron exchange energy level diagram, where HOMO and LUMO represent the highest occupied and lowest unoccupied molecular orbital energy levels of the donor (D) and acceptor (A), respectively. The asterisk indicates the excited state, and dashed arrows represent electron transfer. In the excited state, the donor D exchanges electrons with the ground-state acceptor A, resulting in the donor returning to the ground state (D) and the acceptor being in the excited state (A^*). Figure 11.6b, c show two types of PET processes. In the first type, the excited-state donor D^* transfers an electron to the ground-state acceptor A, forming an excited-state complex $[D^+A^-]$. In the second type, the ground-state donor D transfers an electron to the excited-state acceptor A, forming an excited-state complex $[D^+A^-]^*$. The excited-state complexes formed through the PET process can return to the ground state through both radiative and nonradiative fluorescence processes, and eventually, the accepted electron by the acceptor will return to the donor. It is noteworthy that the fluorescent agent in the excited state can act as both the donor and the acceptor.

Different collisional quenchers have different mechanisms of action on fluorescent agents, and it is often difficult to determine the specific quenching mechanism. Typically, quenching is a combination of three mechanisms: intersystem crossing, Dexter electron exchange, and PET. Among them, oxygen is one of the most common quenchers and can quench almost all known fluorescent agents. The quenching mechanism of oxygen is still a subject of debate, but intersystem crossing is considered the most likely mechanism. Heavy atoms of halogens such as iodine and bromine mainly undergo intersystem crossing as their quenching mechanism. For aromatic and aliphatic amines, the quenching mechanism is PET.

Collisional quenching can only occur at short distances (~10 Å) where there is overlap between the electron clouds of the donor and acceptor molecules. This is because electrons can move between molecules only at sufficiently short distances. The electron transfer rate k_E, exponentially decreases with an increase in the distance between the donor and acceptor, and it can be described as follows

$$k_E = A_E \exp[-\beta(r - r_c)], \tag{11.28}$$

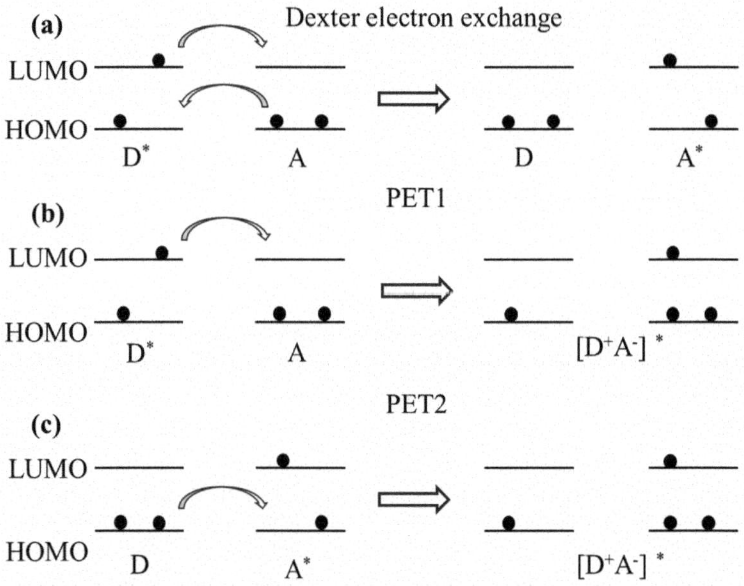

Fig. 11.6 Energy diagrams of dynamic quenching mechanisms. **a** Dexter electron exchange, **b**, **c** two ways of photoinduced electron transfer

where r represents the distance between the centers of the fluorophore and quencher molecules, and r_c is the minimum distance between the molecules. A_E is typically on the order of 10^{13} s^{-1}, and β is usually around 1 Å^{-1}.

The collisional quenching process is commonly studied using the Stern–Volmer equation

$$\frac{F_0}{F} = 1 + k_r \tau_0 [Q] = 1 + K_D [Q], \tag{11.29}$$

where F_0 and F represent the fluorescence intensities in the absence and presence of the quencher, respectively. k_r is the quenching rate constant, τ_0 is the fluorescence lifetime in the absence of the quencher, $[Q]$ is the quencher concentration, and K_D is the Stern–Volmer quenching constant. The relationship between F_0/F and $[Q]$ should be linear. By plotting F_0/F as a function of $[Q]$, K_D and k_r can be determined. The value of k_r reflects the quenching efficiency and the ability of the fluorophore to interact with the quencher. For example, when a fluorophore molecule is attached inside a protein or a thin film polymer, if the protein or thin film is impermeable to the quencher, collisional quenching cannot occur. Measuring k_r indirectly provides information about the permeability of the protein or thin film to the quencher.

Based on the collisional quenching principle, it is possible to quantitatively measure the microenvironment information of fluorescent molecules, including pH values, viscosity, temperature, Ca^{2+}, Mg^{2+}, Na^+, K^+ ion concentrations, O_2 concentration, etc. Quantitative measurement of the microenvironment provides essential

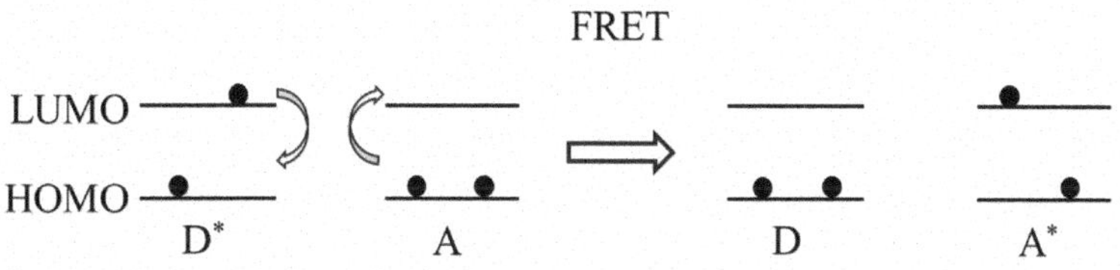

Fig. 11.7 Energy diagrams of Förster resonance energy transfer

information for studying cellular metabolism [6], neural cell signaling processes [7], as well as distinguishing normal and cancerous cells, aiding in cancer surgery [8, 9], and validating anticancer drugs.

(2) Förster Resonance Energy Transfer

Förster Resonance Energy Transfer (FRET) is the phenomenon where a donor molecule in its excited state (D*) transfers its energy to an acceptor molecule (A) in its ground state, causing A to transition to an excited state while D returns to its ground state, as shown in Fig. 11.7. FRET occurs through a dipole–dipole long-range electromagnetic interaction, and it requires spectral overlap between the donor's emission spectrum and the acceptor's absorption spectrum. As FRET is a result of long-range forces, it can occur over distances of 1–10 nm.

The energy transfer rate k_{ET}, depends on (1) the spectral overlap between the donor's emission spectrum and the acceptor's absorption spectrum, and (2) the distance between the donor and acceptor, which can be expressed as follows

$$k_{ET} = \left(\frac{R_0}{\tau_{DA}}\right)^6 \frac{1}{\tau_D}, \tag{11.30}$$

where τ_{DA} and τ_D are the donor's fluorescence lifetimes in the presence and absence of the acceptor, respectively. R_0 is the Förster distance, at which the FRET efficiency is 50%. R_0 is calculated using the following equation

$$R_0^6 = \frac{2.07}{128\pi^5 N_A} \frac{\kappa^2 \cdot QE}{n^4} J(\lambda), \tag{11.31}$$

where κ^2 is the orientation factor that describes the relative orientation of the absorption and emission dipoles in three-dimensional space. If the dipole orientation is random, typically equals the statistical average of 2/3, n is the refractive index, QE is the fluorescence quantum efficiency in the absence of an acceptor, and $J(\lambda)$ is the integral of the overlap region between the donor's emission spectrum and the acceptor's absorption spectrum.

FRET efficiency represents the probability of energy transfer from the excited state donor, and it can be determined by measuring the donor's fluorescence lifetime in the presence and absence of the acceptor

$$E = 1 - \frac{\tau_{DA}}{\tau_D} = \frac{k_{ET}}{\sum_i k_i - k_{ET}}, \tag{11.32}$$

by substituting Eq. (11.30) into Eq. (11.32), the distance between the donor and acceptor can be obtained as follows

$$r = R_0 \left[\frac{1}{E} - 1 \right]^{1/6}, \tag{11.33}$$

the donor and acceptor involved in FRET are paired together. By using donor and acceptor labeling to study the target, and measuring the donor's fluorescence lifetime, the FRET efficiency can be obtained using Eq. (11.32). Then, based on Eq. (11.33), the distance between the labeled target and the acceptor can be quantified.

11.4.3 Factors Affecting Fluorescence Lifetime Imaging Speed

(1) Data acquisition speed

The data acquisition speed determines two key metrics of the FLIM (Fluorescence Lifetime Imaging Microscopy) technique: single frame exposure time and frame rate. FLIM with short single frame exposure time can freeze transient information of the target, while FLIM with a high frame rate can capture the dynamic evolution process of the target. Typically, the single frame exposure time is determined by the acquisition time of a set of FLIM data, and the frame rate is determined by the time interval between adjacent sets of FLIM data. Each set of FLIM data can be used to reconstruct a FLIM image using lifetime algorithms.

There are several time-domain FLIM data acquisition methods, including Time-Correlated Single Photon Counting (TCSPC), gated camera, and streak camera. Traditional TCSPC systems use scanning imaging and single-photon counting data acquisition mode, with single frame exposure time in the order of minutes and frame rates at ~1 frame per minute. Traditional gated camera FLIM uses wide-field imaging and gated selection imaging mode, which has shorter single frame exposure time in the sub-second range, and frame rate depends on the gated camera's frame rate (~10 frames per second). Streak camera FLIM uses line-scanning imaging and integration imaging mode, with single frame exposure time in the order of minutes and frame rates at ~1 frame per minute.

It can be seen that FLIM data acquisition methods have different speeds. Although gated camera FLIM has the shortest single frame exposure time and the highest frame rate, the limited number of gated images obtained leads to lower accuracy in lifetime estimation. TCSPC and streak camera FLIM have slower data acquisition speeds, but they can record the entire time-resolved fluorescence signal curve with higher time resolution, enabling high-precision fluorescence lifetime estimation.

(2) The speed of lifetime algorithms

The speed of lifetime algorithms refers to the speed at which a set of FLIM data is processed to estimate the fluorescence signal decay parameters. Methods with high lifetime algorithm speed can rapidly reconstruct and analyze FLIM images.

In practical applications, fluorescence signals often follow a multi-exponential decay model, which can be represented as follows

$$f(t) = A\sum_{a=1}^{P} q_a \exp(-t/\tau_a), \tag{11.34}$$

where A is the signal amplitude, q_a and τ_a are the proportions and lifetimes, respectively, of the P components of the fluorescence signal. P represents the total number of fluorescence signal components. The fluorescence signal $h(t)$ is the convolution of the decay model $f(t)$ and the instrument response function (IRF) $irf(t)$

$$h(t) = f(t) * irf(t) + \varepsilon(t), \tag{11.35}$$

where $\varepsilon(t)$ is noise and $irf(t)$ is determined by excited light and detection devices. The data obtained from the fluorescence signal $h(t)$ measured by TCSPC and streak camera FLIM system are discrete histograms

$$h_m = \sum_{k=0}^{m} irf_{k-m} f_m + \varepsilon_m, m = 0, \ldots, M - 1, \tag{11.36}$$

$$irf_m = \int_{m\Delta t}^{(m+1)\Delta t} irf(t)\mathrm{d}t, f_m = \int_{m\Delta t}^{(m+1)\Delta t} f(t)\mathrm{d}t, \tag{11.37}$$

among them, h_m and irf_m are the signal strength and IRF strength of the m time bars, M is the number of time bars, and Δt is the time resolution (width of time bars).

For convenience, we use the formal expressions of Eqs. (11.36) and (11.37).

$$h = [h_0, \ldots, h_{M-1}]^{\mathrm{T}}, irf = [irf_0, \ldots, irf_{M-1}]^{\mathrm{T}}, f = [f_0, \ldots, f_{M-1}]^{\mathrm{T}}. \tag{11.38}$$

In this paper, we use bold italics to represent vectors. Fluorescence lifetime algorithm. The function of LDA is to process the h and irf measured by experiments, and to estimate the signal attenuation parameters A, q_a and τ_a.

The most commonly used LDA is the iterative convolution least square method (IC-LSM). Firstly, an exponential attenuation model is selected, that is, the P value in the Eq. (11.34) is selected, and then the attenuation parameters in the selected model are estimated. However, the speed of IC-LSM is very slow at 1 frame per hour (100 × 100 pixel), which is not conducive to the realization of real-time FLIM. Real-time FLIM refers to the rapid completion of FLIM image reconstruction and processing after the completion of data acquisition, so that users can monitor the changes of the target state in real time, which is urgently needed in clinical medicine and combustion state monitoring applications.

The accuracy of life estimation in different applications also limits the speed of FLIM. Define τ and A. Estimation accuracy evaluation numbers F_τ and F_A, the formula is as follows

$$F_\tau(\tau) = \sqrt{N_{tot}}\frac{\sigma_\tau}{\tau},\ F_A(A) = \sqrt{N_{tot}}\frac{\sigma_A}{A}, \tag{11.39}$$

where N_{tot} is the total number of photons collected by the system, σ_τ and σ_A are the standard deviations of τ and A estimates, σ_τ / τ and σ are the relative standard deviations of τ and A estimates, respectively. The lower the value of F_τ and F_A, the higher the estimation accuracy of τ and A, and for the signal containing only Poisson noise, the ideal value of F_τ and F_A is 1. It can be seen that under a certain F_τ and F_A, the relative standard deviation of τ and A estimates decreases with $\sqrt{N_{tot}}$ increase of the signal-to-noise ratio (SNR). Therefore, the higher the precision, the higher the signal-to-noise ratio, which requires a longer data acquisition time to ensure a high signal-to-noise ratio [10].

11.4.4 Method for FLIM Data Acquisition

(1) Time-correlated single photon counting FLIM (TCSPC)

TCSPC is the most commonly used FLIM technique [11]. The conventional TCSPC system consists of a high-repetition-rate ultra-short pulse excitation light source, a point-scanning confocal microscope, a point detector, and a TCSPC module. Figure 11.8 illustrates the principle of TCSPC-FLIM data acquisition.

A large number of pulses are used to excite the sample, and each excitation results in the production of at most one photon in the excitation region. This photon is detected by a single-photon detector and its arrival time is recorded by a time-correlated circuit. When a sufficient number of photons are detected by the single-point detector, a histogram with the fluorescence signal $h(t)$ profile can be obtained.

Traditional TCSPC-FLIM employs a single-pixel single-photon detector (such as a photomultiplier tube, PMT, or single-photon avalanche diode, SPAD) coupled with counting circuitry (including time-to-analog and analog-to-digital converters, TAC-ADC) to perform point-scanning imaging of the sample. This method only allows counting and timing of a large number of photons in a single timing channel, resulting in slow data acquisition speed. Another important factor that limits the data acquisition speed is the dead time, which is the time required for the circuit to reset after detecting a photon in preparation for detecting the next photon. The dead time

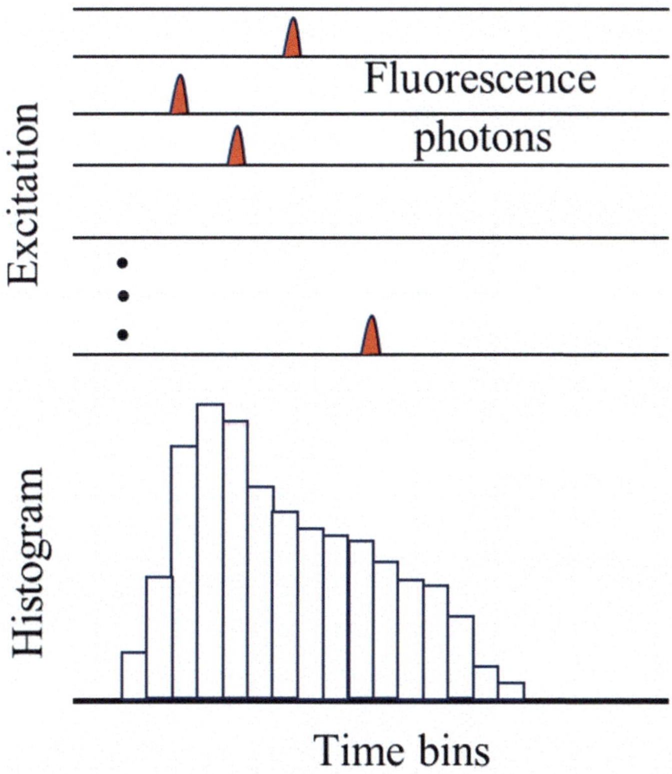

Fig. 11.8 Data-acquisition principle of TCSPC-FLIM systems

of single-photon detectors is around 10–25 ns, and the TAC-ADC counting circuit has a dead time of 25–200 ns. During the dead time, the instrument cannot respond to arriving photons. To ensure that the histogram is not distorted, it is necessary to ensure that at most one photon arrives at the detector during each dead time, otherwise, photon pile-up phenomenon may occur. Photon pile-up refers to the situation where multiple photons arrive at the detector during the dead time, and only the earliest arriving photon is recorded, while the later ones are discarded. Photon pile-up can lead to a shift of the histogram towards shorter lifetimes. Therefore, the upper limit of photon counting rate should be 40 MHz (corresponding to a 25 ns dead time). In practical applications, a lower photon counting rate is required to avoid photon pile-up effects, typically in the order of detecting one photon every 100 excitation events. Thus, for a 40 MHz pulsed laser, the photon counting rate is around 0.4 MHz, and it takes approximately 40 s to acquire a set of FLIM data (256 $\times$ 256 pixels), as shown in Fig. 11.9a. It is evident that single-channel TCSPC-FLIM is not suitable for observing dynamic targets [12].

The TCSPC-FLIM data acquisition speed can be improved by the following methods: shortening the dead time, increasing the number of counting channels,

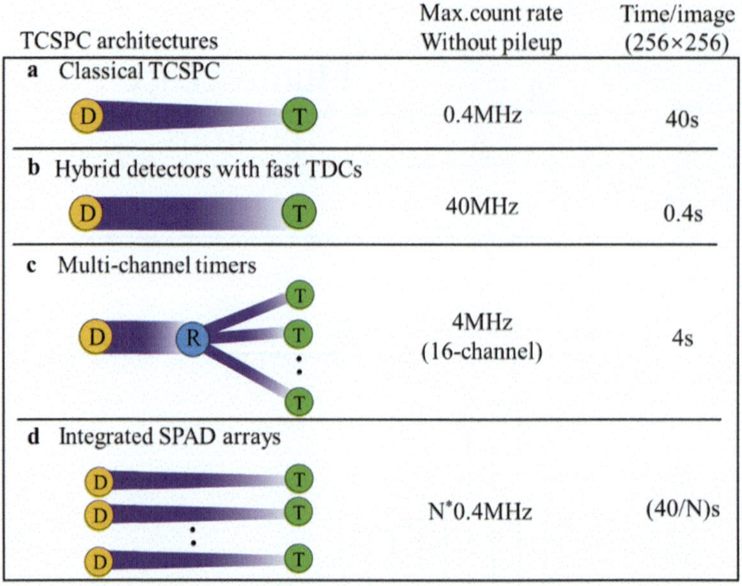

Fig. 11.9 Working principles and data acquisition times for four TCSPC architectures **a** Classical TCSPC with one detector (D) and one timer (T). **b** TCSPC with hybrid detectors and fast TDCs. **c** Multi-channel TCSPC with a router (R) to send detected photons into multiple timing channels. **d** TCSPC with integrated SPAD arrays

using detector arrays for scanning or wide-field imaging, etc. Figure 11.9b illustrates a scheme to increase data acquisition speed by compressing the dead time. The use of a hybrid PMT (Photomultiplier Tube) that combines PMT and SPAD (Single-Photon Avalanche Diode) can reduce the detector's dead time to less than 1 ns, while a Time-to-Digital Converter (TDC) device can shorten the dead time of the counting circuit to less than 1 ns. By combining hybrid PMT and TDC, the photon counting rate can be increased to 40 MHz, and the acquisition time for a single set of data (256 × 256 pixels) is reduced to 0.4 s. Commercial systems like 'rapid FLIMHiRes' (PicoQuant) and 'SP8 FALCON' (Leica) are based on this approach, achieving photon counting rates of ~80 MHz.

The multi-channel TCPSC is shown in Fig. 11.9c, where the detected photons are distributed across multiple channels (usually 8 or 16 channels) for parallel timing. A commercial 16-channel system can increase the photon counting rate to ~4 MHz, and the acquisition time for a single set of data (256 × 256 pixels) is reduced to 4 s.

The N-pixel SPAD array TCPSC are demonstrated in Fig. 11.9d, which increases the counting rate N times compared to single-channel TCSPC. Currently, frame rates at the video level for FLIM data acquisition have been achieved using SPAD arrays of 32 × 32 (50 fps), 64 × 64 (100 fps), and 256 × 256 (10 fps).

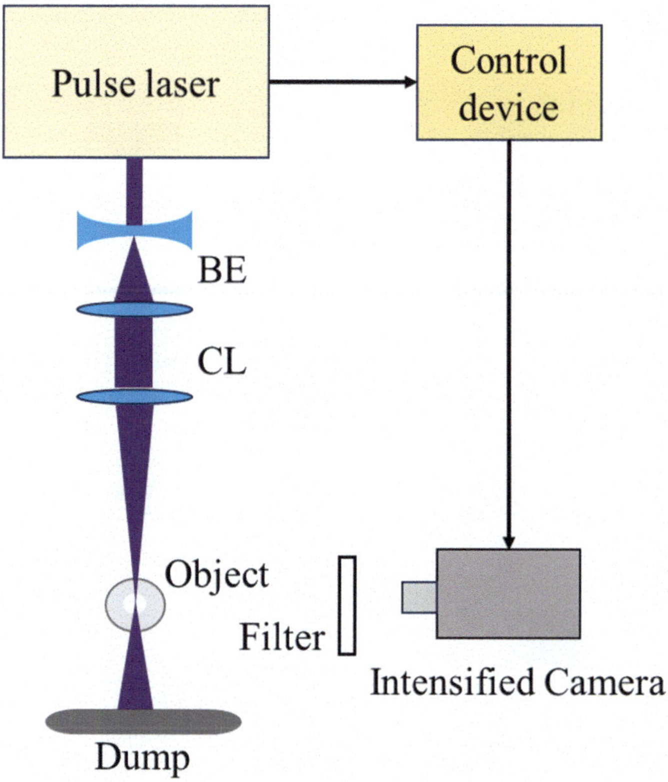

Fig. 11.10 Schematic diagram of a gated camera FLIM system

(2) Gated camera FLIM

A schematic diagram of a typical Gated Camera FLIM system shown in Fig. 11.10.

Gated FLIM requires at least two gated images to reconstruct a fluorescence lifetime image, which means the sample needs to be excited at least twice. To achieve single-shot fluorescence lifetime imaging, D. S. Elson et al. proposed the use of a Segmented Gated Optical Intensifier (SGOI) method [13]. A schematic diagram of a four-channel SGOI is showed in Fig. 11.11, where the time gates of each channel can be independently controlled, allowing simultaneous acquisition of four gated sub-images in a single excitation.

Ehn et al., proposed a method to achieve single-shot FLIM using two Intensified Charge-Coupled Device (ICCD) cameras [14]. The system diagram of this method used for fluorescence lifetime imaging of airflow fields is shown in Fig. 11.12. The fluorescence signals generated by picosecond excitation illumination of the airflow field are split by a beam splitter, and the two sub-signals are detected by two ICCD cameras with different time gates.

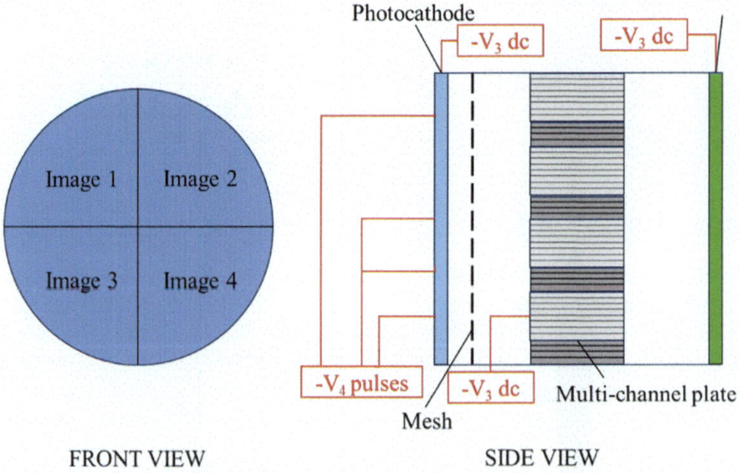

Fig. 11.11 Schematic representation of the four-channel segmented GOI

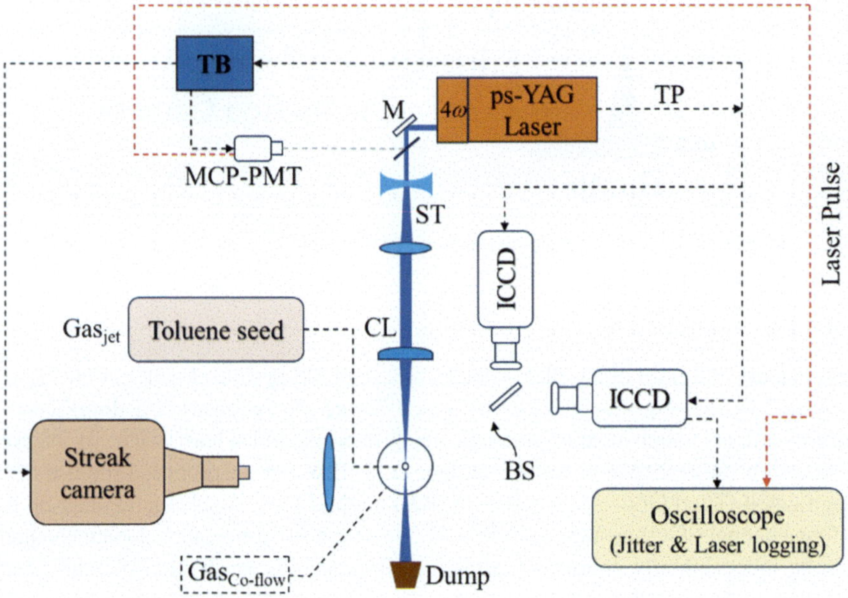

Fig. 11.12 Single-shot time-gated FLIM with two ICCDs system setup

(4) Stripe camera FLIM

Stripe Camera FLIM utilizes a stripe camera to record one-dimensional time-resolved fluorescence intensity curves with sub-picosecond time resolution. Compared to traditional single-channel scanning TCSPC-FLIM, it offers higher time resolution

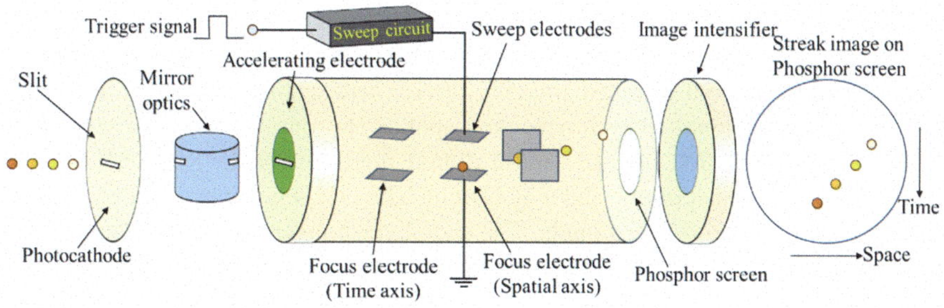

Fig. 11.13 Working principle of a streak camera

and data acquisition speed. The working principle of the stripe camera is shown in Fig. 11.13.

The one-dimensional incident light signal is focused on the photocathode through a slit. The photocathode converts the light signal into an electron stream. The accelerating electrodes in the streak tube accelerate the electron stream.

A pair of deflection electrodes with high-speed scanning voltage scan the electron stream from top to bottom, causing electrons at different moments to strike different positions on the phosphor screen. Finally, the phosphor screen converts the electrons back into photons, which are then imaged onto the imaging device at the end, such as an ICCD camera. The stripe camera image includes a time axis and a spatial axis, corresponding to the time and spatial distribution of the incident light signal, respectively (Fig. 11.14).

Stripe Camera FLIM requires a line scanning process to achieve two-dimensional FLIM, which still results in relatively long data acquisition time. Gao et al.

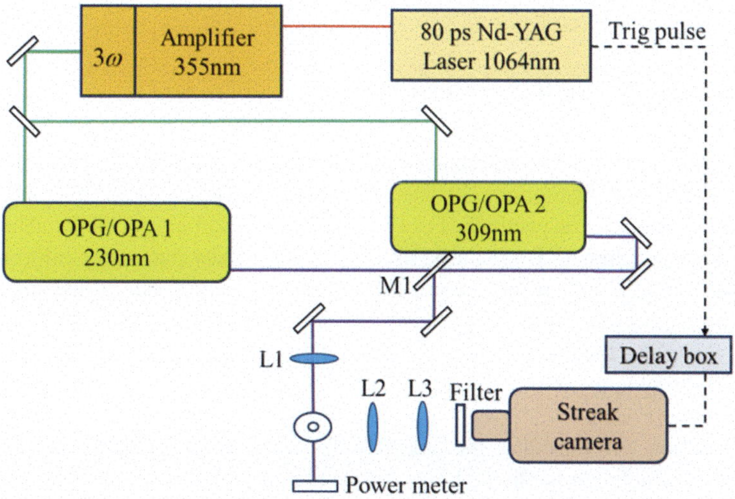

Fig. 11.14 Streak camera FLIM for a CH_4/air flame system setup

proposed the compressed ultrafast photography (CUP) technique based on the stripe camera and compressive sensing algorithm, which extends the stripe camera into a two-dimensional ultrafast imaging device with a time resolution of 10 ps [15]. Three-dimensional (two-dimensional spatial and one-dimensional temporal) dynamic targets are first imaged onto a digital micromirror device (DMD) array through lenses, beam splitters, and a 4f system. The DMD generates random binary coded patterns to encode the signals. The encoded signals are then reflected by the DMD and imaged by the stripe camera. The compressive sensing algorithm is used to reconstruct the three-dimensional target signals from the two-dimensional stripe camera images.

(4) Frequency-domain FLIM

Frequency-domain FLIM utilizes intensity-modulated excitation light to illuminate the sample. By measuring the phase shift and demodulation of the fluorescence signal relative to the excitation light, the fluorescence lifetime can be estimated. A schematic diagram of the traditional frequency-domain FLIM system based on a gated camera is shown in Fig. 11.15. It includes a wide-field microscope, a sinusoidal modulated light source, an image intensifier, a CCD camera, and correlation circuits. The gain of the image intensifier is modulated by an adjustable phase shifter, and each phase shift captures an image. Multiple phase-shifted images can be used to reconstruct a lifetime image.

Traditional frequency-domain FLIM systems require acquiring 12 phase-shifted images to reconstruct a lifetime image to ensure the accuracy of the reconstruction, which limits its imaging speed. To accelerate frequency-domain FLIM, in 2016, Raspe used a new generation of dedicated modulated-FLIM (MEMFLIM) camera to achieve single-image FLIM (siFLIM) [16]. The schematic diagram of the siFLIM system based on the MEMFLIM camera is shown in Fig. 11.16.

11.4.5 FLIM Lifetime Algorithms

The data types for TCSPC, gated camera FLIM, stripe camera FLIM, and frequency-domain FLIM are: histograms, at least two gated images, histograms, and 3–12 phase-shifted images, respectively. For FLIM data of the same type, the same lifetime estimation algorithms (LDA) can be used to extract decay parameter information. Therefore, TCSPC and stripe camera FLIM can utilize per-pixel fitting, per-pixel non-fitting, and global fitting algorithms for data processing. Gated camera FLIM typically employs the rapid lifetime determination (RLD) algorithm, which is a per-pixel non-fitting approach. Frequency-domain FLIM generally uses per-pixel fitting and phasor approach.

This section will introduce the workflow for reconstructing FLIM images from FLIM data, which can be divided into per-pixel and global working modes. It will also present LDAs applicable to both working modes, including fitting and non-fitting LDAs for per-pixel mode and LDAs for global fitting mode. Finally, the estimation

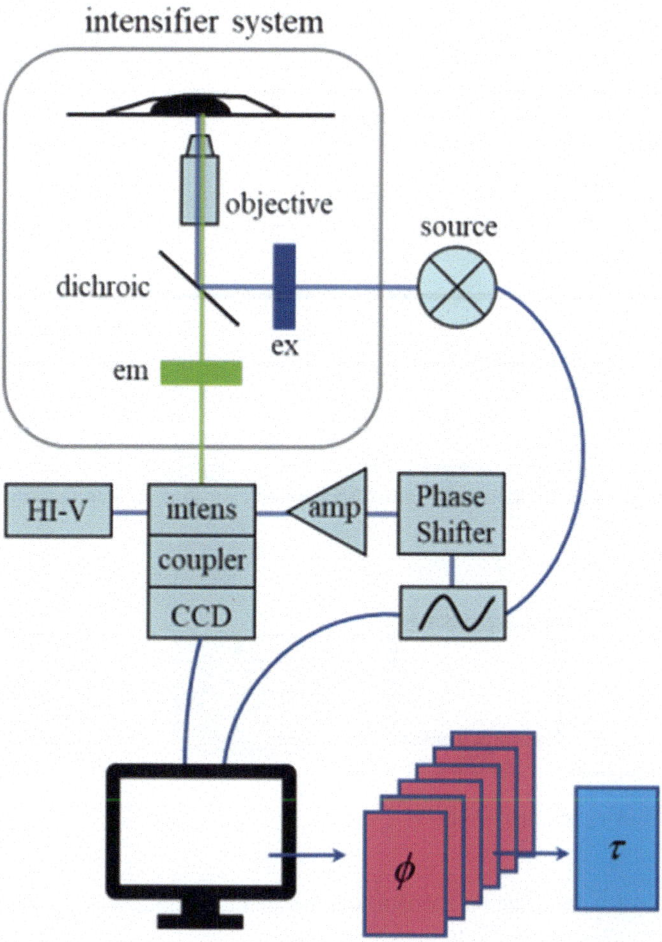

Fig. 11.15 Working principle of a traditional intensifier-based frequency-domain FLIM method system diagram

time and output types of LDAs will be summarized, including decay parameter components and average fluorescence lifetimes, and the application scenarios for different output types will be analyzed.

(1) Lifetime algorithm workflow

The workflow for the per-pixel (pixel-wise, PW) reconstruction mode is illustrated in Fig. 11.17a, where N_{vP} represents the number of valid pixel points. Valid pixels are those with fluorescence integrated intensity higher than the intensity threshold. *h(s)* represents the s-th histogram, where $s = 1, \ldots, N_{vP}$. The (LDA) processes $h(s)$ and the instrument response function (IRF) data, and outputs the decay parameters for the s-th pixel. By performing per-pixel calculations, the decay parameter image

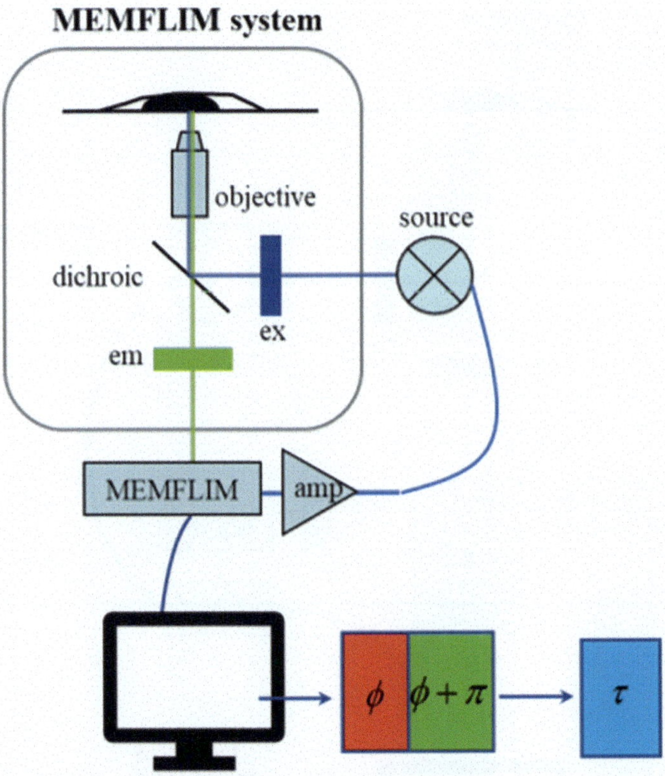

Fig. 11.16 Working principle of siFLIM system

can be reconstructed. The total execution time is $t_{exe}^{PW} = N_{vp} \times t_{LDA}^{PW}$, where t_{LDA}^{PW} is the running time of the LDA in the PW mode.

The workflow for the global fitting for all pixels (GF-P) reconstruction mode is shown in Fig. 11.17b. Unlike the PW mode, the global fitting (GF) assumes that all pixels have the same fluorescence lifetime component τ_a, with only the parameters A and q_a varying for each pixel. The TCSPC data is fed into the LDA capable of performing GF, and the LDA outputs constant τ_a, as well as the A and q_a images for each pixel. The total execution time is $t_{exe}^{GF-P} = t_{LDA}^{GF}$, where t_{LDA}^{GF} is the running time of the LDA capable of implementing GF.

In the PW reconstruction mode, the following minimization problem is constructed

$$\left[\hat{A}^{(s)}, \hat{q}^{(s)}, \hat{\tau}^{(s)}\right] = \arg\min \left\| h^{(s)} - \hat{h}^{(s)} \right\|^2, \tag{11.39}$$

where $\hat{h}(s)$ is the histogram estimated by the LDA from the s-th histogram $h(s)$, and $\|\cdot\|$ represents the l_2 norm. The fitting LDA estimates the decay parameters

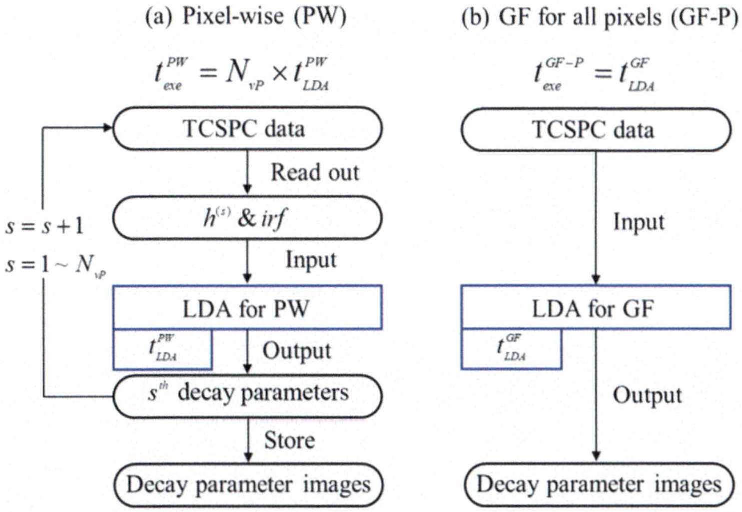

Fig. 11.17 Work flow of FLIM image reconstructions. **a** The pixel-wise (PW) mode and **b** the global fitting mode for all pixels (GF-P)

$\hat{A}^{(s)} = [\hat{A}_1^{(s)}, \ldots, \hat{A}_a^{(s)}]^T$, $\hat{q}^{(s)} = [\hat{q}_1^{(s)}, \ldots, \hat{q}_a^{(s)}]^T$ and $\hat{\tau}^{(s)} = [\hat{\tau}_1^{(s)}, \ldots, \hat{\tau}_a^{(s)}]^T$ of $h(s)$ by solving the minimization problem.

In the GF reconstruction mode, the following minimization problem is constructed

$$\left[\hat{A}^{(s)}, \hat{q}^{(s)}, \hat{\tau}\right] = \arg\min \sum_s \left\| h^{(s)} - \hat{h}^{(s)} \right\|^2, \tag{11.40}$$

the LDA capable of implementing GF estimates the decay parameter $\hat{A}^{(s)} = [\hat{A}_1^{(s)}, \ldots, \hat{A}_a^{(s)}]^T$,$\hat{q}^{(s)} = [\hat{q}_1^{(s)}, \ldots, \hat{q}_a^{(s)}]^T$ and $\hat{\tau} = [\hat{\tau}_1, \ldots, \hat{\tau}_a]^T$ of $h(s)$ by solving the minimization problem.

(2) Per-pixel fitting algorithm

Per-pixel fitting algorithms usually consist of two processes: deconvolution and exponential fitting. Traditional Iterative Convolution Least Squares Method (IC-LSM) takes about 1 h to reconstruct a FLIM image (100 × 100 pixels), making it impractical for rapid data analysis and real-time FLIM implementation. Therefore, many deconvolution algorithms have been proposed to accelerate the deconvolution process and achieve faster LDAs, such as Fourier Transform, Laplace Transform, Moment-Based Deconvolution, Prony Method, Expectation–Maximization, and Laguerre Expansion, among others. Among them, Laguerre Expansion deconvolution is a faster and more effective algorithm. This section will briefly introduce its principles.

Laguerre expansion (LE) uses a set of discrete laguerre basis functions (LBF) to reconstruct the fluorescence decay signal $f^{(s)}$

$$\hat{f}_m^{(s)} = \sum_{l=0}^{L-1} \hat{c}_l^{(s)} b_l(m; \alpha) \tag{11.41}$$

where L and α are parameters that determine the LBF basis. $\hat{c}_l^{(s)}$ represents the coefficient of the l-th LBF and can be expressed as

$$b_l(m; \alpha) = \alpha^{(m-l)/2}(1-\alpha)^{1/2} \sum_{i=0}^{l} (-1)^i \binom{m}{i} \binom{l}{i} \alpha^{l-i}(1-\alpha)^i \tag{11.42}$$

where $l = 0, \ldots, L-1$, $0 < \alpha < 1$, and L is the number of LBF.

(3) Per-pixel non-fitting algorithm

Per-pixel non-fitting algorithms include the centre-of-mass method (CMM) [17], integral extraction method (IEM) [18], phasor method [19], and rapid lifetime determination (RLD) [20], which can achieve rapid estimation of average lifetime. The commonly used average fluorescence lifetimes are of two types: intensity-weighted average lifetime

$$\tau_I = \sum_{p=1}^{P} q_p \tau_p^2 \Big/ \sum_{p=1}^{P} q_p \tau_p, \tag{11.43}$$

as well as amplitude-weighted average lifetime

$$\tau_A = \sum_{p=1}^{P} q_p \tau_p \Big/ \sum_{p=1}^{P} q_p = \sum_{p=1}^{P} q_p \tau_p. \tag{11.44}$$

The formula for estimating fluorescence lifetime using the centre-of-mass method (CMM) is

$$\tau_{CMM}^{(s)} = \frac{\sum_{m=0}^{M-1} t_m \cdot h_m^{(s)}}{\sum_{m=0}^{M-1} h_m^{(s)}} - \frac{\sum_{m=0}^{M-1} t_m \cdot irf_m}{\sum_{m=0}^{M-1} irf_m}, \tag{11.45}$$

the derivation of Eq. (11.45) is as follows: The fluorescence lifetime τ_{CMM} calculated by CMM is the centroid of $f(t)$

$$\tau_{CMM}^{(s)} = \frac{\int_0^\infty t \cdot f(t)\mathrm{d}t}{\int_0^\infty f(t)\mathrm{d}t} = \frac{\sum_p q_p \int_0^\infty t e^{-t/\tau_p} dt}{\sum_p q_p \tau_p} = \frac{\sum_p q_p \tau_p^2}{\sum_p q_p \tau_p}, \tag{11.46}$$

as observed, τ_{CMM} represents the intensity-weighted average lifetime τ_I.

However, in practice, the acquired data are $h(t)$ and $irf(t)$. To obtain Eq. (11.46), integration of t and $h(t)$ is necessary, that is

$$\begin{aligned}
&\int_0^\infty t \cdot h(t)\mathrm{d}t \\
&= \int_0^\infty t \int_0^\infty irf(t-t') \cdot f(t')\mathrm{d}t'\mathrm{d}t \\
&= \int_0^\infty \int_0^\infty (t-t') \cdot irf(t-t') \cdot f(t')\mathrm{d}t'\mathrm{d}t + \int_0^\infty \int_0^\infty irf(t-t') \cdot t' \cdot f(t')\mathrm{d}t'\mathrm{d}t \\
&= \int_0^\infty \left[t \cdot irf(t)\right] * f(t)\mathrm{d}t + \int_0^\infty irf(t) * \left[t \cdot f(t)\right]\mathrm{d}t \\
&= \int_0^\infty t \cdot irf(t)dt \int_0^\infty f(t)\mathrm{d}t + \int_0^\infty f(t)\mathrm{d}t \int_0^\infty t \cdot irf(t)\mathrm{d}t,
\end{aligned} \tag{11.47}$$

$$\int_0^\infty h(t)\mathrm{d}t = \int_0^\infty irf(t)\mathrm{d}t \int_0^\infty f(t)\mathrm{d}t, \tag{11.48}$$

$$\frac{\int_0^\infty t \cdot h(t)\mathrm{d}t}{\int_0^\infty h(t)\mathrm{d}t} = \frac{\int_0^\infty t \cdot irf(t)\mathrm{d}t}{\int_0^\infty irf(t)\mathrm{d}t} + \frac{\int_0^\infty t \cdot f(t)\mathrm{d}t}{\int_0^\infty f(t)\mathrm{d}t}. \tag{11.49}$$

The centroid of the signal $h(t)$ is equal to the sum of the centroids of $irf(t)$ and $f(t)$ are indicated in Eq. (11.49). Therefore, using the centroid of $h(t)$ minus the centroid of $irf(t)$ yields the centroid of $f(t)$, which is τ_{CMM}, the fluorescence lifetime estimated by the centre-of-mass method

$$\tau_{CMM} = \frac{\int_0^\infty t \cdot f(t)\mathrm{d}t}{\int_0^\infty f(t)\mathrm{d}t} = \frac{\int_0^\infty t \cdot h(t)\mathrm{d}t}{\int_0^\infty h(t)\mathrm{d}t} + \frac{\int_0^\infty t \cdot irf(t)\mathrm{d}t}{\int_0^\infty irf(t)\mathrm{d}t}. \tag{11.50}$$

The formula for estimating fluorescence lifetime using the integral extraction method (IEM) is

$$\tau_{IEM}^{(s)} = \frac{\Delta t \sum_{m=0}^{M-1} S_m \cdot \hat{f}_m^{(s)}}{\hat{f}_m^{(s)} - \hat{f}_m^{(s)}}, \tag{11.51}$$

where $S_m = (1/3, 4/3, 2/3, \ldots, 4/3, 1/3)$ represents the coefficients of the Simpson numerical integration formula. The derivation of Eq. (11.51) is as follows: Define the function $g(t)$, such that

$$g(m\Delta t) = \int_{m\Delta t}^{(m+1)\Delta t} f(t)\mathrm{d}t = A\sum_p q_p \tau_p \exp(-m\Delta t/\tau_p)\left[1 - \exp(-\Delta t/\tau_p)\right], \tag{11.52}$$

where $m = 0, 1, \ldots, M - 1$. By making the discrete Eq. (11.52) replacing $m\Delta t$ with t, we obtain

$$g(t) = A\sum_p q_p \tau_p \exp(-t/\tau_p)\left[1 - \exp(-t/\tau_p)\right], \tag{11.53}$$

the integral of $g(t)$ divided by the integral of the first derivative $g'(t)$ yields τ_{IEM}

$$\begin{aligned}\tau_{IEM} &= -\frac{\int_0^\infty g(t)\mathrm{d}t}{\int_0^\infty g'(t)\mathrm{d}t} = \frac{\int_0^\infty g(t)\mathrm{d}t}{g(0) - g(\infty)} = \frac{A\sum_p q_p \tau_p^2\left[1 - e^{-\Delta t/\tau_p}\right]}{A\sum_p q_p \tau_p\left[1 - e^{-\Delta t/\tau_p}\right]} \\ &\approx \frac{\sum_p q_p \tau_p^2\left[1 - (1 - \tau_p^{-1}\Delta t)\right]}{\sum_p q_p \tau_p\left[1 - (1 - \tau_p^{-1}\Delta t)\right]} = \frac{\sum_p q_p \tau_p}{\sum_p q_p},\end{aligned} \tag{11.54}$$

in this process, an exponential function is approximated to the first order. It can be seen that τ_{IEM} represents the amplitude-weighted average fluorescence lifetime τ_A. By approximating the integral of $g(t)$ in Eq. (11.54) using Simpson numerical integration, we obtain Eq. (11.51). It should be noted that IEM cannot directly process the signal h as CMM does. It requires a deconvolution process to reconstruct f before estimating τ_{IEM}.

Phasor is a fluorescence lifetime visualization tool that can analyze multi-exponential decay fluorescence signals using phasor plots. It can also estimate the average fluorescence lifetime using phasor coordinates $(g(s), s(s))$

$$\tau_{Phasor}^{(s)} = \frac{s^{(s)}}{g^{(s)}\omega}, \tag{11.55}$$

the phasor coordinates $(g(s), s(s))$ can be calculated using the following formula

$$g^{(s)} = \frac{R_h + s^{(s)} I_{irf}}{R_{irf}}, \quad s^{(s)} = \frac{R_{irf} I_h + R_h I_{irf}}{R_{irf}^2 + I_{irf}^2}, \tag{11.56}$$

$$R_h = \frac{\sum_{m=0}^{M-1} h_m^{(s)} \cdot \cos(\omega t_m)}{\sum_{m=0}^{M-1} h_m^{(s)}}, \quad I_h = \frac{\sum_{m=0}^{M-1} h_m^{(s)} \cdot \sin(\omega t_m)}{\sum_{m=0}^{M-1} h_m^{(s)}}, \tag{11.57}$$

$$R_{irf} = \frac{\sum_{m=0}^{M-1} irf_m \cdot \cos(\omega t_m)}{\sum_{m=0}^{M-1} irf_m}, \quad I_{irf} = \frac{\sum_{m=0}^{M-1} irf_m \cdot \sin(\omega t_m)}{\sum_{m=0}^{M-1} irf_m}, \tag{11.58}$$

the derivation process of Eq. (11.57) is as follows

$$\begin{aligned} F(\omega) &= \frac{\int_0^\infty f(t) \cdot e^{i\omega t} dt}{\int_0^\infty f(t) dt} = \frac{\sum_p \int_0 q_p e^{-t/\tau_p} e^{i\omega t} dt}{\sum_p q_p \tau_p} \\ &= \frac{\sum_p \left(\frac{q_p \tau_p}{1 - i\omega\tau_p}\right)}{\sum_p q_p \tau_p} = \frac{\sum_p q_p \tau_p \left[\frac{1}{1+\omega^2\tau_p^2} + i\frac{\omega\tau_p}{1+\omega^2\tau_p^2}\right]}{\sum_p q_p \tau_p}, \end{aligned} \tag{11.59}$$

expressing the real and imaginary parts of Eq. (11.59) as g and s, respectively

$$g = \frac{\sum_p \frac{q_p \tau_p}{1+\omega^2\tau_p^2}}{\sum_p q_p \tau_p}, \quad s = \frac{\sum_p \frac{q_p \tau_p^2 \omega}{1+\omega^2\tau_p^2}}{\sum_p q_p \tau_p}, \tag{11.60}$$

if the signal follows a single-exponential decay model, the phasor coordinates (g, s) of the signal with a lifetime τ satisfy

$$(g - \frac{1}{2})^2 + s^2 = (\frac{1}{1+\omega^2\tau^2} - \frac{1}{2})^2 + \left(\frac{\omega\tau}{1+\omega^2\tau^2}\right)^2 = \frac{1}{4}, \tag{11.61}$$

the average fluorescence lifetime calculated using the phasor coordinates (g, s) is

$$\tau_{Phasor} = \frac{s}{g\omega} = \sum_a \frac{q_a \tau_a^2}{1 + \omega^2 \tau_a^2} \Big/ \sum_a \frac{q_a \tau_a}{1 + \omega^2 \tau_a^2}, \tag{11.62}$$

where, τ_{Phasor} is weighted average lifetime, with the weight given by $q_a \tau_a / (1 + \omega^2 \tau_a^2)$. When τ_a / T approaches 0, τ_{Phasor} approximates the intensity-weighted average lifetime τ_I.

The Rapid Lifetime Determination (RLD) algorithm, the integrated signal within two time gates is

$$D_0 = \int_0^{\Delta t_{TG}} A \exp(-t/\tau) dt = A\tau\left[1 - \exp(-t_{TG}/\tau)\right], \tag{11.63}$$

$$D_1 = \int_{\Delta t_{TG}}^{2\Delta t_{TG}} A \exp(-t/\tau) dt = D_0 \exp(-\Delta t_{TG}/\tau), \tag{11.64}$$

where D_0 and D_1 are the integrated signal intensities within time gate 1 and time gate 2, respectively, and Δt_{TG} is the width of the two time gates. The fluorescence lifetime τ can be solved using the following equation

$$\tau = -\Delta t_{TG} / \ln(D_1 / D_0), \tag{11.65}$$

for RLD using different gate widths, D_0 and D_1 can be represented as

$$D_0 = \int_0^{\Delta t_{TG}} A \exp(-t/\tau) dt, \tag{11.66}$$

$$D_1 = \int_{Y\Delta t_{TG}}^{(Y+W)\Delta t_{TG}} A \exp(-t/\tau) dt, \tag{11.67}$$

where Δt_{TG} is the width of time gate 1, Y and W are the starting time factor and width factor of time gate 2. Taking the ratio of D_1 and D_0 yields

$$DR = \frac{D_1}{D_0} = \frac{\exp(-Y\Delta t_{TG}/\tau) - \exp\left[-(Y + W)\Delta t_{TG}/\tau\right]}{1 - \exp(-\Delta t_{TG}/\tau)}. \tag{11.68}$$

(4) Global fitting algorithm

The global fitting algorithm mainly includes two methods: Iterative convolution (IC) and variable projection (VP).

In the IC method, it is assumed that all signal lifetime components τ_a are the same, and only the amplitude A and the lifetime component ratio q_a differ. For the s-th histogram, the decay curve f reconstructed by IC can be expressed as

$$\hat{f}_m^{(s)} = \hat{A}^{(s)} \sum_{a=1}^{A} \hat{q}_a^{(s)} e^{-t_m / \hat{\tau}_a}, \tag{11.69}$$

where $\hat{\tau}_a$ is the estimated value of the lifetime constant, $\hat{A}(s)$ and $\hat{q}_a^{(s)}$ are the estimated values of the amplitude and lifetime ratio, respectively, for the s-th histogram. The estimated value of h can be obtained by convolving with the *irf*

$$\hat{h}_m^{(s)} = \sum_{k=0}^{m} irf_{k-m} \cdot \hat{f}_m^{(s)}, m = 0, 1, 2, \ldots, M - 1. \tag{11.70}$$

The VP method for global fitting is implemented in two steps: The first step involves minimizing a projection function that depends solely on the nonlinear variable (τ_a). The second step involves estimating the linear variables ($A(s)$ and $q_a(s)$) using the projection function as a basis.

A M × P matrix is constructed, where

$$\Phi(\hat{\tau}) = \left[\varphi_1(\hat{\tau}_1), \ldots, \varphi_p(\hat{\tau}_p)\right], \tag{11.71}$$

where $\hat{\tau} = [\hat{\tau}_1, \ldots, \hat{\tau}_a]^T$, is the vector of estimated lifetime constants,$\varphi_a(\hat{\tau}_a) = [\varphi_a(\hat{\tau}_a; t_0), \ldots, \varphi_a(\hat{\tau}_a; t_{M-1})]^T$ is a vector depending only on $\hat{\tau}_a$, representing an exponential decay function convolved with *irf*, and can be expressed as follows

$$\varphi_p(\hat{\tau}_p; t_m) = \sum_{k=0}^{m} irf_{k-m} \cdot \exp(-t_m / \hat{\tau}_p), p = 1, \ldots, P. \tag{11.72}$$

Then, the signal $\hat{h}^{(s)}$ can be estimated from the following equation

$$\hat{h}^{(s)} = \Phi \hat{a}^{(s)}, \tag{11.73}$$

$$\hat{h}_m^{(s)} = \sum_{p=1}^{P} \hat{a}_p^{(s)} \varphi_p(\hat{\tau}_p; t_m), \tag{11.74}$$

$$\hat{a}^{(s)} = \arg\min \sum_{s} \left\| h^{(s)} - \Phi(\hat{\tau}) \hat{a}^{(s)} \right\|^2, \tag{11.75}$$

under a fixed set of $\hat{\tau}$, when the basis $\Phi(\hat{\tau})$ is fixed, the solution to Eq. (11.74) is given by $\hat{a}^{(s)} = \Phi^{-}(\hat{\tau}) h^{(s)}$, where $\Phi^{-}(\hat{\tau})$ is the symmetric generalized inverse of $\Phi(\hat{\tau})$.

Therefore, we can be rewritten as:

$$\hat{\tau} = \arg\min \sum_{s} \left\| P_{\Phi(\tau)}^{\perp} h^{(s)} \right\|, \tag{11.76}$$

the matrix $P_{\Phi(\tau)}^{\perp} = I - \Phi(\hat{\tau})\Phi^{-}(\hat{\tau})$ represents the projection matrix onto the orthogonal complement of the column space of $\Phi(\hat{\tau})$.

In summary, the global fitting algorithm using the variable projection (VP) method consists of two steps:

Step 1: Solve problem to estimate $\hat{\tau} = [\hat{\tau}_1, \ldots, \hat{\tau}_a]^{\mathrm{T}}$.

Step 2: Solve the linear equation $h^{(s)} = \Phi\hat{a}^{(s)}$ to estimate $\hat{a}^{(s)}$.

11.5 General Conclusions

This chapter delves into the intricate realm of fluorescence imaging and microscopy, exploring the fundamental mechanisms, diverse imaging modalities, and complex technologies that drive this field to the forefront of biological and medical research. This chapter provides a comprehensive overview of fluorescence imaging and microscopy, highlighting the continuous development and integration of advanced techniques. From understanding fundamental mechanisms to exploring cutting-edge methodologies such as two-photon microscopy and FLIM, researchers are equipped with powerful tools to unravel the complexities of biological systems at the molecular level. As technology advances, fluorescence imaging plays a pivotal role in shaping our understanding of cellular dynamics, holding immense potential for future discoveries in computational imaging and the life sciences.

Acknowledgements This work was supported by the National Natural Science Foundation of China (Nos. 11874132, 61975044, 12074094, 12374271, and 62375065); Interdisciplinary Research Foundation of HIT (No. IR2021237).

References

1. Crimaldi JP. Planar laser induced fluorescence in aqueous flows. Exp Fluids. 2008;44(6):851–63.
2. Ehn A, Zhu J, Li X, Kiefer J. Advanced laser-based techniques for gas-phase diagnostics in combustion and aerospace engineering. Appl Spectrosc. 2017;71(3):341–66.
3. Ehn A, Bood J, Li Z, Berrocal E, Aldén M, Kristensson E. FRAME: femtosecond videography for atomic and molecular dynamics. Light Sci Appl. 2017;6: e17045.
4. Helmchen F, Denk W. Deep tissue two-photon microscopy. Nat Methods. 2005;2(12):932–40.
5. Mikhnenko V, Blom WM, Nguyen T. Exciton diffusion in organic semiconductors. Energy Environ Sci. 2015;8(7):1867–88.
6. Bower AJ, Li J, Chaney EJ, Marjanovic M, Spillman DR, Boppart SA. High-speed imaging of transient metabolic dynamics using two-photon fluorescence lifetime imaging microscopy. Optica. 2018;5(10):1290–6.

7. Zheng K, Jensen TP, Rusakov DA. Monitoring intracellular nanomolar calcium using fluorescence lifetime imaging. Nat Protoc. 2018;13(3):581–97.
8. Kittle DS, Vasefi F, Patil CG, Mamelak A, Black KL, Butte PV. Real-time optical biopsy: time-resolved fluorescence spectroscopy instrumentation and validation. Sci Rep. 2016;6:38190.
9. Nobis M, McGhee EJ, Morton JP, Schwarz JP, Karim SA, Quinn J, Anderson KI. Intravital FLIM-FRET imaging reveals dasatinib-induced spatial control of Src in pancreatic cancer. Cancer Res. 2013;73(15):4674–86.
10. Zhang Y, Khan AA, Vigil GD, Howard SS. Investigation of signal-to-noise ratio in frequency-domain multiphoton fluorescence lifetime imaging microscopy. J Opt Soc Am A. 2016;33(7):B1–11.
11. Becker W. Advanced time-correlated single photon counting techniques. Springer;2005, p. 20–4
12. Poudel C, Mela I, Kaminski CF. High-throughput, multi-parametric, and correlative fluorescence lifetime imaging. Methods Appl Fluoresc. 2020;8: 024005.
13. Elson DS, Munro I, Requejo-Isidro J, McGinty J, Dunsby C, Galletly N, Stamp GW, Neil MAAA, Lever MJ, Kellett PA, Dymoke-Bradshaw A, Hares J. Real time time domain fluorescence lifetime imaging including single-shot acquisition with a segmented optical image intensifier. New J Phys. 2004;6:180.
14. Ehn A, Johansson O, Arvidsson A, Aldén M, Bood J. Single-laser shot fluorescence lifetime imaging on the nanosecond timescale using a dual image and modeling evaluation algorithm. Opt Express. 2012;20(3):3043–56.
15. Gao L, Liang J, Li C, Wang LV. Single-shot compressed ultrafast photography at one hundred billion frames per second. Nature. 2014;516:74–7.
16. Raspe M, Kedziora KM, Broek B, Zhao Q, Jong S, Herz J, Mastop M, Goedhart J, Gadella J, Young IT, Jalink K. SiFLIM: single-image frequency domain FLIM provides fast and photon-efficient lifetime data. Nat Methods. 2016;13(6):501–4.
17. Li DU, Arlt J, Tyndall D, Walker R, Richardson J, Stoppa D, Charbon E, Henderson RK. Video-rate fluorescence lifetime imaging camera with CMOS single-photon avalanche diode arrays and high-speed imaging algorithm. J Biomed Opt. 2011;16(9): 096012.
18. Li DU, Arlt J, Richardson J, Walker R, Buts A, Stoppa D, Charbon E, Henderson R. Real-time fluorescence lifetime imaging system with a 32 × 32 0.13 μm CMOS low dark-count single-photon avalanche diode array. Opt Express. 2010;18(10):10257–69
19. Sancataldo G, Anselmo S, Vetri V. Phasor-FLIM analysis of thioflavin T self-quenching in concanavalin amyloid fibrils. Microsc Res Tech. 2020;83(7):811–6.
20. Evers M, Salma N, Osseiran S, Casper M, Birngruber R, Evans CL, Manstein D. Enhanced quantification of metabolic activity for individual adipocytes by label-free FLIM. Sci Rep. 2018;8:8757.

Chapter 12
Synthetic Aperture Imaging

Longhui Fu and Shutian Liu

12.1 Introduction

With the development of space technology, the resolution of optical system is required more and more in space observation and space remote sensing. According to the Rayleigh criterion, when it is necessary to improve the resolution of the optical system, it can only be achieved by increasing its aperture [1]. It is not practical to blindly increase the size of the system, which can only increase the complexity and weight of the system, and due to the limitations of processing technology, materials, processing costs and other factors, simply increasing the caliber is no longer the optimal solution. As a result, synthetic aperture technology arises at the historic moment, which overcomes the restrictive factors such as complex processing technology, high cost, large volume, mass and so on [2]. Specifically:

(1) the large aperture system synthesized by a certain number of small aperture systems breaks through the traditional single aperture system and effectively improves the resolution of the system.
(2) the synthetic aperture system tends to be small in size and light in mass. The general system is designed as a folding system, which is convenient to be used in the aerospace field.
(3) the synthetic aperture system is generally arranged by sub-apertures of the same size according to a certain spatial model, which can greatly reduce the processing difficulty and cost.

In summary, synthetic aperture imaging is becoming an indispensable technology in the field of imaging.

L. Fu (✉) · S. Liu
School of Physics, Harbin Institute of Technology, Harbin 150001, China
e-mail: 1224572137@qq.com

S. Liu
e-mail: stliu@hit.edu.cn

Z. Liu et al. (eds.), *Computational Optical Imaging*, Advances in Optics and Optoelectronics, https://doi.org/10.1007/978-981-97-1455-1_12

The main purpose of synthetic aperture is to achieve the coherence and resolution of large aperture system by using several small aperture synthesis systems according to a certain degree of coherence and structural order.

At present, the main imaging methods for synthetic aperture are Michelson interference method, coherent coding method and aperture imaging combination method [2].

Interferometry is mainly a calculation method to directly obtain the radiation field information of objects in the aperture by directly measuring the frequency complex coherence and resolution of the radiation field of the object in the aperture Michelson interferometer is the earliest wide application of this method.

The coherent coding method illuminates the object with coherent photosynthetic echo, records the amplitude and phase of each echo, and uses the recorded images and data to reproduce the echo illuminated object we observed [3]. One of the essential functions of this coherent coding method is active holographic imaging, and one of its main practical applications is coherent optical radiation synthetic aperture radar.

Simple combined aperture system is the most direct optical synthesis method to realize high frequency optical synthetic aperture at present. It mainly uses the coherent comprehensive action of several small aperture synthetic components to calculate the high-frequency aperture information of the object.

The synthetic aperture optical system is formed by combining a certain number of small holes with the same aperture according to a certain structure, and then compensating and adjusting the phase of each sub-aperture optical path to a certain extent, so that each sub-aperture beam meets the same phase, and then coherent superposition is carried out on the same focal plane. The beams emitted by each sub-aperture must be superimposed with the same phase intensity, so that the image fields between them can enhance each other in the receiving and imaging center of the Airy disk, thus narrowing the center of the Airy disk. So that the Airy disk to improve the intensity and resolution of the role of receiving imaging. Otherwise, the image field intensity of each sub-aperture is only equal to the sum of the same phase intensity, which can only improve the imaging efficiency of the Airy disk.

The secondary mirrors of synthetic aperture optical systems can be divided into two categories according to their structural characteristics [3], the first is the structure of shared synthetic aperture secondary mirrors, and the second is the structure of multi-aperture telescopes, as shown in Fig. 12.1. In the former, multiple sub-lenses are used to replace the main mirror structure in the traditional telescope these sub-mirrors are arranged in a specific order so that the receiving light source is projected directly onto the common sub-mirror and then superimposed on the focal plane. It not only makes full use of the traditional technology, but also reduces the difficulty of lens processing.

At present, the common synthetic aperture arrangement structures are: Ring structure (Circle), Golay structure (Golay), Three-arm structure (Tri-arm) [2].

In the system of Ring structure (Circle), many identical sub-holes are evenly distributed on a smaller circumference. With the increase of the number of holes, the distance between each sub-hole decreases relatively and is closely arranged into a

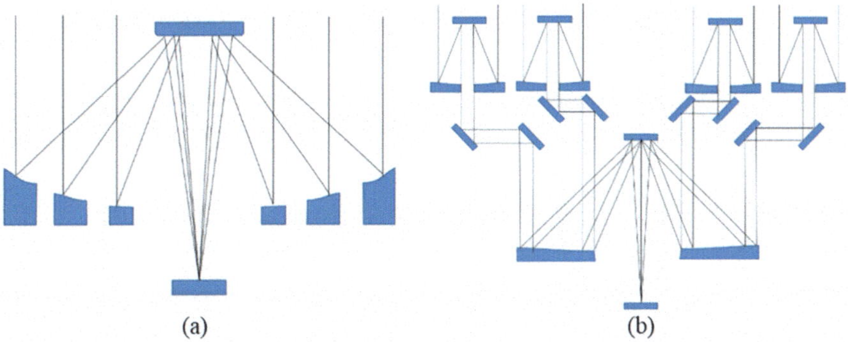

Fig. 12.1 Structure of synthetic aperture imaging system: **a** common mirror structure; **b** multi-telescope structure

ring. Meanwhile, the modulation transfer function of the ring structure will be more and more similar to the structure of the torus as shown in Fig.12.2(a).

The Golay structure is also a widely used synthetic aperture structure. The sub-apertures of the Golay structure are arranged on the grid points of triangles, quadrilaterals and hexagons whose sides are the maximum circumscribed circles of each mirror of the system, so there is no relative position between any two holes in the system as shown in Fig.12.2(b).

Three-arm (Tri-arm) arrangement structure is a common synthetic aperture structure, which arranges three equal sub-apertures at equal intervals in three different directions. The distance between the three axes is 120, forming a Y shape, so the Three-arm (Tri-arm) shape is often called the Y shape (see Fig. 12.2(c)).

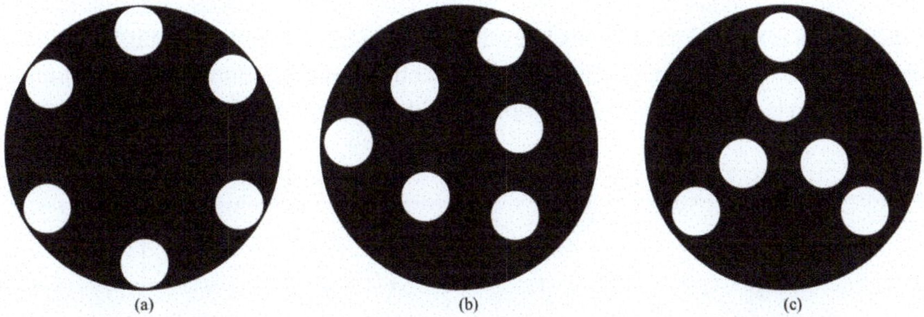

Fig. 12.2 Three pore size distributions: **a** Circle; **b** Golay; **c** Tri-arm

12.2 Basic Principles of Synthetic Aperture System Imaging

12.2.1 Simplified Imaging Model of Synthetic Aperture System

The simplified model of synthetic aperture system is often used to study the imaging characteristics of an optical system and to simulate its imaging process. The synthetic aperture system images a remote target. In this process, the parallel light emitted by the target is divided into multiple sub-beams by the aperture when entering the system, and the sub-beams are coherent and synthesized into the image plane of the system. According to the theory of Fourier optics, the synthetic aperture optical imaging system can be simplified to the imaging model shown in Fig. 12.3. The imaging target is located in the front focal plane of the lens, and the light is illuminated on the object, and the synthetic aperture system is imitated through lens L1, L2 and synthetic aperture diaphragm [4].

According to the theory of physical optics, under the condition of incoherent monochromatic light irradiation, when a diffraction-limited optical system images an infinite distant target, it can be described by

$$I_g(x_i, y_i) = I(x_i, y_i) \otimes \mathrm{PSF}(x_i, y_i) \tag{12.1}$$

That is, the intensity distribution of the image plane is equal to the convolution of the light intensity distribution of the object and the point spread function of the system.

In the spatial domain, to evaluate the imaging quality of an optical system, the quality of the point spread function of the system is generally used to evaluate the imaging quality; in the frequency domain, the Modulation Transfer Function (MTF) of the system is an important index to evaluate the imaging quality. The larger the area covered by the modulation function is, the larger the frequency domain range obtained by the system is and the better the imaging quality is. From the basic theory of Fourier

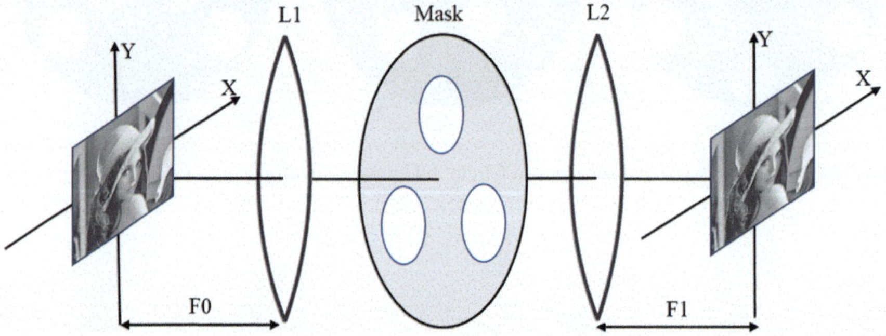

Fig. 12.3 Simplified model of synthetic aperture imaging

optics, it is known that the point modulation function is equal to the Fourier transform of the point spread function. Because the modulation transfer function of an optical system is determined by the number of apertures, the effect equivalent to the full-aperture imaging system can be achieved by changing these physical quantities [2].

If the cutoff frequency of a synthetic aperture optical system is the same as that of a single aperture system, it can be regarded as a large circular aperture system.

In the full-aperture imaging system, assuming that the pupil is circular and the diameter is D, the pupil function is shown in

$$p(x, y) = \frac{\sqrt{x^2 + y^2}}{D/2}. \tag{12.2}$$

The point spread function of the circular aperture can be obtained from the Fourier optics theory as follows [5]:

$$\mathrm{PSF}(x, y) = |\mathcal{F}\{P(x, y)\}|^2, \tag{12.3}$$

In the optical synthetic aperture, the pupil is no longer a single round hole, but is arranged by several small round holes, so the transmittance of the whole diffraction screen can be obtained by the transmittance of a single aperture and the convolution of a function two-dimensional array the position of each function represents the position of a single aperture, and the pupil function of the system can be expressed as [7]

$$P_N(x, y) = \mathrm{circ}(\frac{\sqrt{x^2 + y^2}}{D/2}) \times \sum \delta(x - x_i, y - y_i), \tag{12.4}$$

where (x_i, y_i) is the central coordinate of the pupil function of the n-th keyhole, $P_N(x, y)$ is the pupil function of the sub-aperture.

12.2.2 The Relationship Between PSF and MTF

We can regard the synthetic aperture imaging system as an equivalent low-pass filter composed of several small holes. In the spatial domain, the imaging quality of an optical system is often determined by its own point spread function (PSF). According to the Fourier optics theory, the spot energy distribution function of the system to the ideal target can be reflected by the point spread function, which can measure the imaging quality of an optical system. This function determines the resolving power of the optical system. In the frequency domain, we usually use the optical transfer function to reflect the imaging ability of an optical system [6].

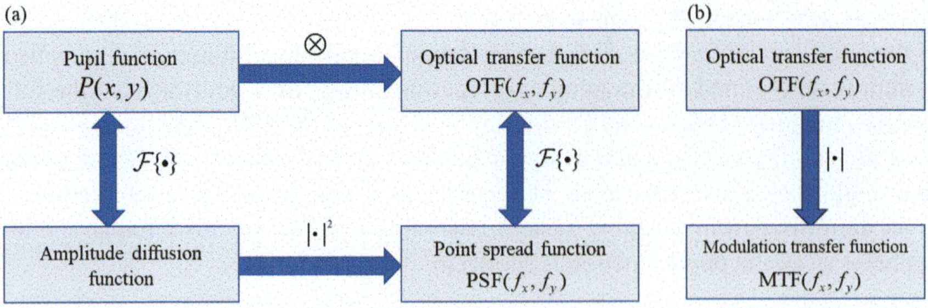

Fig. 12.4 The relationship among pupil function, PSF, OTF and MTF: **a** the relationship among OTF and PSF; **b** the relation among OTF and MTF

According to the theory of Fourier optics, we obtain the conversion relationship among the point spread function, optical transfer function and modulation transfer function of the optical system, as shown in Fig. 12.4.

According to the pupil function, PSF, OTF, MTF relationship and Eq. (12.4) shown in Fig. 12.4, the MTF relation of the synthetic aperture imaging system can be obtained by calculation, as shown in [2]

$$\begin{aligned}\mathrm{MTF}_{\mathrm{syn}}(f_x, f_y) &= \mathrm{MTF}_{\mathrm{sub}}(f_x, f_y) + \frac{1}{N}\mathrm{MTF}_{\mathrm{sub}}(f_x, f_y)\\ &\times \sum_{i=1}^{N(N-1)/2} \delta\left(f_x \pm \frac{x_i}{\lambda d}, f_y \pm \frac{y_i}{\lambda d}\right),\end{aligned} \tag{12.5}$$

$$\mathrm{MTF}_{\mathrm{sub}}(f_x, f_y) = \begin{cases} \frac{2}{\pi}[\cos^{-1}(\frac{\lambda f}{D}) - \frac{\lambda f}{D}\sqrt{1-(\frac{\lambda f}{D})^2}],\ 0 < \rho < \frac{\lambda f}{D}, \\ 0,\ \rho \geq \frac{\lambda f}{D}. \end{cases} \tag{12.6}$$

where, $\mathrm{MTF}_{\mathrm{syn}}(f_x, f_y)$ is the MTF of synthetic aperture. $\mathrm{MTF}_{\mathrm{sub}}(f_x, f_y)$ is the MTF of sub-aperture.

12.2.3 Intermediate Frequency Compensation

The synthetic aperture system also has its inherent defects, because the optical system is partially filled with its equivalent large aperture system, so that the area of the optical system is reduced. According to the analysis, synthetic aperture imaging is the intermittent sampling of its sub-aperture pupil function, which is partially filled by the area, resulting in the reduction or loss of medium and low frequency sampling, resulting in the decrease of image signal-to-noise ratio and blurring. Therefore, it is necessary to improve the medium and low frequency information of the image obtained by synthetic aperture directly, so that we can get high resolution and clear imaging results. The method of improving the low-frequency information in the synthetic aperture optical system has been focused on the optimization of its aperture arrangement, by changing its filling factor and arrangement structure to find a relatively balanced structure to meet the requirements of the medium-and low-frequency transfer function.

The filling factor is defined as the ratio of the area of all the sub-apertures that make up the synthetic aperture system to the area of the circumscribed circular aperture that surrounds the sub-aperture. The specific calculation method is as follows:

$$\text{Filling factor} = \frac{\text{The sum of the areas of each subaperture}}{\text{Subaperture external circle area}}.$$

12.2.4 Middle and Low Frequency Compensation for Smaller Filling Factor

With the decrease of the filling factor and the increase of the interval between the sub-apertures, we can obtain a larger field of view, and the coverage range of the modulation transfer function is also relatively enlarged at the same time, the smaller filling factor is more beneficial to reduce the volume and quality of the optical system, reduce the cost and machining difficulty, and it is easier to correct the arch phase error. Because the pupil function of the synthetic aperture interference array telescope is a discontinuous sampling function of the pupil surface, when its filling factor is small, the light area is insufficient and the signal-to-noise ratio decreases. When the filling factor of the synthetic aperture system is less than the critical filling factor, the interval between the sub-apertures increases, which leads to the interval between the transfer functions of each sub-aperture, that is, the zero-frequency phenomenon of the transfer function of the synthetic aperture system [2].

Generally speaking, the image restoration algorithm is to deconvolution according to the degraded model of the image to remove the blur and noise of the image. However, for the synthetic aperture system, due to the lack of filling factor, the modulation function is missing in the middle and low frequency, which makes the general image restoration algorithm produce a blind area, and it is difficult to restore this part of the information missing in the low frequency. As a result, a scheme of low-frequency compensation for synthetic aperture arises at the historic moment. Through the transfer function analysis of the optical system, the response of the single-aperture optical system to the medium and low frequency is better than that of the synthetic aperture system, and the ability of the synthetic aperture system to obtain high-frequency information is higher than that of the single-aperture system. Therefore, we can use their advantages to complement each other, using a set of single aperture system and a set of synthetic aperture system to sample medium and low frequency and high frequency information respectively. Then, by means of image fusion, the high frequency information and the middle and low frequency information of the two images are extracted respectively, and the full frequency imaging results are obtained, which effectively compensates the medium and low frequency information of the synthetic aperture system.

12.2.5 Middle and Low Frequency Compensation Model

According to the Fourier optical imaging theory, the imaging system should obtain different information of the target object from high frequency to low frequency. On the other hand, the single aperture system has better ability to obtain middle and low frequency information, and synthetic aperture has better ability to obtain high frequency information. Therefore, single aperture and synthetic aperture are used to image the target, and then different frequency information is extracted for image fusion. The compensation principle is shown in Fig. 12.5.

According to Fourier optics, the spot intensity distribution of the object plane has a certain periodicity. According to the Fourier transform, the spot can be regarded as the result of multiple cosine functions with different frequencies and phases. We regard aperiodic functions as functions with infinite periods. That is to say, for one-dimensional (two-dimensional image is similar to one-dimensional discrete image), the imaging process can be expressed [7] as

$$g(x, y) = \sum_{0}^{f_{\text{cut}}} \text{MTF}(f) A(f) \cos(f x), \tag{12.7}$$

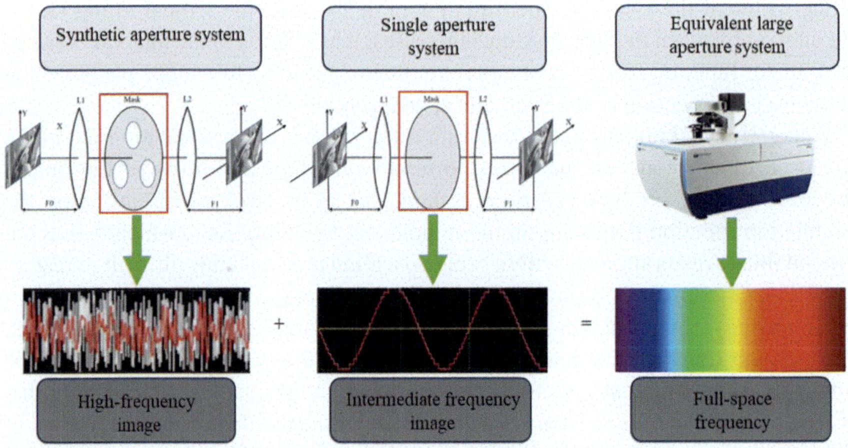

Fig. 12.5 Schematic diagram of low frequency compensation

The optical synthetic aperture with missing intermediate frequency of modulation transfer function is expressed as

$$g_{OSA}(x) = \sum_{0}^{f_{\text{low}}} \text{MTF}(f)A(f)\cos(fx) + \sum_{0}^{f_{\text{high}}} \text{MTF}(f)A(f)\cos(fx), \quad (12.8)$$

where, $A(f)$ is the spectrum of the resulting image, f_{low} is the intermediate frequency in the spatial frequency of the resulting image, f_{high} is the high frequency information of the spatial frequency of the obtained image.

The corresponding single aperture optical system compensating the intermediate frequency is expressed as

$$g_{\text{single}}(x) = \sum_{0}^{f} \text{MTF}(f)A(f)\cos(fx). \quad (12.9)$$

According to Eqs. (12.8) and (12.9), it can be seen that the high frequency information and low frequency information of the corresponding image information are weighted and fused to obtain the whole frequency segment of the image, so as to compensate for the loss of low frequency in the synthetic aperture.

12.2.6 Modulation Transfer Function Analysis of Synthetic Aperture System

The modulation functions of annular six-hole array synthetic aperture system and single aperture structure optical system are analyzed. By taking the square of the pupil function, the modulation transfer function can be obtained [2].

Its transfer function is

$$\sum_{i=1}^{6}\sum_{j=1}^{6} \text{MTF}_d\left(f_x - \frac{x_i - x_j}{\lambda f}, f_y - \frac{y_i - y_j}{\lambda f}\right). \quad (12.10)$$

The filling factor of the annular 6-aperture system is

$$F = \frac{6\pi d^2}{\pi D^2}, \quad (12.11)$$

$$d = \sqrt{\frac{F}{6}}D, \quad (12.12)$$

$$\frac{D - d}{\lambda f} = \left(1 - \sqrt{\frac{F}{6}}\right)\frac{D}{\lambda f} = \left(1 - \sqrt{\frac{F}{6}}\right)f_{Dc}, \quad (12.13)$$

in the above equations, f_{Dc} is the cutoff frequency of the equivalent aperture.

The missing spatial frequency range in the horizontal and vertical directions $[f_{x1}, f_{x2}]$ is

$$f_{x1} = \frac{d}{\lambda f} = \sqrt{\frac{F}{6}} \times \frac{D}{\lambda f} = \sqrt{\frac{F}{6}} f_{Dc}, \tag{12.14}$$

$$f_{x2} = \frac{D - 2d}{\lambda f} = (1 - \frac{1}{3}\sqrt{6F})\frac{D}{\lambda f} = (1 - \frac{1}{3}\sqrt{6F}) f_{Dc}. \tag{12.15}$$

12.2.7 *Transfer Function Analysis*

When the synthetic aperture system and the single aperture system image the same object, and the imaging distance, focal length and aperture of the system are not the same, in order to obtain a complementary spatial frequency between the two systems, the spatial frequency relationship between the two optical systems can be obtained as follows [8]

$$f_0 = \frac{f}{H} f_i, \tag{12.16}$$

in the above equations, f_0 is the object space frequency, f_i is the image spatial frequency, f is the focal length of optical system, H is the object distance of optical system.

The minimum spatial frequency of the missing image in the synthetic aperture system is

$$f_{x2} = (1 - \frac{1}{3}\sqrt{6F})\frac{D_s}{\lambda f_s}, \tag{12.17}$$

among them, D_s is the equivalent aperture size of synthetic aperture system, f_s is the focal length of synthetic aperture system.

The object space frequency of the single aperture system corresponding to the frequency of the missing image space is

$$f_o = (1 - \frac{1}{3}\sqrt{6F})\frac{D_s}{\lambda f_p}\frac{f_s}{H_s} = (1 - \frac{1}{3}\sqrt{6F})\frac{D_s}{\lambda H_s}, \tag{12.18}$$

where, H_s is the object distance of synthetic aperture system, f_p is the focal length of single aperture system.

The image space frequency of the single aperture system corresponding to the frequency of the missing object space is

$$f_i = (1 - \frac{1}{3}\sqrt{6F})\frac{D_s}{\lambda H_p}\frac{H_s}{f_s}, \tag{12.19}$$

where H_p is the focal length of a single aperture.

The cutoff frequency of the single aperture system is

$$f_o = \frac{D_x}{\lambda f_s}, \tag{12.20}$$

where D_x is the aperture size of the single aperture system.

$$\mathrm{MTF}(f_i) = \frac{2}{\pi}[\arccos(\frac{f_i}{f_o}) - (\frac{f_i}{f_o})\sqrt{1 - x^2}]. \tag{12.21}$$

When the transfer function corresponding to the above equation is 0.3, there is

$$(1 - \frac{1}{3}\sqrt{6F})\frac{D_s}{H_p}\frac{H_s}{D_x} = 0.5851. \tag{12.22}$$

The above equation gives the general relationship of the aperture size of the two systems when the spatial frequencies of the two systems are complementary [2].

12.2.8 Middle and Low Frequency Compensation

According to the above analysis, the medium and low frequency information of the large agricultural economy and the high frequency information of the synthetic aperture system are extracted, and the full spatial frequency image can be obtained by simple fusion processing. Weighted fusion is proposed, and the fusion process is expressed as follows [2]

$$F(x, y) = \omega_1 A(x, y) + \omega_2 B(x, y), \tag{12.23}$$

where ω_1, ω_2 is the weighting coefficient, $\omega_1 + \omega_2 = 1$.

The main purpose of this method is to compensate the low frequency loss in the synthetic aperture system by image weighted fusion. For the original image, some spatial frequency segments may not exist. Through the general average weighted image fusion, it is difficult to maximize the amount of information of the image, so it is necessary to determine the weight through the single aperture system and the synthetic aperture system as

$$m = \int_{f_{x2}}^{f_{x1}} A(f)\mathrm{d}f, \tag{12.24}$$

$$n = \int_{f_{x2}}^{f_{x1}} B(f)\mathrm{d}f, \tag{12.25}$$

$$\omega_1 = \frac{a}{a+b}, \ \omega_2 = \frac{b}{a+b}, \tag{12.26}$$

where, m is the sum of high frequency information intensity of imaging results of synthetic aperture system, n is the sum of the medium and low frequency intensity of the imaging result of the single aperture imaging system compensating the synthetic aperture system, A is the spectral information of synthetic aperture imaging results, B is the spectral information of imaging results of single aperture imaging system.

12.3 Shortcomings and Countermeasures

Because the passing area of the synthetic aperture is smaller than that of the actual single aperture, the information of some frequencies of the image cannot be obtained accurately. When a synthetic aperture system is imaging a remote target, the light emitted or reflected by the target passes through the aperture to form multiple sub-beams. Due to the difference in the position of the sub-aperture, there is a certain phase difference among the sub-beams, which cannot meet the condition of coherent superposition. The direction and phase of each beam must be adjusted differently to make it coherent on the same focal plane and then the target image can be obtained. The point spread function of the system will be expanded to a certain extent, and the response of the system to the middle and low spatial frequency will be reduced, so that the imaging results will be more blurred compared with its equivalent aperture imaging results this phenomenon is called image degradation.

In order to obtain better imaging results, we can first use the direct method of optimizing the synthetic aperture system as far as possible to improve the imaging quality of the system; another method is to repair and improve the image information by using some means in the field of image processing to improve the image quality. Image restoration is designed to repair damaged or contaminated images. In the actual imaging process, there are many factors that affect the imaging quality to some extent. These factors are roughly divided into internal and external factors. Internal causes: the unavoidable aberration of the optical system itself, the installation and adjustment error between sub-lenses, etc.; external factors: atmospheric turbulence in the air, noise of CCD devices, as well as the influence of machine vibration, the relative motion between the measured target and the lens and other factors will cause the imaging results to be blurred. There are many methods for image quality restoration, such as Wiener filter restoration, constrained least square restoration and other common image restoration methods [2]. It is not too much to introduce here because it is related to the topic.

12.4 Two-Dimensional Synthetic Aperture Imaging Algorithm

The traditional synthetic aperture radar uses large wideband signals to achieve range resolution and one-dimensional synthetic aperture to achieve azimuth resolution to obtain two-dimensional images. In some near imaging situations, it is difficult to meet the requirements depending on the signal bandwidth because of high imaging resolution. Therefore, a two-dimensional synthetic aperture imaging system is proposed, which uses two-dimensional synthetic aperture to achieve two-bit resolution, the resolution depends on the synthetic aperture length, is not restricted by the signal bandwidth, and can achieve ultra-high resolution. The signal uses a simple single-frequency signal, which can greatly simplify the system. Two-dimensional synthetic aperture of single-frequency signal is a high-resolution imaging technique.

12.4.1 Two-Dimensional Planar Synthetic Aperture Imaging System

The model of the relationship between signal and geometry is established. As shown in Fig. 12.6, the position coordinate of the target is (x, y, z), the synthetic aperture plane is perpendicular to the Y axis, the distance from the origin to the synthetic aperture plane is R_0, and the position of the antenna in the aperture plane is (x_a, R_0, z_a). The distance between the antenna and the target is [9]

$$R(x_a, R_0, z_a) = \sqrt{(x - x_a)^2 + (y - R_0)^2 + (z - z_a)^2}. \tag{12.27}$$

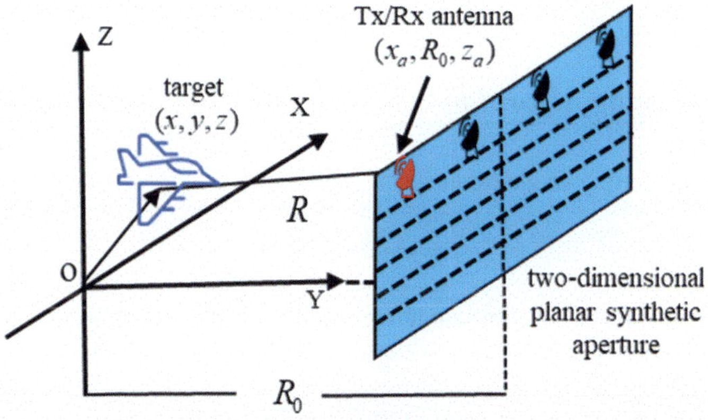

Fig. 12.6 Schematic diagram of two-dimensional planar synthetic aperture imaging system

12.4.2 Aperture Implementation

Two-dimensional plane synthetic aperture is needed to realize two-dimensional imaging of single-frequency signal. There are generally three ways to realize two-dimensional planar synthetic aperture [9]:

(1) single antenna, two-dimensional scanning. Advantages: the system is simple. Disadvantages: long scanning time, it is difficult to meet the requirements of imaging time in practical applications.
(2) one-dimensional antenna array, one-dimensional scanning. The complexity and cost of the system are moderate, and the scanning time is short.
(3) two-dimensional planar array. Advantages: no mechanical scanning is required. Disadvantages: the system is complex, the cost is high, and the consistency of antenna elements is difficult to be guaranteed.

Comprehensive consideration, one-dimensional linear array, one-dimensional scanning is a compromise scheme, which is relatively easy to implement, and this scheme can be given priority in the actual system.

12.4.3 Selection of Frequency

Through the analysis of the above signals and transverse parameters, it is known that there is a contradiction between penetration and resolution in the in-wall target detection system. Penetration requires lower frequency, longer signal wavelength at lower frequency, and larger accumulation angle in the synthetic aperture plane. The wavelength of the higher frequency signal is shorter, and in the case of the same resolution, the accumulation angle can be very small, but the penetration of the higher frequency is poor [9].

12.4.4 Imaging Algorithm

Consider transmitting single frequency signal [9]

$$s_t = \exp(-\mathrm{i}2\pi f_c t). \tag{12.28}$$

Omitting the amplitude constant term, with the target echo at the distance R, the function s_r is expressed as

$$s_r = s(x, y, z) \exp[-\mathrm{i}2\pi f_c (t - 2R(x_a, R_0, z_a)/c)], \tag{12.29}$$

where $R(x_0, R_0, z_a)$ is shown in the Eq. (12.27).

The mixing reference signal is

$$s_{r0} = \exp[-i2\pi f_c(t - 2R_0/c)]. \tag{12.30}$$

By mixing the transmitted signal with the received signal, we can get:

$$d(x_a, R_0, z_a) = s(x, y, z)\exp(ik_r R_0)\exp(-ik_r R), \tag{12.31}$$

where $s(x, y, z)$ is the scattering coefficient at (x, y, z) and $k_r = 4\pi f_c/c$ is constant.

The method of coherent accumulation can be used to realize the reconstruction of scattering points [10] as

$$\begin{aligned} s(x, y, z) &= \iint \mathrm{d}x_a \mathrm{d}R_0 \mathrm{d}z_a \exp(-ik_r R_0)\exp(ik_r R_0)\exp(ik_r R) \\ &= \iiint_A \mathrm{d}x_a \mathrm{d}R_0 \mathrm{d}z_a \exp(-ik_r R_0)\exp[ik_r\sqrt{(x - x_a)^2 + (y - R_0)^2 + (z - z_a)^2}] \end{aligned} \tag{12.32}$$

Rewrite the Eq. (12.34) to get

$$\begin{aligned} s(x, y, z) =& \exp(-ik_r R_0)\iint_A \mathrm{d}y_a \\ & \exp[ik_r\sqrt{(x - x_a)^2 + (y - R_0)^2 + (z - z_a)^2}]\mathrm{d}x_a \mathrm{d}z_a \end{aligned} \tag{12.33}$$

The following two-dimensional integral in Eq. (12.35) is about the two-dimensional convolution of plane aperture coordinates, that is, the two-dimensional convolution of $d(x_a, y_a, z_a)$ and $\exp\{ik_r[(x - x_a)^2 + (y - R_0)^2 + (z - z_a)^2]^{1/2}\}$ with respect to x_a, z_a. The two-dimensional convolution can be calculated in the frequency domain, which is similar to the frequency domain calculation of one-dimensional convolution. First, the two-dimensional Fourier transform of $d(x_a, y_a, z_a)$ about $x_a z_a$ and $\exp\{ik_r[(x - x_a)^2 + (y - R_0)^2 + (z - z_a)^2]^{1/2}\}$ about $x_a z_a$ are calculated respectively, then the two-dimensional frequency spectrum is multiplied, and the result of two-dimensional inverse Fourier transform is obtained [11, 12].

So, the key problem is computing

$$\begin{aligned} E(k_x, k_z) = \iint & \exp[ik_r\sqrt{(x - x_a)^2 + (y - R_0)^2 + (z - z_a)^2}] \\ & \exp(-ik_x x - ik_z z)\mathrm{d}x\mathrm{d}z \end{aligned} \tag{12.34}$$

From the principle of stationary point, it can be obtained as

$$E(k_x, k_z) = \frac{2\pi k_r}{ik_y^2}\exp[-ik_y(y - R_0)] \tag{12.35}$$

$$k_y = \sqrt{k_r^2 - k_x^2 - k_z^2} \tag{12.36}$$

So,

$$s(x, y, z) = \exp(-\mathrm{i}k_r R_0) \iint_{k_x k_z} D(k_x, R_0, k_z) \frac{2\pi k_r}{\mathrm{i}k_y^2} \exp[-\mathrm{i}k_y(y - R_0)] \times \exp(\mathrm{i}k_x x + \mathrm{i}k_z z)\mathrm{d}k_x \mathrm{d}k_z \tag{12.37}$$

Phase compensation for reference point out, that is, matched filtering is

$$H_{mf} = \exp[-\mathrm{i}k_r R_0 + \mathrm{i}k_y(y - R_0)]. \tag{12.38}$$

It can be seen that the matched filter must know the distance y of the target, that is, the matched filter focus can only focus on a fixed distance y and defocus for other distances. In practice, different y can be selected to focus on targets with different distances. This characteristic determines that the two-dimensional synthetic aperture of single-frequency signal can have a certain range resolution.

After matched filtering, the signal becomes

$$s(x, y, z) = \iint_{k_x, k_z} D(k_x, R_0, k_z) \frac{2\pi k_r}{\mathrm{i}k_y^2} \exp(\mathrm{i}k_x x + \mathrm{i}k_z z)\mathrm{d}k_x \mathrm{d}k_z. \tag{12.39}$$

Because of $1/k_y^2 \approx 1/k_r$, its effect can be ignored in the imaging process. Therefore, the Eq. (12.41) is a two-dimensional Ifft process, so the focused image can be obtained by two-dimensional IFFT.

The above two-dimensional synthetic aperture imaging algorithms only need to use one-dimensional interpolation and FFT or IFFT operation, and the computational complexity of the algorithm is small. Through interpolation processing, the accurate matching of the original signal can be realized in the frequency domain, so the focusing accuracy of the algorithm is high [11].

12.5 Synthetic Aperture Imaging Algorithm

12.5.1 Overview

Light field imaging is a process of collecting light field information in space and processing the collected data. The information includes not only the intensity information of the light but also the direction information of the light. Because the traditional single imaging device loses the light direction information when recording the light intensity information, when the target is obscured, it is difficult for the single angle of view to effectively image the occluded object. Therefore, occlusion is a challenging task for many computer vision applications. As an important branch in the field of light field imaging, synthetic aperture imaging can overcome the defect of single-view imaging and image the occluded object. Synthetic aperture imaging using camera array to capture multi-view images can obtain light intensity information and direction information in space at the same time. According to the synthetic aperture imaging algorithm, the multi-view image is synthesized into a virtual large aperture image. Because of its extremely shallow depth of field, when focusing to a certain plane, the non-focusing plane will be seriously virtual. Synthetic aperture imaging takes advantage of the sharp characteristic that only the object on the focal plane is sharp by averaging the images from different perspectives, the image composed of fuzzy objects far away from the focal plane and sharp objects on the focal plane can be obtained. Therefore, synthetic aperture imaging can realize the "penetration" imaging of the occluded object [12].

However, in practical application, when the depth difference between the occluded object and the occluded object is small, due to the serious occlusion in the scanning direction of the camera array and the limitation of imaging space, the linear array detection mode is easy to appear that the measured object is beyond the imaging field of vision the serious lack of sampling information of the occluded object affects the reconstruction of the target. According to the principle of synthetic aperture imaging, the reconstruction of the occluded object is averaged by the multi-view image the more perfect the effective data contained in the multi-view image, the better the reconstruction effect. Therefore, a nonlinear array can be formed to detect the occluded object by changing the camera arrangement position and the direction of the camera optical axis, so as to make up for the detected occluded information. Although the nonlinear array can improve the acquisition of the information of the occluded object, in the process of synthetic aperture imaging, the pixels of the foreground occlusion participate in the image averaging, resulting in blurring caused by occlusion and defocusing in the reconstruction results of the occluded object.

As a result, an algorithm for occluded object reconstruction based on nonlinear array arises at the historic moment, which aims to enhance the ability of camera array to capture the target scene information and ensure the integrity of the occluded object information. Furthermore, through the rearrangement of the nonlinear multi-view image, the changed image sequence can meet the requirements of synthetic aperture imaging and realize the reconstruction of the target. At the same time, in order to solve the problem of reconstructed image blur caused by the occlusion participating in synthetic aperture imaging, the un-occlusion imaging method is studied. Combined with the synthetic aperture imaging mechanism, the occlusion and the occluded objects in the scene are identified by the defocus function, and then the multi-view image without occlusion is obtained, and the high quality reconstruction results are obtained.

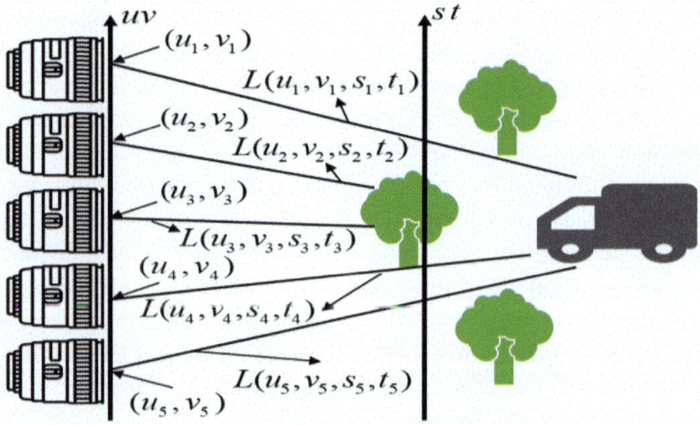

Fig. 12.7 The camera array collects the light field information of the scene

12.5.2 Basis of Synthetic Aperture Imaging Technology

Synthetic aperture imaging technology uses a camera array to simulate an equivalent large aperture camera to obtain a focus.

A synthetic aperture image at one depth. Because of the extremely shallow depth of field of the large aperture camera, the foreground occlusion can be effectively obscured when focusing on the occluded object, and the perspective imaging of the occluded object can be realized.

When the object is blocked by an occluder, the occluder information obtained from only one direction is difficult to reconstruct the object. However, it is possible to obtain target information from multiple angles by building camera arrays. As shown in Fig. 12.7, when the car is blocked by the tree, the target information cannot be obtained completely from the perspective of a single camera and cannot be reconstructed. Through the camera array, the target information can be obtained from multiple perspectives, so as to make up for the shortage of information obtained by a single camera [12].

12.5.3 Perspective Imaging Principle of Large Aperture Camera

The resolution of all optical imaging devices is limited, so it is difficult to form the same point as the object surface in the image plane, and an allowable value should be specified according to the characteristics of the optical imaging device. When a pupil is given, the depth of the imaging space is also determined, and the image is specified to be a clear image within this depth range. Generally speaking, the depth range of object space in which a clear image can be obtained on the imaging surface is called depth of field [1].

The imaging within a certain distance before and after the scene is within the prescribed speckle diameter, and at this time, the depth of the image point mapped to the object space within the focal depth is the depth of field. The depth of field is the sum of the spatial range of objects that are allowed to be clearly imaged in front of the focal plane. The depth of field is closely related to the size of the incident pupil. The larger the incident pupil is, the smaller the depth of field is. On the contrary, the smaller the incident pupil is, the greater the depth of field is. For the same point, when the incident pupil is larger, the focal depth is smaller, and the corresponding depth of field is also small. Adding an aperture diaphragm in front of the incident pupil to limit the incident light is equivalent to reducing the size of the incident pupil. Within the specified size of the dispersion spot, the imaging increases, the focal depth increases, and the corresponding depth of field increases. According to this relationship, when the scene is photographed by a large aperture camera, the depth of field is small and can be clearly imaged only in the plane near the focal plane. With the increase of the aperture, the range of depth of field allowed by the camera will become smaller and smaller, until each scene can only focus on one depth plane, except for the scene on that depth plane, the other scene cannot be clearly imaged. According to this principle, when the target in the scene is partially obscured, the large aperture camera can disperse the defocused object, thus highlighting the occluded object in the formed image and completing the imaging of the occluded object.

However, in reality, the super-large aperture camera does not exist, even if it can make expensive super-large lens, the vertical aberration and geometric aberration of the lens are difficult to be corrected [1]. In this case, synthetic aperture imaging arises at the historic moment. The scene information is captured by a camera array composed of multiple single cameras, and the synthetic aperture imaging algorithm is used to synthesize the multi-view image to simulate the image captured by the large aperture camera. In the synthetic aperture image, the occlusion is virtual, the target is highlighted, and the occlusion is reconstructed.

12.5.4 Principle of Synthetic Aperture Algorithm

For a linear array of N cameras, the camera optical axis is perpendicular to the baseline [12], the distance between the two adjacent cameras is B, and any two adjacent cameras are at the same distance, as shown in Fig. 12.8.

In Fig. 12.8, the distance between an object point M and the linear array is z, and the order of the camera from left to right is 1, 2... $N-1$, N. n and R_e represent the number of digits sorted by the camera, in which $n = 1, 2 \ldots N-1, N$, R_e is the reference camera, and the general reference camera is the intermediate viewing angle camera. Assuming that the internal parameters of all cameras are the same, the focal length is f. For camera n and reference camera R_e, the line between the light center and M intersects with the image plane, which is the projection point of

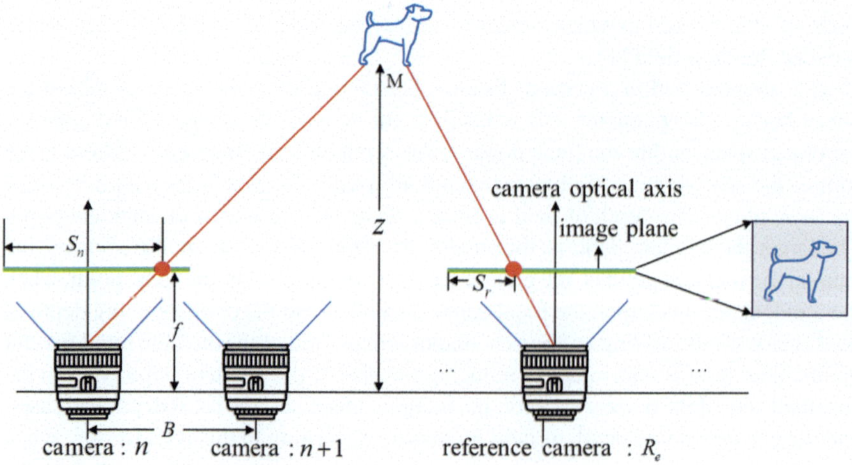

Fig. 12.8 Principle of linear array synthetic aperture imaging

M on the image plane, and the distance from the left end of the image is S_n and S_r, respectively. The light center and the object point M form a triangle, the projection point in the image plane and M can also form a triangle, and the two triangles are similar triangles. According to the similar triangle relation, the Eq. (12.42) can be obtained as

$$\frac{(R_e - n) \cdot B}{z} = \frac{(R_e - n) \cdot B + S_r - S_n}{z - f}, \tag{12.40}$$

where, d_n is defined as the parallax of the M point in the image of camera n and reference camera R_e as

$$d_n = S_n - S_r, \tag{12.41}$$

According to Eqs. (12.42) and (12.43), we can obtain

$$d_n = \frac{(R_e - n) \times B \times f}{z}. \tag{12.42}$$

According to the Eq. (12.44), the parallax of a point in space in the image of the two cameras and the depth of the point.

The greater the depth value, the smaller the parallax; on the contrary, the smaller the depth value, the larger the parallax, the two are inversely proportional. In synthetic aperture imaging, in order to highlight the target in the synthetic image, it is necessary to eliminate the parallax of all the objects in the image. Assuming that the target object is M, the image of camera n after parallax elimination is

$$I_{n,z}(s') = I_n(s - d), \tag{12.43}$$

where s represents the position of all the object points in the original image, s' represents the new position of the transformed object point, I_n is the image of the camera, I_{nz} is the image under the camera n focus to the depth z, and the parallax d_n is replaced by the depth as

$$I_{n,z}(s') = I_n(s - \Delta B_n \times \tau). \tag{12.44}$$

Equation (12.44) is the equation for focusing the image n to the depth z, which can be focused on objects of different depths by adjusting the size of the threshold value. It should be noted that the focus here is not the focus of the camera on a scene, but the translation operation on the multi-view image to eliminate the parallax of the target in order to ensure that the synthetic aperture image is focused on a target. The same method carries out the same operation for cameras with other viewing angles, and the images focused to M point at all viewing angles can be obtained. Finally, the multi-view image after focus is summed and averaged, and the synthetic aperture imaging equation with focus to M point or depth to z plane is obtained as

$$S_z = \frac{1}{N}\sum_{n=1}^{N} I_{n,z}. \tag{12.45}$$

For a two-dimensional surface array, simply change the Eq. (12.46) into

$$I_{n,z}(s', t') = I_n(s - \Delta B_{n,x} \times \tau, t - \Delta B_{n,y} \times \tau). \tag{12.46}$$

The synthetic aperture imaging results of two-dimensional planar arrays can also be obtained by using the Eq. (12.48). The synthetic aperture imaging algorithm synthesizes the multi-view image obtained by the camera array into a synthetic aperture image captured by a virtual large aperture camera. In the traditional imaging equipment, the light intensity of a certain point on the detector comes from the light information in all directions received by the pupil. In the camera array, the pixel value of a certain point in the synthetic aperture image comes from the superposition of the pixel value of a certain point in the multi-view image. Due to the different camera arrangement position, the image under the corresponding visual angle has parallax, and the position of the same object mapping in different camera detectors is also different. Reflected in the coordinate space, the same point has different positions on the s-axis of different cameras, as shown in Fig. 12.9.

In Fig. 12.9, shadow and gray are located under depth Z_1 and depth Z_2, respectively, and the parallax is reflected in the image because of their different depths. When the synthetic aperture imaging is focused to a certain depth, objects at other depths will not be able to focus.

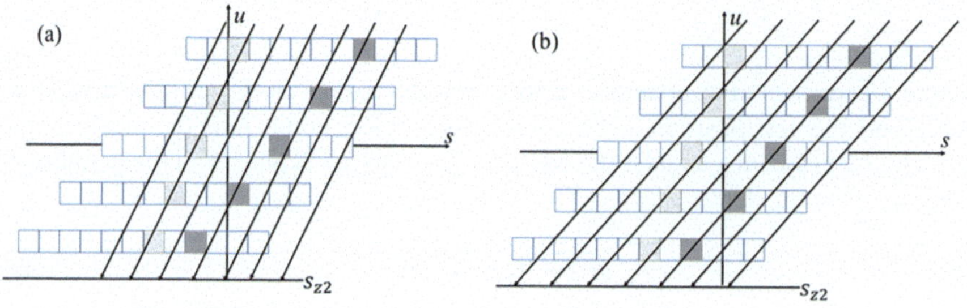

Fig. 12.9 Imaging coordinate characterization of synthetic aperture at different depths: **a** synthetic aperture imaging in focus to depth Z_1 (shadow); **b** synthetic aperture imaging in focus to depth Z_2 (gray)

12.5.5 Synthetic Aperture Imaging Algorithm

The prerequisite for the reconstruction of the target is that the target information is obtained by the camera array when the distance between the target and the occlusion in the optical axis direction is small or the space of the detection array is limited, the detection effect of the nonlinear array is better than that of the linear array. Although nonlinear array is widely used in 3D reconstruction, panoramic view imaging and other fields, it is rarely used in the field of synthetic aperture imaging. In order to solve the problem that the current linear array light field information processing and reconstruction algorithm cannot directly adapt to the nonlinear array light field information processing, this chapter proposes a synthetic aperture imaging algorithm based on nonlinear array. Arc array is a typical nonlinear array.

In synthetic aperture imaging, the baseline length of the dense camera array can reflect the size of the synthetic aperture. The larger the effective aperture diameter is, the more complete the target information detected by the camera array is. For a linear array [12] with a baseline length of B_1, an arc array with a chord length of B_1 can be placed at the same space size, as shown in Fig. 12.10.

In Fig. 12.10, the baseline length of the arc array is B_2. Assuming that the center angle of the arc array is θ and the radius is r there is a relation as follows

$$B_1 = 2r \sin\frac{\theta}{2}, \tag{12.47}$$

$$B_2 = r\theta. \tag{12.48}$$

It can be obtained according to the above two equations

$$B_2 = B_1 \frac{\theta}{2\sin\frac{\theta}{2}}. \tag{12.49}$$

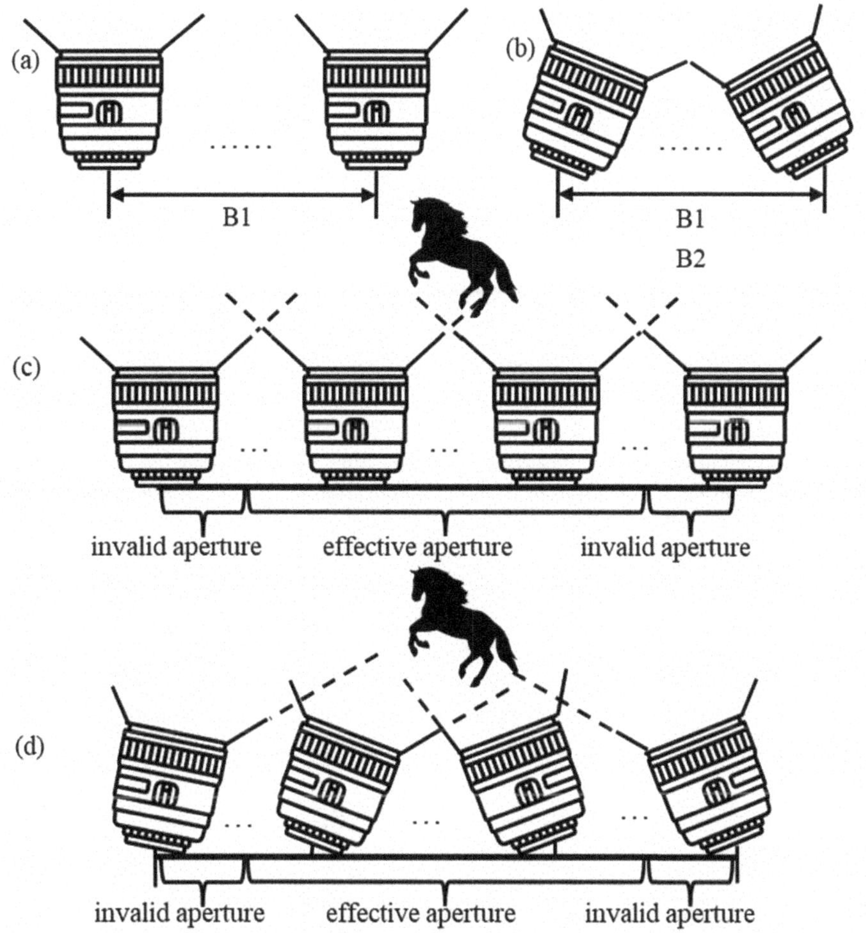

Fig. 12.10 Camera array in the same space: **a** linear array; **b** arc array; **c** a linear array of large baselines; **d** rotate the array of camera lenses

Because $\theta/2 \geq \sin\theta/2$, $B_2 \geq B_1$, for curved arrays, the equal sign holds when $\theta = 0$.

Therefore, for the two kinds of arrays in the same space, the equivalent aperture of the arc array is larger. If the equivalent aperture is enlarged by increasing the baseline length of the linear array, the target will exceed part of the camera imaging field of vision, resulting in a large number of invalid aperture. One of the ways to solve the invalid aperture is to rotate the camera lens and change the detection direction to change the invalid aperture into an effective aperture. Although this method can increase the effective aperture, it will occupy much more space than the arc array in the same detection range. In addition, due to the great difference in the distance between the camera and the target in different positions, it is easy to cause the target beyond the camera depth of field, so it is necessary to adjust the aperture of each camera according to the actual situation, which brings great inconvenience. Therefore, both the equivalent aperture size and the space occupied by the arc array are better than the linear array [12].

When the distance between targets in the direction of the optical axis is small or the arrangement space of the detection array is limited, it is difficult for the line array to reconstruct the occluded object. Taking advantage of the large equivalent aperture of the nonlinear array, the target information can be collected from more locations around the occlusion.

Because the optical axis of the camera on the nonlinear array is not parallel and the image is not on the same plane, the synthetic aperture imaging algorithm using linear array will not be able to reconstruct the target. Therefore, the first step of the algorithm is to rearrange the multi-view image. As shown in Fig. 12.11, keep the reference camera still and set the straight line perpendicular to the optical axis of the reference camera as the baseline. By calibrating the camera array, the position relationship between cameras and the internal parameters of the camera are obtained [13–15] before the image rearrangement, the external parameters of the camera are assumed to be known conditions. According to computer vision, the one-to-one mapping between two camera images can be described by homography matrix H [16]. According to this transformation relationship, the original image is rearranged so that it has the same baseline as the reference camera, and its optical axis is parallel to the optical axis of the reference camera.

It is assumed that the camera array needs a point with z distance from the reference camera and M coordinates for reconstruction, and the coordinate of M point mapping

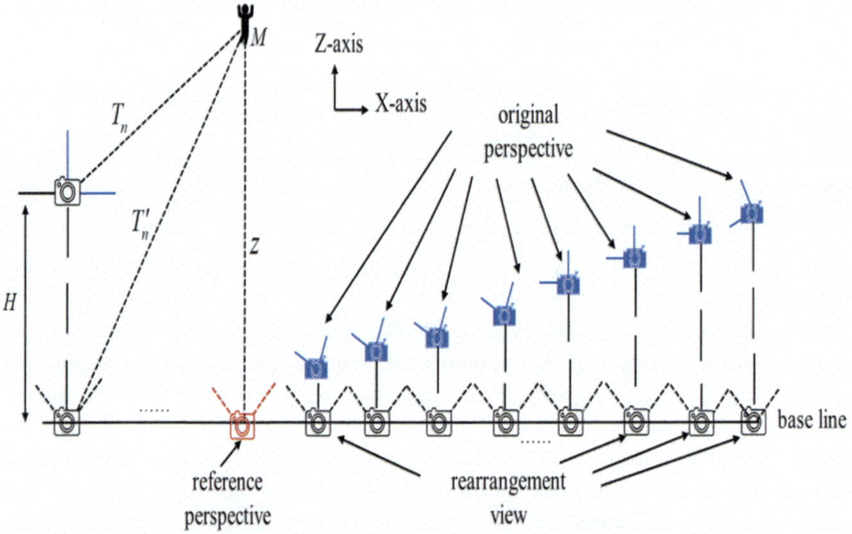

Fig. 12.11 Schematic diagram of nonlinear array image rearrangement

in image n is m_n, then the two are connected by projection matrix P_n.

$$m_n = P_n M = K[R_n|T_n]M. \tag{12.50}$$

In the Eq. (12.50), K is the internal reference matrix of the camera. In the system built, the cameras on the camera array are all the same, so the K value can be assumed to be the same for all cameras. R_n is the rotation matrix from the camera n coordinate system to the world coordinate system, and T_n is the translation matrix from the camera n coordinate system to the world coordinate system. Equation (12.50) connects the world coordinate system, the camera coordinate system and the pixel coordinate system.

Through calibration, the position relationship between the reference camera and other cameras can be obtained. Let $R_{n,r}$ be the rotation matrix from the camera n coordinate system to the reference camera coordinate system. $T_{n,r}$ is the translation vector from the camera n coordinate system to the reference camera coordinate system. $T_{n,r}$ can be expressed as

$$T_{n,r} = [T_{n,x}, T_{n,y}, T_{n,z}]^T, \tag{12.51}$$

where T_{nx}, T_{ny}, T_{nz} represents the amount by which the camera n coordinate system translates to the reference camera coordinate system in three directions under the camera n coordinate system. For the hypothetical reconstruction point M, the translation vector from the camera n coordinate system to the world coordinate system is

$$T_n = R_{n,r}^{-1} T_r + T_{n,r}, \tag{12.52}$$

where T_r is the translation vector that the reference camera translates to M under the reference camera coordinates, where the size is equal to z. In the reference camera coordinate system, the optical center position of the camera n rearranged is only different from that of the reference camera on the x axis, and the other positions are zero. Therefore, the translation vector of the camera n rearranged coordinate system to the world coordinate system is

$$T_n' = [T_{n,x}', 0, z]^T, \tag{12.53}$$

where $T_{n,x}'$ is the position of the optical center of the camera after n rearrangement and the position of the optical center of the reference camera, which can be obtained by

$$T_{n,x}' = [1, 0, 0] R_{n,r} T_{n,r}. \tag{12.54}$$

It can be seen from the Eq. (12.54) that $T_{n,x}'$ is the distance between the camera n and the reference camera on x axis in the reference camera coordinate system,

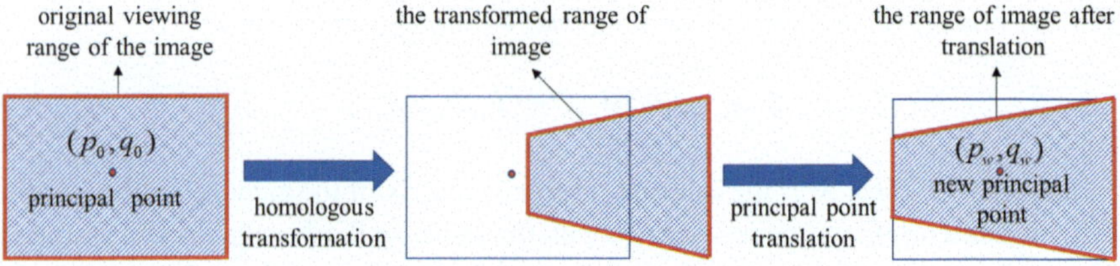

Fig. 12.12 Homologous transformation and principal point translation

which is the same as the distance between the camera and the reference camera after rearrangement.

As can be seen from Fig. 12.12, the direction of the coordinate axis of the rearranged camera is the same as that of the reference camera, but the difference is its translation vector to M. According to the Eqs. (12.52) and (12.53), the projection matrix P_n of camera n and the projection matrix P_n' after camera rearrangement can be obtained.

$$P_n = K[R_{n,r}|T_n], \tag{12.55}$$

$$P_n' = K[E|T_n'], \tag{12.56}$$

where E represents the unit matrix of 3 to 3. Assuming that H_n is the matrix of camera n and its rearranged camera, you can contact the point on the image of camera n and the point m_n' on the rearranged image, as

$$m_n' = H_n m_n \tag{12.57}$$

According to the above equations, we can get

$$H_n = P_n' P_n^+ \tag{12.58}$$

where P_n^+ is the pseudo inverse matrix of P_n.

When the multi-view image is rearranged to the same plane, a lot of effective information is lost after homography transformation, so the utilization value of the transformed image is not high. To solve this problem, we transform the principal point (PP) [17] of the image to ensure that the multi-view image retains as much effective scene information as possible. The image information after homography transformation is not really lost, but because the position of a lot of image information after image transformation has exceeded the current range of view, so it is necessary to readjust the observation range of the image adjust the invisible image information back to the observable range of the image.

In Fig. 12.12, the shadow part represents the image information, the black box is the visible range of the image, and the center point of the black box is the main point of the image. Because the main point of the original image is in the center of the image, which coincides with the central position of the visible range of the image, all the scene information can be seen in the original image, but some of the image information is not in the visible range after homography transformation.

Suppose that the four vertex positions of the image are stored in the set (p_w, q_w) after the homography transformation. For an image after homography transformation, the new principal point coordinate position of the image can be obtained by

$$\begin{cases} p_0 = \dfrac{\max(p_w) + \min(p_w)}{2}, \\ q_0 = \dfrac{\max(q_w) + \min(q_w)}{2}. \end{cases} \quad (12.59)$$

In the Eq. (12.59), (p_w, q_w) represents the new principal point coordinates of the image.

In Fig. 12.12, the PP of the image is translated again to coincide with the center of the image, and more effective information of the scene is back out of the current visual range, so that the multi-view image can be used for synthetic aperture imaging as

$$I_{n,z}(s') = I_n(s - \Delta s - \Delta B_n \times \tau). \quad (12.60)$$

For a rearranged image in which the PPs are translated in both directions, the Eq. (12.60) can be extended to

$$I_{n,z}(s') = I_n(s - \Delta s - \Delta B_{n,x} \times \tau, t - \Delta t - s - \Delta s - \Delta B_{n,y} \times \tau). \quad (12.61)$$

The flow chart of synthetic aperture imaging algorithm based on nonlinear array [12] is shown in Fig. 12.13. For the original image, it is necessary to use nonlinear camera array to shoot from different angles to obtain pictures of different angles, according to the relative position of the image under different angles, the image is reshot, and then the main point translation is carried out to shift the image into the observable area, and finally the focus is carried out to obtain the image of synthetic aperture imaging.

12.6 General Conclusions

Synthetic aperture imaging can solve the problem that the resolution of the optical system is limited by the limited aperture. The system based on synthetic aperture imaging can reduce the processing difficulty of optical devices, and has the advantages of small size and light weight. While ensuring the portability of the system, it can greatly improve the resolution of the system and obtain high-quality and clear images.

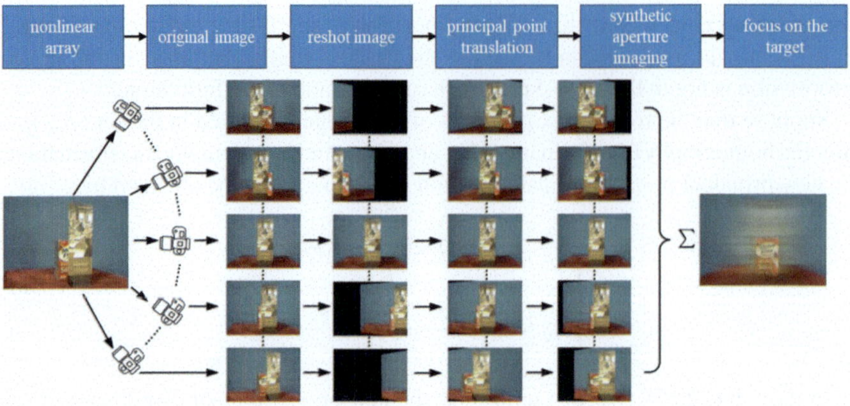

Fig. 12.13 Flow chart of synthetic aperture imaging algorithm based on nonlinear array

However, the resolution of one-dimensional synthetic aperture system is limited to some extent due to the limitation of signal bandwidth in close range. Therefore, two-dimensional synthetic aperture technology comes into being, its resolution depends on the synthetic aperture length, is not limited by the signal bandwidth, and can achieve ultra-high resolution.

At the same time, when the single-aperture system is imaging, it is usually limited by obstacles, and it is difficult to image the object behind the obstacle. However, the system based on synthetic aperture can overcome this defect, image the object being blocked, and obtain the light intensity information and direction information in the space, so as to obtain high-quality image information.

Synthetic aperture imaging also has its disadvantages. Because the passing area of synthetic aperture is smaller than that of actual single aperture, some frequency information in the image cannot be accurately obtained. The point expansion function of the system will be expanded to a certain extent, and the response of the system to low and medium spatial frequencies will be reduced, making the imaging results fuzzier than those of the equivalent aperture imaging.

All in all, although some shortcomings exist, synthetic aperture imaging can still solve the problems that single-aperture imaging cannot solve, and has great technical advantages.

Acknowledgements This work was supported by the National Natural Science Foundation of China (Nos. 11874132, 61975044, 12074094, 12374271, and 62375065); Interdisciplinary Research Foundation of HIT (No. IR2021237).

References

1. Lizuka K. Engineering optics. Springer Cham, Shanghai. 2019;125–132:146–57.
2. Tang J, Wu J, Wang K, Ren Z, Wu X, Hu L, Di J, Liu G, Zhao J. Restore Net-Plus: image restoration via deep learning in optical synthetic aperture imaging system. Opt Lasers Eng. 2021;146(8): 106707.
3. Nguyen L, Le C. 3D imaging for millimeter-wave forward-looking synthetic aperture radar (SAR). In: Passive and active millimeter-wave imaging conference;2020
4. Zhi X, Jiang S, Zhang L, Gong J. Multi-frame image restoration method for novel rotating synthetic aperture imaging system. Results Phys. 2021;23: 103991.
5. Chen J, Lqbal M, Yang W, Wang P, Sun B. Mitigation of azimuth ambiguities in spaceborne stripmap SAR images using selective restoration. IEEE T Geosci Remote. 2014;52(7):4038–45.
6. Yamaki R, Hirose A. Singular unit restoration in interferograms based on complex-valued Markov random field model for phase unwrapping. IEEE GRSL. 2009;6(1):18–22.
7. Chang J, Song T. A new synthetic aperture focusing method to suppress the diffraction of ultrasound. T-UFFC. 2011;58(2):327–37.
8. Viswannath S, Ghulyani M, Beco S, Dahan M, Arigovindan M. Image restoration by combined order regularization with optimal spatial adaptation. IEEE Trans Image Process. 2020;29:6315–29.
9. Chen H, Deng Y, Wang R, Tian W, Zeng T. Two-dimensional deformation measurement based on multiple aperture interferometry in GB-SAR. IEEE Geosci Remote Sens Lett. 2016;14(2):208–12.
10. Ulander L, Hellsten H, Stenstorm G. Synthetic-aperture radar processing using fast factorized back-projection. IEEE Trans Aerosp Electron Syst. 2003;39(3):760–76.
11. Sheen D, McMakin D, Hall T. Three-dimensional millimmeter-wave imaging for concealed weapon detection. IEEE Trans Microw Theory Tech. 2001;49(9):1581–92.
12. Lopez J, Fortuny J. 3-D radar imaging using range migration techniques. IEEE Trans Antennas Propag. 2000;48(5):728–37.
13. Zhao W, Wen C, Yuan Q, Zheng R. Efficient rotating synthetic aperture radar imaging via robust sparse array synthesis. IEEE Trans Geosci Remote Sens. 2023;61:1–12.
14. Zhang Z. A flexible new technique for camera calibration. IEEE Trans Pattern Anal Mach Intell. 2000;22(11):1330–4.
15. Zhang Z. Camera calibration with one-dimensional object. IEEE Trans Pattern Anal Mach Intell. 2004;26(7):892–9.
16. Prince S. Computer vison: models, learning, and inference. London: Cambridge University Press; 2012. p. 236–57.
17. Bradski G, Kaehler A. Learning OpencCV. Beijing: O' Reilly Media Inc.; 2008. p. 324–8.

Chapter 13
Mask-Modulated Lensless Imaging

Zhengyu Wu and Zhengjun Liu

Abstract The progression of human society has always been intertwined with scientific exploration. Visual information, as the most intuitive and explicit form of data, has become one of humanity's primary ways of acquiring information. Nevertheless, human visual capabilities have always been limited. To overcome this, people have combined specific optical principles and developed a range of optical instruments to explore the unknown world.

13.1 Introduction

The progression of human society has always been intertwined with scientific exploration. Visual information, as the most intuitive and explicit form of data, has become one of humanity's primary ways of acquiring information. Nevertheless, human visual capabilities have always been limited. To overcome this, people have combined specific optical principles and developed a range of optical instruments to explore the unknown world. From Galileo's use of telescopes to observe the vast cosmos to Leeuwenhoek's observations of the microscopic world using microscopes, humanity has managed to surpass its own limitations and explore information that was previously unobservable.

For the past centuries, lens-based optical imaging systems have followed to a point-to-point imaging pattern. Designers have tended to create optical imaging systems with compact point spread functions. This ensures that even within the constraints of physical laws, sensors can still capture sufficiently clear images. However, this imaging approach also restricts the system's degrees of freedom. With

Z. Wu · Z. Liu (✉)
School of Physics, Harbin Institute of Technology, Harbin 150001, China
e-mail: zjliu@hit.edu.cn

Z. Wu
School of Instrumentation Science and Engineering, Harbin Institute of Technology, Harbin 150001, China

Z. Liu et al. (eds.), *Computational Optical Imaging*, Advances in Optics and Optoelectronics, https://doi.org/10.1007/978-981-97-1455-1_13

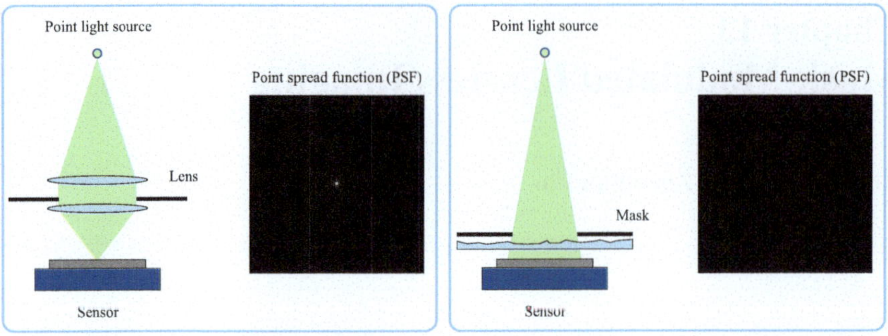

Fig. 13.1 Comparison of Point Spread Function (PSF) between lens imaging system and lensless imaging system

the advancement of computer science and the semiconductor industry, the computational power of computers has significantly increased. This has enabled the integration of computational techniques into the imaging process. By combining computational technology, imaging systems have seen remarkable enhancements in their ability to reconstruct images across various dimensions. Furthermore, the computational burden of imaging has shifted from massive and expensive hardware to the realm of computation itself, making new architectures for cost-effective cameras a possibility.

In recent years, lensless imaging has gained increasing attention due to its compact design and easy setup characteristics. Unlike traditional lens-based imaging, lensless imaging replaces the lens component with a specially designed mask, resulting in a larger Point Spread Function (PSF) scale for the imaging system. This indicates that information from point light sources on the target plane will be projected onto a significant portion of the sensor (Fig. 13.1). Consequently, images captured by sensors in lensless imaging systems often lack the visual features of natural scenes. Therefore, for lensless systems, efficient and accurate reconstruction algorithms are also indispensable. In Fig. 13.2, the imaging modes, processes and characteristics of lens imaging and lensless imaging are compared in detail.

13.2 Development of Mask-Modulated Lensless Imaging Systems

As a compact imaging method, lensless imaging is gaining increasing attention. The earliest form of lensless imaging is pinhole imaging. During the Spring and Autumn Period and the Warring States Period in ancient China, Mozi discovered that placing a board with a small hole between a wall and an object would create an inverted image of the object on the wall. The "Mo Jing" recorded and explained this phenomenon, stating, "The scenery, which is illuminated by light, is radiant and shining. The person

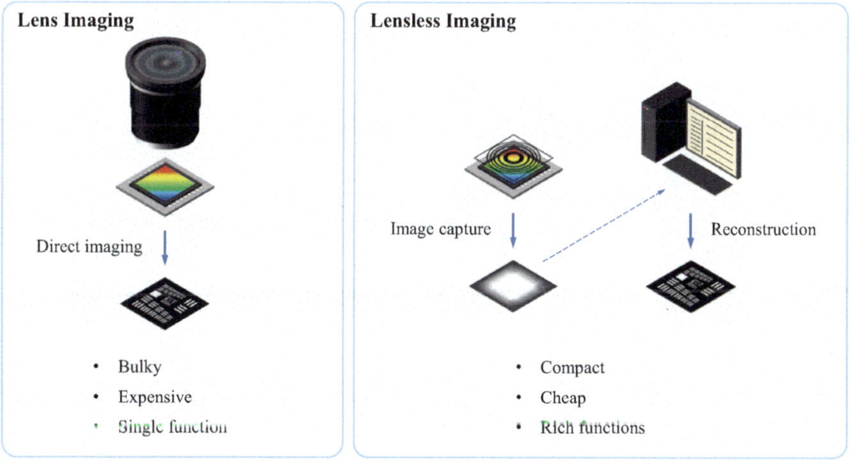

Fig. 13.2 Comparison between lens imaging and lensless imaging

standing lower appears taller, and the person standing taller appears lower. The feet obstruct the light from below, thus forming the image above. The head obstructs the light from above, thus forming the image below. There are boundaries in terms of distance, related to light, so the images form in the storage area of the scenery."

However, from the imaging perspective, single pinhole imaging reduces the amount of incoming light, resulting in a lower signal-to-noise ratio and poorer image quality. To address this issue, the Arab physicist Ibn al-Haytham introduced the classical lens imaging system. By utilizing the focusing function of lenses to converge light rays, the optical system's light input is enhanced, thereby improving image quality. Another approach to increasing light input is to replace the single pinhole with multiple pinholes and enhance light input through aperture coding. This method was initially applied in imaging with electromagnetic waves in the invisible light spectrum. In X-ray and gamma-ray imaging, the wavelengths of these types of rays are short, and they possess high photon energy, often passing through lenses directly without deflection. As a result, the lenses or spherical mirrors required for imaging in such wavelength ranges are usually expensive. Through aperture coding, it is possible to encode the radiation source, and then, through corresponding decoding algorithms, recover the original imaging target [1, 2].

In recent years, lensless imaging techniques for the visible light spectrum have also been continuously proposed (Fig. 13.3). In comparison to lens imaging systems that directly capture images, lensless imaging system replaces the lens with a modulation device, enabling modulation of the imaging scene of the target. The process involves combining the system's PSF with specifically designed recovery algorithms to decode the acquired pseudo-random patterns, thereby restoring the original scene. Depending on the type of modulation, modulation masks can be categorized into two types: amplitude modulation and phase modulation.

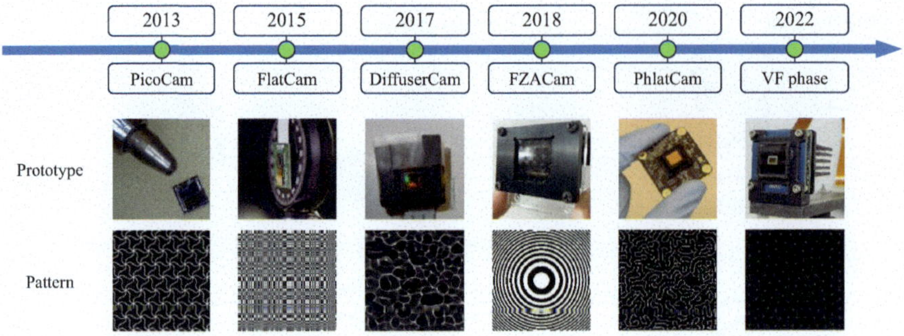

Fig. 13.3 Proposed prototypes of lensless camera

13.2.1 Amplitude Modulation

Amplitude modulation masks primarily consist of an array of transparent and opaque basic units, selectively controlling the passage, obstruction, or attenuation of light to achieve modulation. In 2015, Asif et al. designed the FlatCam lensless camera based on a binary phase mask [3]. Utilizing a geometric optics model, the authors proposed an imaging model and its corresponding computational methods. Mathematically, the mask of this imaging system can be decomposed into the inner product of two one-dimensional vectors, greatly reducing the computational memory burden.

Researchers Nakamura et al. from Japan employed Fresnel zone aperture plates as modulation devices and drew from stripe scanning techniques to introduce the FZA lensless camera [4]. By capturing images modulated with zone aperture plates in four different phase patterns and applying corresponding mathematical calculations, this method effectively eliminates twin images. Building upon this imaging system, Wu et al. further extended its capabilities. They introduced regularization techniques from compressive sensing, which effectively eliminated twin images and enabled single-frame image reconstruction [5].

13.2.2 Phase Modulation

Amplitude modulation still achieves modulation by blocking a portion of the light, which in turn reduces the incoming light and leads to a lower signal-to-noise ratio in the reconstructed results. In contrast, phase modulation, unlike amplitude modulation, utilizes modulation of the phase of the light field for imaging. This approach effectively increases the number of photons, thereby enhancing the imaging results. In 2013, Stork et al. designed a lensless camera based on spiral phase gratings and provided corresponding computational methods, resulting in a low-cost and compact imaging system [6].

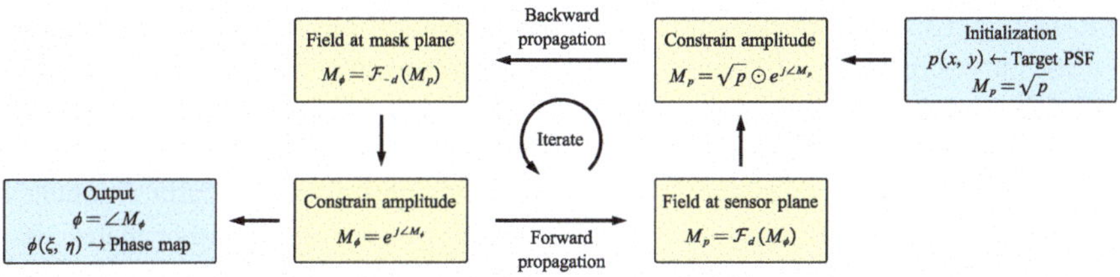

Fig. 13.4 Illustration of the phase mask design

In 2017, Antipa et al. utilized a diffuser as modulation device and developed the DiffuserCam [7]. Before capturing images, calibration of the point spread functions for different depths of imaging is required. Subsequently, they combined the Alternating Direction Method of Multipliers (ADMM) reconstruction algorithm to restore the original scene [8]. By integrating the core idea of compressive sensing, this method achieved three-dimensional imaging.

In 2020, Boominathan et al. introduced a phase mask design method (Fig. 13.4) based on wave optics analysis and phase retrieval techniques [9]. This method effectively calculates the thickness of a mask that possesses a sharp point spread function. The paper also analyzed and summarized the characteristics of well-performing masks. Based on these features, the authors proposed a phase mask based on the Perlin noise and verified its performance.

Fu et al. proposed a Fourier-domain evaluation metric called Modulation Transfer Function volume (MTFv) [10]. This method objectively assesses the mask's ability to transmit spatial frequency information. Leveraging this evaluation metric, the authors introduced a phase modulation function reminiscent of insect compound eyes, namely the Voronoi-Fresnel phase. This phase function is composed of first-order Fresnel phases arranged in a Voronoi pattern, achieving phase modulation. Through continuous optimization of the Voronoi structure and in combination with the MTFv function, the final phase function exhibits excellent optical performance.

13.2.3 Comparison of Modulation Mask

In the previous sections, we discussed the architectures of several classical lensless imaging systems. It's important to note that the performance of these lensless systems varies. A well-designed mask can yield better results, whereas a suboptimal mask might lead to a decrease in performance.

In optical systems, higher system transmission frequencies indicate richer high-frequency information on the sensor. In the context of lensless imaging systems, a more capable mask in terms of transmitting information corresponds to richer details in the reconstructed image. Modulation Transfer Function (MTF) is a fundamental concept that characterizes how different spatial frequencies are captured or transmitted by an imaging system. It is obtained through the Fourier transform of the

PSF. In Fig. 13.5, we present the MTF curves for multiple masks. The variations in these curves indicate differences in the transmission capabilities of the respective masks. However, it is important to note that MTF is a two-dimensional function that describes spatial frequencies, while MTF curve represents a one-dimensional function. Therefore, to provide a more intuitive evaluation of the overall spatial frequency transmission ability, we introduce the MTFv to serve as a single-number metric that quantifies the comprehensive transmission ability across all spatial frequencies. Its expression is defined as follows:

$$MTFv = \frac{1}{|\Omega|} \iint MTF(\mathrm{f_X},\ \mathrm{f_Y})\mathrm{df_X}d\mathrm{f_Y}. \tag{13.1}$$

To assess the transmission ability of the modulation elements in terms of frequencies, we employed the decorrelation analysis method proposed in [11] to estimate the cut-off frequency of the corresponding sensor recording. In the analyses presented in Fig. 13.6 (b1)-(b4) and (d1)-(d4), we considered 500 sampling points within the spatial frequency range of 0 to 1, and the estimated cut-off frequencies are marked by black vertical lines. According to the Power Spectral Density (PSD) curve of the sensor recording, the decorrelation analysis method successfully estimates the cut-off frequency. We summarized the data of the cut-off frequency of sensor recording and MTFv in Table 13.1.

As illustrated in Fig. 13.7, a logarithmic relationship can be observed between the cut-off frequency and MTFv. This relationship suggests that as the MTFv increases, the high-frequency information captured by the sensor becomes richer. However, the rate of growth in high-frequency information weakens as the MTFv continues to increase.

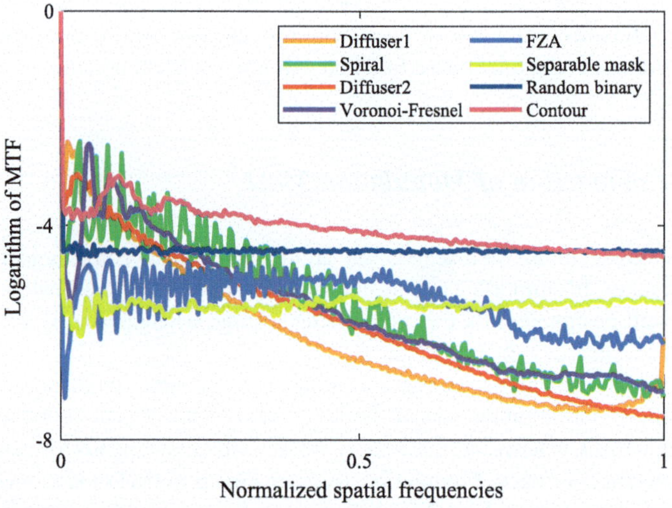

Fig. 13.5 MTF curves of lensless systems with different masks

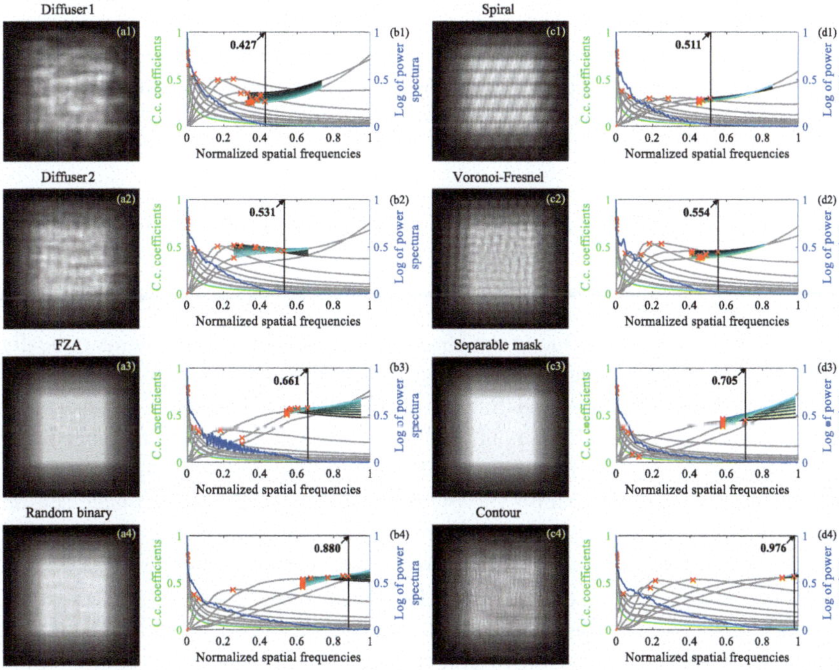

Fig. 13.6 (a1)-(a4), (c1)-(c4) Sensor recordings of different lensless imaging systems. (b1)-(b4), (d1)-(d4) Decorrelation analyses of corresponding sensor measurement. Green line, decorrelation function before high-pass filtering; blue line, PSD curve; gray lines, all high-pass filtered decorrelation functions; black to cyan lines, decorrelation functions with refined high-pass filtering range. Red crosses, all local maxima. Black vertical line, cut-off frequency of sensor measurement. C.c., cross-correlation

Table 13.1 Relationship of MTFv and cut-off frequency of sensor recording

MTFv	160	187	216	284	366	421	1363	1639
$f_{cut-off}$	0.427	0.511	0.531	0.554	0.661	0.705	0.880	0.976

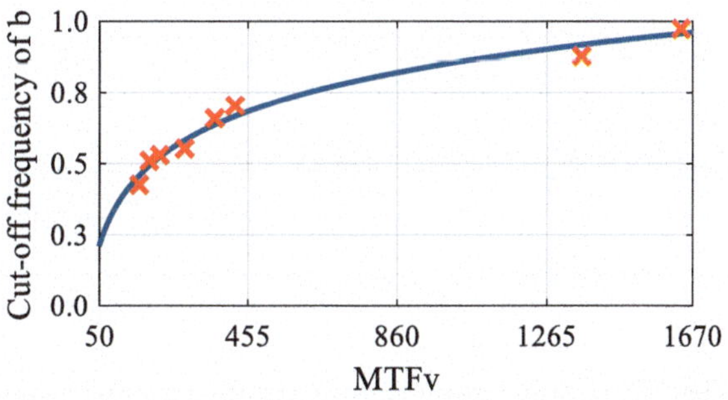

Fig. 13.7 Relationship of cut-off frequency of sensor recording and MTFv. Here b is sensor recording

Therefore, an optimally designed mask should possess excellent transmission capabilities across all spatial frequencies, aiming to maximize the cut-off frequency on the sensor. This ensures the preservation of rich high-frequency information and ultimately leads to enhanced reconstruction performance.

13.3 Image Reconstruction

Compared to lens-based imaging, lensless imaging employs modulating devices in place of lens components. Due to the absence of lenses, imaging systems typically do not directly generate images of the target scene. Instead, they establish a mapping from the scene to sensor patterns. Subsequently, the original scene is reconstructed using solving algorithms. Each point of light in the scene projects the shadow of an encoded mask onto the image sensor. Projections generated by light sources at different spatial positions exhibit varying positions and sizes on the image sensor. The resulting encoded pattern is essentially a superposition of shifted and scaled versions of the mask's projection. As a result, the lensless imaging model based on encoded masks is fundamentally akin to a linear regression model. Recovering the scene image from the encoded pattern constitutes a linear inverse problem. The primary challenge in solving this inverse problem lies in its ill-posed nature. In the context of encoded mask imaging, model errors and noise present in the encoded pattern can lead to significant artifacts in the reconstructed image.

13.3.1 Formation Model

Figure 13.8 demonstrates the schematic of general lensless imaging system. The whole system comprises of sensor and modulation device (mask). The simplest case is imaging for a single planar plane with sparse spatial intensity. For brevity, we will first analyze the imaging model of the system in a simplified one-dimensional scenario.

Consider the modulation function of the mask is $h(x)$, the imaging result of placing a point light source with brightness $v(x_0)$ at (z_0, x_0) is

$$b(x) = h(x + d\frac{x_0 - x}{z_0 + d})v(x_0), \tag{13.2}$$

which is also the scaled version of point spread function. It should be noted that this equation is derived under the assumption of a small angle approximation. If we expand the input scene from a singular point to a plane situated at the depth of z_0, the intensity at the sensor pixel x can be formulated as follows:

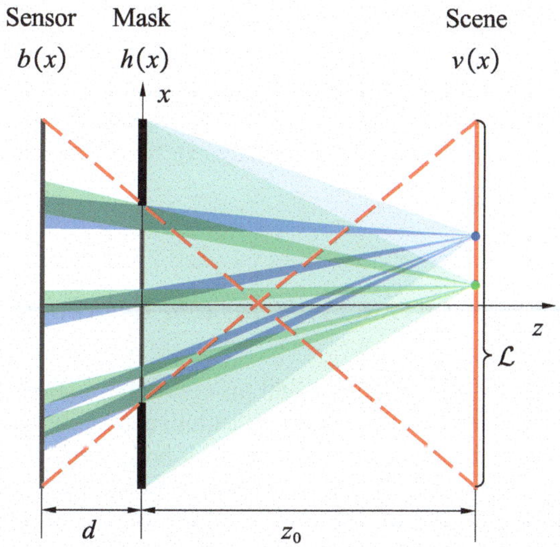

Fig. 13.8 Schematic of lensless imaging system

$$b(x) = \int_{x_0 \in \mathcal{L}} h(x + d\frac{x_0 - x}{z_0 + d})v(x_0)\mathrm{d}x_0. \tag{13.3}$$

Here, $\mathcal{L}$ represents the lensless camera's field of view, within which a point light source projects a complete camera modulation pattern onto the sensor. This formation can also be expressed using the convolution model:

$$b(x) = h(\frac{z_0}{z_0 + d}) * \frac{z_0}{d}v(-\frac{z_0}{d}x). \tag{13.4}$$

The convolution model illustrates that the imaging system functions as a linear system with shift invariance. Discretize and extend the equations into a two-dimensional format, the formation model can be formulated as follows:

$$\mathbf{b} = \mathbf{H}\mathbf{v}. \tag{13.5}$$

Here, **b** and **v** represent the scene and sensor recording vectors respectively, while **H** is a Toeplitz matrix that signifies a linear operator of the imaging process.

All the aforementioned models can be expanded to encompass 3D imaging. The important factors to consider include: (a) the point spread function should be depth dependent and (b) light from different planes should be linear additive at the sensor. With these in mind, the three-dimensional linear model can all be prolonged to incorporate the three-dimensional context as:

$$b = \sum_{z} Hv. \tag{13.6}$$

In this formation, z signifies the depth-related characteristics of scene elements or appropriate transfer functions, and Σ_z indicates a summation across the depth dimension. The computational and memory demands of these forward models grow linearly with the quantity of depth layers employed in the model. Nonetheless, apart from this aspect, these models exhibit comparable limitations to the 2D models discussed earlier.

13.3.2 Inverse Problems and Regularization

In the 1940s, Tikhonov and his colleagues extensively investigated ill-posed problems [12], and introduced a theoretical approach called *regularization*. The aim of regularization is to transform an ill-posed problem into a well-posed one by modifying it, ensuring that the solution to the problem is both physically sound and stable. Of course, the process of constructing a well-posed problem doesn't involve unconditional modifications to the ill-posed problem; rather, it requires adherence to prior knowledge of the actual problem. Therefore, the fundamental idea behind regularization is to utilize a priori knowledge about the solution, introduce additional constraints, to stabilize the solution of the inverse problem. Currently, researchers have developed various regularization techniques and methods that are widely applied to solving diverse inverse problems.

In lensless imaging system, the whole process can be viewed as a linear process. For any linear system, it can be expressed using the following system of linear equations:

$$b = Hv + N. \tag{13.7}$$

Convolution model states that the equation can also be similarly written as a mulitplication in the Foutier domain:

$$\mathrm{b} = \mathcal{F}^{-1}\{\mathcal{F}\{\mathrm{h}\} \odot \mathcal{F}\{\mathrm{v}\}\} + \mathrm{n}, \tag{13.8}$$

where $\odot$ is the Hadamard product. The vector v represents the discrete samples of the original data, while vector b represents the discrete samples of the measured data. H is a linear operator that encapsulates the characteristics of the entire linear system; essentially, it's a measurement matrix. Vector N represents additive noise. When the operator H in Eq. 13.7 is known and in the absence of noise, the original data can be obtained by taking the inverse of the measurement matrix H:

$$\hat{v} = H^{-1}b, \tag{13.9}$$

or from the aspect of frequency domain:

$$\tilde{v}=\mathcal{F}^{-1}\{\mathcal{F}\{\mathrm{b}\}./\mathcal{F}\{\mathrm{c}\}\}, \tag{13.10}$$

where the division should be performed element-wisely.

In practical situations, the measurement matrix H may not necessarily be a square matrix. In such cases, the inverse matrix might not exist, but we can substitute it with a pseudoinverse matrix. In this scenario, we can employ the method of least squares to solve the problem as

$$\min_{v}\|Hv - b\|_2^2. \tag{13.11}$$

For the given optimization problem, the approach involves first expanding the expres sion, then taking the derivative with respect to the vector **v**, and setting the derivative to zero to obtain the solution. When the number of rows in matrix **H** is greater than the number of columns, the system of equations is overdetermined. In this situation, the pseudoinverse matrix is given by:

$$H_{\text{pinv}}^{-1} = (H^{\mathrm{T}}H)^{-1}H^{\mathrm{T}}. \tag{13.12}$$

When the number of rows in matrix **H** is less than the number of columns, the system of equations is underdetermined. In this case, the pseudoinverse matrix is:

$$H_{\text{pinv}}^{-1} = H^{\mathrm{T}}(HH^{\mathrm{T}})^{-1} \tag{13.13}$$

However, when there are zero or very small values in the content of $\mathcal{F}\{\mathrm{c}\}$, dividing by a value close to zero will significantly amplify the noise generated in the measurement results, leading to the failure of reconstruction. Figure 13.9 illustrates the deconvolution results of measurement images with varying degrees of noise. The measurement results are obtained from the convolution calculation of the original target image with a 9×9 Gaussian blur kernel. Reconstruction of the original image can be achieved by directly using Eq. 13.10 when there is no noise in the measurement results. However, when noise is added to the measurement image, reconstruction fails due to noise of varying degrees. From a numerical computation perspective, the calculation method of the formula introduces unstable factors, manifested specifically as slight changes in the measurement results leading to substantial variations in the solution.

From an algebraic perspective, the imaging process establishes a mapping from object space to image space, while the result of deconvolution indicates that a very small image space corresponds to a very large object space. Therefore, to obtain accurate results, it's necessary to constrain the object space corresponding to the image space, which involves creating a space containing natural images. This can be achieved by finding the intersection of the object space and the natural image space to satisfy the requirements of data fitting and natural image priors as

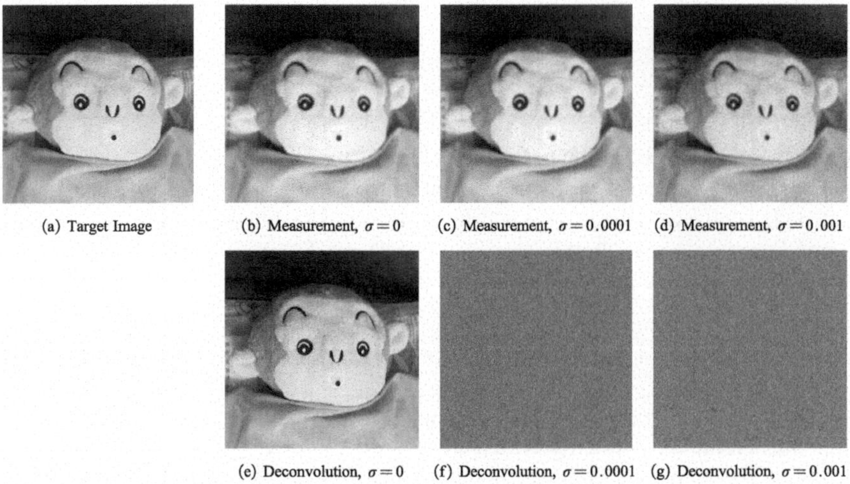

Fig. 13.9 Deconvolution result for the toy monkey image; σ is the standard deviation of the zero-mean Gaussian noise added to the corresponding measurements. The original measurement is derived from the convolutional result of target image and Gaussian blur kernel

$$\min_{v} \frac{1}{2} \|Hv - b\|_2^2 + \tau \mathcal{R}(v). \tag{13.14}$$

Common prior information includes smoothness, sparsity, sparse gradients, self-similarity enhancement, and many other types of information. For an image $\mathbf{v}$, the sparsity of the image itself can be mathematically represented using the ℓ_1-norm

$$\min_{v} \frac{1}{2} \|Hv - b\|_2^2 + \tau \|v\|_1. \tag{13.15}$$

or in the case of the smoothness prior:

$$\min_{v} \|Hv - b\|_2^2 + \tau \|\Delta v\|_2^2. \tag{13.16}$$

where Δ represents the finite difference approximation of the Laplacian operator. This represents the second derivative that models the smoothness of the image. The most widely used regularization term in image processing is Total Variation (TV). TV describes the primary characteristic of natural images, which is that most of the image should be smooth, with sharp transitions occurring only at the edges between smooth regions. This is exactly what TV attempts to capture as a feature of natural images: sharp transitions between regions of constant intensity.

In the process of solving inverse problems, we aim for sparse gradients, meaning that the gradients of solution should contain a significant number of zero elements. The ℓ_0 norm is primarily used to indicate the count of non-zero elements in a vector.

However, the solution space of the ℓ_0 norm is generally discontinuous and non-convex, making it difficult to solve. To expedite the solution of sparse representations, it's common to apply convex relaxation to the optimization problem.

For instance, transforming an ℓ_0 problem into a minimum ℓ_2 norm or ℓ_1 norm problem. When using the ℓ_2 norm constraint on the solution space, as depicted in Fig. 13.10, the ℓ_2 norm geometrically represents an inflating sphere starting from the origin. The optimal solution to the minimum ℓ_2 norm problem is the first intersection point of the solution space with the expanded sphere. However, the optimal solution of the minimum ℓ_2 norm is typically not sparse. Therefore, optimization and reconstruction algorithms based on the ℓ_1 norm have gained attention. Compared to the ℓ_2 norm, the ℓ_1 norm allows the optimal solution to be as close to the coordinate axes as possible, promoting sparse solutions. Unlike the ℓ_0 norm, problems constrained by the ℓ_1 norm tend to have stronger convexity, facilitating rapid search for optimal solutions.

To compute the TV, it's necessary to initially calculate the differentials in two directions of the image: the vertical differentiation is denoted as $\mathbf{G}_x$, and the horizontal differentiation is denoted as $\mathbf{G}_y$ (Fig. 13.11). In this context, $\mathbf{G}_x$ and $\mathbf{G}_y$ represent differential operators used for expressing the image's discrete differences. As shown in Fig. 13.12, natural image has obvious sparsity in the gradient domain.

Similarly, gradient calculations can also be transformed into the frequency domain for computation as

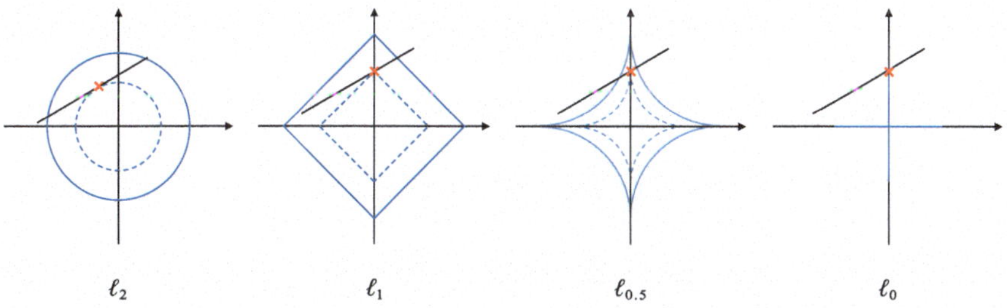

Fig. 13.10 Isosurfaces under different norm metrics in two dimensional space

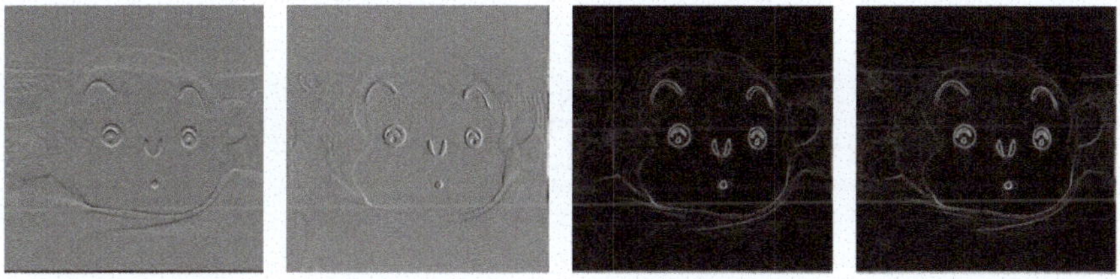

(a) Horizontal gradients $\mathbf{G}_x\mathbf{x}$ (b) Vertical gradients $\mathbf{G}_y\mathbf{x}$ (c) Anisotropic TV (d) Isotropic TV

Fig. 13.11 Gradients and TV of natural image

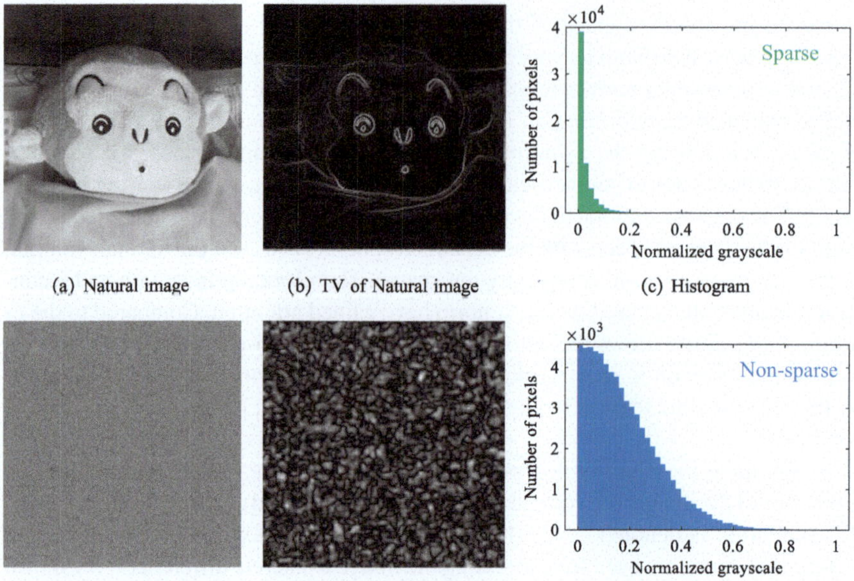

(a) Natural image (b) TV of Natural image (c) Histogram

(d) Non-natural image (e) TV of Non-nature image (f) Histogram

Fig. 13.12 Comparison of gradients of natural and non-natural images

$$G_{\mathrm{x}}x \Leftrightarrow \mathrm{x} * \mathrm{g_x},\ \mathrm{g_x}=\begin{bmatrix} 0 & 0 & 0 \\ 0 & -1 & 1 \\ 0 & 0 & 0 \end{bmatrix}. \tag{13.17}$$

$$G_{\mathrm{y}}x \Leftrightarrow \mathrm{x} * \mathrm{g_y},\ \mathrm{g_y}=\begin{bmatrix} 0 & 0 & 0 \\ 0 & -1 & 0 \\ 0 & 1 & 0 \end{bmatrix}. \tag{13.18}$$

The TV pseudonorm employs the ℓ_1-norm of image gradients to enhance sparse gradients. For this purpose, the anisotropic TV adds up the ℓ_1-norms of the two gradient components

$$\begin{aligned} \mathrm{TV}_{anisotropic}(x) &= \|G_{\mathrm{x}}x\|_1 + \left\|G_{\mathrm{y}}x\right\|_1 \\ &= \sum_{\mathrm{i}=1}^{\mathrm{N}} |(G_{\mathrm{x}}x)_{\mathrm{i}}| + \left|(G_{\mathrm{y}}x)_{\mathrm{i}}\right| = \sum_{\mathrm{i}=1}^{\mathrm{N}} \sqrt{(G_{\mathrm{x}}x)_{\mathrm{i}}^2 + \sqrt{(G_{\mathrm{y}}x)_{\mathrm{i}}^2}}. \end{aligned} \tag{13.19}$$

Isotropic total variation, on the other hand, is computed in a slightly different manner as

$$\mathrm{TV}_{isotropic}(x) = \|Gx\|_{2,1} = \sum_{\mathrm{i}=1}^{\mathrm{N}} \left\| \begin{bmatrix} (G_{\mathrm{x}}x)_{\mathrm{i}} \\ (G_{\mathrm{y}}x)_{\mathrm{i}} \end{bmatrix} \right\|_2 = \sum_{\mathrm{i}=1}^{\mathrm{N}} \sqrt{(G_{\mathrm{x}}x)_{\mathrm{i}}^2 + (G_{\mathrm{y}}x)_{\mathrm{i}}^2}. \tag{13.20}$$

13.3.3 Separability Model

In Eq. 13.5, we establish a linear transformation model for lensless imaging. From the perspective of computational complexity, when the input and output are images of size $\mathrm{N} \times \mathrm{N}$ and $\mathrm{M} \times \mathrm{M}$ respectively, H should be an $\mathrm{M}^2 \times \mathrm{N}^2$ matrix. Given a mask and a sensor, the specific values of H can be obtained either by simulating light propagation from the scene to the sensor or through a calibration process. Clearly, even for moderately sized systems, estimating (calibrating) the system or performing inversion calculations is very challenging.

To address the issue of excessive computation and achieve rapid reconstruction, Asif et al. proposed a separable model in [3]. Through specially designed separable templates, fast reconstruction of the original scene can be achieved. In the architecture of FlatCam, the template for the mask can be represented as the outer product of two one-dimensional vectors. Thus, the imaging model proposed in Eq. 13.5 can be further expressed as follows:

$$b = \Phi_{\mathrm{L}} v \Phi_{\mathrm{R}}^{\mathrm{T}} + N. \tag{13.21}$$

Here, Φ_{L} and $\Phi_{\mathrm{R}}^{\mathrm{T}}$ represent separable operations on the rows and columns of the image, respectively. The original scene is v an $\mathrm{N} \times \mathrm{N}$ matrix, the imaging result b is an $\mathrm{M} \times \mathrm{M}$ matrix, and N accounts for sensor noise and errors introduced by model mismatches. At this point, Φ_{L} is an $\mathrm{M} \times \mathrm{N}$ matrix, and $\Phi_{\mathrm{R}}^{\mathrm{T}}$ is a matrix with size of $\mathrm{N} \times \mathrm{M}$. Additionally, since separable matrix operations are highly parallelizable, they can be applied row-wise and then column-wise, reducing the computational complexity to $\mathcal{O}(\mathrm{N})$, where N is the number of pixels. Compared to the linear imaging model, the separable model can significantly conserve memory, reduce computation, and thereby enhance processing speed.

The most direct approach to solving the problem in the Eq. 13.5 is through the method of least squares, which seeks to find the solution to the equation. However, this method is also susceptible to the problem of error amplification. Therefore, an additional ℓ_2 constraint can be added to help solve for the original scene as

$$\hat{v} = \arg\min_{v} \left\| \Phi_{\mathrm{L}} v \Phi_{\mathrm{R}}^{\mathrm{T}} - b \right\|_2^2 + \tau \|v\|_2^2, \tag{13.22}$$

where τ is a adjustable parameter used to control the constraint of the prior term. Taking the derivative of the equation and setting it to zero, we can further simplify as follows:

$$\Phi_{\mathrm{L}}^{\mathrm{T}}(\Phi_{\mathrm{L}} v \Phi_{\mathrm{R}}^{\mathrm{T}} - b)\Phi_{\mathrm{R}} + \tau v = 0, \tag{13.23}$$

$$\Phi_{\mathrm{L}}^{\mathrm{T}}\Phi_{\mathrm{L}} v \Phi_{\mathrm{R}}^{\mathrm{T}}\Phi_{\mathrm{R}} + \tau v = \Phi_{\mathrm{L}}^{\mathrm{T}} b \Phi_{\mathrm{R}}. \tag{13.24}$$

Applying singular value decomposition (SVD) to the matrix Φ_{L} and $\Phi_{\mathrm{R}}^{\mathrm{T}}$:

$$V_{\mathrm{L}}\Sigma_{\mathrm{L}}^{2}V_{\mathrm{L}}^{\mathrm{T}} v V_{\mathrm{R}}\Sigma_{\mathrm{R}}^{2}V_{\mathrm{R}}^{\mathrm{T}} + \tau v = V_{\mathrm{L}}\Sigma_{\mathrm{L}}U_{\mathrm{L}}^{\mathrm{T}} b U_{\mathrm{R}}\Sigma_{\mathrm{R}}V_{\mathrm{R}}^{\mathrm{T}}. \tag{13.25}$$

Further simplification leads to:

$$\Sigma_{\mathrm{L}}^{2}V_{\mathrm{L}}^{\mathrm{T}} v V_{\mathrm{R}}\Sigma_{\mathrm{R}}^{2} + \tau V_{\mathrm{L}}^{\mathrm{T}} v V_{\mathrm{R}} = \Sigma_{\mathrm{L}}U_{\mathrm{L}}^{\mathrm{T}} b U_{\mathrm{R}}\Sigma_{\mathrm{R}}. \tag{13.26}$$

Representing the diagonal matrix Σ_{L}^{2} and Σ_{R}^{2} as vectors σ_{L} and σ_{R}, the equation simplifies to

$$V_{\mathrm{L}}^{\mathrm{T}} v V_{\mathrm{R}} \odot (\sigma_{\mathrm{L}}\sigma_{\mathrm{R}}^{\mathrm{T}}) + \tau V_{\mathrm{L}}^{\mathrm{T}} v V_{\mathrm{R}} = \Sigma_{\mathrm{L}}U_{\mathrm{L}}^{\mathrm{T}} b U_{\mathrm{R}}\Sigma_{\mathrm{R}}, \tag{13.27}$$

$$V_{\mathrm{L}}^{\mathrm{T}} v V_{\mathrm{R}} \odot (\sigma_{\mathrm{L}}\sigma_{\mathrm{R}}^{\mathrm{T}} + \tau \mathbf{1}\mathbf{1}^{\mathrm{T}}) = \Sigma_{\mathrm{L}}U_{\mathrm{L}}^{\mathrm{T}} b U_{\mathrm{R}}\Sigma_{\mathrm{R}}, \tag{13.28}$$

$$V_{\mathrm{L}}^{\mathrm{T}} v V_{\mathrm{R}} = \Sigma_{\mathrm{L}}U_{\mathrm{L}}^{\mathrm{T}} b U_{\mathrm{R}}\Sigma_{\mathrm{R}}./(\sigma_{\mathrm{L}}\sigma_{\mathrm{R}}^{\mathrm{T}} + \tau \mathbf{1}\mathbf{1}^{\mathrm{T}}), \tag{13.29}$$

where $\odot$ and ./ denote element-wise multiplication and division operations. **1** denotes the all-one vector. The solution to the equation can be expressed as:

$$\hat{v} = V_{\mathrm{L}}[\Sigma_{\mathrm{L}}U_{\mathrm{L}}^{\mathrm{T}} b U_{\mathrm{R}}\Sigma_{\mathrm{R}}./(\sigma_{\mathrm{L}}\sigma_{\mathrm{R}}^{\mathrm{T}} + \tau \mathbf{1}\mathbf{1}^{\mathrm{T}})]V_{\mathrm{R}}^{\mathrm{T}} \tag{13.30}$$

13.3.4 Regularized Image Reconstruction with ADMM

The essential nature of the non-coherent lensless imaging model based on coded masks is essentially that of a linear regression model. Recovering the scene image from the encoded image belongs to a linear inverse problem. For the equation in Eq. 13.5, numerically, when the modulation matrix H is known, the original scene v can be directly computed using the inverse matrix H^{-1}. However, in the context of practical imaging systems, the modulation matrix H might not be square. In this case, the inverse matrix H^{-1} may not really exist, and the equation generally represents an underdetermined or overdetermined system of equations. This implies that the equation might have infinitely many solutions or no solutions at all. Additionally, noise is generated during the imaging process, which gets amplified during the inverse matrix computation and significantly affects the recovery results. In the case of lensless imaging, the number of columns in H is much larger than the number of rows, meaning that the equation's solution is underdetermined. However, images generated from natural scenes should exhibit sparsity in their gradients. Therefore, this sparsity can act as a prior constraint in solving linear problems. In this scenario, solving the Eq. 13.5 transforms into a regularized non-negative ℓ_1 norm constrained optimization problem:

$$\hat{v} = \arg\min_{v \geq 0} \frac{1}{2} \|b - Hx\|_2^2 + \tau_1 \|\Psi v\|_1. \tag{13.31}$$

Here, the operator Ψ signifies the operation that maps the image to the gradient domain, and τ_1 acts as a tunable parameter to adjust the strength of the gradient sparsity constraint. The inverse problem of lensless imaging is of considerable scale, typically involving millions of input and output parameters. Conventional gradient descent methods for solving such problems often require a significant amount of time.

ADMM (Alternating Direction Method of Multipliers) [8] is an iterative optimization method that splits the objective function and solves it alternately by utilizing proximal operators for both the data fidelity term and the regularization term. As a result, the ADMM method in conjunction with variable separation [13–15] can be utilized to address the problem. Equation 13.31 can be equivalently transformed into:

$$\begin{aligned} v = \arg\min_{w,x,u} \frac{1}{2} \|b - x\|_2^2 + \tau_1 \|u\|_1 + \mathbb{I}_+(w), \\ \text{s.t.} x = Hv, u = \Psi v, w = v, \end{aligned} \tag{13.32}$$

in which

$$\mathbb{I}_+(\mathrm{x}) = \begin{cases} 0, \mathrm{x} \geq 0, \\ \infty, \ \mathrm{x} < 0, \end{cases} \tag{13.33}$$

acts as a penalty function to constrain the non-negative of result.

By using the augmented Lagrangian method to process the equation, the problem is transformed into

$$\begin{aligned} \max_{\xi,\eta,\rho} \Bigg[\min_{u,v,w,x} \frac{1}{2} \|b - x\|_2^2 + \frac{\mu_1}{2} \left\| Hv - x + \frac{\xi}{\mu_1} \right\|_2^2 \\ + \tau_1 \|u\|_1 + \frac{\mu_2}{2} \left\| \Psi v - u + \frac{\eta}{\mu_2} \right\|_2^2 \\ + \mathbb{I}_+(w) + \frac{\mu_3}{2} \left\| v - w + \frac{\rho}{\mu_3} \right\|_2^2 \Bigg], \end{aligned} \tag{13.34}$$

where ξ, η and ρ are the Lagrange multipliers of variables x, u and v respectively. In addition, μ_1, μ_2 and μ_3 are adjustable parameters to adjust the degree of constraint of the penalty term separately during the optimization solution process.

Firstly, derive the optimal solution for each original variable, and assume that the other variables are fixed and invariant. In the kth iteration, there should be

$$
\begin{aligned}
u^{k+1} &\leftarrow \arg\min_{u^k}\left[\tau_1\|u\|_1+\frac{\mu_2}{2}\left\|\Psi v^k-u^k+\frac{\eta^k}{\mu_2}\right\|_2^2\right],\\
x^{k+1} &\leftarrow \arg\min_{x^k}\left[\frac{1}{2}\|b^k-x^k\|_2^2+\frac{\mu_1}{2}\left\|Hv^k-x^k+\frac{\xi^k}{\mu_1}\right\|_2^2\right],\\
w^{k+1} &\leftarrow \arg\min_{w^k}\left[\mathbb{I}_+(w^k)+\frac{\mu_3}{2}\left\|v^k-w^k+\frac{\rho^k}{\mu_3}\right\|_2^2\right],\\
v^{k+1} &\leftarrow \arg\min_{v^k}\left[\frac{\mu_1}{2}\left\|Hv^k-x^k+\frac{\xi^k}{\mu_1}\right\|_2^2\right.\\
&\quad+\frac{\mu_2}{2}\left\|\Psi v^k-u^k+\frac{\eta^k}{\mu_2}\right\|_2^2\\
&\quad\left.+\frac{\mu_3}{2}\left\|v^k-w^k+\frac{\rho^k}{\mu_3}\right\|_2^2\right].
\end{aligned}
\tag{13.35}
$$

By solving each sub optimization problem, it can be concluded that

$$
\begin{aligned}
u^{k+1} &\leftarrow \boldsymbol{T}_{\frac{\tau_1}{\mu_2}}\left(\Psi v^k+\frac{\eta^k}{\mu_2}\right),\\
x^{k+1} &\leftarrow \frac{\xi^k+\mu_1 Hv+b}{1+\mu_1},\\
w^{k+1} &\leftarrow \max\left(\frac{\rho^k}{\mu_3}+v^k,0\right),\\
v^{k+1} &\leftarrow [\mu_1 H^{\mathrm{T}}H+\mu_2\Psi^{\mathrm{T}}\Psi+\mu_3 E]^{-1}r^k,\\
\xi^{k+1} &\leftarrow \xi^k+\mu_1(Hv^{k+1}-x^{k+1}),\\
\eta^{k+1} &\leftarrow \eta^k+\mu_2(\Psi v^{k+1}-u^{k+1}),\\
\rho^{k+1} &\leftarrow \rho^k+\mu_3(v^{k+1}-w^{k+1}),
\end{aligned}
\tag{13.36}
$$

where

$$r^k \leftarrow H^T(\mu_1 x^{k+1} - \xi^k) + \Psi^T(\mu_2 u^{k+1} - \eta^k) + (\mu_3 w^{k+1} - \rho^k). \quad (13.37)$$

The operator $\mathcal{T}_v$ represents a soft threshold operation with a threshold of v [16], which is the identity matrix.

13.3.5 Data Driven Methods

The integration of deep learning techniques with optical imaging methods effectively enhances both the speed and quality of imaging, with successful applications in various fields. In recent years, deep learning has also been gradually applied to the realm of lensless imaging. When compared to traditional optimization-based algorithms, data-driven techniques offer several notable advantages. Unlike optimization algorithms, deep learning can be specifically tailored to scenarios within specialized applications, presenting a more flexible framework. Additionally, data-driven approaches leverage extensive prior statistical information from natural scenes, resulting in better performance and faster reconstruction speed compared to optimization methods.

Lensless cameras rely on computational algorithms to reconstruct scenes from complex measurements. Current iterative optimization-based reconstruction methods tend to generate images that are noisier and visually less appealing. Khan et al. introduce a non-iterative reconstruction approach based on deep learning, which leads to a remarkable improvement in image quality for lensless reconstructions [17]. The method, named "FlatNet" (Fig. 13.13), establishes a framework for producing high-quality, realistic images from mask-based lensless cameras where the camera's forward model is known. FlatNet consists of two key stages:

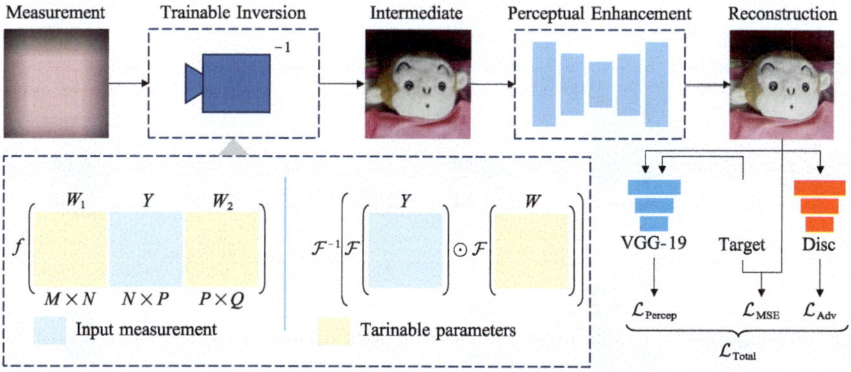

Fig. 13.13 Overall architecture of the FlatNet

(1) An inversion stage that maps measurements into an intermediate reconstruction space by learning parameters within the forward model.
(2) A perceptual enhancement stage that enhances the perceptual quality of this intermediate reconstruction.

These stages are trained simultaneously in an end-to-end manner. Experiments show that the end-to-end approach is speedy, produces photorealistic reconstructions, and can be easily applied to other mask-based lensless cameras.

Monakhova et al. introduced a computationally bounded, trainable neural network for image reconstruction [18] (Fig. 13.14). Authors leveraged understanding of the physical system by unrolling a conventional model-based optimization algorithm and employed a specialized variable splitting technique tailored for lensless imaging [7, 8]. This approach integrates insights into the image creation process directly into the neural network, while also enables the learning of network parameters from the available data. Three distinct architectures were introduced: Le-ADMM, Le-ADMM*, and Le-ADMM-U, each progressively featuring more trainable parameters. Compared to conventional methods, this architecture achieves superior perceptual image quality and operates at a speed 20 times faster, allowing for interactive scene previews. The authors explore a spectrum between model-based and deep learning approaches, highlighting the advantages of adopting an intermediate strategy.

In lensless imaging with masks, iterative reconstruction methods based on the geometric optics model lead to artifacts and demand significant computational resources. Wu et al. introduce a lensless camera prototype that leverages a deep neural network (DNN) to achieve rapid reconstruction for Fresnel zone aperture (FZA) imaging [19]. This system includes a deep back-projection network (DBPN) connected to a U-Net, creating an error feedback mechanism that enables self-correction of features for image detail recovery (Fig. 13.15). Training data are

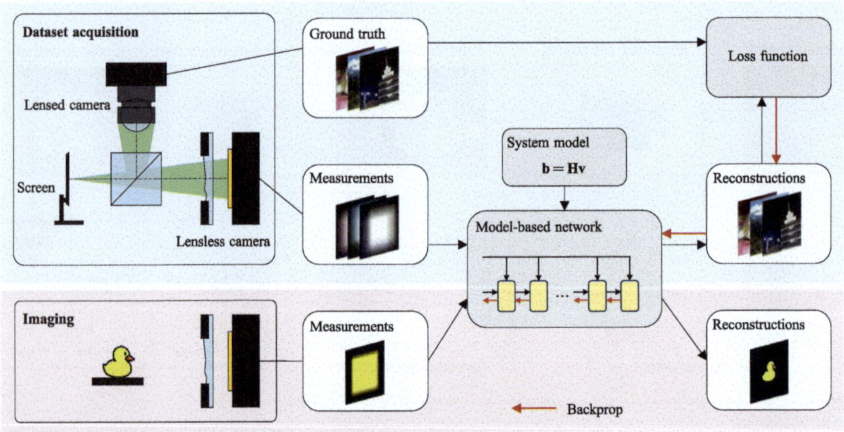

Fig. 13.14 Overview of learned reconstructions for lenless imaging

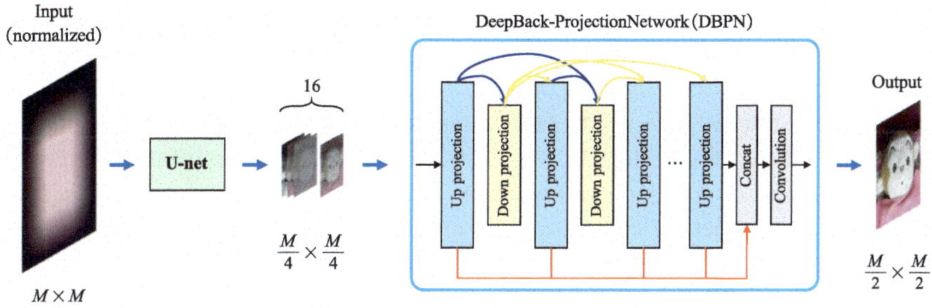

Fig. 13.15 Architecture of the DNN-FZA camera

generated using a diffraction model, simulating conditions of broadband incoherent imaging. The reconstructed results demonstrate a reduction in blur caused by diffraction, while the computational time is two orders of magnitude faster compared to traditional iterative image reconstruction methods.

13.4 Applications

13.4.1 Video from Single Capture

Traditional lens imaging operates on a point-to-point basis, resulting in limited bandwidth for image sensor chip pixel readouts. When recording videos, there is usually a trade-off between frame rate and pixel count. On the other hand, lensless imaging projects information from point light sources in the imaging scene onto a large plane. This suggests that even with insufficient sampling, scene reconstruction can be achieved through compressive sensing methods. Antipa et al. employed multiplexing optics to spatially compress the scene, allowing the entire scene's information to be sampled from a row of sensor pixels [20]. This information can be rapidly read out using a rolling-shutter CMOS sensor. This specific functionality is achievable through DiffuserCam. Utilizing sparse recovery methods, the proposed imaging system can reconstruct 140 video frames at a rate exceeding 4500 frames per second, all from a single image captured using a rolling-shutter sensor.

13.4.2 Light-Field Imaging

In lensless imaging, when dealing with scenes characterized by dense textures spread across a considerable depth range, the resulting inverse problem is highly unstable. This means that the number of unknowns significantly surpasses the number of measurements, making the problem difficult to solve. Drawing inspiration from light field cameras, Hua et al. devised a strategy involving programmable masks to enhance the conditioning of depth estimation and image reconstruction [21]. This method involves shifting a single mask pattern to generate encoded images from various viewing perspectives. The fundamental steps used for reconstruction are akin to generating a coded focus stack of the scene. Subsequently, straightforward deconvolution techniques are employed to recover the volumetric textures of the scene. Through this approach, a lensless camera can achieve focusing at different depths within a scene, effectively enabling refocusing capabilities.

13.4.3 Hyperspectral Imaging

Hyperspectral imaging finds significant applications in various fields such as agriculture and military. However, conventional scanning-based hyperspectral imaging devices are hindered by their slow speed and high cost, making widespread adoption challenging. Monakhova et al. propose an innovative computational camera for snapshot hyperspectral imaging based on the DiffuserCam system [22]. This system comprises a flat spectral filter array directly placed on the image sensor and a diffuser positioned near the sensor. The authors introduce the concept of the "Spectral DiffuserCam", a lensless imaging setup equipped with a spectral filter array. This array consists of multiple spectral filters arranged in a grid pattern, each designed to transmit light within a specific wavelength range. By positioning this spectral filter array in front of the light-sensitive sensor, the system can simultaneously capture spectral information for each spatial location in the scene. Subsequently, the authors utilize sparse-constraint inverse problem solving to recover the hyperspectral volume at sub-pixel resolution.

13.5 General Conclusions

In this chapter, we begin by highlighting the similarities and differences between lens-based imaging and lensless imaging. Lensless imaging, in contrast to lens-based imaging, projects information from each point in the scene onto a plane. This implies that the captured images often lack the visual characteristics of natural images and require effective algorithms for reconstruction. Additionally, we introduce several classical architectures of lensless imaging and, by incorporating objective evaluation parameters, further analyze the performance of different mask patterns from the frequency transmission perspective.

Next, we provide a mathematical model of lensless imaging systems. For the most basic two-dimensional planar imaging under paraxial approximation, the imaging outcome can be regarded as the convolution of the target imaging scene and the mask pattern. The convolution imaging model also reveals that lensless imaging is fundamentally a linear process, and solving for the original scene is tantamount to addressing a linear inverse problem. By pre-calibrating, establishing a clear understanding of the system's optical propagation, and calculating the point spread function, the original scene can be obtained through the methods for solving linear inverse problems.

Subsequently, we introduce the methods for solving lensless imaging, wherein the task of recovering the original scene from the images captured by the sensor is fundamentally an inverse problem of a linear nature. However, for a linear inverse problem, slight errors introduced during measurements and mismatches between the actual optical process and the propagation model can be amplified dramatically during the reconstruction process, potentially leading to reconstruction failure. To mitigate this issue, it is common to incorporate prior information about gradient sparsity during the solution process to enhance stability. By jointly constraining data fitting and regularization terms, we can obtain results that fulfill both the accuracy requirements of the data and the characteristics of natural images. Furthermore, data-driven approaches can be employed for scene reconstruction. By collecting a large dataset of images and their corresponding lensless imaging outcomes, constructing a training dataset, and training a model, we can achieve rapid and high-quality reconstruction.

Lastly, we introduced some innovative applications of lensless imaging. Currently, lensless imaging still cannot compete with lens-based imaging in terms of real-time imaging and high-quality imaging. Nevertheless, the distinctive point-to-plane imaging mode of lensless imaging has shown unique advantages in acquiring high-dimensional data such as 3D measurements and hyperspectral imaging. The advanced applications in lensless imaging present significant promise, introducing an exciting new dimension to imaging system design.

Acknowledgements This work was supported by the National Natural Science Foundation of China (Nos. 11874132, 61975044, 12074094, 12374271, and 62375065); Interdisciplinary Research Foundation of HIT (No. IR2021237).

References

1. Dicke RH. Scatter-hole cameras for x-rays and gamma rays. Astrophys J. 1968;153:L101.
2. Cannon TM, Fenimore EE. Coded aperture imaging: many holes make light work. Opt Eng. 1980;19:283–9.
3. Asif MS, Ayremlou A, Sankaranarayanan A, Veeraraghavan A, Baraniuk RG. Flatcam: thin, lensless cameras using coded aperture and computation. IEEE Trans Comput Imag. 2016;3:384–97.
4. Shimano T, Nakamura Y, Tajima K, Sao M, Hoshizawa T. Lensless light-field imaging with Fresnel zone aperture: quasi-coherent coding. Appl Opt. 2018;57:2841–50.
5. Wu J, Zhang H, Zhang W, Jin G, Cao L, Barbastathis G. Single-shot lensless imaging with fresnel zone aperture and incoherent illumination. Light Sci Appl. 2020;9:53.
6. Gill PR, Stork DG (2013) Lensless ultra-miniature imagers using odd-symmetry spiral phase gratings. Imag Appl Opt CW4C
7. Antipa N, Kuo G, Heckel R, Mildenhall B, Bostan E, Ng R, Waller L. DiffuserCam: lensless single-exposure 3D imaging. Optica. 2018;5:1–9.

8. Boyd S, ParikhN,Chu E, PeleatoB, Eckstein J.Distributed optimization and statistical learning via the alternating direction method of multipliers. Found Trends Mach Learn. 2011;3:1–122.
9. Boominathan V, Adams JK, Robinson JT, Veeraraghavan A. Phlatcam: Designed phase-mask based thin lensless camera. IEEE Trans Pattern Anal Mach Intell. 2020;42:1618–29.
10. Fu Q, Yan DM, Heidrich W. Diffractive lensless imaging with optimized Voronoi-Fresnel phase. Opt Express. 2022;30:45807–23.
11. Descloux A, Grußmayer KS, Radenovic A. Parameter-free image resolution estimation based on decorrelation analysis. Nat Methods. 2019;16:918–24.
12. Willoughby RA. Solutions of ill-posed problems (AN Tikhonov and VY Arsenin). SIAM Rev. 1979;21:266–7.
13. Matakos A, Ramani S, Fessler JA. Accelerated edge-preserving image restoration without boundary artifacts. IEEE Trans Image Process. 2013;22:2019–29.
14. Almeida MSC, Figueiredo M. Deconvolving images with unknown boundaries using the alternating direction method of multipliers. IEEE Trans Image Process. 2013;22:3074–86.
15. Afonso MV, Bioucas-Dias JM, Figueiredo MAT. Fast image recovery using variable splitting and constrained optimization. IEEE Trans Image Process. 2010;19:2345–56.
16. Wang Y, Yang J, YinW, Zhang Y. A new alternating minimization algorithm for total variation image reconstruction. SIAM J Imag Sci. 2008;1:248–72.
17. Khan SS, Sundar V, Boominathan V, Veeraraghavan A, Mitra K. Flatnet: towards photorealistic scene reconstruction from lensless measurements. IEEE Trans Pattern Anal Mach Intell. 2020;44:1934–48.
18. Monakhova K, Yurtsever J, Kuo G, Antipa N, Yanny K,Waller L. Learned reconstructions for practical mask-based lensless imaging. Opt Express. 2019;27:28075–90.
19. Wu J, Cao L, Barbastathis G. DNN-FZA camera: a deep learning approach toward broadband FZA lensless imaging. Opt Lett. 2021;46:130–3.
20. Antipa N, Oare P, Bostan E, Ng R, Waller L (2019) Video from stills: lensless imaging with rolling shutter. In: The International conference on computational photography 2019, pp 1–8
21. Hua Y, Nakamura S, Asif MS, Sankaranarayanan AC. Sweepcam—depth-aware lensless imaging using programmable masks. IEEE Trans Pattern AnalMach Intell. 2020;42:1606–17.
22. Monakhova K, Yanny K, Aggarwal N, Waller L. Spectral DiffuserCam: lensless snapshot hyperspectral imaging with a spectral filter array. Optica. 2020;7:1298–307.

Chapter 14
Polarization Imaging

Ziling Qiao and Zhengjun Liu

Abstract Polarization imaging is a technique that involves capturing and analyzing the polarization properties of light to obtain additional information about a scene or an object.

14.1 Introduction

Light exhibits the characteristics of transverse wave whose electromagnetic field oscillates perpendicular to the direction of propagation, which creates a polarization state of light in which the direction of the field is perpendicular to the direction of propagation. Polarization imaging is a technique that involves capturing and analyzing the polarization properties of light to obtain additional information about a scene or an object [1, 2].

In traditional imaging, the intensity of light is recorded, which provides information about the brightness or color of the scene. However, polarization imaging goes beyond intensity and takes into account the polarization state of light. The polarization state describes the specific orientation and amplitude of the electric field vector as the light wave propagates.

Polarization imaging can be achieved by using polarizing elements such as polarizers, wave plates, or retarders. These elements selectively transmit or alter the polarization state of incident light. By placing them in front of a camera or an image sensor, the light's polarization can be measured and recorded.

The primary motivation behind polarization imaging is to reveal hidden or subtle information that is not visible in conventional intensity-based imaging. Polarized light interacts with objects or materials in a manner that depends on their physical

Z. Qiao · Z. Liu (✉)
School of Physics, Harbin Institute of Technology, Harbin 150001, China
e-mail: zjliu@hit.edu.cn

Z. Qiao
e-mail: 18831306020@163.com

Z. Liu et al. (eds.), *Computational Optical Imaging*, Advances in Optics and Optoelectronics, https://doi.org/10.1007/978-981-97-1455-1_14

properties, surface characteristics, and internal structures. By analyzing the polarization changes that occur when light interacts with these objects, polarization imaging can provide valuable insights.

Polarization imaging provides an additional dimension of information by exploiting the polarization properties of light. It offers enhanced capabilities for material characterization, surface analysis, target detection, and other applications where polarization effects play a significant role [3–5].

14.2 Polarization Theory and Expression Method

14.2.1 The Polarization State of the Light

Light has wave-particle duality. As a high frequency vector wave belongs to the electromagnetic wave category, so light has all the attributes of electromagnetic wave, including wavelength, amplitude, frequency, etc. The propagation of light is accompanied by the propagation of electric field and magnetic field, and the direction of electric field and magnetic field are orthogonal to each other and perpendicular to the optical axis. In view of the spatial relationship between the two, usually only the electric field is identified, that is, the light vector (electric field intensity) to characterize the characteristics of the light wave, when the light vector maintains a certain vibration direction in the vertical plane of the light wave propagation direction, the polarization phenomenon occurs. Polarization is one of the inherent properties of light. The electric field vector of light has different vibration states in the process of medium transmission, which makes the light show different vibration states. According to the polarization state of light, it can be mainly divided into natural light (non-polarized light), partially polarized light and fully polarized light. According to the vibration trajectory of the end of the electric vector in the plane perpendicular to the propagation direction, it can be divided into different polarization states of polarized light, such as linear polarized light, circular polarized light and elliptically polarized light.

(1) **Non-polarized Light**

Non-polarized light, also known as natural light, is light whose light vector is evenly distributed in the vertical plane of the light wave propagation directions. The vibration of non-polarized light exists in all directions, which means that the vibration state of non-polarized light is random in time, and the vibration state of the light vector at the next moment cannot be inferred from the current moment. From the perspective of space, in the vertical plane of the optical axis, the vibration direction of the light vector is randomly distributed, there may be light vectors in any direction, and non-polarized light can be represented by two mutually orthogonal and equal amplitude light vectors.

(2) **Partial Polarized Light**

When the light vector is not evenly distributed and the amplitude is different in the vertical plane of the light wave propagation direction, the light generated at this time is partially polarized light, partially polarized light is a mixed polarized light, it can be decomposed into non-polarized light and fully polarized light. The degree of polarization of partially polarized light is determined by the amount of fully polarized light doped, and is usually quantified by the degree of polarization P as

$$P = \frac{I_P}{I_t} = \frac{I_P}{I_P + I_n} = \frac{I_{\max} - I_{\min}}{I_{\max} + I_{\min}}, \tag{14.1}$$

where $I_{\max}$ is the maximum value of light intensity in the vibration plane, and $I_{\min}$ is the minimum value of light intensity in the vibration plane. According to the definition of polarization degree, when $P = 0$, it is non-polarized light; When $0 < P < 1$, it is partially polarized light; when $P = 1$, it is completely polarized light.

Partially polarized light contains a mixture of light waves with some preferred orientation of the electric field vector. The preferred orientation can be due to scattering or reflection from surfaces. It has a partial alignment of its electric field vectors in a particular direction, but there is still some randomness in the orientation. This means that some portions of the light are polarized, while others remain unpolarized.

(3) **Fully Polarized Light**

Physical phenomena such as reflection, scattering and refraction of light exist in nature. When the non-polarized light is reflected and refracted on the surface of the material, the vibration state of the light vector changes, and if the vibration direction of the outgoing light vector has obvious regularity, we define the outgoing light as fully polarized light. The vibration direction of fully polarized light is clear, and it can be regarded as the synthesis of two orthogonal light vectors. Assuming that light propagates along the z axis and the light vector (electric field vector E) vibrates in the *xoz* plane, the propagation of the light vector can be represented by Fig. 14.1. To understand fully polarized light quantitatively, we introduce a system of equations, as shown in Eq. (14.2).

$$\begin{cases} E_x(z,t) = E_{ox}(t)\cos[\omega t - kz + \delta_x(t)], \\ E_y(z,t) = E_{ox}(t)\cos[\omega t - kz + \delta_y(t)] \\ E = iE_x(z,t) + jE_y(z,t), \end{cases}, \tag{14.2}$$

where E represents the projection of the electric field vector on the x and y axes, and δ is the phase of the electric field vector in the x and y directions. In order to more intuitively analyze the vibration trajectory of electric field vector, Eq. (14.2) is simplified and t is eliminated to obtain

$$\begin{cases} \left(\frac{E_x}{E_{ox}}\right)^2 + \left(\frac{E_y}{E_{oy}}\right)^2 - \frac{2E_xE_y\cos\delta}{E_{ox}E_{oy}} = \sin^2\delta, \\ \delta = \delta_x - \delta_y, \end{cases} \tag{14.3}$$

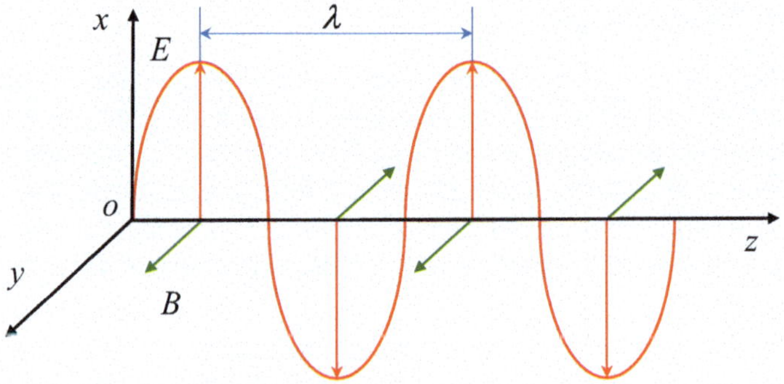

Fig. 14.1 Diagram of light vector propagation

$$\begin{vmatrix} E_{ox}^2 & -\frac{\cos\delta}{E_{ox}E_{oy}} \\ -\frac{\cos\delta}{E_{ox}E_{oy}} & E_{oy}^2 \end{vmatrix} = \frac{\sin^2\delta}{E_{ox}^2 E_{oy}^2} \geqslant 0, \tag{14.4}$$

where δ is the phase difference between x and y directions. It can be seen from Eq. (14.4) that the value of the coefficient determinant in Eq. (14.3) is greater than 0, so it can be judged that Eq. (14.3) is an elliptically polarized light expression.

When $\delta = \delta_x - \delta_y = m\pi (m = 0, \pm 1, \pm 2, \cdots)$, the Eq. (14.3) will be simplified to:

$$\frac{E_x}{E_y} = (-1)^m \frac{E_{ox}}{E_{oy}}. \tag{14.5}$$

It can be seen that the elliptically polarized light becomes linearly polarized light at this time.

When $\delta = \delta_x - \delta_y = m\pi/2 (m = \pm 1, \pm 3, \pm 5, \cdots)$, and $E_{ox} = E_{oy}$, the Eq. (14.3) will be simplified to:

$$\left(\frac{E_x}{E_{ox}}\right)^2 + \left(\frac{E_y}{E_{oy}}\right)^2 = 1. \tag{14.6}$$

As can be seen from Eq. (14.6), at this time, elliptically polarized light degenerates into circularly polarized light. From the above analysis, it can be found that the generation of linearly polarized light and circularly polarized light is closely related to the phase difference of the beam in the x and y directions, and their appearance is affected by the δ value. The polarization state of the ellipse changes with δ, as shown in Fig. 14.2.

Fully polarized light can be achieved through various methods, such as passing unpolarized light through a polarizing filter, which selectively transmits light waves with a specific polarization direction and blocks the others. Polarized sunglasses or

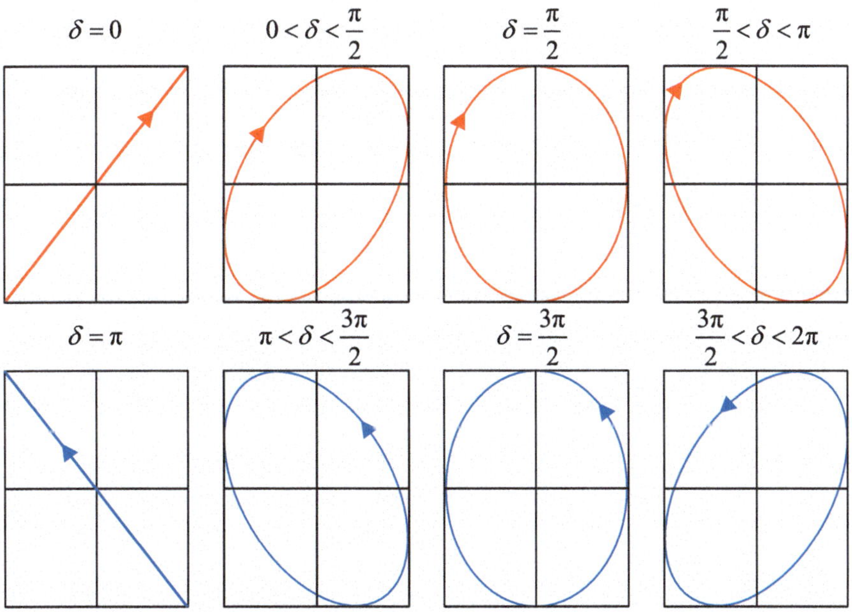

Fig. 14.2 Elliptically polarized light corresponding to different δ

polarizing filters used in photography are examples of devices that can produce or manipulate fully polarized light.

14.2.2 Representation of Polarized Light

After people discovered the properties of polarized light, they only had a qualitative understanding of them. In order to accurately distinguish polarized light in these different states, researchers used mathematical and geometric tools to quantitatively describe the state of polarized light, and proposed many descriptive methods, including trigonometric function method, Jones matrix method, Stokes vector method and Bongar sphere diagram method. The following is mainly to introduce the last three description methods.

(1) **Jones Matrix Method**

The Jones vector method is a mathematical approach used to represent the polarization state of light. It was developed by Jones and is based on the use of complex numbers and matrices to describe the amplitude and phase of the electric field vector associated with the light wave.

Jones matrix uses a matrix with two rows and one column to represent the components of the light vector in the x and y directions. The polarization state of light is represented using a two-element column vector known as the Jones vector. The elements of the vector are complex numbers and correspond to the amplitudes of the electric field components in specific polarization directions. To fully characterize the polarization state of light using the Jones vector method, it is necessary to consider both the amplitude and phase information of the electric field components. The phase information is incorporated by representing the complex numbers in terms of their magnitudes and phase angles as

$$J = \begin{bmatrix} E_x \\ E_y \end{bmatrix} = \begin{bmatrix} E_{ox} e^{i\delta_1} \\ E_{oy} e^{i\delta_2} \end{bmatrix}. \tag{14.7}$$

In the above formula, E_{ox} and δ_1, E_{ox} and δ_2 are the amplitudes and phases of the light vector in the x and y directions, respectively. In order to more directly reflect the relationship between the amplitude and phase of each component, we simplify the Eq. (14.7) and extract the common factor to obtain:

$$J = E_{ox} e^{\delta_x} \begin{bmatrix} 1 \\ E_0 E^{i\delta} \end{bmatrix}. \tag{14.8}$$

Due to the limitations of the Jones matrix, it can only be used to describe fully polarized light, not unpolarized light and partially polarized light. The Jones vector method allows for the representation of a wide range of polarization states, including linear, circular, and elliptical polarization. By applying appropriate mathematical operations on the Jones vectors, such as rotations and linear transformations, it is possible to analyze and manipulate the polarization properties of light.

Several typical fully polarized lights are normalized, and their Jones matrix is shown in Table 14.1.The Jones vector method is widely used in various fields, including optics, telecommunications, and quantum optics. It provides a convenient and intuitive representation of the polarization properties of light and facilitates the design and analysis of polarization-based optical systems.

(2) **Stokes Vector Method**

Stokes vector method is a mathematical method to describe the polarization state of light. Developed by Stokes, it provides a comprehensive representation of the polarization properties of light waves.

Table 14.1 Jones matrix of fully polarized light

Typical fully polarized light	Jones matrix
0° line polarized light	$E = [1 \quad 0]^T$
±45° line polarized light	$E = \frac{1}{\sqrt{2}[1 \quad \pm 1]^T}$
90° line polarized light	$E = [0 \quad 1]^T$
Right-handed circularly polarized light	$E = \frac{1}{\sqrt{2}[1 \quad -i]^T}$
Left-handed circularly polarized light	$E = \frac{1}{\sqrt{2}[1 \quad i]^T}$

In the Stokes vector method, the polarization state of light is described using a four-element vector known as the Stokes vector. The elements of the vector are calculated based on the intensity and polarization characteristics of the light wave. The four elements of the Stokes vector are denoted as S. Compared with Jones matrix, Stokes vector has two more parameters, but there is no exponential and complex operation, and these four parameters are real numbers, as long as supplemented by simple photoelectric detection equipment, these parameters can be measured. The specific function definition and physical meaning are as follows

$$S_0 = \langle E_{ox}^2(t)\rangle + \langle E_{oy}^2(t)\rangle, \tag{14.9}$$

$$S_1 = \langle E_{ox}^2(t)\rangle - \langle E_{oy}^2(t)\rangle, \tag{14.10}$$

$$S_2 = 2\langle E_{ox}(t)E_{oy}(t)\cos\delta(t)\rangle, \tag{14.11}$$

$$S_3 = 2\langle E_{ox}(t)E_{oy}(t)\sin\delta(t)\rangle, \tag{14.12}$$

where S_0 represents the total intensity of the light wave and is equal to the sum of the intensities of the vertically and horizontally polarized components. S_1 represents the difference in intensity between the horizontally and vertically polarized components of the light. S_2 represents the difference in intensity between the 45° and 135° linearly polarized components. S_3 represents the difference in intensity between the right-handed and left-handed circularly polarized components.

By calculating the following four elements, the Stokes vector fully characterizes the polarization state of light. The Stokes vector method allows for the representation of various types of polarization states, including linear, circular, and elliptical polarization, as well as unpolarized light. In the process of polarization detection, if only $[\,S_0\ S_1\ S_2\,]^{\mathrm{T}}$ are obtained, the system is called linear polarization imaging system. If the four components of $[\,S_0\ S_1\ S_2\ S_3]^{\mathrm{T}}$ are obtained, then the system is a full-polarization imaging system.

$$\begin{aligned} S &= \left[\,S_0\ S_1\ S_2\ S_3\,\right]^{\mathrm{T}} \\ &= \left[\,I_{0^\circ} + I_{90^\circ}\ I_{0^\circ} + I_{90^\circ}\ I_{45^\circ} - I_{135^\circ}\ I_R - I_L\,\right]^{\mathrm{T}}. \end{aligned} \tag{14.13}$$

In order to further accurately describe the properties of polarized light, the concepts of degree of polarization and Angle of polarization are introduced. Polarization reflects the proportion of polarized light intensity in the total light intensity, defined as

$$D_{op} = \frac{\sqrt{Q^2 + U^2 + V^2}}{I} = \frac{\sqrt{(I_{0^\circ} - I_{90^\circ})^2 + (I_{45^\circ} - I_{135^\circ})^2 + (I_R - I_L)^2}}{I_{0^\circ} + I_{90^\circ}}. \tag{14.14}$$

When the circular polarization component S_3 is ignored, the degree of polarization becomes linear polarization, and the expression is:

$$D_{olp} = \frac{\sqrt{Q^2 + U^2}}{I} = \frac{\sqrt{(I_{0^\circ} - I_{90^\circ})^2 + (I_{45^\circ} - I_{135^\circ})^2}}{I_{0^\circ} + I_{90^\circ}}. \quad (14.15)$$

Polarization Angle, also known as polarization azimuth, represents the Angle between the major half axis of the ellipse and the X-axis in elliptically polarized light. It can characterize the spatial orientation of elliptically polarized light, and is expressed as

$$A_{op} = \frac{1}{2}\arctan\left(\frac{U}{Q}\right). \quad (14.16)$$

The Stokes vector method is widely used in fields such as optics, polarimetry, and remote sensing. It provides a compact and quantitative representation of the polarization properties of light, enabling detailed analysis and manipulation of polarized light waves.

Table 14.2 shows the calculation formula and expression significance of each polarization parameter. These polarization parameters can reflect the inherent properties of material such as medium characteristics, structural characteristics, roughness and moisture content [6–8].

(3) **Poincaré sphere**

Poincaré used a sphere with a radius of $S_0 = 1$ to describe the polarization state of light waves, which was named the Poincaré sphere, as shown in Fig. 14.3. Poincaré sphere is a graphical expression of Stokes vector, which can visually reflect the polarization state of light waves, and polarized light of any polarization state can find corresponding points on the sphere.

Use the equator as the dividing line for analysis. The equator of the Poincaré sphere represents linear polarization states. The North Pole represents right circular polarization, and the South Pole represents left circular polarization. The points on the sphere between the poles represent different elliptical polarization states, with

Table 14.2 Expression of polarization parameters and information representation

Polarization parameter	Expression	Information representation
Degree of polarization	$D_{op} = \frac{\sqrt{S_1^2+S_2^2+S_3^2}}{S_0}$	Polarization information, image contrast
Polarization azimuth	$A_{op} = \frac{1}{2}\tan^{-1}\left(\frac{S_2}{S_1}\right)$	Edge information
Degree of circular polarization	$D_{ocp} = \frac{S_3}{S_0}$	Circularly polarized light content
Polarization ellipticity	$\varepsilon = \frac{1}{2}\tan^{-1}\left(\frac{S_3}{\sqrt{S_1^2+S_2^2}}\right)$	Roughness, reflectivity

Fig. 14.3 Poincaré sphere

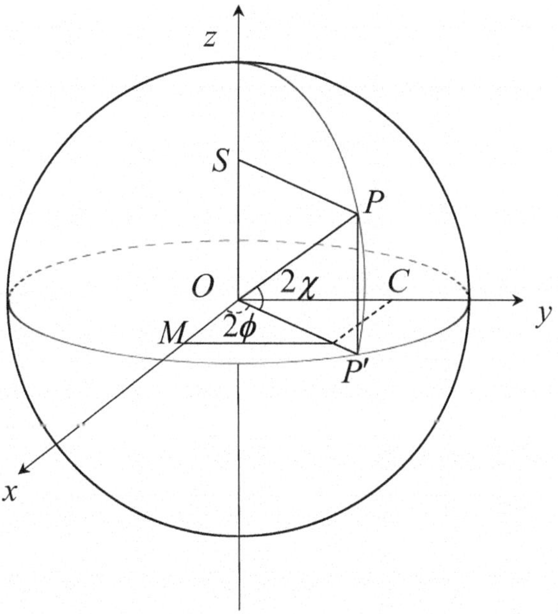

varying degrees of ellipticity and tilt angles. Unpolarized light is represented by the center of the sphere. The distance from the center of the sphere to a particular point represents the degree of polarization. Points on the surface of the sphere represent fully polarized light, while points inside the sphere represent partially polarized light.

Points on the equator correspond to $S_3 = 0$, and according to Stokes vector formula, equator describes linearly polarized light; The north and south poles of the equator correspond to $S_1 = S_2 = 0$, so the poles describe circularly polarized light; None of the other three components on the sphere is zero, so these points describe elliptically polarized light, and left-polarized light or right-polarized light depends on its position above and below the equator. The relationship between the Stokes vector and the Poincaré sphere is shown as

$$\begin{bmatrix} S_0 \\ S_1 \\ S_2 \\ S_3 \end{bmatrix} = I \begin{bmatrix} 1 \\ P\cos 2\chi \cos 2\phi \\ P\cos 2\chi \sin 2\phi \\ P\sin 2\chi \end{bmatrix}. \tag{14.17}$$

In the above formula, P is the degree of polarization (D_o), and the position of P can represent the polarization state of the light wave. According to the geometric relation, when $S_0^2 = S_1^2 + S_2^2 + S_3^2$, P on the sphere corresponds to fully polarized light; When $S_0^2 > S_1^2 + S_2^2 + S_3^2$, P is located in the sphere, which corresponds to partially polarized light; When P is at the center of the sphere, it describes unpolarized light.

By using the Poincaré sphere, it is possible to visualize various polarization transformations, such as rotations, phase shifts, and conversions between different polarization states. It also provides a useful tool for analyzing polarization components, such as polarizers, waveplates, and retardation plates, and their effects on the polarization state of light.

The Poincaré sphere is extensively employed in various fields, including optics, telecommunications, quantum optics, and polarimetry. It simplifies the understanding and analysis of polarization phenomena and facilitates the design and optimization of polarization-based devices and systems.

14.2.3 Polarization Measurement Method

(1) **Müller Matrix**

The Müller matrix, also known as Muller matrix, named after physicist Hans Müller, is a mathematical representation of the polarization properties of an optical system. It describes the transformation of an incident polarized light state into an output polarized light state when interacting with various optical elements, such as wave plates, polarizers, and birefringent materials [9–12].

When the light beam passes through the optical device, the optical device will have a certain effect on the incident light, causing the polarization state of the outgoing light to change. Because there are sometimes multiple optical devices in a system, in order to describe the changes of polarized light represented by Stokes vector in the transmission process of the system, a 4 × 4 matrix was introduced to reflect the transmission characteristics of the optical devices, which was named Muller matrix. Let the Stokes vector of the incident light be S_{in}, the Stokes vector of the outgoing light be S_{out}, and M be the Muller matrix of the optical device, then the relationship among them is as follows

$$S_{out} = M \times S_{in} = \begin{bmatrix} m_{11} & m_{12} & m_{13} & m_{14} \\ m_{21} & m_{22} & m_{23} & m_{24} \\ m_{31} & m_{32} & m_{33} & m_{34} \\ m_{41} & m_{42} & m_{43} & m_{44} \end{bmatrix} \begin{bmatrix} S_0 \\ S_1 \\ S_2 \\ S_3 \end{bmatrix}. \tag{14.18}$$

They quantify how the system modifies the incident light's intensity and polarization components. For example, m_{11} represents the output intensity when the input light is unpolarized, m_{12} and m_{21} describe how the system transforms linearly polarized light, and m_{44} characterizes the circular polarization response.

The Muller matrix can capture depolarization effects caused by scattering, random fluctuations, or multiple optical paths within a system. A depolarizing system will have non-zero elements off the diagonal [m_{12} m_{13} m_{14} m_{23} m_{24} m_{34}], indicating how it mixes or attenuates the polarization components.

The Muller matrix can be decomposed into a combination of basic polarization effects. This decomposition provides insights into the specific optical processes occurring within a system. Common decompositions include the Pauli decomposition, the spectral decomposition.

The Muller matrix can be used to describe the state change in the beam propagation process intuitively. As a common polarizing light modulator, the Muller matrix of polarizer and liquid crystal adjustable phase retarder is shown as

$$M_\alpha = \frac{1}{2}\begin{bmatrix} 1 & \cos 2\alpha & \sin 2\alpha & 0 \\ \cos 2\alpha & \cos^2 2\alpha & \cos 2\alpha \sin 2\alpha & 0 \\ \sin 2\alpha & \cos 2\alpha \sin 2\alpha & \sin^2 2\alpha & 0 \\ 0 & 0 & 0 & 0 \end{bmatrix}, \tag{14.19}$$

$$M_{\beta,\delta} = \begin{bmatrix} 1 & 0 & 0 & 0 \\ 0 & \cos^2 2\beta + \sin^2 2\beta \cos\delta & \cos 2\beta \sin 2\beta(1-\cos\delta) & -\sin 2\beta \sin\delta \\ 0 & \cos 2\beta \sin 2\beta(1-\cos\delta) & \sin^2 2\beta + \cos^2 2\beta \cos\delta & \cos 2\beta \sin\delta \\ 0 & \sin 2\beta \sin\delta & -\cos 2\beta \sin\delta & \cos\delta \end{bmatrix}. \tag{14.20}$$

In Eq. (14.19), α is the transmittance direction of the polarizer. In Eq. (14.20), β is the fast axis direction of the liquid crystal tunable phase retarder, and δ is the phase delay of the liquid crystal tunable phase retarder.

To measure or calculate the Muller matrix, a known input polarization state is applied to the system, and the resulting output polarization state is measured or determined. This process is repeated for multiple input polarization states to obtain the full Muller matrix.

The Muller matrix provides valuable information about how an optical system affects the polarization state of light. It can be used to analyze and predict the behavior of polarized light when passing through or reflecting off various optical components or materials. The matrix elements can reveal the effects of birefringence, dichroism, depolarization, and other polarization phenomena.

The Muller matrix can be used to study the behavior of individual polarization components. By applying the Muller matrix to specific input states (e.g., linearly or circularly polarized light), the resulting output Stokes vector can be analyzed to determine the polarization characteristics of the system, such as the degree of polarization, the orientation of polarization, or the presence of polarization-dependent effects.

Accurate measurement of the Muller matrix requires careful calibration of the measurement setup. This involves calibrating the detectors, compensating for instrumental artifacts, and accounting for any polarization alterations introduced by the measurement system itself. Calibration methods can include using known polarized references or employing polarimetric calibration standards.

In some cases, it is desirable to invert the Muller matrix to retrieve the input Stokes vector from the measured output Stokes vector. This inversion process can be challenging due to the potential presence of noise, non-idealities, and limitations in measurement precision. Various techniques, such as least-squares fitting or optimization algorithms, can be employed for matrix inversion and reconstruction.

Overall, the Muller matrix provides a comprehensive description of the polarization properties of an optical system. It is a versatile tool for understanding and analyzing the polarization response of various optical elements and materials. By utilizing the Muller matrix, researchers and engineers can design, optimize, and characterize optical systems with specific polarization requirements, as well as develop advanced polarization imaging and measurement techniques.

(2) **Linear Polarization Measurement of Stokes Vector**

In engineering practice, polarization detection sometimes does not need to obtain all four components of Stokes vector according to different research objects. For example, in the process of infrared polarization imaging, the energy of the circular polarization component S_3 is very small and is often ignored. This kind of detection, which ignores S_3 component and only aims at acquiring $[\, S_0\ S_1\ S_2 \,]$ components, is called linear polarization measurement of Stokes vector (Fig. 14.4).

Suppose that the initial polarization state of the incident light after collimating by the collimating lens is $S_{in} = \left[\, S_0\ S_1\ S_2\ S_3 \,\right]^{\mathrm{T}}$, and the polarization state of the outgoing light after modulating by the polarization plate is $S_{out} = \left[\, S_0'\ S_1'\ S_2'\ S_3' \,\right]^{\mathrm{T}}$. The outgoing light can be expressed as $S_{out} = M_\alpha \times S_{in}$ by incident light, which is expanded as

$$\begin{bmatrix} S_0' \\ S_1' \\ S_2' \\ S_3' \end{bmatrix} = \frac{1}{2} \begin{bmatrix} 1 & \cos 2\alpha & \sin 2\alpha & 0 \\ \cos 2\alpha & \cos^2 2\alpha & \cos 2\alpha \sin 2\alpha & 0 \\ \sin 2\alpha & \cos 2\alpha \sin 2\alpha & \sin^2 2\alpha & 0 \\ 0 & 0 & 0 & 0 \end{bmatrix} \begin{bmatrix} S_0 \\ S_1 \\ S_2 \\ S_3 \end{bmatrix}, \tag{14.21}$$

$$S_0' = \frac{S_0 + S_1 \cos 2\alpha + S_2 \sin 2\alpha}{2}. \tag{14.22}$$

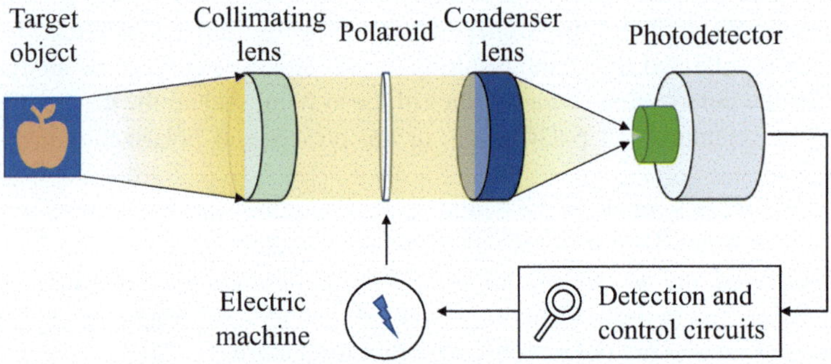

Fig. 14.4 Principle of linear polarization measurement

From the Eq. (14.22), we can see that in order to obtain the three unknowns of (S_0 S_1 S_2), we need to construct three equations. For the solution of these three components, there are two main methods in engineering practice:

The polarization Angle of the polarizer is 0°, 60° and 120° respectively, and the three components can be obtained as follows:

$$\begin{cases} S_0 = \frac{2}{3}\left[S_0'(0^\circ) + S_0'(60^\circ) + S_0'(120^\circ)\right], \\ S_1 = \frac{2}{3}[2S_0'(0^\circ) - S_0'(60^\circ) - S_0'(120^\circ)], \\ S_2 = \frac{2\sqrt{3}}{3}[S_0'(60^\circ) - S_0'(120^\circ)]. \end{cases} \tag{14.23}$$

The polarization angles of the polarizer are 0°, 45°, 90° and 135° respectively, and the three components can be obtained as follows

$$\begin{cases} S_0 = \frac{1}{2}[S_0'(0^\circ) + S_0'(45^\circ) + S_0'(90^\circ) + S_0'(135^\circ)], \\ S_1 = S_0'(0^\circ) - S_0'(90^\circ), \\ S_2 = S_0'(45^\circ) - S_0'(135^\circ). \end{cases} \tag{14.24}$$

Method 1 and method 2 have the same principle, both construct three equations to solve the unknowns simultaneously, but method 2 requires one more rotation of the polarizer than method 1. After the linear polarization component of the Stokes vector is obtained, the degree of polarization and Angle of polarization of the incident light can then be calculated.

(3) **Full Polarization Measurement of Stokes Vector**

The linear polarization measurement of Stokes vector simplifies the measurement scheme by sacrificing parameters, which has great limitations in practical application. In order to fully understand the properties of the measured object in the process of polarization imaging, a phase retarder ($\lambda/4$ wave plate) was added to the on-line polarization measurement for full polarization measurement. The measurement principle is shown in Fig. 14.5.

Here $\lambda/4$ plate is a traditional phase retarder, which can make the phase difference between two components with different vibration directions in the light vector

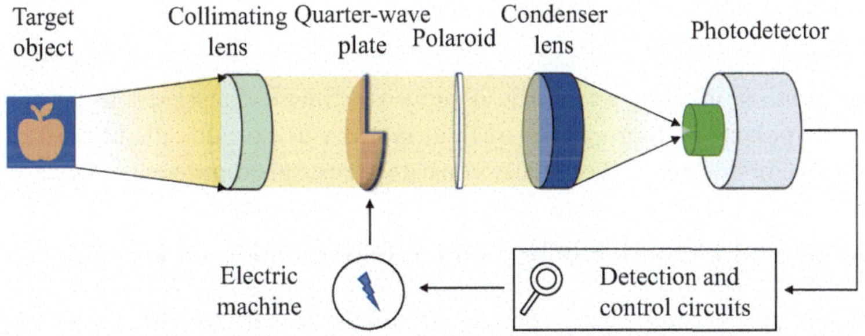

Fig. 14.5 Principle of full polarization measurement

become $\lambda/4$, which is generally used to produce circular polarized light, and its Muller matrix can be expressed as

$$M_\beta = \frac{1}{2}\begin{bmatrix} 1 & 0 & 0 & 0 \\ 0 & \cos^2 2\beta & \cos 2\alpha \sin 2\alpha & -\sin 2\beta \\ 0 & \cos 2\beta \sin 2\beta & \sin^2 2\beta & \cos 2\beta \\ 0 & \sin 2\beta & -\cos 2\beta & 0 \end{bmatrix}. \tag{14.25}$$

In the above formula, β is the Angle between the main axis of the $\lambda/4$ wave plate and the reference direction. The outgoing light in full polarization measurement can be expressed as:

$$S_{out} = M_a M_\beta S_{in}. \tag{14.26}$$

By analogy with the derivation of the linear polarization measurement method, the Stokes parameters of the outgoing light in the total polarization measurement can be obtained as follows:

$$\begin{aligned} S_0'(\alpha, \beta) = &\frac{1}{2}S_0 + \frac{1}{2}\cos 2\alpha(S_1 \cos^2 2\beta + S_2 \cos 2\beta \sin 2\beta - S_3 \sin 2\beta) \\ &+ \frac{1}{2}\sin 2\alpha(S_1 \cos 2\beta \sin 2\beta + S_2 \sin^2 2\beta + S_3 \cos 2\beta). \end{aligned} \tag{14.27}$$

When $\beta = 0°$ this is linear polarization measurement, then the three components of (S_0, S_1, S_3) can be obtained according to method 1 or method 2 mentioned above; When $\beta = 45°$ and $\alpha = 0°$, the Eq. (14.27) can be simplified to:

$$S_0'(0°, 45°) = \frac{S_0 - S_3}{2}. \tag{14.28}$$

The S_3 component can be obtained through the Eq. (14.28), and the solution of all four components in Stokes vector has been completed.

14.3 Polarization Imaging System

The polarization imaging system is a setup specifically designed to capture and analyze the polarization properties of light in order to extract valuable information about a scene or an object. It typically consists of optical components, detectors, and data analysis techniques tailored for polarization measurements.

Typical components of a polarization imaging system fall into the following categories:

Light Source: A controlled and polarized light source is essential for polarization imaging. It can be achieved using lasers, light-emitting diodes (LEDs), or specialized polarized light sources. The source can provide a specific polarization state (e.g., linear, circular) or a combination of states.

Polarizers: Polarizers are optical elements that transmit light with specific polarization orientations while blocking light with other orientations. They are often used to control the polarization state of incident light or to analyze the polarization properties of the transmitted or reflected light.

Wave Plates: Wave plates, also known as retarders or phase shifters, are birefringent optical components that introduce controlled phase delays between the polarization components of light. They are employed to manipulate and transform the polarization state of light.

Imaging Optics: These include lenses, mirrors, or other optical elements used to focus and direct the light onto the imaging sensor. The imaging optics should be designed to minimize polarization distortions and aberrations.

Imaging Sensor: The imaging sensor captures the intensity and polarization information of the light. It can be a charge-coupled device (CCD) sensor, a complementary metal–oxide–semiconductor (CMOS) sensor, or a specialized polarization-sensitive sensor. Each pixel of the sensor records the intensity and polarization state of the incident light.

Data Acquisition: The polarization information captured by the imaging sensor needs to be acquired and processed. This can be done using analog-to-digital converters (ADCs) to convert the analog signal from the sensor into digital data. The acquired data is then ready for further analysis.

Data Analysis: Various algorithms and techniques are employed to process and analyze the acquired polarization data. This can involve computing Stokes parameters, Mueller matrices, or other polarization descriptors to extract relevant information about the scene or object. Image processing techniques, machine learning algorithms, and statistical analysis methods may be applied to enhance the extracted polarization information.

Display and Visualization: The polarization imaging system may include a display or visualization module to present the processed polarization information in a meaningful way. This can involve color-coded maps, polarization images, or other visualization techniques to highlight polarization variations or features.

The specific design and configuration of a polarization imaging system can vary depending on the application requirements, the desired accuracy, and the level of complexity needed for polarization analysis. Customizations can be made by incorporating additional optical elements, polarization modulators, or specialized sensors to suit specific research or industrial needs. According to different polarization imaging principles, polarization imaging systems can be divided into the following categories.

14.3.1 Linear Polarization Information Detection Technology

At present, there are three imaging methods for linear polarization information measurement:

(1) **Division-of-time Polarimeter**

Division-of-time polarimeter is used to obtain different polarization state images of the same scene by rotating polarization device or liquid crystal phase variable delay device plus polarizer. Because the system structure is simple and easy to implement, it is the earliest and most mature polarization imaging system. The system has the advantages of simple design, small size and easy debugging, but it also has some disadvantages such as registration error due to mechanical rotation and not suitable for dynamic scenes (Fig. 14.6).

The division-of-time polarimeter is based on the principle of rotating polarizers and uses a modulation technique to measure the polarization state. It consists of a series of polarizing elements, such as polarizers or wave plates, along with detectors to capture the intensity of the light after each polarization element. The polarizers are typically arranged in a rotating fashion.

Division-of-time polarimeters are commonly used in research and industrial applications where precise polarization measurements are required. They find applications in areas such as material science, optics, telecommunications, and biophysics. By accurately characterizing the polarization properties of light, division-of-time polarimeters contribute to the understanding and advancement of various fields related to light-matter interactions.

(2) **Division-of-amplitude Polarimeter**

The Division-of-amplitude polarimeter, also known as a Mach–Zehnder interferometer, is an optical device that combines polarization analysis with interferometry. It is used to measure the polarization state of light and can provide information about the amplitude and phase changes induced by a sample.

The amplitude-based polarization information detection system, as shown in Fig. 14.7, first divides the incident light into three or four optical paths by using the spectral element, and each light path is incident into its corresponding detector through the polarizer of different polarization angles. Then the polarization information obtained by each detector is collected and fused by the computer to solve the problem, and finally the polarization image is obtained. To achieve the purpose of simultaneous imaging.

The split-amplitude imaging method makes up for the timeliness of the time-sharing polarization imaging system, but the complexity of the system increases,

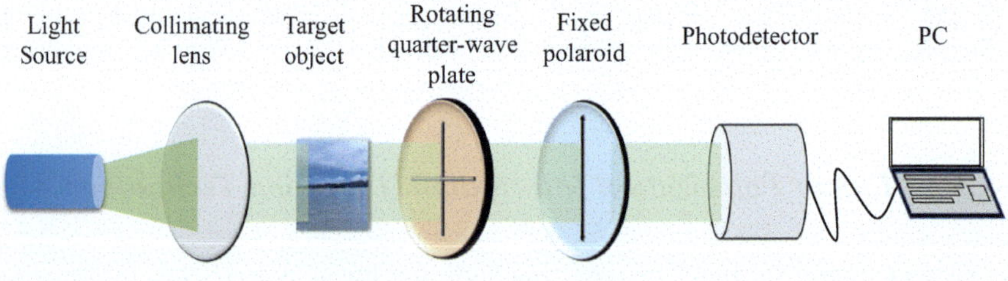

Fig. 14.6 Principle diagram of division-of-time polarization imaging system

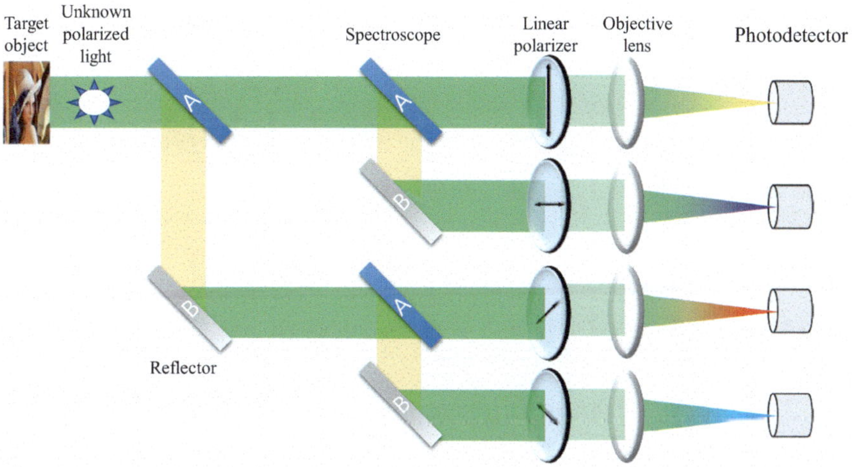

Fig. 14.7 Principle diagram of division-of-amplitude polarimeter imaging system

and multiple sets of lenses, spectroscopic elements and detectors are added, making the volume, weight and manufacturing cost of the system greatly increase, and the optical elements and detectors of each spectroscopic path are different, there is inevitably non-ideal noise, which makes the imaging conditions of each optical path different. Error correction and image registration are necessary to achieve the accurate calculation of polarization characteristics.

Division-of-amplitude polarimeters are widely used in various fields, including optics, microscopy, metrology, and optical characterization of materials. They provide valuable insights into the polarization properties of light and help in studying polarization-dependent phenomena and characterizing samples based on their polarization properties.

(3) **Division-of-aperture Polarimeter**

As Fig. 14.8, the scene incident light in this system is divided into four channels after passing through the micro-lens array, and polarization elements with different polarization directions are placed in the four channels respectively. After the light passes through the polarization element, it is imaged into four equal regions in the same detector, and finally through the computer registration to achieve simultaneous linear polarization imaging.

In order to overcome the real-time problem of the mechanical rotation polarization imaging system and the complex structure of the split-amplitude polarization imaging system, a new polarization information detection form has been developed. With the innovation of detector technology and the emergence of micro-lens array, a polarizing camera with splitter aperture was proposed in the late 1990s.

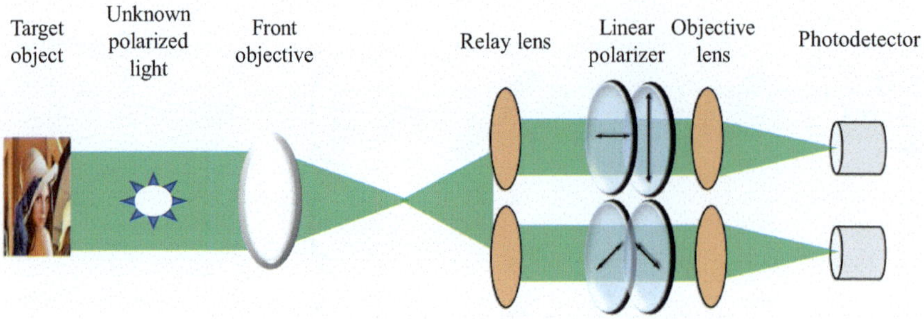

Fig. 14.8 Principle diagram of division-of- aperture polarization imaging system

14.3.2 Full Polarization Information Detection Technology

Compared with linear polarization information detection technology, full polarization information detection technology can obtain more comprehensive polarization information about the target scene. With the increasingly extensive application of polarization technology in the field of target detection, the relevant research work on circular polarization characteristics is also increasing, especially in the aspects of cloud and fog penetration and underwater detection. At present, there are three imaging methods for full polarization information measurement.

(1) **Polarization Modulation Technology Based on Liquid Crystal**

The traditional polarization modulation usually adopts wave plate, which needs manual or motor to drive the wave plate to rotate during the experiment. Although this method is simple to operate, due to mechanical rotation, the system will be unstable in the measurement, and the obtained data is not accurate. With the development of liquid crystal technology, liquid crystal based variable phase retarder came into being. Its appearance effectively makes up for the shortage of wave plate in polarization modulation.

Liquid crystal modulation polarization information detection technology has two forms, one is to use voltage control liquid crystal wave plate to produce different phase delay, through the combination of two phase delay of two liquid crystal wave plates to obtain four phase delay states of the beam. The other is to use the voltage to change the optical axis direction of the liquid crystal element, and the two optical axis direction changes of the two liquid crystal wave plates are combined to obtain the polarization image of four different polarization angles.

Both of these forms reflect the polarization parameters of the incident light by repeatedly measuring the intensity of the emitted light. This detection technology can measure all Stokes vectors and calculate the full polarization information. Figure 14.9 is the optical schematic diagram of the liquid crystal modulated polarization camera.

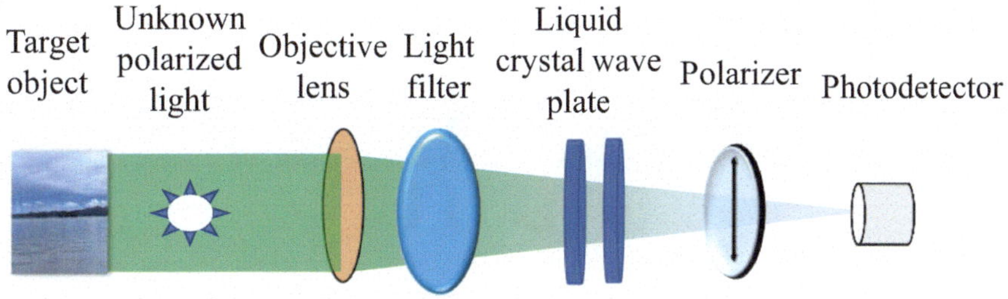

Fig. 14.9 Principle of liquid crystal modulated polarization imaging system

(2) **Division-of-focal-plane polarimeter**

The division-of-focal-plane polarimeter, also known as a DFP polarimeter, is an optical instrument used to measure the polarization properties of light with high spatial resolution. Unlike traditional polarimeters that analyze the polarization state of the entire beam simultaneously, DFP polarimeters divide the incident light into sub-beams using an array of polarization-sensitive elements located at the focal plane of the optical system. Each sub-beam is then independently analyzed for its polarization state. The imaging mode of the focal plane polarization imaging system is shown in Fig. 14.10. In order to reduce the instantaneous field error, the Stokes vector of each pixel is generally obtained by interpolation in the calculation process of this method, so the spatial resolution is relatively reduced [13].

Division-of-focal-plane polarimeters find applications in various fields, including remote sensing, polarized imaging, microscopy, astronomy, and biomedical research. They enable the characterization of the polarization properties of light with high spatial resolution, providing valuable insights into polarization-dependent phenomena and enhancing our understanding of the interaction between light and matter in different contexts. In addition, the non-uniformity of the polarizer array needs to be corrected due to the difference in the marking accuracy and transmittance of the polarizer unit. At the same time, due to many difficulties in working mechanism, optimal design and accurate registration with detector pixels, circular

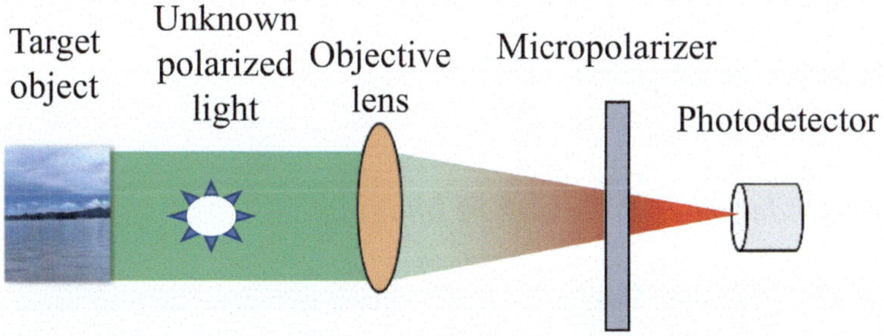

Fig. 14.10 Principle of division-of-focal-plane polarimeter imaging system

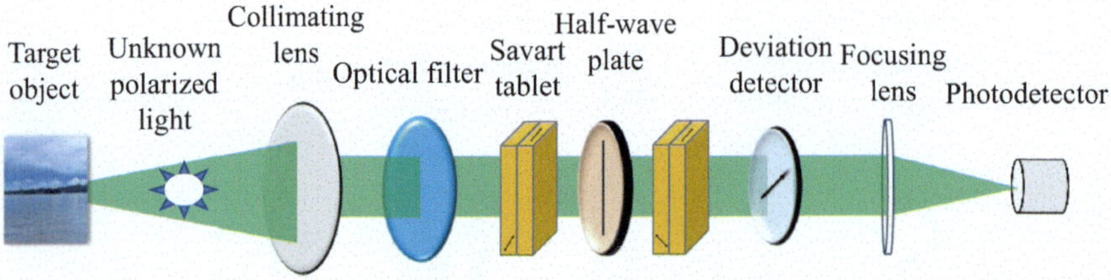

Fig. 14.11 Principle of Channeled linear polarimeter imaging system

polarizer arrays are mostly in the research and development stage and have not been widely used.

(3) **Channel Modulated Polarization Imaging**

Channel-modulated polarization information detection technology uses optical splitting elements to divide the incident light into multiple beams with modulated polarization information, which are simultaneously imprinted on the focal plane of the detector and interfere. The obtained interference image with the polarization information of the target is solved by Fourier transform to calculate each Stokes vector, so as to achieve full polarization information detection. According to the different spectral components of this detection technology, it is further subdivided into birefringent prism spectral type, Savart plate spectral type and grating spectral type. Figure 14.11 shows a Savart plate split modulated polarization detection system. Incident light is modulated into four light paths with different polarization directions after passing through two Savart plates, and converges to the detector focal plane after passing through the focusing lens. Fourier analysis is used to calculate each Stokes vector from the interference pattern.

14.4 Application of Polarization Imaging

Applications of polarization imaging are found in various fields, including remote sensing, material characterization, biomedical imaging, and computer vision. The following is a brief introduction to the three areas where polarization imaging applications are more excellent [14, 15].

14.4.1 Underwater Polarization Imaging

Underwater polarization imaging is a specialized technique used to capture and analyze the polarization properties of light in aquatic environments. It aims to overcome challenges such as light scattering, absorption, and surface reflections, which can degrade image quality and visibility in underwater scenes. By selectively analyzing polarized light, underwater polarization imaging offers advantages such as improved contrast, reduced glare, and enhanced object detection and recognition in underwater applications.

Polarized light in water: When light interacts with water, it becomes partially polarized due to scattering and reflection processes. The polarization state of light in water varies with factors such as depth, water composition, and the angle of incidence. Polarization imaging systems for underwater use typically employ polarizing filters and analyzers to selectively filter or analyze polarized light. These optical components allow specific polarization states to pass while blocking others. They can be used to reduce glare and enhance contrast in underwater scenes.

Polarization imaging can enhance object detection and recognition in underwater environments. By exploiting the polarization differences between objects and the surrounding water, polarization imaging can help differentiate objects from the background and enhance their visibility. This is particularly useful for underwater robotics, marine biology research, and underwater surveillance.

Polarization imaging can aid in underwater navigation and mapping tasks. By leveraging polarization cues, such as the polarization patterns of the sky or underwater landmarks, polarization imaging systems can assist in navigation and orientation in environments where traditional visual cues may be limited. Polarization imaging techniques can be valuable in underwater archaeology and the study of submerged cultural heritage. By enhancing image quality and reducing surface reflections, polarization imaging can assist in the identification and documentation of underwater archaeological sites, submerged structures, and artifacts (Table 14.3).

Underwater polarization imaging continues to be an active area of research and development, with ongoing efforts to improve imaging techniques, hardware, and image analysis algorithms. By exploiting the polarization properties of light underwater, researchers and practitioners aim to overcome the challenges associated with underwater imaging and expand our understanding of underwater ecosystems, underwater structures, and underwater cultural heritage.

14.4.2 Infrared Polarization Imaging

Infrared polarization imaging refers to the process of capturing and analyzing the polarization properties of infrared light. It involves using infrared cameras or sensors that are sensitive to the infrared wavelength range, typically beyond what is visible to the human eye. Infrared polarization imaging has gained significant attention and applications in various fields.

Infrared Light and Polarization: Infrared (IR) light has longer wavelengths than visible light and falls within the electromagnetic spectrum between microwaves and visible light. Similar to visible light, infrared light can also exhibit polarization characteristics. By analyzing the polarization state of infrared light, valuable information about the objects, materials, and phenomena being imaged can be obtained.

Table 14.3 Comparison of different polarization imaging methods

Type	Time type	Advantages	Disadvantages
Division-of-time	Time division	Mature technology; Simple structure; Low cost	Poor real-time performance; The measurement accuracy of dynamic target is low
Division-of-amplitude	Simultaneous	Good real-time performance; High spatial resolution	Complex structure; The accuracy of inter-channel calibration is high
Division-of-aperture	Simultaneous	Good real-time performance; Compact structure	Loss of spatial resolution; Difficult to ensure consistency between channels
Based on liquid crystal	Time division	Small size; Low power consumption; Large aperture through light; High detection accuracy	Poor real-time performance; Narrow band
Division-of-focal-plane	Time division	Good real-time performance; Small size; Simple optical structure	Loss of spatial resolution; There is registration error; Complex processing technology
Channel modulated	Simultaneous	Good real-time performance; Compact structure; Less energy loss	Loss of spectral resolution; The solution is complex; Information is easily aliased

Infrared polarization imaging can be combined with thermal imaging techniques. Thermal imaging captures the distribution of heat emitted by objects, while polarization imaging adds polarization information. This combination enables the detection of polarization-dependent thermal effects, which can provide additional insights into the thermal properties and behavior of objects.

Polarization imaging is also widely used in the field of material analysis and detection. It can reveal important information about the composition, structure, and thermal properties of materials. By analyzing the polarization state of infrared light reflected, transmitted, or emitted by materials, infrared polarization imaging can assist in non-destructive testing, quality control, and defect detection in industries such as manufacturing, electronics, and aerospace.

By combining the advantages of infrared imaging with polarization analysis, it can provide additional information about tissues and biological structures. It may help in identifying anomalies, characterizing tumors, and studying the polarization properties of biological tissues for diagnostic purposes.

Infrared polarization imaging offers a valuable tool for extracting additional information from the infrared spectrum. By combining infrared imaging with polarization analysis, it provides insights into the polarization properties of objects and materials, enhancing various applications in fields such as material science, remote sensing, biomedical imaging, and astronomy.

14.4.3 Polarization Imaging Defogging

Polarization imaging defogging, also known as polarization-based defogging or polarimetric defogging, is a technique used to improve the visibility of outdoor scenes affected by fog, haze, or atmospheric scattering. It leverages the polarization properties of light to estimate and remove the atmospheric scattering component, resulting in clearer and sharper images.

Atmospheric Scattering and Fog: When light travels through a foggy or hazy atmosphere, it undergoes scattering due to the presence of suspended particles and molecules. This scattering causes a loss of contrast, reduced visibility, and degraded image quality in outdoor scenes. Scattered light exhibits specific polarization patterns that can be exploited for defogging. Polarization imaging techniques capture and analyze the polarization state of light to estimate the amount of scattering and recover the underlying scene information.

Polarization Difference Imaging: One common approach in polarization imaging defogging is to capture multiple images with different polarization states of light, such as horizontal and vertical polarizations. These images are then used to compute the polarization difference, which corresponds to the polarization-dependent scattering component. From the polarization difference images, various algorithms can be employed to estimate the atmospheric parameters, such as the scattering coefficients or the degree of polarization. These parameters characterize the scattering properties of the foggy atmosphere and are essential for the defogging process.

Using the estimated atmospheric parameters, the polarization imaging defogging algorithm aims to separate the scene information from the scattering component. By utilizing inverse scattering models or statistical estimation techniques, the algorithm reconstructs a clearer and less-scattered image, effectively removing the fog or haze.

Polarization imaging defogging offers advantages over traditional defogging methods that rely solely on intensity information. By exploiting the polarization properties of light, it can better distinguish between the scene and the scattering component, resulting in more accurate and robust defogging. However, polarization imaging defogging methods require specialized hardware, such as polarizing filters or polarimeters, and may have limitations in certain scenarios with complex or non-uniform fog conditions.

Polarization imaging defogging techniques have been applied in various fields, including surveillance, transportation, remote sensing, and outdoor imaging applications. By effectively reducing the impact of atmospheric scattering, polarization imaging defogging enhances image visibility, improves object detection, and enables clearer visualization of outdoor scenes affected by fog or haze.

14.5 General Conclusions

In this chapter, we first introduce the imaging principle of polarization imaging. The polarization state of light is regarded as an important optical information in modern optical technology and is applied to many scenarios. By using the polarization state of light, we can obtain the desired image in a number of scenes that cannot be properly imaged by conventional optics.

Secondly, we systematically introduce the mechanism of light polarization. According to the degree of polarization, light can be divided into: unpolarized light, partially polarized light, and fully polarized light; According to the direction of polarization, polarized light can be divided into linear polarized light, circular polarized light and elliptically polarized light. The polarization state of light can be expressed effectively by using Jones matrix, Stokes matrix and Poincaré sphere. In practical applications, we use a variety of measurement methods to obtain the polarization state of light. Such as Muller matrix, Stokes vector measurement method and so on.

In the third part, we introduce several polarization imaging systems in application. For the measurement of linear polarization information, there are division of time polarization imaging systems, division-of-amplitude polarization imaging systems and division-of-aperture polarization measurement systems. For the measurement of full polarization information, there are based on liquid crystal modulation polarization imaging systems, division-of-plane polarization imaging systems and channel modulated polarization imaging systems. These polarization imaging methods have their advantages and disadvantages in practice, and their characteristics are summarized.

In the last part, we briefly introduce the application of polarization imaging, in modern imaging methods, polarization system in the military field, biomedical imaging, infrared imaging, image processing and other fields have a very surprising advantage. In particular, there are three aspects of underwater polarization imaging, infrared polarization imaging and polarization defogging. Polarization imaging technology allows us to use more information about light, to achieve imaging tasks under more demanding conditions, and to improve the clarity of the image and increase the visual effect of the image. We believe there will be more and better applications in the future!

Acknowledgements This work was supported by the National Natural Science Foundation of China (Nos. 11874132, 61975044, 12074094, 12374271, and 62375065); Interdisciplinary Research Foundation of HIT (No. IR2021237).

References

1. Solomon JE. Polarization imaging. Appl Opt. 1981;20(9):1537–44.
2. Song Q, Baroni A, Sawant R, Ni P, Brandli V, Chenot S, Vézian S, Damilano B, de Mierry P, Khadir S. Ptychography retrieval of fully polarized holograms from geometric-phase metasurfaces. Nat Commun. 2020;11(1):2651.
3. Tang J, Wang YB, Zhao DH, Guo XT, Zhao J, Shen C, Liu J. Application of polarized light compass system on solar position calculation. Optik. 2019;187:135–47.
4. Yang FC, Tang XJ, Hu BL, We RY, Kong L, Li Y. A method of removing reflected highlight on images based on polarimetric imaging. J Sens. 2016;2016:9537320.
5. Goudail F, Dai J. Optimal polarimeter structures for estimating polarization degree, angle, and ellipticity in the presence of additive noise. Opt Lett. 2020;45(12):3264–7.
6. Goudail F, Li X, Boffety M, Roussel S, Liu T, Hu H. Precision of retardance autocalibration in full-Stokes division-of-focal-plane imaging polarimeters. Opt Lett. 2019;44(22):5410–3.
7. Rubin NA, D'Aversa G, Chevalier P, Shi Z, Chen WT, Capasso F (2019) Matrix Fourier optics enables a compact full-Stokes polarization camera. Science 365(6448): eaax1839
8. Arbabi E, Kamali SM, Arbabi A, Faraon A. Full-Stokes imaging polarimetry using dielectric metasurfaces. ACS Photon. 2018;5(8):3132–40.
9. Qi J, Elson DS. Mueller polarimetric imaging for surgical and diagnostic applications: a review. J Biophoton. 2017;10(8):950–82.
10. Ku S, Mahato KK, Mazumder N. Polarization-resolved Stokes-Mueller imaging: a review of technology and applications. Lasers Med Sci. 2019;34:12814–21293.
11. Gao S, Gruev V. Gradient-based interpolation method for division-of-focal-plane polarimeters. Opt Express. 2013;21(1):1137–51.
12. He C, He H, Chang J, Chen B, Ma H, Booth MJ. Polarisation optics for biomedical and clinical applications: a review. Light Sci Appl. 2021;10(1):194.
13. Gruev V, Perkins R, York T. CCD polarization imaging sensor with aluminum nanowire optical filters. Opt Express. 2010;18(18):19087–94.
14. Nomura T, Javidi B, Murata S, Nitanai E, Numata T. Polarization imaging of a 3D object by use of on-axis phase-shifting digital holography. Opt Lett. 2007;32(5):481–3.
15. Tong L, Huang X, Wang P, Ye L, Peng M, An L, Sun Q, Zhang Y, Yang G, Li Z, Zhong F, Wang F, Wang Y, Motlag M, Wu W, Cheng GJ, Hu W. Stable mid-infrared polarization imaging based on quasi-2D tellurium at room temperature. Nat Commun. 2020;11(1):2308.

GPSR Compliance

The European Union's (EU) General Product Safety Regulation (GPSR) is a set of rules that requires consumer products to be safe and our obligations to ensure this.

If you have any concerns about our products, you can contact us on ProductSafety@springernature.com

In case Publisher is established outside the EU, the EU authorized representative is:

Springer Nature Customer Service Center GmbH
Europaplatz 3
69115 Heidelberg, Germany

The manufacturer's authorised representative in the EU is Springer Nature Customer Service Centre GmbH, Europaplatz 3, 69115 Heidelberg, Germany. If you have any concerns regarding our products, please contact ProductSafety@springernature.com

Printed and bound by CPI Group (UK) Ltd, Croydon, CR0 4YY
01/10/2026
02225935-0015